Ralf Kories · Heinz Schmidt-Walter

Electrical Engineering – A Pocket Reference

Springer

*Berlin
Heidelberg
New York
Hong Kong
London
Milan
Paris
Tokyo*

Ralf Kories · Heinz Schmidt-Walter

Electrical Engineering

A Pocket Reference

With 610 Figures

Springer

Prof. Dr. Ralf Kories

Deutsche Telekom AG, Fachhochschule Leipzig
Gustav-Freytag-Str. 43-45, 04277 Leipzig
Germany

Prof. Dr. Heinz Schmidt-Walter

FH Darmstadt, FB Elektrotechnik und Informationstechnik
Schöfferstr. 3, 64295 Darmstadt
Germany

Library of Congress Cataloging-in-Publication Data

Kories, Ralf
 Electrical engineering : a pocket reference / Ralf Kories, Heinz Schmidt-Walter.
    p. cm.
 Includes bibliographical references and index.
 ISBN 3-540-43965-X (alk. paper)
 1. Electric engineering--Handbooks, manuals, etc. I. Schmidt-Walter, Heinz. II. Title.

TK151.K583 2003
621.3--dc21
                                                                                2003044162

Based on the 5th German edition of Kories/Schmidt-Walter "Taschenbuch der Elektrotechnik", 2003. Published by Wissenschaftlicher Verlag Harri Deutsch GmbH, Frankfurt am Main.

ISBN 3-540-43965-X Springer-Verlag Berlin Heidelberg New York

This work is subject to copyright. All rights are reserved, whether the whole or part of the material is concerned, specifically the rights of translation, reprinting, reuse of illustrations, recitation, broadcasting, reproduction on microfilm or in other ways, and storage in data banks. Duplication of this publication or parts thereof is permitted only under the provisions of the German Copyright Law of September 9, 1965, in its current version, and permission for use must always be obtained from Springer-Verlag. Violations are liable for prosecution act under German Copyright Law.

Springer-Verlag Berlin Heidelberg New York
a member of BertelsmannSpringer Science + Business Media GmbH

http://www.springer.de

© Springer-Verlag Berlin Heidelberg 2003
Printed in Germany

The use of general descriptive names, registered names, trademarks, etc. in this publication does not imply, even in the absence of a specific statement, that such names are exempt from the relevant protective laws and regulations and therefore free for general use.

Typesetting and final processing: PTP-Berlin Protago-TeX-Production GmbH, Berlin
Cover-Design: design & production GmbH, Heidelberg

Printed on acid-free paper   62/3141Yu – 5 4 3 2 1 0

Preface

Purpose

The purpose of the *Electrical Engineering – A Pocket Reference* is to provide the basics of electrical engineering and electronics in a single handy volume.

The book addresses university students in electrical engineering, telecommunications, computer engineering as well as other engineering disciplines with a minor in electrical engineering.

The book is a lot more than a collection of equations. It provides concise explanation of fundamental principles and their application.

The appendix collects useful reference data on standards, electrical, physical and chemical data on materials, etc. An extensive list of commonly used acronyms is included.

We hope that students will find this book helpful for reviewing classroom material, prepare for exams and understanding fundamental principles.

From the German edition we learned that quite a number of practising engineers keep the book on their shelves for ready reference.

Organisation

The book is organized the way most electrical engineering curricula are taught. We tried to avois extensive cross-referencing between chapters. Need-to-know facts are provided in the context where and how you need them.

An extensive number of figures illustrates basic facts and principles. Numerous tables are provided to summarize facts and relations.

To make sure each symbol in an equation you look up for reference is well understood, each chapter provides a list of symbols with their meanings and physical units.

Useful mathematical relations in a notation suitable for their application in electrical engineering are listed in the Appendix. Fourier transforms are given in both (ω and f) notations to avoid confusion and mistakes by converting one into another.

An extensive index helps to find subject matters easily, entries for overview tables are highlighted in bold print.

No electrical device functions without a power supply. The last chapter gives an insight into state-of-the-art power supply technology drawing on all previous chapters. This is a useful source of information for novices and practising engineers.

Acknowledgement

The book is the result of collaborative work of faculty from three European Universities. We are greatfully indebted to Elmar Jung and Paul O'Leary from the Dublin Institute of Technology for translating the German edition. Our thanks extend to Christopher Bruce, Frank Duignan and Eugene Coyle for initial proof reading and many useful comments.

The book wouldn't have come into beeing without Elmar Jung's efforts in typesetting. We gratefully acknowledge the meticulous copy editing work by Ms. Tracey Wilbourn. The final form of the book owes much to their collective efforts.

Finally we would like to thank hundreds of readers who over the years provided us with tips and hints how to improve the book to make it even more useful to our readership.

The Authors

Contents

1 DC Systems   1
1.1 Basic Quantities, Basic Laws   1
- 1.1.1 Electric Charge   1
- 1.1.2 Electric Current   1
- 1.1.3 Voltage and Potential   2
- 1.1.4 Ohm's Law   2
- 1.1.5 Resistance and Conductance   3
- 1.1.6 Temperature Dependence of Resistance   3
- 1.1.7 Inductance   4
- 1.1.8 Capacitance   4
- 1.1.9 Ideal Voltage Source   5
- 1.1.10 Ideal Current Source   5
- 1.1.11 Kirchhoff's Law   6
  - 1.1.11.1 Kirchhoff's First Law (Current Law)   6
  - 1.1.11.2 Kirchhoff's Second Law (Voltage Law)   6
- 1.1.12 Power and Energy   7
  - 1.1.12.1 Energy and Power in a Resistor   7
  - 1.1.12.2 Energy in an Inductor   8
  - 1.1.12.3 Energy in a Capacitor   9
- 1.1.13 Efficiency   9
- 1.1.14 Maximum Power Transfer   10

1.2 Basic Circuits   11
- 1.2.1 Real Voltage and Current Sources   11
  - 1.2.1.1 Real Voltage Source   11
  - 1.2.1.2 Real Current Source   12
  - 1.2.1.3 Voltage–Current Source Conversion   13
- 1.2.2 Circuit Elements in Series and Parallel   13
  - 1.2.2.1 Series Combination of Resistors   13
  - 1.2.2.2 Parallel Combination of Resistors   13
  - 1.2.2.3 Series Combination of Conductances   14
  - 1.2.2.4 Parallel Combination of Conductances   14
  - 1.2.2.5 Series Combination of Inductances   15
  - 1.2.2.6 Parallel Combination of Inductances   15
  - 1.2.2.7 Series Combination of Capacitances   16
  - 1.2.2.8 Parallel Combination of Capacitances   16
- 1.2.3 Star–Delta Transformation   17
- 1.2.4 Voltage and Current Divider   18
  - 1.2.4.1 Voltage Divider   18
  - 1.2.4.2 Current Divider   18
  - 1.2.4.3 Capacitive and Inductive Dividers   19
- 1.2.5 *RC* and *RL* Combinations   19
  - 1.2.5.1 Series Combination of R and C Driven by a Voltage Source   21
  - 1.2.5.2 Series Combination of R and C Driven by a Current Source   21
  - 1.2.5.3 Parallel Combination of R and C Driven by a Current Source   22
  - 1.2.5.4 Parallel Combination of R and C Driven by a Voltage Source   22

		1.2.5.5	Series Combination of R and L Driven by a Voltage Source	23

           1.2.5.5   Series Combination of R and L Driven by a Voltage Source 23
           1.2.5.6   Series Combination of R and L Driven by a Current Source 23
           1.2.5.7   Parallel Combination of R and L Driven by a Voltage Source 24
           1.2.5.8   Parallel Combination of R and L Driven by a Current Source 24
   1.2.6 RLC Combinations 25
           1.2.6.1   Series Combination of R, L and C 26
1.3 Calculation Methods for Linear Circuits 29
   1.3.1 Rules for Signs 29
   1.3.2 Circuit Calculation with Mesh and Node Analysis 30
   1.3.3 Superposition 31
   1.3.4 Mesh Analysis 32
   1.3.5 Node Analysis 33
   1.3.6 Thévenin's and Norton's Theorem 33
           1.3.6.1   Calculating a Load Current by Thévenin's Theorem ... 34
           1.3.6.2   Calculating a Current Within a Network 36
1.4 Notation Index 37
1.5 Further Reading 38

2 Electric Fields 39

2.1 Electrostatic Fields 39
   2.1.1 Coulomb's Law 39
   2.1.2 Definition of Electric Field Strength 40
   2.1.3 Voltage and Potential 41
   2.1.4 Electrostatic Induction 42
   2.1.5 Electric Displacement 43
   2.1.6 Dielectrics 44
   2.1.7 The Coulomb Integral 44
   2.1.8 Gauss's Law of Electrostatics 45
   2.1.9 Capacitance 46
   2.1.10 Electrostatic Field at a Boundary 47
   2.1.11 Overview: Fields and Capacitances of Different Geometric Configurations 48
   2.1.12 Energy in an Electrostatic Field 49
   2.1.13 Forces in an Electrostatic Field 50
           2.1.13.1 Force on a Charge 50
           2.1.13.2 Force at the Boundary 50
   2.1.14 Overview: Characteristics of an Electrostatic Field 52
   2.1.15 Relationship between the Electrostatic Field Quantities ... 53
2.2 Static Steady-State Current Flow 53
   2.2.1 Voltage and Potential 53
   2.2.2 Current 54
   2.2.3 Electric Field Strength 54
   2.2.4 Current Density 55
   2.2.5 Resistivity and Conductivity 56
   2.2.6 Resistance and Conductance 57
   2.2.7 Kirchhoff's Laws 58
           2.2.7.1   Kirchhoff's First Law (Current Law) 58
           2.2.7.2   Kirchhoff's Second Law (Mesh Law) 59
   2.2.8 Static Steady-State Current Flow at Boundaries 60

		2.2.9	Overview: Fields and Resistances of Different Geometric Configurations	61
		2.2.10	Power and Energy in Static Steady-State Current Flow	62
		2.2.11	Overview: Characteristics of Static Steady-State Current Flow ...	63
		2.2.12	Relationship Between Quantities in Static Steady-StateCurrent Flow	64
	2.3	Magnetic Fields		64
		2.3.1	Force on a Moving Charge	65
		2.3.2	Definition of Magnetic Flux Density.....................	66
		2.3.3	Biot–Savart's Law	68
		2.3.4	Magnetic Field Strength	69
		2.3.5	Magnetic Flux	70
		2.3.6	Magnetic Voltage and Ampere's Law	71
		2.3.7	Magnetic Resistance, Magnetic Conductance, Inductance	73
		2.3.8	Materials in a Magnetic Field	74
			2.3.8.1 Ferromagnetic Materials	75
		2.3.9	Magnetic Fields at Boundaries	77
		2.3.10	The Magnetic Circuit	78
		2.3.11	Magnetic Circuit with a Permanent Magnet	80
		2.3.12	Overview: Inductances of Different Geometric Configurations ...	82
		2.3.13	Induction	83
			2.3.13.1 Induction in a Moving Electrical Conductor	83
			2.3.13.2 Faraday's Law of Induction.....................	84
			2.3.13.3 Self-Induction................................	87
		2.3.14	Mutual Induction	88
		2.3.15	Transformer Principle	90
		2.3.16	Energy in a Magnetic Field............................	90
			2.3.16.1 Energy in a Magnetic Circuit with an Air Gap	91
		2.3.17	Forces in a Magnetic Field	92
			2.3.17.1 Force on a Current-Carrying Conductor	92
			2.3.17.2 Force at the Boundaries	93
		2.3.18	Overview: Characteristics of a Magnetic Field	94
		2.3.19	Relationship between the Magnetic Field Quantities	95
	2.4	Maxwell's Equations		95
	2.5	Notation Index		96
	2.6	Further Reading		98
3	**AC Systems**			**99**
	3.1	Mathematical Basics of AC		99
		3.1.1	Sine and Cosine Functions	99
			3.1.1.1 Addition of Sinusoidal Waveforms	100
		3.1.2	Complex Numbers	101
			3.1.2.1 Complex Arithmetic	102
			3.1.2.2 Representation of Complex Numbers	103
			3.1.2.3 Changing Between Different Representations	105
		3.1.3	Complex Calculus	105
			3.1.3.1 Complex Addition and Subtraction	105
			3.1.3.2 Multiplication of Complex Numbers	106
		3.1.4	Overview: Complex Number Arithmetic....................	107
		3.1.5	The Complex Exponential Function	107
			3.1.5.1 Exponential Function with Imaginary Exponents......	108
			3.1.5.2 Exponential Function with Complex Exponents	108
		3.1.6	Trigonometric Functions with Complex Arguments	109
		3.1.7	From Sinusoidal Waveforms to Phasors	109
			3.1.7.1 Complex Magnitude	109

		3.1.7.2	Relationship Between Sinusoidal Waveforms and Phasors 110
		3.1.7.3	Addition and Subtraction of Phasors 111

3.2 Sinusoidal Waveforms 112
    3.2.1 Characteristics of Sinusoidal Waveforms 113
    3.2.2 Characteristics of Nonsinusoidal Waveforms 115
3.3 Complex Impedance and Admittance 116
    3.3.1 Impedance 116
    3.3.2 Complex Impedance of Passive Components 118
        3.3.2.1 Resistor 118
        3.3.2.2 Inductor 118
        3.3.2.3 Capacitor 118
    3.3.3 Admittance 119
    3.3.4 Complex Admittance of Passive Components 120
    3.3.5 Overview: Complex Impedance 121
3.4 Impedance of Passive Components 122
3.5 Combinations of Passive Components 123
    3.5.1 Series Combinations 123
        3.5.1.1 General Case 123
        3.5.1.2 Resistor and Inductor in Series 124
        3.5.1.3 Resistor and Capacitor in Series 125
        3.5.1.4 Resistor, Inductor and Capacitor in Series 126
    3.5.2 Parallel Combinations 128
        3.5.2.1 General Case 128
        3.5.2.2 Resistor and Inductor in Parallel 129
        3.5.2.3 Resistor and Capacitor in Parallel 130
        3.5.2.4 Resistor, Inductor and Capacitor in Parallel 132
    3.5.3 Overview of Series and Parallel Circuits 134
3.6 Network Transformations 135
    3.6.1 Transformation from Parallel to Series Circuits and Vice Versa ... 135
    3.6.2 Star–Delta (Wye–Delta) and Delta–Star (Delta–Wye) Transformations 137
    3.6.3 Circuit Duality 139
3.7 Simple Networks 140
    3.7.1 Complex Voltage and Current Division 140
    3.7.2 Loaded Complex Voltage Divider 142
    3.7.3 Impedance Matching 143
    3.7.4 Voltage Divider with Defined Input and Output Resistances 145
    3.7.5 Phase-Shifting Circuits 146
        3.7.5.1 RC Phase Shifter 146
        3.7.5.2 Alternative Phase-Shifting Circuits 148
    3.7.6 AC Bridges 149
        3.7.6.1 Balancing Condition 149
        3.7.6.2 Application: Measurement Technique 150
3.8 Power in AC Circuits 151
    3.8.1 Instantaneous Power 151
        3.8.1.1 Power in a Resistance 151
        3.8.1.2 Power in a Reactive Element 151
    3.8.2 Average Power 152
        3.8.2.1 Real Power 153
        3.8.2.2 Reactive Power 154
        3.8.2.3 Apparent Power 155
    3.8.3 Complex Power 155
    3.8.4 Overview: AC Power 156

		3.8.5	Reactive Current Compensation 156

	3.8.5	Reactive Current Compensation	156
3.9	Three-Phase Supplies		158
	3.9.1	Polyphase Systems	158
	3.9.2	Three-Phase Systems	159
		3.9.2.1 Properties of the Complex Operator $\underline{a}$	160
	3.9.3	Delta-Connected Generators	161
	3.9.4	Star-Connected Generators	162
3.10	Overview: Symmetrical Three-Phase Systems		164
	3.10.1	Power in a Three-Phase System	165
3.11	Notation Index		166
3.12	Further Reading		167

4 Current, Voltage and Power Measurement — 169

4.1	Electrical Measuring Instruments		169
	4.1.1	Moving-Coil Instrument	169
	4.1.2	Ratiometer Moving-Coil Instrument	169
	4.1.3	Electrodynamic Instrument	170
	4.1.4	Moving-Iron Instrument	171
	4.1.5	Other Instruments	171
	4.1.6	Overview: Electrical Instruments	173
4.2	Measurement of DC Current and Voltage		174
	4.2.1	Moving-Coil Instrument	174
	4.2.2	Range Extension for Current Measurements	174
	4.2.3	Range Extension for Voltage Measurements	175
	4.2.4	Overload Protection	176
	4.2.5	Systematic Measurement Errors in Current and Voltage Measurement	176
4.3	Measurement of AC Voltage and AC Current		177
	4.3.1	Moving-Coil Instrument with Rectifier	177
	4.3.2	Moving-Iron Instruments	179
	4.3.3	Measurement Range Extension Using an Instrument Transformer	179
	4.3.4	RMS Measurement	180
4.4	Power Measurement		181
	4.4.1	Power Measurement in a DC Circuit	181
	4.4.2	Power Measurement in an AC Circuit	182
		4.4.2.1 Three-Voltmeter Method	183
		4.4.2.2 Power Factor Measurement	184
	4.4.3	Power Measurement in a Multiphase System	185
		4.4.3.1 Measurement of the Real Power in a Multiphase System	185
		4.4.3.2 Measurement of the Reactive Power in a Multiphase System	186
4.5	Measurement Errors		187
	4.5.1	Systematic and Random Errors	187
	4.5.2	Guaranteed Error Limits	188
4.6	Overview: Symbols on Measurement Instruments		188
4.7	Overview: Measurement Methods		190
4.8	Notation Index		190
4.9	Further Reading		191

5 Networks at Variable Frequency — 192

5.1	Linear Systems		192
	5.1.1	Transfer Function, Amplitude and Phase Response	192
5.2	Filters		194

	5.2.1	Low-Pass Filter		195
	5.2.2	High-Pass Filter		195
	5.2.3	Bandpass Filter		196
	5.2.4	Stop-Band Filter		197
	5.2.5	All-Pass Filter		197
5.3	Simple Filters			197
	5.3.1	Low-Pass Filter		197
		5.3.1.1	Rise Time	198
	5.3.2	Frequency Normalisation		199
		5.3.2.1	Approximation of the Magnitude Response	200
	5.3.3	High-Pass Filter		200
		5.3.3.1	Approximation of the Magnitude Response	202
	5.3.4	Higher-Order Filters		202
	5.3.5	Bandpass Filter		204
	5.3.6	Filter Realisation		206
5.4	Notation Index			206
5.5	Further Reading			207

6 Signals and Systems — 208

6.1	Signals			208
	6.1.1	Definitions		208
	6.1.2	Symmetry Properties of Signals		209
6.2	Fourier Series			210
	6.2.1	Trigonometric Form		210
		6.2.1.1	Symmetry Properties	211
	6.2.2	Amplitude–Phase Form		211
	6.2.3	Exponential Form		212
		6.2.3.1	Symmetry Properties	213
	6.2.4	Overview: Fourier Series Representations		213
	6.2.5	Useful Integrals for the Calculation of Fourier Coefficients		214
	6.2.6	Useful Fourier Series		215
	6.2.7	Application of the Fourier Series		217
		6.2.7.1	Spectrum of a Rectangular Signal	217
		6.2.7.2	Spectrum of a Sawtooth Signal	218
		6.2.7.3	Spectrum of a Composite Signal	219
6.3	Systems			220
	6.3.1	System Properties		220
		6.3.1.1	Linear Systems	220
		6.3.1.2	Causal Systems	221
		6.3.1.3	Time-Invariant Systems	221
		6.3.1.4	Stable Systems	222
		6.3.1.5	LTI Systems	222
	6.3.2	Elementary Signals		222
		6.3.2.1	The Step Function	222
		6.3.2.2	The Rectangular Pulse	222
		6.3.2.3	The Triangular Pulse	223
		6.3.2.4	The Gaussian Pulse	223
		6.3.2.5	The Impulse Function (Delta Function)	224
	6.3.3	Shifting and Scaling of Time Signals		225
	6.3.4	System Responses		226
		6.3.4.1	Impulse Response	226
		6.3.4.2	Step Response	227
		6.3.4.3	System Response to Arbitrary Input Signals	228
		6.3.4.4	Rules of Convolution	228

		6.3.4.5	Transfer Function 230
		6.3.4.6	System Response Calculation in the Frequency Domain 230
	6.3.5	Impulse and Step Response Calculation 231
		6.3.5.1	Normalisation of Circuits 231
		6.3.5.2	Impulse and Step Response of First-Order Systems 232
		6.3.5.3	Impulse and Step Response of Second-Order Systems .. 234
	6.3.6	Ideal Systems 236
		6.3.6.1	Distortion-Free Systems 236
		6.3.6.2	Ideal Low-Pass Filter 238
		6.3.6.3	Ideal Bandpass Filter 240

6.4 Fourier Transforms
- 6.4.1 Principle 241
- 6.4.2 Definition 242
- 6.4.3 Representation of the Fourier Transform 243
  - 6.4.3.1 Symmetry Properties 244
- 6.4.4 Overview: Properties of the Fourier Transform 244
- 6.4.5 Fourier Transforms of Elementary Signals 245
  - 6.4.5.1 Spectrum of the Delta Function 245
  - 6.4.5.2 Spectrum of the Signum and the Step Functions 246
  - 6.4.5.3 Spectrum of the Rectangular Pulse 247
  - 6.4.5.4 Spectrum of the Triangular Pulse 247
  - 6.4.5.5 Spectrum of the Gaussian Pulse 248
  - 6.4.5.6 Spectrum of Harmonic Functions 249
- 6.4.6 Summary of Fourier Transforms 250

6.5 Nonlinear Systems
- 6.5.1 Definition 253
- 6.5.2 Characterisation of Nonlinear Systems 253
  - 6.5.2.1 Characteristic Equation 253
  - 6.5.2.2 Total Harmonic Distortion 254
  - 6.5.2.3 Signal-to-Intermodulation Ratio 255

6.6 Notation Index 258
6.7 Further Reading 259

7 Analogue Circuit Design   261

7.1 Methods of Analysis
- 7.1.1 Linearisation at the Operating Point 261
- 7.1.2 AC Equivalent Circuit 262
- 7.1.3 Input and Output Impedance 263
  - 7.1.3.1 Determination of the Input Impedance 263
  - 7.1.3.2 Determination of the Output Impedance 263
  - 7.1.3.3 Combination of Two-Terminal Networks 264
- 7.1.4 Two-Port Networks 265
  - 7.1.4.1 Two-Port Network Equations 265
  - 7.1.4.2 Hybrid Parameters (h-Parameters) 265
  - 7.1.4.3 Admittance Parameters (y-Parameters) 266
- 7.1.5 Block Diagrams 267
  - 7.1.5.1 Calculation Rules for Block Diagrams 268
- 7.1.6 Bode Plot 269

7.2 Silicon and Germanium Diodes
- 7.2.1 Current–Voltage Characteristic of Si and Ge Diodes 270
- 7.2.2 Temperature Dependency of the Threshold Voltage 270
- 7.2.3 Dynamic Resistance (Differential Resistance) 271

7.3 Small-Signal Amplifier with Bipolar Transistors
- 7.3.1 Transistor Characteristics 272

		7.3.1.1 Symbols, Voltages and Currents for Bipolar Transistors . 272
		7.3.1.2 Output Characteristics 272
		7.3.1.3 Transfer Characteristic 273
		7.3.1.4 Input Characteristic 273
		7.3.1.5 Static Current Gain β_{DC} 274
		7.3.1.6 Differential Current Gain β 274
		7.3.1.7 Transconductance g_m 275
		7.3.1.8 Thermal Voltage Drift 275
		7.3.1.9 Differential Input Resistance r_{BE} 275
		7.3.1.10 Differential Output Resistance r_{CE} 275
		7.3.1.11 Reverse Voltage-Transfer Ratio A_r 276
		7.3.1.12 Unity Gain and Critical Frequencies 276
	7.3.2	Equivalent Circuits...................................... 276
		7.3.2.1 Static Equivalent Circuit 276
		7.3.2.2 AC Equivalent Circuit........................... 277
		7.3.2.3 The Giacoletto Equivalent Circuit 278
	7.3.3	Darlington Pair.. 278
		7.3.3.1 Pseudo-Darlington Pair.......................... 279
	7.3.4	Basic Circuits with Bipolar Transistors 280
	7.3.5	Common-Emitter Circuit 280
		7.3.5.1 Common-Emitter Circuit Two-Port Network Equations . 281
		7.3.5.2 Common-Emitter AC Equivalent Circuit 282
		7.3.5.3 Common-Emitter Circuit Input Impedance 283
		7.3.5.4 Common-Emitter Circuit Output Impedance 284
		7.3.5.5 Common-Emitter Circuit AC Voltage Gain 285
		7.3.5.6 Operating Point Biasing 286
		7.3.5.7 Operating Point Stabilisation 288
		7.3.5.8 Load Line 290
		7.3.5.9 Common-Emitter Circuit at High Frequencies 291
	7.3.6	Common-Collector Circuit (Emitter Follower) 291
		7.3.6.1 Common-Collector AC Equivalent Circuit 292
		7.3.6.2 Common-Collector Circuit Input Impedance 293
		7.3.6.3 Common-Collector Circuit Output Impedance 293
		7.3.6.4 Common-Collector Circuit AC Current Gain......... 294
		7.3.6.5 Common-Collector Circuit at High Frequencies 294
	7.3.7	Common-Base Circuit 294
		7.3.7.1 Common-Base AC Equivalent Circuit 295
		7.3.7.2 Common-Base Circuit Input Impedance 295
		7.3.7.3 Common-Base Circuit Output Impedance 295
		7.3.7.4 Common-Base Circuit AC Voltage Gain 296
		7.3.7.5 Common-Base Circuit at High Frequencies.......... 296
	7.3.8	Overview: Basic Bipolar Transistor Circuits 296
	7.3.9	Bipolar Transistor Current Sources 296
	7.3.10	Bipolar Transistor Differential Amplifier 298
		7.3.10.1 Differential Mode Gain.......................... 300
		7.3.10.2 Common-Mode Gain 301
		7.3.10.3 Common-Mode Rejection Ratio 301
		7.3.10.4 Differential Amplifier Input Impedance 302
		7.3.10.5 Differential Amplifier Output Impedance 302
		7.3.10.6 Offset Voltage of the Differential Amplifier 302
		7.3.10.7 Differential Amplifier Offset Current................ 302
		7.3.10.8 Input Offset Voltage Drift 302
		7.3.10.9 Differential Amplifier Examples 303
	7.3.11	Overview: Bipolar Transistor Differential Amplifiers 304

	7.3.12	Current Mirror		304
		7.3.12.1	Current Mirror Variations	305
7.4	Field-Effect Transistor Small-Signal Amplifiers			305
	7.4.1	Transistor Characteristics and Ratings		305
		7.4.1.1	Symbols, Voltages and Currents for Field-Effect Transistors	305
		7.4.1.2	JFET Characteristic Curves	307
		7.4.1.3	IGFET Characteristic Curves	307
		7.4.1.4	Transconductance	308
		7.4.1.5	Dynamic Output Resistance	309
		7.4.1.6	Input Impedance	309
	7.4.2	Equivalent Circuit		309
		7.4.2.1	Equivalent Circuit for Low Frequencies	309
		7.4.2.2	Equivalent Circuit for High Frequencies	310
		7.4.2.3	Critical Frequency of Transconductance	310
	7.4.3	Basic Circuits using Field-Effect Transistors		310
	7.4.4	Common-Source Circuit		310
		7.4.4.1	Common-Source Two-Port Parameters	311
		7.4.4.2	AC Equivalent Circuit of the Common-Source Circuit	312
		7.4.4.3	Input Impedance of the Common-Source Circuit	313
		7.4.4.4	Output Impedance of the Common-Source Circuit	313
		7.4.4.5	AC Voltage Gain	314
		7.4.4.6	Operating-Point Biasing	314
		7.4.4.7	Common-Drain Circuit, Source Follower	316
		7.4.4.8	AC Equivalent Circuit of the common-drain Circuit	316
		7.4.4.9	Input Impedance of the Common-Drain Circuit	317
		7.4.4.10	Output Impedance of the Common-Drain Circuit	317
		7.4.4.11	Voltage Gain of the Common-Drain Circuit	317
		7.4.4.12	Common-Drain Circuit at High Frequencies	317
	7.4.5	Common-Gate Circuit		317
		7.4.5.1	Input Impedance of the Common-Gate Circuit	318
		7.4.5.2	Output Impedance of the Common-Gate Circuit	318
		7.4.5.3	Voltage Gain of the Common-Gate Circuit	318
	7.4.6	Overview: Basic Circuits using Field-Effect Transistors		318
	7.4.7	FET Current Source		319
	7.4.8	Differential Amplifier with Field-Effect Transistors		319
		7.4.8.1	Differential Mode Gain	320
		7.4.8.2	Common-Mode Gain	320
		7.4.8.3	Common-Mode Rejection Ratio	321
		7.4.8.4	Input Impedance	321
		7.4.8.5	Output Impedance	321
	7.4.9	Overview: Differential Amplifier with FETs		321
	7.4.10	Controllable Resistor FETs		321
7.5	Negative Feedback			322
	7.5.1	Feedback Topologies		324
	7.5.2	Influence of Negative Feedback on the Input and Output Impedance		326
		7.5.2.1	Input and Output Impedance of the Four Kinds of Feedback	327
	7.5.3	Influence of Negative Feedback on Frequency Response		327
	7.5.4	Stability of Systems with Negative Feedback		328
7.6	Operational Amplifiers			329
	7.6.1	Characteristics of the Operational Amplifier		330
		7.6.1.1	Output Voltage Swing	330

		7.6.1.2	Offset Voltage 330
		7.6.1.3	Offset Voltage Drift 331
		7.6.1.4	Common-Mode Input Swing 331
		7.6.1.5	Differential Mode Gain 331
		7.6.1.6	Common-Mode Gain 331
		7.6.1.7	Common-Mode Rejection Ratio 332
		7.6.1.8	Power Supply Rejection Ratio. 332
		7.6.1.9	Input Impedance 332
		7.6.1.10	Output Impedance 332
		7.6.1.11	Input Bias Current 332
		7.6.1.12	Gain–Bandwidth Product (Unity Gain Frequency). 333
		7.6.1.13	Critical Frequency 333
		7.6.1.14	Slew Rate of the Output Voltage 333
		7.6.1.15	Equivalent Circuit of the Operational Amplifier. 333
	7.6.2	Frequency Compensation 334	
	7.6.3	Comparators 335	
	7.6.4	Circuits with Operational Amplifiers 335	
		7.6.4.1	Impedance Converter (follower) 336
		7.6.4.2	Noninverting Amplifier 336
		7.6.4.3	Inverting Amplifier 337
		7.6.4.4	Summing Amplifier 338
		7.6.4.5	Difference Amplifier 339
		7.6.4.6	Instrumentation Amplifier 340
		7.6.4.7	Voltage-Controlled Current Source 341
		7.6.4.8	Integrator 341
		7.6.4.9	Differentiator 342
		7.6.4.10	AC Voltage Amplifier with Single-Rail Supply 343
		7.6.4.11	Voltage Setting with Defined Slew Rate 343
		7.6.4.12	Schmitt Trigger 344
		7.6.4.13	Triangle- and Square-Wave Generator 345
		7.6.4.14	Multivibrator 346
		7.6.4.15	Sawtooth Generator 346
		7.6.4.16	Pulse-Width Modulator 346
7.7	Active Filters 348		
	7.7.1	Low-Pass Filters. 349	
		7.7.1.1	Theory of Low-Pass Filters 349
		7.7.1.2	Low-Pass Filter Calculations 356
		7.7.1.3	Low-Pass Filter Circuits 357
	7.7.2	High-Pass Filters 359	
		7.7.2.1	Theory of High-Pass Filters 359
		7.7.2.2	High-Pass Filter Circuits 359
	7.7.3	Bandpass Filters 361	
		7.7.3.1	Second-Order Bandpass Filter 361
		7.7.3.2	Second-Order Bandpass Filter Circuit 362
		7.7.3.3	Fourth- and Higher-Order Bandpass Filters 362
	7.7.4	Universal Filter 363	
	7.7.5	Switched-Capacitor Filter 363	
7.8	Oscillators 364		
	7.8.1	RC Oscillators 365	
		7.8.1.1	Phase-Shift Oscillator 365
		7.8.1.2	Wien Bridge Oscillator 366
	7.8.2	LC Tuned Oscillators 367	
		7.8.2.1	Meissner Oscillator 367
		7.8.2.2	Hartley Oscillator 367

		7.8.2.3	Colpitts Oscillator	368

- 7.8.3 Quartz/Crystal Oscillators ... 368
  - 7.8.3.1 Pierce Oscillator ... 369
  - 7.8.3.2 Quartz Oscillator with TTL Gates ... 370
- 7.8.4 Multivibrators ... 370
- 7.9 Heating and Cooling ... 370
  - 7.9.1 Reliability and Lifetime ... 371
  - 7.9.2 Temperature Calculation ... 373
    - 7.9.2.1 Thermal Resistance ... 373
    - 7.9.2.2 Thermal Capacity ... 374
    - 7.9.2.3 Transient Thermal Impedance ... 375
- 7.10 Power Amplifiers ... 376
  - 7.10.1 Emitter Follower ... 376
  - 7.10.2 Complementary Emitter Follower in Class B Operation ... 379
  - 7.10.3 Complementary Emitter Follower in Class C Operation ... 382
  - 7.10.4 The Characteristic Curves of the Operation Classes ... 383
  - 7.10.5 Complementary Emitter Follower in Class AB Operation ... 383
    - 7.10.5.1 Biasing for Class AB Operation ... 384
    - 7.10.5.2 Complementary Emitter Follower with Darlington Transistors ... 385
    - 7.10.5.3 Current-Limiting Complementary Emitter Follower ... 386
  - 7.10.6 Input Signal Injection to Power Amplifiers ... 386
    - 7.10.6.1 Input Signal Injection using a Differential Amplifier ... 386
    - 7.10.6.2 Input Signal Injection Using an Op-Amp ... 388
  - 7.10.7 Switched-Mode Amplifiers ... 388
- 7.11 Notation Index ... 389
- 7.12 Further Reading ... 390

8 Digital Electronics   392

- 8.1 Logic Algebra ... 392
  - 8.1.1 Logic Variables and Logic Gates ... 392
    - 8.1.1.1 Inversion ... 392
    - 8.1.1.2 AND Function ... 392
    - 8.1.1.3 OR Function ... 393
  - 8.1.2 Logic Functions and their Symbols ... 393
    - 8.1.2.1 Inverter (NOT) ... 394
    - 8.1.2.2 AND Gate ... 394
    - 8.1.2.3 OR Gate ... 394
    - 8.1.2.4 NAND Gate ... 395
    - 8.1.2.5 NOR Gate ... 395
    - 8.1.2.6 XOR Gate, Exclusive OR ... 396
  - 8.1.3 Logic Transformations ... 396
    - 8.1.3.1 Commutative Laws ... 396
    - 8.1.3.2 Associative Laws ... 397
    - 8.1.3.3 Distributive Laws ... 397
    - 8.1.3.4 Inversion Laws (DeMorgan's Rules) ... 398
  - 8.1.4 Overview: Logic Transformations ... 398
  - 8.1.5 Analysis of Logic Circuits ... 399
  - 8.1.6 Sum of Products and Product of Sums ... 400
    - 8.1.6.1 Sum of Products ... 400
    - 8.1.6.2 Product of Sums ... 401
  - 8.1.7 Systematic Reduction of a Logic Function ... 402
    - 8.1.7.1 Karnaugh Map ... 402
    - 8.1.7.2 The Quine–McCluskey Technique ... 406

	8.1.8	Synthesis of Combinational Circuits 408
		8.1.8.1 Implementation Using only NAND Gates 408
		8.1.8.2 Implementation Using only NOR Gates 408
8.2	Electronic Realisation of Logic Circuits 409
	8.2.1	Electrical Specification 409
		8.2.1.1 Voltage Levels 409
		8.2.1.2 Transfer Characteristic 409
		8.2.1.3 Loading 410
		8.2.1.4 Noise Margin 410
		8.2.1.5 Propagation Delay Time 411
		8.2.1.6 Rise Times 411
		8.2.1.7 Power Loss 412
		8.2.1.8 Minimum Slew Rate 412
		8.2.1.9 Integration 412
	8.2.2	Overview: Notation in Data Sheets 412
	8.2.3	TTL Family 414
		8.2.3.1 TTL Devices 414
		8.2.3.2 Basic TTL Gate Circuit 416
	8.2.4	CMOS Family 417
	8.2.5	Comparison of TTL and CMOS 418
		8.2.5.1 Other Logic Families 418
	8.2.6	Special Circuit Variations 420
		8.2.6.1 Outputs with Open Collector..................... 420
		8.2.6.2 Wired AND/OR 420
		8.2.6.3 Tri-State Outputs 422
		8.2.6.4 Schmitt Trigger Inputs 422
8.3	Combinational Circuits and Sequential Logic 423
	8.3.1	Dependency Notation 423
		8.3.1.1 Overview: Dependency Notation.................. 425
	8.3.2	Circuit Symbols for Combinational and Sequential Logic 425
8.4	Examples of Combinational Circuits 426
	8.4.1	1-to-n Decoder 426
	8.4.2	Multiplexer and Demultiplexer 426
		8.4.2.1 Overview of Circuits 428
8.5	Latches and Flip-Flops 428
	8.5.1	Flip-Flop Applications 428
	8.5.2	SR Flip-Flop 429
		8.5.2.1 SR Flip-Flop with Clock Input 430
	8.5.3	D Flip-Flop 430
	8.5.4	Master–Slave Flip-Flop 431
	8.5.5	JK Flip-Flop 432
	8.5.6	Flip-Flop Triggering 432
	8.5.7	Notation for Flip-Flop Circuit Symbols 433
	8.5.8	Overview: Flip-Flops 434
	8.5.9	Overview: Edge-Triggered Flip-Flops....................... 434
	8.5.10	Synthesis of Edge-Triggered Flip-Flops 436
	8.5.11	Overview: Flip-Flop Circuits 438
8.6	Memory 439
	8.6.1	Memory Construction 439
	8.6.2	Memory Access 440
	8.6.3	Static and Dynamic RAMs 441
		8.6.3.1 Variations of RAM 442
	8.6.4	Read-Only Memory 443
	8.6.5	Programmable Logic Devices.............................. 444

			8.6.5.1	Principle of Operation	444
			8.6.5.2	PLD Types	445
			8.6.5.3	Output Circuits	446
	8.7	Registers and Shift Registers			448
	8.8	Counters			449
		8.8.1	Asynchronous Counters		450
			8.8.1.1	Binary Counter	450
			8.8.1.2	Decimal Counter	450
			8.8.1.3	Down Counter	453
			8.8.1.4	Up/Down Counter	454
			8.8.1.5	Programmable Counter	454
		8.8.2	Synchronous Counters		455
			8.8.2.1	Cascading Synchronous Counters	456
		8.8.3	Overview: TTL and CMOS Counters		458
			8.8.3.1	TTL Counters	459
			8.8.3.2	CMOS Counters	459
	8.9	Design and Synthesis of Sequential Logic			460
	8.10	Further Reading			467
9	**Power Supplies**				**469**
	9.1	Power Transformers			469
	9.2	Rectification and Filtering			470
		9.2.1	Different Rectifier Circuits		472
	9.3	Analogue Voltage Stabilisation			473
		9.3.1	Voltage Stabilisation with Zener Diode		473
		9.3.2	Analogue Stabilisation with Transistor		474
		9.3.3	Voltage Regulation		475
			9.3.3.1	Integrated Voltage Regulators	475
	9.4	Switched Mode Power Supplies			476
		9.4.1	Single-Ended Converters, Secondary Switched SMPS		477
			9.4.1.1	Buck Converter	477
			9.4.1.2	Boost Converter	479
			9.4.1.3	Buck-Boost Converter	481
		9.4.2	Primary Switched SMPS		482
			9.4.2.1	Flyback Converter	482
			9.4.2.2	Single-Transistor Forward Converter	486
			9.4.2.3	Push–Pull Converters	489
			9.4.2.4	Resonant Converters	491
		9.4.3	Overview: Switched-Mode Power Supplies		494
		9.4.4	Control of Switched-Mode Power Supplies		496
			9.4.4.1	Voltage-Mode Control	496
			9.4.4.2	Current-Mode Control	497
			9.4.4.3	Comparison: Voltage-Mode vs. Current-Mode Control	498
			9.4.4.4	Design of the PI Controller	498
		9.4.5	Design of Inductors and High-Frequency Transformers		499
			9.4.5.1	Calculation of Inductors	499
			9.4.5.2	Calculation of High-Frequency Transformers	500
		9.4.6	Power Factor Control		504
			9.4.6.1	Currents, Voltages and Power of the PFC	505
			9.4.6.2	Controlling the PFC	506
		9.4.7	Radio-Frequency Interference Suppression of Switched-Mode Power Supplies		507
			9.4.7.1	Radio-Frequency Interference Radiation	507
			9.4.7.2	Mains Input Conducted-Mode Interference	508

XX Contents

		9.4.7.3	Suppression of Common-Mode Radio-Frequency Interference 509
		9.4.7.4	Suppression of Differential-Mode Radio Frequency Interference 509
		9.4.7.5	Complete Radio-Frequency Interference Filter 510
	9.5	Notation Index 511	
	9.6	Further Reading 512	

A Mathematical Basics — 513

- A.1 Trigonometric Functions 513
  - A.1.1 Properties 513
  - A.1.2 Sums and Differences of Trigonometric Functions 514
  - A.1.3 Sums and Differences in the Argument 515
  - A.1.4 Multiples of the Argument 515
  - A.1.5 Weighted Sums of Trigonometric Functions 516
  - A.1.6 Products of Trigonometric Functions 516
  - A.1.7 Triple Products 516
  - A.1.8 Powers of Trigonometric Functions 517
  - A.1.9 Trigonometric Functions with Complex Arguments 517
- A.2 Inverse Trigonometric Functions (Arc Functions) 517
- A.3 Hyperbolic Functions 518
- A.4 Differential Calculus 518
  - A.4.1 Basics of Differential Calculus 518
  - A.4.2 Derivatives of Elementary Functions 519
- A.5 Integral Calculus 519
  - A.5.1 Basics of Integral Calculus 519
    - A.5.1.1 Integrals of Elementary Functions 520
  - A.5.2 Integrals Involving Trigonometric Functions 521
  - A.5.3 Integrals Involving Exponential Functions 523
  - A.5.4 Integrals Involving Inverse Trigonometric Functions 524
  - A.5.5 Definite Integrals 524
- A.6 The Integral of the Standard Normal Distribution 527

B Tables — 530

- B.1 The International System of Units (SI) 530
  - B.1.1 Decimal Prefixes 531
  - B.1.2 SI Units in Electrical Engineering 532
- B.2 Naturally Occurring Constants 533
- B.3 Symbols of the Greek Alphabet 533
- B.4 Units and Definitions of Technical–Physical Quantities 534
- B.5 Imperial and American Units 535
- B.6 Other Units 537
- B.7 Charge and Discharge Curves 540
- B.8 IEC Standard Series 541
- B.9 Resistor Colour Code 542
- B.10 Parallel Combination of Resistors 543
- B.11 Selecting Track Dimensions for Current Flow 544
- B.12 American Wire Gauge 545
- B.13 Dry Cell Batteries 546
- B.14 Notation of Radio-Frequency Ranges 548
- B.15 Ratios 549
  - B.15.1 Absolute Voltage Levels 549
    - B.15.1.1 Conversion of Power and Voltage Level Ratios 550
  - B.15.2 Relative Levels 551

    B.16 V.24 Interface 552
    B.17 Dual-Tone Multi-Frequency 553
    B.18 ASCII Coding 554
    B.19 Resolution and Coding for Analogue-to-Digital Converters 555
    B.20 Chemical Elements... 556
    B.21 Materials 559

C   Acronyms   **561**

D   Circuit Symbols   **595**

Index   **601**

1 DC Systems

1.1 Basic Quantities, Basic Laws

1.1.1 Electric Charge

Système International (SI) unit of charge: C = As (coulomb)

Electricity is based upon the existence of electric **charges**, which are positive or negative. A force exists between electric charges, which is described by Coulomb's law (Sect. 2.1.1). Like charges repel each other, and unlike charges attract each other.

From the physical point of view, every charge is a multiple of the **elementary charge** e.
Elementary charge $e = \pm 1.602 \cdot 10^{-19}$ coulomb
Electrons carry a negative charge, and protons carry a positive charge. A lack of electrons in a body means the body is positively charged. Similarly, an excess of electrons means it is negatively charged.

1.1.2 Electric Current

SI unit of current: A (ampere)

The directed motion of electric charge carriers is called an **electric current**.

$$I = \frac{dQ}{dt} \tag{1.1}$$

The electric current I in a conductor is the charge dQ passing through the conductor cross-sectional area during the time interval dt. The current is a Direct current if the charge passing the conductor per time interval is constant.

$$\text{DC current:} \quad I = \frac{dQ}{dt} = \text{constant} \tag{1.2}$$

Technical direction of current:

The **positive current direction** is the motion of the positive charge carriers. This is equivalent to the opposite motion of negative charge carriers. In metal conductors electrons are the charge carriers. From the physical point of view, the electrons therefore move opposite the positive current flow (Fig. 1.1).

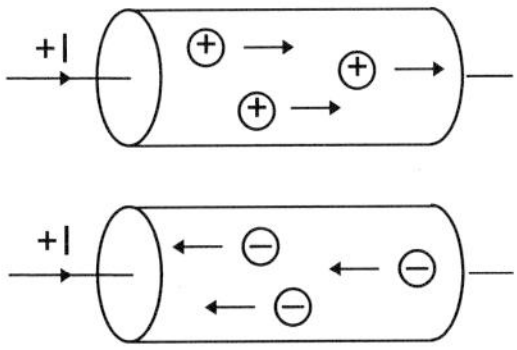

Fig. 1.1. Definition of the positive current direction

Electric charges always move in a closed loop. This means:

- The electric current always flows in a closed circuit.

1.1.3 Voltage and Potential

SI unit of voltage: V (volt)

The electric **voltage** is the force that causes the movement of the charge carriers.

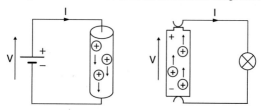

Fig. 1.2. Electrical circuits showing the direction of the voltage and the current

The electric current always flows from the positive terminal to the negative terminal of the voltage source. Since the current flows in a closed loop, *inside* the voltage source (e.g. a battery) the current flows from the negative to the positive terminal (Fig. 1.2).

The **potential** φ is a scalar quantity. Given that one point in space has the potential $\varphi = 0$, then all other points in space can be assigned an absolute potential. This potential is obtained from the energy that has to be provided to move the elementary charge from the point with $\varphi = 0$ to the given point. In this physical model, the voltage V is the difference between two potentials (Fig. 1.3). For this reason voltage is often referred to as potential difference.

$$V_{21} = \varphi_2 - \varphi_1 \tag{1.3}$$

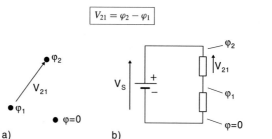

Fig. 1.3. Relationship between voltage and potential: **a** for arbitrary points; **b** in a circuit

1.1.4 Ohm's Law

The current flowing through a load is dependent on the driving voltage. Provided the properties of the load are independent of the current flowing through it and the voltage applied to it, **Ohm's law** holds:

$$\begin{aligned} V &\propto I, \\ \text{or} \quad V &= R \cdot I \end{aligned} \tag{1.4}$$

The current changes proportionally with the voltage. The constant R relating current and voltage is called the **electric resistance**.

1.1.5 Resistance and Conductance

SI unit of resistance: Ω (ohm), $1\,\Omega = 1\,\dfrac{V}{A}$

SI unit of conductance: S (siemens), $1\,S = 1\,\dfrac{A}{V}$

The relationship between current and voltage is described by the quantities **resistance** R and **conductance** G (Fig. 1.4).

$$\begin{aligned} V &= R \cdot I, \quad \text{or} \quad R = \frac{V}{I} \\ I &= G \cdot V, \quad \text{or} \quad G = \frac{I}{V} \end{aligned} \tag{1.5}$$

Fig. 1.4. Resistance and conductance as electrical circuit symbols

1.1.6 Temperature Dependence of Resistance

For real resistors a change in the temperature causea a change in resistance. The relationship between both values is linear to a first approximation. The relationship is described by the **temperature coefficient** α (K^{-1}).

If the resistor R_1 is heated from temperature ϑ_1 to temperature ϑ_2, then the change in resistance is given by:

$$\Delta R = R_1 \alpha (\vartheta_2 - \vartheta_1) \tag{1.6}$$

At a temperature ϑ_2 the resistance is:

$$R_2 = R_1 \left[1 + \alpha(\vartheta_2 - \vartheta_1)\right] \tag{1.7}$$

The temperature coefficient α is often given for a temperature of $\vartheta = 20°C$. This value is sufficient for calculations for temperatures up to approximately 200°C. For most resistive materials (apart from certain semiconductors), α has a positive value. This means that the resistance increases with temperature.

EXAMPLE: For aluminium and copper is $\alpha = 0.004$ K^{-1}. For a temperature change of $\Delta \vartheta = 100$ K the resistance of aluminium or copper wire therefore changes by 40%.

For calculations over larger temperature ranges, the nonlinearity of $R = f(\vartheta)$ can be taken into account by including a squared term with the coefficient β. In this case, $R = f(\vartheta)$ is represented by:

$$R_2 = R_1 \left[1 + \alpha(\vartheta_2 - \vartheta_1) + \beta(\vartheta_2 - \vartheta_1)^2\right] \tag{1.8}$$

1.1.7 Inductance

SI unit of inductance: H (henry), $1\,\text{H} = 1\,\dfrac{\text{Vs}}{\text{A}}$

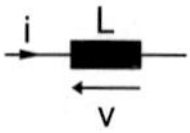

Fig. 1.5. Inductance as a circuit symbol

- For an **inductance** L the voltage v is proportional to the rate of change of the current i.

$$v = L\frac{di}{dt}, \qquad i = \frac{1}{L}\int_{t_0}^{t_1} v\,dt + I_0, \qquad L = \frac{v\,dt}{di} \tag{1.9}$$

The current flowing at the beginning of the integration interval is called I_0. If a constant voltage is applied to an inductance, the current increases linearly (Fig. 1.6).

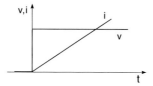

Fig. 1.6. Time progression of the current in an inductance when a constant voltage has been applied

- In an inductance the current cannot change instantaneously, while the voltage can change instantaneously.

- The current in an inductance is proportional to the time-integral of the applied voltage.

The inductive component is called the **inductor**, **choke** or **coil**.

1.1.8 Capacitance

SI unit of capacitance: F (farad), $1\,\text{F} = 1\,\dfrac{\text{As}}{\text{V}}$

- In a **capacitance** C the current i is proportional to the rate of change of the voltage v.

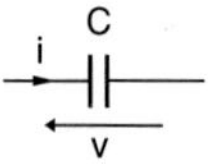

Fig. 1.7. Capacitance as a circuit symbol

$$i = C\frac{dv}{dt}, \qquad v = \frac{1}{C}\int_{t_0}^{t_1} i\, dt + V_0, \qquad C = \frac{i\, dt}{dv} \qquad (1.10)$$

The voltage applied across the capacitance at the beginning of the integration interval is V_0. If a capacitance is supplied with a constant current, the voltage increases linearly (Fig. 1.8).

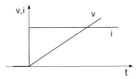

Fig. 1.8. Time progression of the voltage across a capacitance with a constant current flowing through it

- The voltage across a capacitance is continuous (cannot change instantaneously), while the current can change instantaneously.

The capacitive component is called the **capacitor**. When a current flows into a capacitance it can be said that the capacitor is being charged.

1.1.9 Ideal Voltage Source

A **voltage source** drives an electric current (Fig. 1.9).

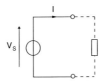

Fig. 1.9. Ideal voltage source

- The ideal **voltage source** supplies a voltage V_S, which is independent of the current I.

1.1.10 Ideal Current Source

- The ideal **current source** supplies a current I_S, which is independent of the applied voltage V (Fig. 1.10).

Fig. 1.10. Ideal current source

1.1.11 Kirchhoff's Law

Kirchhoff's laws describe the behaviour of current and voltage in electrical circuits. An electrical circuit can be represented by an **equivalent circuit diagram**. A circuit consists of **branches, nodes** and **loops*** (Fig. 1.11). Connection points are referred to as nodes. A branch joins two nodes, and a closed loop is formed with individual branches.

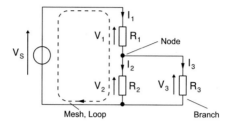

Fig. 1.11. Typical circuit of branches, nodes and loops

1.1.11.1 Kirchhoff's First Law (Current Law)

- The sum of all currents at a node is always equal to zero.

$$\sum_n I_n = 0 \tag{1.11}$$

Expressed differently, this means that the sum of currents flowing into a node is equal to the sum of currents flowing out of the node. This yields for the circuit given in Fig. 1.11:

$$I_1 - I_2 - I_3 = 0$$

Kirchhoff's current law is easier to understand if it is remembered that current always flows in a closed loop. This means that no extra current can 'join' the current path.

1.1.11.2 Kirchhoff's Second Law (Voltage Law)

- The sum of all voltages in a loop is always equal to zero.

$$\sum_m V_m = 0 \tag{1.12}$$

For the circuit given in Fig. 1.11 this yields:

$$\begin{aligned} -V_S + V_1 + V_2 &= 0, \quad \text{and} \\ -V_S + V_1 + V_3 &= 0, \quad \text{and} \\ V_2 - V_3 &= 0 \end{aligned}$$

---

* The term mesh may be used instead of loop.

1.1.12 Power and Energy

SI unit of power: W (watt), 1 W = 1 VA
SI unit of energy: J (joule), 1 J = 1 Ws

The **instantaneous power** is defined as:

$$p(t) = i(t) \cdot v(t) \tag{1.13}$$

In most technical applications the **average power** P is important. For example, the average power loss in a diode yields the heat dissipation in the diode.

$$P = \frac{1}{T} \int_0^T i(t) \, v(t) \, \mathrm{d}t \tag{1.14}$$

For DC this simplifies to:

$$P = V \cdot I \tag{1.15}$$

The **electrical energy** W is the integral of the power over time:

$$W = \int_{t_1}^{t_2} p(t) \, \mathrm{d}t = \int_{t_1}^{t_2} i(t) \, v(t) \, \mathrm{d}t \tag{1.16}$$

For DC this simplifies to:

$$W = P \cdot (t_2 - t_1) = V \cdot I \cdot (t_2 - t_1) \tag{1.17}$$

NOTE: Power and energy relate the electrical SI units with the mechanical and thermodynamic SI units, respectively. All calculations in systems with mechanical and thermodynamic quantities on one side and electrical quantities on the other side are done via this relationship.

EXAMPLE: What current is necessary to heat 1 l of water in 10 min from 0°C to 100°C using 230 V? (Remember, 1 J/s = 1 V·A.)

$$W = 100 \text{ kcal} = 418.7 \text{ kJ} = 0.116 \text{ kWh}$$
$$W = VIt$$
$$I = \frac{W}{V \cdot t} = \frac{418.7 \text{ kJ}}{230 \text{ V} \cdot 600 \text{ s}} = 3.0 \text{ A}$$

1.1.12.1 Energy and Power in a Resistor

In a resistor electric energy is converted into thermal energy. For resistors $v \propto i$, therefore:

$$p(t) = v(t) \, i(t) = i(t)^2 R = \frac{v(t)^2}{R} \tag{1.18}$$

The resulting change in resistance caused by the heating is neglected here.

The average power is:

$$P = \frac{1}{T} \int_0^T v(t)\,i(t)\,dt = \frac{1}{T} \int_0^T i(t)^2 R\,dt = \frac{1}{T} \int_0^T \frac{v(t)^2}{R}\,dt \tag{1.19}$$

For DC this simplifies to:

$$P = V \cdot I = I^2 \cdot R = \frac{V^2}{R} \tag{1.20}$$

EXAMPLE: A motor delivers mechanical energy of $P = 500$ W at 230 V. What is the value of the equivalent resistor that represents the power consumption of this motor, assuming that the motor is loss-free?

$$P = \frac{V^2}{R} \quad \Rightarrow \quad R = \frac{(230 \text{ V})^2}{500 \text{ W}} = 106\,\Omega$$

The energy W that is converted into heat in a time interval can be calculated as:

$$W = \int_{t_1}^{t_2} p(t)\,dt \tag{1.21}$$

For DC it holds that:

$$W = V \cdot I \cdot (t_2 - t_1) = I^2 \cdot R \cdot (t_2 - t_1) = \frac{V^2}{R}(t_2 - t_1) \tag{1.22}$$

1.1.12.2 Energy in an Inductor

An ideal inductor absorbs and releases electrical energy. No energy is transformed into heat. The energy is stored in the magnetic field (see Sect. 2.3.16).

For the energy stored in an inductor, it holds in general that:

$$W = \int_{t_0}^{t_1} v(t)\,i(t)\,dt + W_0$$

The starting energy in the time interval under consideration is W_0. With $v = L\,di/dt$ and $W_0 = 0$, it follows that:

$$W = \int L \frac{di}{dt} i\,dt = L \int i\,di = \frac{1}{2} L\,i^2$$

$$W = \frac{1}{2} L\,i^2 \tag{1.23}$$

For DC this is

$$W = \frac{1}{2} L\,I^2 \tag{1.24}$$

- The energy stored in an inductor is proportional to the inductance and to the square of the current flowing through it.

1.1.12.3 Energy in a Capacitor

An ideal capacitor absorbs and releases electrical energy. No energy is transformed into heat. The energy is stored in the electric field (see Sect. 2.1.12).

For the energy stored in a capacitor, it holds in general that:

$$W = \int_{t_0}^{t_1} v(t)\, i(t)\, dt + W_0$$

The starting energy in the time interval under consideration is W_0. With $i = C\, dv/dt$ and $W_0 = 0$, it follows that:

$$W = \int C \frac{dv}{dt} v\, dt = C \int v\, dv = \frac{1}{2} C v^2$$

$$\boxed{W = \frac{1}{2} C v^2} \quad (1.25)$$

For DC this is

$$\boxed{W = \frac{1}{2} C V^2} \quad (1.26)$$

- The energy stored in a capacitor is proportional to the capacitance and to the square of the voltage across it.

1.1.13 Efficiency

The **efficiency** η is defined as the ratio of the effective (useful) power P_{out} to the total power P_{total}.

$$\boxed{\eta = \frac{P_{\text{out}}}{P_{\text{total}}} = \frac{P_{\text{out}}}{P_{\text{out}} + P_{\text{loss}}}} \quad (1.27)$$

EXAMPLE: A motor consumes a power of $P = 230\,\text{V} \cdot 5\,\text{A}$ and delivers a torque of $M = 2.5\,\text{Nm}$ at $n = 3000\,\text{rpm}$ (rounds per minute).

The efficiency is:

$$\eta = \frac{P_{\text{out}}}{P_{\text{total}}} = \frac{M\omega}{VI} = \frac{M \frac{2\pi}{60} n}{VI} = 0.68 = 68\%$$

Next, the efficiency of a real voltage source with a load resistor is calculated. The load resistor R_L corresponds to the effective power, and the source resistor R_S corresponds to the power loss (Fig. 1.12).

$$P_{\text{out}} = V \cdot I, \qquad P_{\text{total}} = V_S \cdot I, \qquad P_{\text{loss}} = I^2 \cdot R_S$$

$$\eta = \frac{VI}{V_S I} = \frac{V_S \dfrac{R_L}{R_S + R_L} \dfrac{V_S}{R_S + R_L}}{V_S \dfrac{V_S}{R_S + R_L}} = \frac{R_L}{R_S + R_L}$$

- The smaller the source resistance, the higher the efficiency! If the source resistance of a voltage source is zero then the efficiency has a value of 1 (Fig. 1.13).

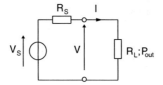

Fig. 1.12. Real voltage source with a load resistor

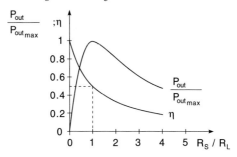

Fig. 1.13. Efficiency and supplied power for a real voltage source

1.1.14 Maximum Power Transfer

In some cases the efficiency is not as important as the voltage source delivering maximum power. This is true, for example, for many sensors and for audio systems, where the signal power is very low and the power loss is unimportant.

The useful power P_{out} that is delivered by a voltage source with source resistance R_S is:

$$P_{out} = VI = V_S \frac{R_L}{R_S + R_L} \frac{V_S}{R_S + R_L} = V_S^2 \frac{R_L}{(R_S + R_L)^2}$$

With $dP_{out}/dR_L = 0$, the load resistance R_L at which the useful power P_{out} reaches a maximum can be determined:

$$\frac{dP_{out}}{dR_L} = 0 = V_S^2 \frac{(R_S + R_L)^2 - 2R_L(R_S + R_L)}{(R_S + R_L)^4}$$

This yields:

$$\boxed{R_L = R_S} \qquad (1.28)$$

This is known as **impedance matching**.

The efficiency is then:

$$\boxed{\eta = \frac{R_L}{R_S + R_L} = \frac{1}{2} = 50\%} \qquad (1.29)$$

- With a load of $R_L = R_S$, a voltage source delivers the maximum power. The efficiency is then 50%.

1.2 Basic Circuits

1.2.1 Real Voltage and Current Sources

1.2.1.1 Real Voltage Source

The terminal voltage of a real voltage source (e.g. a battery) depends on the current being drawn from it. The terminal voltage decreases as the output current increases. A real voltage source can often be described by an equivalent circuit, as shown in Fig. 1.14, and consists of an ideal voltage source V_S and a source resistor R_S in series.

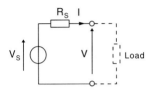

Fig. 1.14. Equivalent circuit of a real voltage source

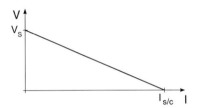

Fig. 1.15. Current–voltage diagram of a voltage source with a source resistance

Calculation of the current–voltage characteristic can be done through application of Kirchhoff's voltage law:

$$-V_S + I \cdot R_S + V = 0$$
$$\boxed{V = V_S - I \cdot R_S} \tag{1.30}$$

This equation describes a linear relationship, that is, the voltage V decreases linearly with increasing current I. Nonlinearities of a real voltage source are not considered in this equivalent circuit. However, in most cases, this equivalent circuit is a good representation of a real voltage source.

- In the **open-circuit** case (i.e. $I = 0$), $V = V_S$ can be measured at the terminals of the equivalent voltage source.

- In the case of a short circuit (i.e. $V = 0$), the current is:

$$I = I_{s/c} = \frac{V_S}{R_S}$$

$I_{s/c}$ is known as the **short-circuit current**.

- The lower the source resistance R_S, the more similar the real voltage source is to an ideal voltage source.

1.2.1.2 Real Current Source

The current delivered by a real current source is dependent on the applied voltage. The current decreases as the resistance of the load increases. For example, a photodiode is a current source for which incoming light causes a current to flow that is almost independent of the applied voltage. A real current source often can be described in the equivalent circuit in Fig. 1.16. It consists of an ideal current source I_S in parallel with a source resistor R_S.

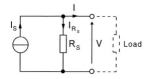

Fig. 1.16. Equivalent circuit of a real current source

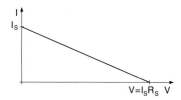

Fig. 1.17. Current–voltage diagram of a current source with internal resistance

If the load has a large resistance, then a large voltage appears at the terminals. The higher the voltage V, the more the source resistance R_S drains the current, which is therefore lost at the terminals.

Calculation of the current-voltage characteristic can be done through application of Kirchhoff's current law:

$$-I_S + \frac{V}{R_S} + I = 0$$

$$\boxed{I = I_S - \frac{V}{R_S}} \tag{1.31}$$

This equation describes a linear relationship in which the current I decreases linearly with increasing voltage V. Nonlinearities of a real current source are not considered in this equivalent circuit. However, in most cases this equivalent circuit is a good representation of a real current source.

- For a short circuit ($V = 0$), the current $I = I_S$.

- For an open circuit the entire current I_S flows through the internal resistance. Then the voltage is:
$$V = V_{o/c} = I_S R_S$$
$V_{o/c}$ is the open-circuit voltage.

- The higher the source resistance R_S, the more similar the real current source is to an ideal current source.

1.2.1.3 Voltage–Current Source Conversion

Current and voltage sources have an identical linear voltage–current behaviour, which is shown as a negatively sloped line on a V–I-graph. Therefore a real current source can be regarded as a voltage source with a high internal resistance, and a real voltage source can be regarded as a current source with a low internal resistance (Fig. 1.18).

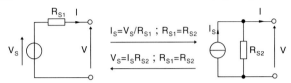

Fig. 1.18. Changing from voltage to current sources and vice versa

1.2.2 Circuit Elements in Series and Parallel

- **Series combination**: Circuit elements in series experience the *same* current flow.
- **Parallel combination**: Circuit elements in parallel experience the *same* applied voltage.

1.2.2.1 Series Combination of Resistors

A series combination of resistors R is shown in Fig. 1.19. Application of Kirchhoff's voltage law yields:

$$V = IR_1 + IR_2 + \cdots + IR_n = I(R_1 + R_2 + \cdots + R_n) = I \cdot R_{\text{total}}$$

$$\boxed{R_{\text{total}} = R_1 + R_2 + \cdots + R_n} \tag{1.32}$$

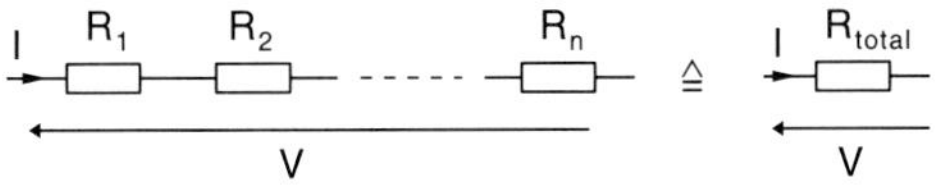

Fig. 1.19. Series combination of resistors

1.2.2.2 Parallel Combination of Resistors

A number of resistors R combined in parallel is shown in Fig. 1.20. Application of Kirchhoff's current law yields:

$$I = \frac{V}{R_1} + \frac{V}{R_2} + \cdots + \frac{V}{R_n} = V\left(\frac{1}{R_1} + \frac{1}{R_2} + \cdots + \frac{1}{R_n}\right) = V\frac{1}{R_{\text{total}}}$$

$$\boxed{\frac{1}{R_{\text{total}}} = \frac{1}{R_1} + \frac{1}{R_2} + \cdots + \frac{1}{R_n}} \tag{1.33}$$

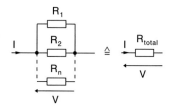

Fig. 1.20. Parallel combination of resistors

For the parallel combination of *two* resistors:

$$\frac{1}{R_{\text{total}}} = \frac{1}{R_1} + \frac{1}{R_2}, \qquad R_{\text{total}} = \frac{R_1 R_2}{R_1 + R_2} \tag{1.34}$$

- The resulting resistance of a parallel combination of resistors is smaller than either of the individual resistances.

1.2.2.3 Series Combination of Conductances

For a number of conuctances G combined in series (Fig. 1.21), the application of Kirchhoff's voltage law yields:

$$V = \frac{I}{G_1} + \frac{I}{G_2} + \cdots + \frac{I}{G_n} = I \left(\frac{1}{G_1} + \frac{1}{G_2} + \cdots + \frac{1}{G_n} \right) = I \frac{1}{G_{\text{total}}}$$

$$\frac{1}{G_{\text{total}}} = \frac{1}{G_1} + \frac{1}{G_2} + \cdots + \frac{1}{G_n} \tag{1.35}$$

For the series combination of *two* conductances:

$$\frac{1}{G_{\text{total}}} = \frac{1}{G_1} + \frac{1}{G_2}, \qquad G_{\text{total}} = \frac{G_1 G_2}{G_1 + G_2} \tag{1.36}$$

- The resulting conductance of a series combination is smaller than either of the individual conductances.

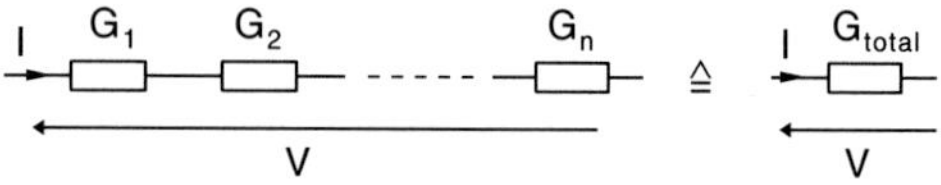

Fig. 1.21. Series combination of conductances

1.2.2.4 Parallel Combination of Conductances

The parallel combination of a number of conductances G is shown in Fig. 1.22. Application of Kirchhoff's current law yields:

$$I = VG_1 + VG_2 + \cdots + VG_n = V(G_1 + G_2 + \cdots + G_n) = V \cdot G_{\text{total}}$$

$$G_{\text{total}} = G_1 + G_2 + \cdots + G_n \tag{1.37}$$

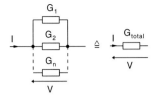

Fig. 1.22. Parallel combination of conductances

1.2.2.5 Series Combination of Inductances

For a number of inductances L combined in series (Fig. 1.23), the application of Kirchhoff's voltage law yields:

$$v = L_1 \frac{\mathrm{d}i}{\mathrm{d}t} + L_2 \frac{\mathrm{d}i}{\mathrm{d}t} + \cdots + L_n \frac{\mathrm{d}i}{\mathrm{d}t} = (L_1 + L_2 + \cdots + L_n) \frac{\mathrm{d}i}{\mathrm{d}t} = L_{\text{total}} \frac{\mathrm{d}i}{\mathrm{d}t}$$

$$\boxed{L_{\text{total}} = L_1 + L_2 + \cdots + L_n} \tag{1.38}$$

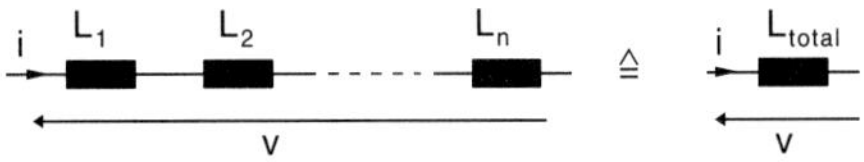

Fig. 1.23. Series combination of inductances

1.2.2.6 Parallel Combination of Inductances

For a parallel combination of inductances L (Fig. 1.24), application of Kirchhoff's current law yields:

$$\begin{aligned} i &= \frac{1}{L_1} \int v \, \mathrm{d}t + \frac{1}{L_2} \int v \, \mathrm{d}t + \cdots + \frac{1}{L_n} \int v \, \mathrm{d}t + I_{01} + I_{02} + \cdots + I_{0n} \\ &= \left(\frac{1}{L_1} + \frac{1}{L_2} + \cdots + \frac{1}{L_n} \right) \int v \, \mathrm{d}t + I_{01} + I_{02} + \cdots + I_{0n} \\ &= \frac{1}{L_{\text{total}}} \int v \, \mathrm{d}t + I_0 \end{aligned}$$

$$\boxed{\frac{1}{L_{\text{total}}} = \frac{1}{L_1} + \frac{1}{L_2} + \cdots + \frac{1}{L_n}} \tag{1.39}$$

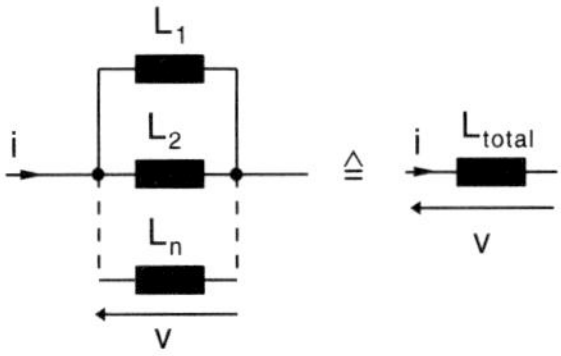

Fig. 1.24. Parallel combination of inductances

For the parallel combination of *two* inductances:

$$\frac{1}{L_{\text{total}}} = \frac{1}{L_1} + \frac{1}{L_2}, \qquad L_{\text{total}} = \frac{L_1 L_2}{L_1 + L_2} \tag{1.40}$$

- The resulting inductance of a parallel combination is smaller than either of the individual inductances.

1.2.2.7 Series Combination of Capacitances

For capacitances C combined in series (Fig. 1.25), the application of Kirchhoff's voltage law yields:

$$\begin{aligned}
v &= \frac{1}{C_1}\int i\,\mathrm{d}t + V_{01} + \frac{1}{C_2}\int i\,\mathrm{d}t + V_{02} + \cdots + \frac{1}{C_n}\int i\,\mathrm{d}t + V_{0n} \\
&= \left(\frac{1}{C_1} + \frac{1}{C_2} + \cdots + \frac{1}{C_n}\right)\int i\,\mathrm{d}t + V_{01} + V_{02} + \cdots + V_{0n} \\
&= \frac{1}{C_{\text{total}}}\int i\,\mathrm{d}t + V_0
\end{aligned}$$

$$\frac{1}{C_{\text{total}}} = \frac{1}{C_1} + \frac{1}{C_2} + \cdots + \frac{1}{C_n} \tag{1.41}$$

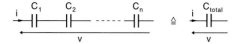

Fig. 1.25. Series combination of capacitances

For the series combination of *two* capacitances:

$$\frac{1}{C_{\text{total}}} = \frac{1}{C_1} + \frac{1}{C_2}, \qquad C_{\text{total}} = \frac{C_1 C_2}{C_1 + C_2} \tag{1.42}$$

- The resulting capacitance of a series combination is smaller than either of the individual capacitances.

1.2.2.8 Parallel Combination of Capacitances

For a parallel combination of capacitances C (Fig. 1.26), the application of Kirchhoff's current law yields:

$$i = C_1\frac{\mathrm{d}v}{\mathrm{d}t} + C_2\frac{\mathrm{d}v}{\mathrm{d}t} + \cdots + C_n\frac{\mathrm{d}v}{\mathrm{d}t} = (C_1 + C_2 + \cdots + C_n)\frac{\mathrm{d}v}{\mathrm{d}t}$$

$$C_{\text{total}} = C_1 + C_2 + \cdots + C_n \tag{1.43}$$

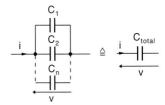

Fig. 1.26. Parallel combination of capacitances

1.2.3 Star–Delta Transformation (Wye–Delta Transformation)

A **star-configuration** can be transformed into an equivalent **delta-configuration** and vice versa (Fig. 1.27).[†] This can be necessary when calculating complex circuits of resistors in order to reduce the calculation to those of series and parallel combinations.

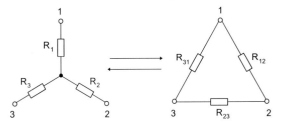

Fig. 1.27. Star–delta transformation

In a star–delta transformation:

$$\begin{aligned} R_{23} &= R_2 + R_3 + \frac{R_2 R_3}{R_1}, \\ R_{31} &= R_1 + R_3 + \frac{R_1 R_3}{R_2}, \\ R_{12} &= R_1 + R_2 + \frac{R_1 R_2}{R_3} \end{aligned} \quad (1.44)$$

In a delta–star transformation:

$$\begin{aligned} R_1 &= \frac{R_{31} R_{12}}{R_{12} + R_{23} + R_{31}}, \\ R_2 &= \frac{R_{23} R_{12}}{R_{12} + R_{23} + R_{31}}, \\ R_3 &= \frac{R_{23} R_{31}}{R_{12} + R_{23} + R_{31}} \end{aligned} \quad (1.45)$$

[†] In American literature the term *Wye* may be used instead of *star*.

1.2.4 Voltage and Current Divider

1.2.4.1 Voltage Divider

If the same current flows through two resistors (Fig. 1.28), then:

$$I = \frac{V_1}{R_1} = \frac{V_2}{R_2} = \frac{V}{R_1 + R_2}$$

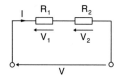

Fig. 1.28. Voltage divider

The **voltage-divider rule** follows from this:

$$\boxed{\frac{V_1}{V_2} = \frac{R_1}{R_2}, \quad \frac{V_1}{V} = \frac{R_1}{R_1 + R_2}, \quad \frac{V_2}{V} = \frac{R_2}{R_1 + R_2}} \tag{1.46}$$

- In a series combination the individual voltages are proportional to the resistances they appear across. This also holds for series combinations of more than two resistors.

1.2.4.2 Current Divider

If the same voltage is applied across two conductances or resistances (Fig. 1.29), then:

$$V = \frac{I_1}{G_1} = \frac{I_2}{G_2} = \frac{I}{G_1 + G_2}$$

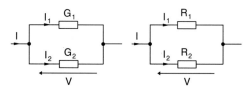

Fig. 1.29. Current divider

The **current-divider rule** follows from this:

$$\boxed{\frac{I_1}{I_2} = \frac{G_1}{G_2}, \quad \frac{I_1}{I} = \frac{G_1}{G_1 + G_2}, \quad \frac{I_2}{I} = \frac{G_2}{G_1 + G_2}} \tag{1.47}$$

- In parallel combinations the individual currents are proportional to the conductances they flow through. This also holds for combinations of more than two conductances.

Replacing the conductances with resistances gives:

$$\boxed{\frac{I_1}{I_2} = \frac{R_2}{R_1}, \quad \frac{I_1}{I} = \frac{R_2}{R_1 + R_2}, \quad \frac{I_2}{I} = \frac{R_1}{R_1 + R_2}} \tag{1.48}$$

- The individual currents behave inversely to the individual resistances.

1.2.4.3 Capacitive and Inductive Dividers

Examples of capacitive and inductive dividers along with their respective voltage and current relations are shown in Fig. 1.30. The specifications given in Fig. 1.30 are valid provided that *before* the voltage v or the current i are applied, the circuit elements were energy-free, i.e. $V_n(t=0) = 0$, and $i_n(t=0) = 0$.

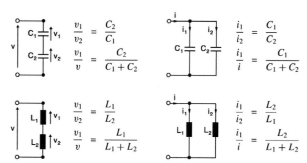

Fig. 1.30. Capacitive (top) and inductive (bottom) dividers

1.2.5 *RC* and *RL* Combinations

This section concerns the settling processes that occur when a DC voltage or a DC current is applied to a circuit containing an inductance or a capacitance next to a resistance. Processes like this can be described with first-order differential equations.

A **linear first-order differential equation** has the following form:

$$q(t) = \tau \cdot \frac{dy}{dt} + y \tag{1.49}$$

The solution of the inhomogeneous differential equation is combined from the solution of the homogeneous equation $\left(0 = \tau \cdot \frac{dy}{dt} + y\right)$ plus any special solution (for example, in the case of the step response, the special solution may be obtained by examining the system behaviour as $t \to \infty$).

The solution of the inhomogeneous differential equation is then:

$$y(t) = y(t)_{\text{homogeneous}} + y(t)_{\text{special}} \tag{1.50}$$

The coefficient τ is called the **time constant**.

The solution of the first-order homogeneous differential equation is

$$y(t)_{\text{homogeneous}} = K_1 \cdot e^{-\frac{t}{\tau}} \tag{1.51}$$

The constant K_1 is evaluated from the starting conditions of the system, that is, $y(t=0)$.

EXAMPLE: Calculation of the step response of an RC low-pass filter:

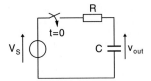

Fig. 1.31. Series combination of R and C as a low-pass filter

Applying Kirchhoff's voltage law to the system shown in Fig. 1.31, it can be seen that

$$V_S = iR + V_{\text{out}}, \quad \text{with} \quad i = C\frac{dV_{\text{out}}}{dt}.$$

The inhomogeneous differential equation follows from this:

$$V_S = \underbrace{RC}_{\tau}\frac{dV_{\text{out}}}{dt} + V_{\text{out}}$$

The solution of the inhomogeneous differential equation is:

$$V_{\text{out}}(t) = V_{\text{out}}(t)_{\text{homogeneous}} + V_{\text{out}}(t)_{\text{special}} = K_1 \cdot e^{-\frac{t}{RC}} + V_S$$

Given the starting condition $V_{\text{out}}(0) = 0$, K_1 can be calculated:

$$0 = K_1 + V_S \quad \Rightarrow \quad K_1 = -V_S$$

The solution of the inhomogeneous differential equation is therefore:

$$V_{\text{out}}(t) = -V_S \cdot e^{-\frac{t}{RC}} + V_S = V_S\left(1 - e^{-\frac{t}{RC}}\right)$$

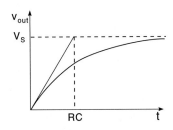

Fig. 1.32. Step response of $V_{\text{out}}(t)$

- Constant τ is called the time constant. At the time τ the function value has reached 63% of its final value. After 5τ, the function value is within 1% of its final value (Fig. 1.32).

1.2.5.1 Series Combination of *R* and *C* Driven by a Voltage Source

The switch is closed at time $t = 0$. The capacitor is assumed to be uncharged at this time. Application of Kirchhoff's voltage law leads to the differential equation:

$$V_S = iR + \frac{1}{C} \int i \, dt$$

The solution of the differential equation is given by:

$$\begin{aligned}
i(t) &= \frac{V_S}{R} \cdot e^{-\frac{t}{RC}}, \\
V_C(t) &= V_S \left(1 - e^{-\frac{t}{RC}}\right), \\
V_R(t) &= V_S \cdot e^{-\frac{t}{RC}}, \quad \tau = RC
\end{aligned} \tag{1.52}$$

The capacitor is charged via the resistor. Since the voltage across the capacitor increases during the charging process, the voltage across the resistor decreases. The current is proportional to the voltage V_R and therefore also decreases (Fig. 1.33).

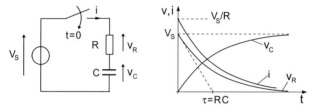

Fig. 1.33. Series combination of *R* and *C* driven by a voltage source

1.2.5.2 Series Combination of *R* and *C* Driven by a Current Source

A series combination of a resistor and a capacitor driven by a current source is shown in Fig. 1.34. The switch is toggled at time $t = 0$. The capacitor is assumed to be uncharged at this time.

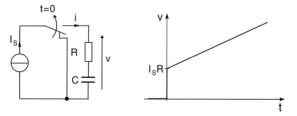

Fig. 1.34. Series combination of *R* and *C* driven by a current source

Application of Kirchhoff's voltage law yields:

$$v = I_S R + \frac{1}{C}\int I_S \, dt$$

Solution:

$$v(t) = I_S R + \frac{1}{C} I_S t \tag{1.53}$$

1.2.5.3 Parallel Combination of R and C Driven by a Current Source

The parallel combination of a resistor and a capacitor is shown in Fig. 1.35. The switch is toggled at time $t = 0$. The capacitor is assumed to be uncharged at this time.

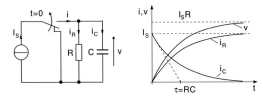

Fig. 1.35. Parallel combination of R and C driven by a current source

Application of Kirchhoff's current law leads to the differential equation:

$$I_S = \frac{v}{R} + C \frac{dv}{dt}$$

Solution of the differential equation gives:

$$\begin{aligned} v(t) &= I_S R \left(1 - e^{-\frac{t}{RC}}\right), \\ i_R(t) &= I_S \left(1 - e^{-\frac{t}{RC}}\right), \\ i_C(t) &= I_S \cdot e^{-\frac{t}{RC}}, \quad \tau = RC \end{aligned} \tag{1.54}$$

1.2.5.4 Parallel Combination of R and C Driven by a Voltage Source

Theoretically, the voltage across the capacitor has to change in an infinitely short time. Therefore the current $i_C = C \cdot dv/dt$ should be an infinitely high value. In practice such circuits lead to the destruction of the switch (Fig. 1.36).

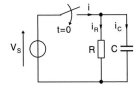

Fig. 1.36. Parallel combination of R and C driven by a voltage source

1.2.5.5 Series Combination of *R* and *L* Driven by a Voltage Source

Application of Kirchhoff's voltage law to a resistor and a capacitor in series driven by a voltage source leads to the differential equation:

$$V_S = iR + L\frac{di}{dt}$$

Solution of the differential equation yields:

$$\boxed{\begin{aligned} i(t) &= \frac{V_S}{R}\left(1 - e^{-\frac{t}{L/R}}\right), \\ V_R(t) &= V_S\left(1 - e^{-\frac{t}{L/R}}\right), \\ V_L(t) &= V_S \cdot e^{-\frac{t}{L/R}}, \quad \tau = \frac{L}{R} \end{aligned}} \qquad (1.55)$$

The voltage $V_L = V_S$ is applied at time $t = 0$. The current i increases at a rate of $di/dt = V_S/L$. Therefore, the voltage drop across R increases, and V_L and di/dt decrease (Fig. 1.37).

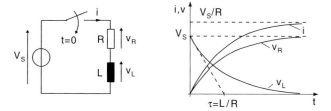

Fig. 1.37. Series combination of *R* and *L* driven by a voltage source

1.2.5.6 Series Combination of *R* and *L* Driven by a Current Source

Toggling the switch shown in Fig. 1.38 can be considered an attempt to create an infinitely large di/dt. This would result in an infinitely high voltage across L, which is not achievable in reality.

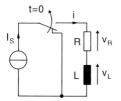

Fig. 1.38. Series combination of *R* and *L* driven by a current source

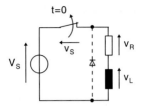

Fig. 1.39. Switching off a resistive–inductive load

Very high values of di/dt result when switching off a resistive–inductive load (Fig. 1.39). At the time $t = 0$ the current V_S/R is flowing. Opening the switch results in a current change of $di/dt \to -\infty$. Therefore $v_L \to -\infty$. Kirchhoff's voltage law applied to the loop gives $V_S = v_{\text{Switch}} + v_R + v_L$. This shows that not only v_L but also v_{Switch} increases greatly, but v_R and V_S have finite values. In practice, this results in the destruction of the switch. To avoid this, a diode can be added to the circuit, called a **free-wheeling diode**.

1.2.5.7 Parallel Combination of R and L Driven by a Voltage Source

A resistor and an inductor connected in parallel and driven by a voltage source are shown in Fig. 1.40. Application of Kirchhoff's current law for this circuit yields:

$$i(t) = \frac{V_S}{R} + \frac{1}{L} \int V_S \, dt$$

Solution:

$$\boxed{i(t) = \frac{V_S}{R} + \frac{V_S t}{L}} \quad (1.56)$$

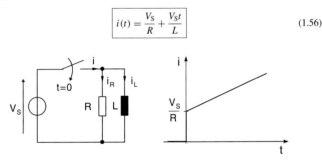

Fig. 1.40. Parallel combination of R and L driven by a voltage source

1.2.5.8 Parallel Combination of R and L Driven by a Current Source

A resistor and an inductor are combined in parallel and driven by a current source are shown in Fig. 1.41. The switch is closed at $t = 0$. At this time the current i_L is assumed to be zero. Application of Kirchhoff's current law leads to the differential equation:

$$I_S = \frac{v}{R} + \frac{1}{L} \int v \, dt$$

Solution of the differential equation yields:

$$v(t) = I_S R \cdot e^{-\frac{t}{L/R}},$$
$$i_L(t) = I_S \left(1 - e^{-\frac{t}{L/R}}\right), \qquad (1.57)$$
$$i_R(t) = I_S \cdot e^{-\frac{t}{L/R}}, \quad \tau = \frac{L}{R}$$

After toggling the switch a current I_S flows through resistor R. The current i_L increases by a factor $d i / d t = I_S R / L$. While i_L increases i_R decreases until the inductance has taken over the entire current I_S. Then $v = 0$ because $i_R = 0$ (Fig. 1.41).

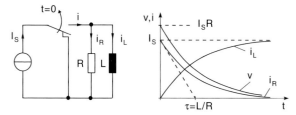

Fig. 1.41. Parallel combination of R and L driven by a current source

1.2.6 RLC Combinations

This section deals with transients that appear when applying a DC voltage or DC current to a circuit containing inductances *and* capacitances. Systems containing two independent energy storage components can oscillate, depending on the damping of the system. Inductances and capacitances are independent energy storage components in this sense. This kind of process is described by second-order differential equations.

A linear second-order differential equation with constant coefficients has the form:

$$q(t) = \frac{1}{\omega_0^2} \frac{d^2 y}{d t^2} + \frac{2D}{\omega_0} \frac{d y}{d t} + y \qquad (1.58)$$

The solution of the inhomogeneous differential equation is combined from the solution of the homogeneous differential equation:

$$\frac{1}{\omega_0^2} \frac{d^2 y}{d t^2} + \frac{2D}{\omega_0} \frac{d y}{d t} + y = 0$$

and any special solution. In order to calculate the step response, it is easiest to regard $y(t \to \infty)$.

The solution of the inhomogeneous differential equation is then:

$$y(t) = y(t)_{\text{homogeneous}} + y(t)_{\text{special}} \qquad (1.59)$$

The coefficient D is called the **damping ratio**, and the coefficient ω_0 is called the **resonant frequency**: $\omega_0 = 2\pi f_0$. Three different cases have to be distinguished for the solution of second-order homogeneous differential equations:

1. **Overdamped case**
   $\mathbf{D > 1}$:

$$y(t) = K_1 e^{\lambda_1 t} + K_2 e^{\lambda_2 t}, \quad \lambda_{1,2} = -D\omega_0 \pm \omega_0 \sqrt{D^2 - 1} \tag{1.60}$$

2. **Critically damped case**
   $\mathbf{D = 1}$:

$$y(t) = (K_1 t + K_2) e^{-D\omega_0 t} \tag{1.61}$$

3. **Underdamped case**
   $\mathbf{D < 1}$:

$$y(t) = e^{-D\omega_0 t}(K_1 \cos \omega t + K_2 \sin \omega t), \quad \omega = \omega_0 \sqrt{1 - D^2} \tag{1.62}$$

The constants K_1 and K_2 are determined from the initial conditions, namely $y(0)$ and $y'(0)$.

The angular frequency ω is called the **natural frequency**, which is the frequency of the fading oscillation of a damped system. Its value is slightly lower than the resonant frequency and depends on the damping ratio.

NOTE: An oscillator with a resonant circuit as a means of frequency determination oscillates at the resonant frequency. This is because the circuit attenuation is compensated with an active component (e.g. a transistor), so that $D = 0$.

In the context of oscillating circuits the following terms are also commonly used:

Loss factor: $\quad d = 2D$

Quality or Q-factor: $\quad Q = \dfrac{1}{2D}$

Bandwidth: $\quad B = \dfrac{\omega_0}{2\pi} \cdot 2D$

1.2.6.1 Series Combination of R, L and C

The entire procedure of solving the differential equation can be explained using the example of the series combination of R, L and C. This circuit forms a low-pass filter (Fig. 1.42), and the step response $v_\text{out}(t)$ is calculated here.

Application of Kirchhoff's voltage law yields: $V_S = L\dfrac{di}{dt} + R \cdot i + v_\text{out}$. With $i = C\dfrac{dv_\text{out}}{dt}$, the inhomogeneous differential equation is:

$$V_S = \underbrace{LC}_{\frac{1}{\omega_0^2}} \frac{d^2 V_\text{out}}{dt^2} + \underbrace{RC}_{\frac{2D}{\omega_0}} \frac{dV_\text{out}}{dt} + V_\text{out}, \quad \omega_0 = \frac{1}{\sqrt{LC}}, \quad D = \frac{R}{2}\sqrt{\frac{C}{L}} \tag{1.63}$$

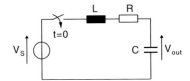

Fig. 1.42. Series combination of R, L and C as a low-pass filter

The homogeneous differential equation is:

$$0 = LC\frac{d^2 v_{\text{out}}}{dt^2} + RC\frac{d v_{\text{out}}}{dt} + v_{\text{out}} \tag{1.64}$$

One special solution of the differential equation is, for example:

$$v_{\text{out, special}} = v_{\text{out}}(t \to \infty) = v_{\text{S}} \tag{1.65}$$

To determine the coefficients K_1 and K_2 in the general solution two specific values are required for $v_{\text{out}}(t)$. The initial conditions usually used are:

$$v_{\text{out}}(t = 0) = 0, \quad \text{and} \quad \left.\frac{d v_{\text{out}}}{dt}\right|_{t=0} = 0 \tag{1.66}$$

The solutions of the inhomogeneous differential equation are:

1. Overdamped case
$D > 1$:

$$v_{\text{out}}(t) = v_{\text{out}}(t)_{\text{homogeneous}} + v_{\text{out, special}}$$
$$= K_1 e^{\lambda_1 t} + K_2 e^{\lambda_2 t} + v_{\text{S}}, \qquad \lambda_{1,2} = -D\omega_0 \pm \omega_0\sqrt{D^2 - 1}$$

K_1 and K_2 can be determined from the initial conditions: first the derivative $d v_{\text{out}}(t)/dt$ is calculated, then $t = 0$ is inserted in $v_{\text{out}}(t)$ and in $d v_{\text{out}}(t)/dt$. This results in two equations for K_1 and K_2.

$$\frac{d v_{\text{out}}(t)}{dt} = \lambda_1 K_1 e^{\lambda_1 t} + \lambda_2 K_2 e^{\lambda_2 t}$$

$$v_{\text{out}}(t = 0) = 0 \Longrightarrow K_1 + K_2 + V_{\text{S}} = 0$$

$$\left.\frac{d v_{\text{out}}}{dt}\right|_{t=0} = 0 \Longrightarrow \lambda_1 K_1 + \lambda_2 K_2 = 0$$

$$\Longrightarrow \quad K_1 = \frac{\lambda_2 V_{\text{S}}}{\lambda_2 - \lambda_1}, \text{ and } K_2 = \frac{\lambda_1 V_{\text{S}}}{\lambda_2 - \lambda_1}$$

- The solution of the differential equation is:

$$v_{\text{out}}(t) = \frac{V_{\text{S}}}{\lambda_1 - \lambda_2}(\lambda_2 e^{\lambda_1 t} - \lambda_1 e^{\lambda_2 t}) + V_{\text{S}} \tag{1.67}$$

2. Critically damped case
$D = 1$:

$$v_{\text{out}}(t) = v_{\text{out}}(t)_{\text{homogeneous}} + v_{\text{out, special}} = (K_1 t + K_2)e^{-\omega_0 t} + V_{\text{S}}, \qquad D\omega_0 = \omega$$

K_1 and K_2 can be calculated from the initial conditions: first the derivative $\mathrm{d}v_{\text{out}}/\mathrm{d}t$ needs to be calculated. Then $t = 0$ is inserted in $v_{\text{out}}(t)$ and in $\mathrm{d}v_{\text{out}}/\mathrm{d}t$. This results in two equations for K_1 and K_2.

$$\frac{\mathrm{d}v_{\text{out}}}{\mathrm{d}t} = K_1 e^{-\omega_0 t} - (K_1 t + K_2)\omega_0 e^{-\omega_0 t}$$

$$v_{\text{out}}(t=0) = 0 \Longrightarrow K_2 + V_S = 0$$

$$\left.\frac{\mathrm{d}v_{\text{out}}}{\mathrm{d}t}\right|_{t=0} = 0 \Longrightarrow K_1 - \omega_0 K_2 = 0$$

$$\Longrightarrow \quad K_1 = -\omega_0 V_S, \text{ and } K_2 = -V_S$$

- The solution of the differential equation for the critically damped case is:

$$\boxed{v_{\text{out}}(t) = -(\omega_0 V_S t + V_S)e^{-\omega_0 t} + V_S} \quad (1.68)$$

3. Underdamped case
D < 1:

$$v_{\text{out}}(t) = v_{\text{out}}(t)_{\text{homogeneous}} + v_{\text{out, special}}$$
$$= e^{-D\omega_0 t}(K_1 \cos \omega t + K_2 \sin \omega t) + V_S, \qquad \omega = \omega_0 \sqrt{1 - D^2}$$

K_1 and K_2 can be calculated from the initial conditions: first the derivative of $\mathrm{d}v_{\text{out}}/\mathrm{d}t$ is calculated. Then $t = 0$ is inserted in $v_{\text{out}}(t)$ and in $\mathrm{d}v_{\text{out}}/\mathrm{d}t$. This results in two equations for K_1 and K_2.

$$\frac{\mathrm{d}v_{\text{out}}}{\mathrm{d}t} = -D\omega_0 e^{-D\omega_0 t}(K_1 \cos \omega t + K_2 \sin \omega t) + e^{-D\omega_0 t}(-K_1 \omega \sin \omega t + K_2 \omega \cos \omega t)$$

$$v_{\text{out}}(t=0) = 0 \Longrightarrow K_1 + V_S = 0$$

$$\left.\frac{\mathrm{d}v_{\text{out}}}{\mathrm{d}t}\right|_{t=0} = 0 \Longrightarrow -D\omega_0 K_1 + \omega K_2 = 0$$

$$\Longrightarrow \quad K_1 = -V_S, \text{ and } K_2 = -\frac{D}{\sqrt{1-D^2}} V_S$$

- The solution for the differential equation for the underdamped case is:

$$\boxed{v_{\text{out}}(t) = -V_S e^{-D\omega_0 t}\left(\cos \omega t + \frac{D}{\sqrt{1-D^2}} \sin \omega t\right) + V_S} \quad (1.69)$$

Figure 1.43 shows the step responses $v_{\text{out}}(t)$ for different system damping ratios.

The level of the step is $V_S = 1$ V in this example, and the resonant angular frequency is $\omega_0 = \frac{1}{s}$. Small attenuation values (underdamped case) result in high overshooting of v_{out}. In the case of $D = 1$, the output voltage reaches the final value quickly without overshooting. For $D > 1$, the output voltage approaches the final value slowly (overdamped case). All functions approach $v_{\text{out}}(t \to \infty) = V_S$.

NOTE: In electronic systems often $D = 1/\sqrt{2}$. This setting lets the output value reach the final value much more quickly than with critical damping and has an overshoot of only 4%.

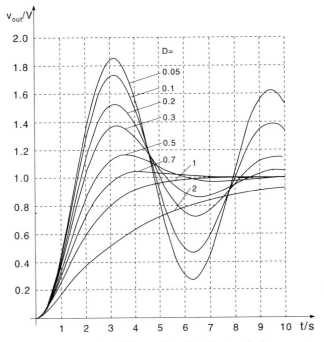

Fig. 1.43. Step responses of an LRC low-pass filter using the damping ratio D as a parameter

1.3 Calculation Methods for Linear Circuits

1.3.1 Rules for Signs

An unknown branch of a circuit can act both as a **generator** and as a **load**. Generators are components that supply energy, which can be voltage or current sources.
Loads are components that absorb energy. These are usually resistors, inductors and capacitors. They may also be components that usually supply energy. A rechargeable battery can become an absorber whilst being charged.

Generators and loads are distinguished in a circuit by assigning voltage and current directions. In the generator the current and the voltage point in the same direction. In the load the current and the voltage have opposite directions (Fig. 1.44).

This convention is used in the following sections for the analysis of circuits. When the exact nature of an element is unknown (that is, whether it is supplying or absorbing energy), a nominal direction for the arrows is chosen arbitrarily. At the end of the analysis, if the solution for the element has a positive sign, then the nominal direction was correct. On the other hand, a negative sign implies a reversing of the arrows.

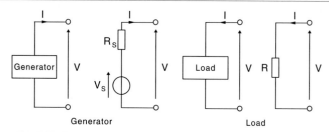

Fig. 1.44. Generator and load

1.3.2 Circuit Calculation with Mesh and Node Analysis

For a known circuit (i.e. all component values are known) Kirchhoff's laws deliver enough independent equations as required to calculate all of the currents present. If one or more values of components are unknown, this lack of information must be compensated for by the same number of known currents or voltages.

A circuit consisting of n nodes and m meshes delivers $(n-1)$ independent node equations (Kirchhoff's first law) and $m - (n-1)$ independent mesh equations (Kirchhoff's second law). There are therefore m independent equations. Equations are independent of each other if they cannot be generated from a linear combination of the other equations.

Mesh equations, including ideal current sources, do not deliver additional information, because the voltage drop across the current source is independent of the respective current source. Branches enclosing current sources are therefore not considered in the number of branches m.

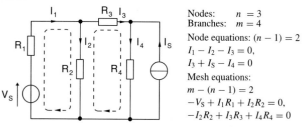

Nodes: $n = 3$
Branches: $m = 4$
Node equations: $(n-1) = 2$
$I_1 - I_2 - I_3 = 0$,
$I_3 + I_S - I_4 = 0$
Mesh equations:
$m - (n-1) = 2$
$-V_S + I_1 R_1 + I_2 R_2 = 0$,
$-I_2 R_2 + I_3 R_3 + I_4 R_4 = 0$

Fig. 1.45. Example for mesh and node equations

Gauss's method:

The solution of a system of m equations with m variables is done by step-by-step elimination of the variables until only one variable remains. The elimination is done by scaling and adding/subtracting two equations. When one variable is known then it can be inserted into an equation with a further variable, and so on. With this method all variables can be solved in a stepwise fashion. To make it easier to keep track of the operations, the calculation is done to a defined scheme (Table 1.1)

EXAMPLE: Calculation of the current I_4 in Fig. 1.45 with Gauss's method:

Solution:
$$I_4 = \frac{R_2(I_S R_1 + V_S) + (R_1 + R_2)I_S R_3}{R_2 R_1 + (R_1 + R_2)(R_3 + R_4)}$$

Table 1.1. Solution with Gauss's method

I_1	I_2	I_3	I_4	Right side	Operation	Eliminated
1	-1	-1	0	0	$+$ 2. line	
0	0	1	-1	$-I_S$	$\times(-R_3) +$ 4. line	I_3
R_1	R_2	0	0	V_S		
0	$-R_2$	R_3	R_4	0		
1	-1	0	-1	$-I_S$	$\times(-R_1) +$ 3. line	
0	$-R_2$	0	$R_3 + R_4$	$I_S R_3$		I_1
R_1	R_2	0	0	V_S		
0	$R_1 + R_2$	0	R_1	$I_S R_1 + V_S$	$\times R_2$	I_2
0	$-R_2$	0	$R_3 + R_4$	$I_S R_3$	$\times (R_1 + R_2)$	
0	0	0	$R_2 R_1 +$ $(R_1+R_2)(R_3+R_4)$	$R_2(I_S R_1 + V_S) +$ $+(R_1+R_2)I_S R_3$		

1.3.3 Superposition

According to the **principle of superposition** in physics it is possible in linear systems (i.e. where cause and effect are proportional) to determine the effect of *one* cause independently from all other causes and effects. The overall resulting cause is then the sum of all individual causes.

In the analysis of linear circuits this means that first all (partial) currents are calculated as caused by the individual voltage and current sources. Then the individual partial currents are summed, keeping their correct signs, in order to determine the resulting current. When calculating the partial currents, all voltage sources *not* under consideration are replaced by short circuits, and all the current sources *not* under consideration are taken out (replaced by open circuits).

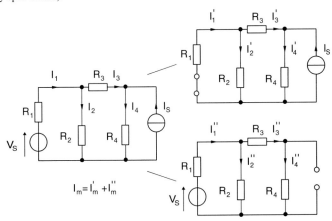

Fig. 1.46. Solution using the principle of superposition

EXAMPLE: Calculation of the current I_4 in Fig. 1.46 with the principle of superposition:

Short circuiting the voltage source V_S and applying the current-divider rule yields for I_4':

$$I_4' = I_S \frac{R_3 + \dfrac{R_1 R_2}{R_1 + R_2}}{R_4 + R_3 + \dfrac{R_1 R_2}{R_1 + R_2}}$$

Removing the current source I_S and applying the current-divider rule yields for I_4'':

$$I_4'' = I_1'' \frac{R_2}{R_2 + R_3 + R_4} = \frac{V_S}{R_1 + \dfrac{R_2(R_3 + R_4)}{R_2 + R_3 + R_4}} \cdot \frac{R_2}{R_2 + R_3 + R_4}$$

The current I_4 is then given as:

$$I_4 = I_4' + I_4''$$

1.3.4 Mesh Analysis

In mesh analysis a ring current is introduced for every independent mesh, and the respective equation is described. This results in a system of as many equations as there are meshes. The branch currents are obtained by adding the ring currents while taking into account their correct signs.

If the mesh contains a current source, the source current can be considered as the mesh current.

- This method is very suitable for calculating the currents in a circuit.

- The equation system becomes very simple when the circuit contains many current sources.

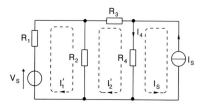

Fig. 1.47. Solution method using mesh analyses

EXAMPLE: Calculation of the current I_4 through R_4 in the circuit shown in Fig. 1.47:

System of equations:

$$-V_S + I_1'(R_1 + R_2) - I_2' R_2 = 0$$
$$-I_1' R_2 + I_2'(R_2 + R_3 + R_4) + I_S R_4 = 0$$

It follows then for I_2':

$$I_2' = \frac{V_S R_2 - I_S R_4 (R_1 + R_2)}{(R_1 + R_2)(R_2 + R_3 + R_4) - R_2^2}$$

The current I_4 is then: $I_4 = I_2' + I_S$.

1.3.5 Node Analysis

In node analysis every node is assigned a potential, where *one* node is assigned the reference potential $\varphi = 0$. Then the independent node equations are formed by expressing the currents with the node potential differences divided by the respective resistors, $I_n = \Delta\varphi/R_m$. This results in a system of as many equations as there are unknown potentials.

- This method is very suitable for calculating the voltages in a circuit.
- The equation system becomes very simple when the circuit contains many voltage sources.

EXAMPLE: Calculation of the voltage V_4 across R_4 in the circuit in Fig. 1.48:

System of equations:

$$I_1 - I_2 - I_3 = 0 \implies \frac{V_S - (\varphi_1 - \varphi_0)}{R_1} - \frac{\varphi_1 - \varphi_0}{R_2} - \frac{\varphi_1 - \varphi_2}{R_3} = 0$$

$$I_3 + I_S - I_4 = 0 \implies \frac{\varphi_1 - \varphi_2}{R_3} + I_S - \frac{\varphi_2 - \varphi_0}{R_4} = 0$$

with $\varphi_0 = 0$, it follows that:

$$\frac{V_S - \varphi_1}{R_1} - \frac{\varphi_1}{R_2} - \frac{\varphi_1 - \varphi_2}{R_3} = 0$$

$$\frac{\varphi_1 - \varphi_2}{R_3} + I_S - \frac{\varphi_2}{R_4} = 0$$

This yields for $\varphi_2 = V_4$:

$$\varphi_2 = V_4 = \frac{R_4[V_S R_2 + I_S(R_1 R_2 + R_2 R_3 + R_1 R_3)]}{R_1 R_2 + R_2 R_3 + R_1 R_3 + R_1 R_4 + R_2 R_4}$$

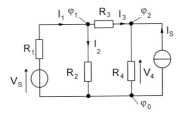

Fig. 1.48. Solution method using node analyses

1.3.6 Thévenin's and Norton's Theorem

By **Thévenin's theorem** every active linear two-terminal network, no matter how many sources and resistors are interconnected, can be represented by an equivalent circuit, which contains only one voltage source and one resistor (Fig. 1.49).

By **Norton's theorem** every active linear two-terminal network, no matter how many sources and resistors are interconnected, can be represented by an equivalent circuit, which contains only one current source and one resistor (Fig. 1.50).

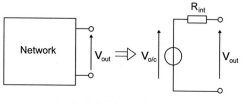

Fig. 1.49. Thévenin's theorem

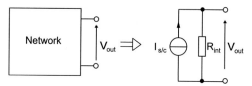

Fig. 1.50. Norton's theorem

Thévenin's and Norton's equivalent circuits, called 'real voltage source' and 'real current source' in Sect. 1.2.1 respectively, have the same voltage–current characteristic (Fig. 1.51).

It is a declining straight line, starting with $V_{o/c}$, $I_{out} = 0$ (o/c stands for open circuit) for no-load operation and ending with $V_{out} = 0$, $I_{s/c}$ (s/c stands for short circuit). The slope of the line is determined by the internal resistance R_{int}.

$$R_{int} = \frac{\Delta V_{out}}{\Delta I_{out}} = \frac{V_{o/c}}{I_{s/c}} \qquad (1.70)$$

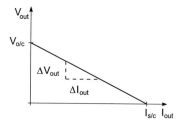

Fig. 1.51. Voltage–current characteristic of Thévenin's and Norton's equivalent circuit

1.3.6.1 Calculating a Load Current by Thévenin's Theorem

By Thévenin's theorem an active two-terminal network can be reduced to its equivalent circuit, consisting of a voltage source and an internal resistor.

The voltage–current characteristic of the equivalent circuit can be determined very simply:

a) by measuring the open circuit voltage $V_{o/c}$ with a voltmeter, and by measuring a second value of the characteristic V_1, I_1 by loading the circuit (Fig. 1.53). Then $R_{int} = \dfrac{V_{o/c} - V_1}{I_1}$.

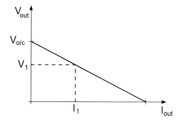

Fig. 1.52. Voltage–current characteristic

b) by calculating the open circuit voltage $V_{o/c}$ and the short circuit current $I_{s/c}$, if the network is known.

Then $R_{int} = \dfrac{V_{o/c}}{I_{s/c}}$.

After $V_{o/c}$ and R_{int} are known, the load current can be determined very simply (Fig. 1.53):

$$I_{out} = \frac{V_{o/c}}{R_{int} + R_{load}}, \qquad V_{out} = V_{o/s}\frac{R_{load}}{R_{int} + R_{load}} \tag{1.71}$$

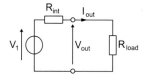

Fig. 1.53. Loaded Thévenin's equivalent circuit

EXAMPLE: Analysing the circuit shown in Fig. 1.54 using Thévenin's theorem:

Step one: calculate $V_{o/c}$ using the voltage-divider rule

$$V_{o/c} = V_{S1} \cdot \frac{R_2}{R_1 + R_2} = 12\text{ V} \cdot \frac{20\ \Omega}{30\ \Omega} = 8\text{ V}$$

Step two: calculate $I_{s/c}$ by shorting the output terminals

$$I_{s/c} = \frac{V_{S1}}{R_1} = \frac{12\text{ V}}{10\ \Omega} = 1.2\text{ A}$$

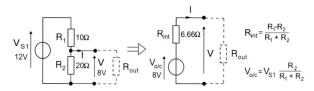

Fig. 1.54. Application of Thévenin's theorem to a circuit

For R_{int} it follows that:

$$R_{int} = \frac{V_{o/c}}{I_{s/c}} = \frac{V_{S1} \cdot \frac{R_2}{R_1+R_2}}{\frac{V_{S1}}{R_1}} = \frac{R_2 R_1}{R_1+R_2} = \frac{10\,\Omega \cdot 20\,\Omega}{30\,\Omega} = 6.66\,\Omega$$

or directly

$$R_{int} = \frac{V_{o/c}}{I_{s/c}} = \frac{8\,V}{1.2\,A} = 6.66\,\Omega$$

1.3.6.2 Calculating a Current Within a Network

Thévenin's theorem can also be used to determine a certain current within a network. Therefore the network must be divided into two parts, where the current is to be calculated (Fig. 1.55).

After this, both parts of the divided network can be reduced to Thévenin equivalent circuits:
Then $I = \dfrac{V_1 - V_2}{R_1 + R_2}$.

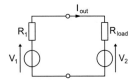

Fig. 1.55. Using Thévenin's theorem to calculate a certain current

Often the right part of the network is passive, which means that it can be reduced to a resistor (or in an AC circuit to an impedance). In this case, the calculation follows that given in Sect. 1.3.6.1.

EXAMPLE: Calculation of the current I_3 for the circuit shown in Fig. 1.56.

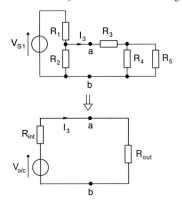

Fig. 1.56. Calculation of I_3 by Thévenin's theorem

First the circuit is divided into two parts at the two points a and b. Then the left part is converted into an equivalent voltage source, and the right part is converted into an equivalent resistor:

$$V_{o/c} = V_{S1}\frac{R_2}{R_1 + R_2}, \qquad R_{int} = \frac{R_1 R_2}{R_1 + R_2}, \qquad R_{load} = R_3 + \frac{R_4 R_5}{R_4 + R_5}$$

I_3 is then:

$$I_3 = \frac{V_{o/c}}{R_{int} + R_{load}} = \frac{V_{S1}\dfrac{R_2}{R_1 + R_2}}{\dfrac{R_1 R_2}{R_1 + R_2} + R_3 + \dfrac{R_4 R_5}{R_4 + R_5}}$$

1.4 Notation Index

C	capacitance (F = As/V)
D	damping ratio
e	elementary or electronic charge ($e = \pm 1.602 \cdot 10^{-19}$ As)
f	frequency (Hz)
G	conductance (S = A/V)
i	time variant current (A)
I	DC current
$I_{s/c}$	short-circuit current (A)
I_S	source current (A)
L	inductance (H = Vs/A)
n	rounds per minute (min^{-1})
M	torque (Nm)
P	power (W = VA)
Q	charge (As)
R	resistance (Ω = V/A)
R_S	source or internal resistance (Ω = V/A)
R_L	load resistance (Ω = V/A)
t	time (s)
T	period length (s)
v	time variant voltage (V)
V	DC voltage (V)
V_S	source voltage (V)
$V_{o/c}$	open-circuit voltage (V)
W	energy (Ws = VAs)
α	temperature coefficient (k^{-1})
β	temperature coefficient (k^{-2})
η	efficiency
ϑ	temperature (°C)
τ	time constant (s)
φ	potential (V)
ω	angular frequency (s^{-1})
ω_0	resonant angular frequency (s^{-1})

1.5 Further Reading

BIRD, J. O.: *Electrical Circuit Theory & Technology, 1st Edition*
Butterworth-Heinemann (1997)

BOYLESTAD, R. L.: *Introductory Circuit Analysis, 9th Edition*
Prentice Hall (1999)

FLOYD, T. L.: *Electric Circuits Fundamentals, 5th Edition*
Prentice Hall (2001)

FLOYD, T. L.: *Electronics Fundamentals: Circuits, Devices, and Applications, 5th Edition*
Prentice Hall (2000)

FLOYD, T. L.: *Electronic Devices, 5th Edition*
Prentice Hall (1998)

GROB, B.: *Basic Electronics, 8th Edition*
McGraw-Hill (1996)

MUNCASTER, R.: *A-Level Physics*
Stanley Thornes Ltd. (1997)

NELKON, M.; PARKER, P.: *Advanced Level Physics*
Heinemann (1995)

TSE, CHI KONG: *Linear Circuit Analysis, 1st Edition*
Addison-Wesley (1998)

2 Electric Fields

Electric fields are caused by electrical charges. The state of motion of the charges is important. The physical phenomena may be divided into those caused by *stationary* charges and those caused by *moving* charges. The former are described by an **electrostatic field**, the latter by an **electric flow field** or a **magnetic field**. Charges in motion cause electric *and* magnetic fields. For an electrostatic field to exist alone, the charges must be stationary.

2.1 Electrostatic Fields

Electrostatics describes the relationships between *stationary* electric charges. The electric field that is created by *stationary* electric charges is known as an electrostatic field.

2.1.1 Coulomb's Law

Electric charges exert forces on each other. Charges with the same sign repel each other, whereas charges with opposite signs attract each other. The force between two stationary point charges Q_1 and Q_2 is defined by **Coulomb's law** (point charges are charges with no spatial volume):

$$|F| = \frac{1}{4\pi\varepsilon} \cdot \frac{Q_1 \cdot Q_2}{r^2}, \quad \text{with} \quad \varepsilon = \varepsilon_0 \cdot \varepsilon_r \tag{2.1}$$

ε_0: permittivity of free space, $\varepsilon_0 = 8.86 \cdot 10^{-12} \, \frac{\text{As}}{\text{Vm}}$;

ε_r: relative permittivity;

r: distance between the charges.

The space between the charges is assumed to be an insulator, whose properties are equal and independent of orientation (that is, isotropic) in all places. In a vacuum $\varepsilon_r = 1$, which is also approximately true in air. The force F_2 on the point charge Q_2 can be represented by a vector:

$$\vec{F}_2 = \frac{1}{4\pi\varepsilon} \cdot \frac{Q_1 \cdot Q_2}{r^2} \cdot \vec{e}_r \tag{2.2}$$

The equation is written in spherical coordinates, where the point charge Q_1 is in the centre of the coordinate system. $\vec{e}_r$ is the unity vector $\frac{\vec{r}}{|\vec{r}|}$, which points radially away from the charge Q_1.

- Coulomb's law is also valid to a good approximation for spheres whose diameters are small with respect to their separation; r is then the distance between the centre points.

2.1.2 Definition of Electric Field Strength

The **electric field strength** may be derived from Coulomb's law:

$$\vec{F}_2 = Q_2 \cdot \underbrace{\frac{Q_1}{4\pi\varepsilon r^2} \cdot \vec{e}_r}_{\vec{E}} = Q_2 \cdot \vec{E} \qquad (2.3)$$

This defines a field for the point charge Q_1, pointing radially away from the charge and decreasing with the distance squared. This leads to a description of the force on Q_2 that would be caused by an electric field, without explicitly knowing the source of the field (the charge Q_1 in the position $r = 0$).

- The SI unit of electric field strength is the Volts per Meter, $\frac{V}{m}$.

- When a force is exerted on an electric charge, then the charge is in an electric field.

In general, the force on a point charge in an electric field is given by:

$$\vec{F} = Q \cdot \vec{E} \qquad (2.4)$$

This equation is generally valid, independently of how the field strength vector $\vec{E}$ was caused.

NOTE: The field of the point charge Q for which one wants to calculate the force is in this model meaningless. It points radially away from (or towards) the point and exercises no force on the charge.

NOTE: To calculate the force on an extended charge, it is divided up into infinitesimal point charges, in order to calculate the resulting force by integration. In the Cartesian coordinate system the calculation is expressed as:

$$\vec{F} = \int_Q \vec{E}(x, y, z) \, \mathrm{d}Q(x, y, z)$$

Of course, the coordinate system that is chosen is suitable to the given problem.

The fields may be visualised by **electric flux lines** (Fig. 2.1). Lines whose direction at any point corresponds to the direction of the force on a point charge at that position are used to represent a field. The density of the field lines is thus a measure of the field strength.

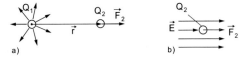

Fig. 2.1. Representation of the force on a charge Q_2: **a)** using Coulomb's law; **b)** using the electric field

- An electrostatic field is a **source field**. Electric flux lines always begin and end on electric charges.

- The **positive direction** of the flux lines is defined from positive charges to negative charges.

NOTE: If an electric field is drawn around a single charge (as in the definition of the electric field strength), then this automatically implies that the opposite charge is at infinity. This visually simplifies the calculation of the electric field. Where there are several charges present, then the resulting electric field can be constructed by the superposition of each individual field at each point in space (Fig. 2.2).

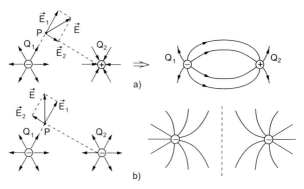

Fig. 2.2. The resulting electric field of two **a)** opposite charges **b)** same-sign charges

2.1.3 Voltage and Potential

The **electric voltage** is a measure of the work required to move a unit charge in an electric field from one place to another.

$$W = \int \vec{F} \, d\vec{s} = \int Q \cdot \underbrace{\vec{E} \, d\vec{s}}_{V} = Q \cdot V$$

- The **electric voltage** between two points in space is equal to the line integral of the electric field strength between the two points. The particular path over which the integration is carried out is *irrelevant*.

$$V_{12} = \int_s \vec{E} \, d\vec{s} \tag{2.5}$$

The **electric potential** φ is an absolute scalar quantity. If a point in space is chosen where the electric potential $\varphi = 0$ (reference potential), then all other points in space can be assigned an absolute potential. The potential $\varphi = 0$ is normally assigned to earth, or is placed in abstract physical models at infinity. The voltage V in this model is given by the difference between two potentials (Fig. 2.3a).

$$V_{12} = \varphi_1 - \varphi_2 \tag{2.6}$$

Areas of equal potential are called **equipotential surfaces**. The electric field strength and the electric flux density vectors are perpendicular to these surfaces. An infinitesimal surface element $d\vec{A}$ of the equipotential surface is a vector that is perpendicular to the surface. The direction of the vector $d\vec{A}$ is the same as the direction of the electric field strength (Fig. 2.3b).

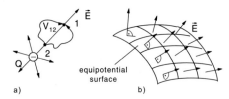

Fig. 2.3. The voltage V and the potential φ in an electrostatic field

2.1.4 Electrostatic Induction

Electrostatic induction is the shift of the mobile charges in a conductor that has been placed in an electric field. The charges shift so that the electric field strength in the electrical conductor remains zero.

- An electrical conductor is always internally field-free. (This is strictly valid only in an electrostatic field, but is also approximately true in low-frequency alternating fields.)

The electric field strength can be experimentally shown with **Maxwell's parallel plates** (Fig. 2.4). When two electrically conducting parallel plates are placed in an electric field, the mobile charges in the plates shift to the outer surfaces. In Fig. 2.4a, the negative charges shift left, and the positive shift right. The space inside the parallel plates is field-free. If the plates are now drawn apart (Fig. 2.4b), the charges remain on the plates and continue to balance out the field strength between the plates. If the parallel plates are removed from the field (Fig. 2.4c), then the charges on the parallel plates create a new electric field that can, for example, be measured in the discharge current.

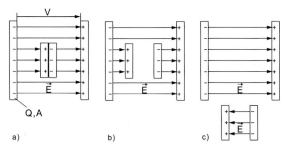

Fig. 2.4. Maxwell's parallel plates to demonstrate electric fields

Electrostatic shielding can be used to prevent electrostatic induction by an external field. A hollow conductor is always internally field-free, as the charges shift to its outer surface so as to prevent an internal electric field. This is also approximately true when the hollow conductor does not have a closed outer surface, but rather has a grating structure. Such a shielding cage is called a **Faraday cage**, after its inventor (Fig. 2.5a). However, it is also possible to surround a charge with a hollow electrical conductor to keep the external space field-free (Fig. 2.5b). An equal and opposite charge to the internal charge gathers on the inner side of the hollow body. The charge on the outer surface of the hollow body flows away to earth. The outer space is thus kept field-free.

NOTE: If there is an alternating field inside the hollow body, then an alternating current flows on the earth conductor, as in this case the charge on the outer surface of

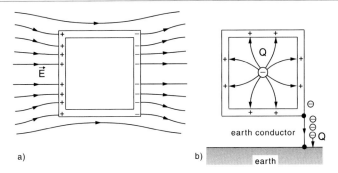

Fig. 2.5. Faraday cage

the hollow body must continuously change. The fact that this alternating current creates a magnetic field, which in turn induces a voltage, need not be taken into account at low frequencies. (The frequency range is not simple to specify, but for a good short earth conductor can be from 1 to 10 MHz.)

2.1.5 Electric Displacement

The **electric displacement** is a measure of the electrostatically induced charge after displacement. It is a vector field quantity.

$$\vec{D} = \frac{dQ}{dA_\perp} \cdot \vec{e}_{A_\perp} \tag{2.7}$$

Thus $dA_\perp$ is the surface element of an equipotential surface. The unity vector $\vec{e}_{A_\perp}$ points in the direction of the electric field strength.

- The SI unit of electric displacement is $\dfrac{As}{m^2}$.

- The electric displacement is equivalent to the charge density on the outer surface of a conductor. In an electrostatic field it is equal to the charge density on an equipotential surface, if an electrically conducting foil is placed on the surface.

- The electric displacement intersects the equipotential surfaces perpendicularly, as does the electric field strength.

EXAMPLE: A point charge Q_+ is placed in the centre of a hollow spherical conductor (Fig. 2.6b). The electric displacement on the inner surface of the hollow sphere amounts to $\vec{D} = \dfrac{Q}{4\pi R^2} \cdot \vec{e}_r$. It is obvious therefore that the opposite charge Q_- is spread over the inner surface of the sphere. Inside the hollow sphere the equipotential surfaces form concentric shells around the point charge Q_+. The electric displacement inside the sphere can be given in general by

$$\vec{D}_{(r)} = \frac{Q}{4\pi r^2} \cdot \vec{e}_r.$$

With spherical coordinates the charge Q_+ lies in the centre of this representation.

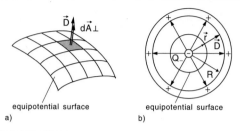

Fig. 2.6. Electric displacement: **a)** general definition; **a)** in a hollow sphere

2.1.6 Dielectrics

The space an electrostatic field fills is known as a dielectric. The dielectric field quantities are the electric field strength $\vec{E}$ and the electric displacement $\vec{D}$. In an electrostatic field:

$$\vec{D} = \varepsilon \cdot \vec{E} \tag{2.8}$$

where ε is the **permittivity** (also known as the dielectric constant). For an isotropic dielectric:

- $\vec{E}$ and $\vec{D}$ point in the same direction and lie perpendicular to the equipotential surfaces.
- The proportionality factor between the electric displacement and the electric field strength is the permittivity ε.

The permittivity is derived from the **free space permittivity** ε_0 and the **relative permittivity** $\varepsilon_r \geq 1$ (relative dielectric constant).

$$\varepsilon = \varepsilon_0 \cdot \varepsilon_r \tag{2.9}$$

The value of the free space permittivity is:

$$\varepsilon_0 = 8.85419 \cdot 10^{-12} \frac{\mathrm{As}}{\mathrm{Vm}} \tag{2.10}$$

The relative permittivity ε_r depends on the material in which the field extends. The values of ε_r for most dielectrics lie between 1 and 100, but ε_r can be up to 10 000.

- The relative permittivity is always $\varepsilon_r \geq 1$.
- The relative permittivity of a vacuum is $\varepsilon_r = 1$.
- The relative permittivity of air is $\varepsilon_r \approx 1$.
- The relative permittivity of insulators usually lies in the range 2–3.

2.1.7 The Coulomb Integral

The electric field strength at an arbitrary point in space can be calculated with the aid of the superposition principle. The total field strength is equal to the vector sum of the individual field strengths from each of the charges.

$$\vec{D} = \sum_i \frac{Q_i}{4\pi r_i^2} \cdot \vec{e}_{ri}, \quad \text{or} \quad \vec{E} = \sum_i \frac{Q_i}{4\pi \varepsilon \, r_i^2} \cdot \vec{e}_{ri} \tag{2.11}$$

A spatially distributed charge may be considered as a number of spatially distributed point charges dQ_i. The resulting field strength is then equal to the integral of the individual field strengths from each of the point charges dQ_i.

$$\vec{D} = \int_Q \frac{dQ_i}{4\pi r^2} \cdot \vec{e}_{ri}, \quad \text{or} \quad \vec{E} = \int_Q \frac{dQ_i}{4\pi\varepsilon r^2} \cdot \vec{e}_{ri} \tag{2.12}$$

This is known as the **Coulomb integral**.

EXAMPLE: The calculation of the electric field strength around a straight line charge $\lambda \left(\dfrac{As}{m}\right)$ by using the Coulomb integral:

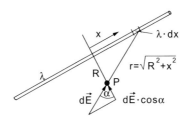

Fig. 2.7. Calculation of the electric field strength using the Coulomb integral

It can be assumed that the field spreads out radially and symmetrically from the line charge. If the line charge is rotated about its longitudinal axis, the field at a fixed point in space will not change for symmetrical reasons. The calculation can thus be reduced to a planar (two-dimensional) problem. The spatially distributed point charges may be considered as $dQ = \lambda\,dx$. The distance from the charge dQ to the point P is given by $r = \sqrt{R^2 + x^2}$. The cosine of the angle α can be represented in Cartesian coordinates as $\dfrac{R}{\sqrt{R^2 + x^2}}$ (Fig. 2.7). The calculation of the field strength E at the point P:

$$E = \frac{D}{\varepsilon} = \int_{-\infty}^{+\infty} \frac{\lambda}{4\pi\varepsilon} \cdot \frac{dx}{R^2 + x^2} \cdot \underbrace{\frac{R}{\sqrt{R^2 + x^2}}}_{\cos\alpha} = 2 \cdot \frac{\lambda R}{4\pi\varepsilon} \int_{0}^{+\infty} \frac{dx}{(R^2 + x^2)^{3/2}} = \frac{\lambda}{2\pi\varepsilon R}$$

The field components in the x-direction balance each other out. The resulting field strength points radially away from the line charge.

2.1.8 Gauss's Law of Electrostatics

Gauss's law of electrostatics states that the surface integral of the electric displacement over a closed surface is equal to the charge enclosed.

$$\oint_A \vec{D}\,d\vec{A} = Q \tag{2.13}$$

The vector $d\vec{A}$ points out from the surface area.

EXAMPLE: The calculation of the electric field strength around a straight line charge λ using Gauss's law:

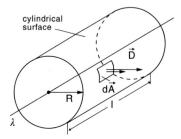

Fig. 2.8. Field calculation for a line charge

A field calculation using Gauss's law essentially depends on a suitable choice of the coordinate system. In this particular case a cylindrical coordinate system is suitable for symmetrical reasons, since the field points radially away from a line charge. A cylindrical surface is therefore placed around the line charge, as shown in Fig. 2.8. Gauss's law states that:

$$\oint_A \vec{D} \, d\vec{A} = \vec{D} \cdot 2\pi R \cdot l = \lambda l \quad \Rightarrow \quad \vec{D}(R) = \frac{\lambda}{2\pi R} \cdot \vec{e}_r$$

$$\Rightarrow \quad \vec{E}(R) = \frac{\lambda}{2\pi \varepsilon R} \cdot \vec{e}_r$$

2.1.9 Capacitance

- In a configuration of two electrodes, the ratio of the charge on the two electrodes to the voltage between the electrodes is constant. This ratio only depends on the geometry of the configuration and the dielectric constant of the space between the electrodes.

- The ratio of the charge to the voltage is called the **capacitance**.

$$C = \frac{Q}{V} \tag{2.14}$$

The SI unit of capacitance is the farad, $1 \text{ F} = 1 \dfrac{\text{As}}{\text{V}}$

If field quantities are used to describe Q and V, then C can be calculated as:

$$C = \frac{Q}{V} = \frac{\oint_A \vec{D} \, d\vec{A}}{\int_s \vec{E} \, d\vec{s}}, \quad \text{with} \quad \vec{e}_A \| \vec{e}_s \tag{2.15}$$

Equation (2.15) is valid for the case where the path element $d\vec{s}$ is perpendicular to the area element $d\vec{A}$. Therefore, in order to evaluate this integral, the field qualities must be known qualitatively, i.e. the direction of the field strength and the electric displacement must be known.

EXAMPLE: Calculation of the capacitance of a coaxial conductor of length l (Fig. 2.9):

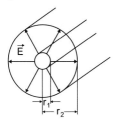

Fig. 2.9. Calculation of the capacitance of a coaxial conductor

Knowledge of the electric field is necessary for the calculation of the capacitance. Equation (2.15) is used, as either a voltage V is applied to, or a charge Q is placed on, the electrodes. In this case a charge Q is placed on the electrodes, and the field $\vec{E}(r)$ is expressed as a function of this charge. As can be seen, the charge Q cancels out to yield an expression for C that only depends on the geometry of the configuration and the material characteristics of the dielectric.

$$C = \frac{Q}{V} = \frac{\oint_A \vec{D} \, \mathrm{d}\vec{A}}{\int_s \vec{E} \, \mathrm{d}\vec{s}} = \frac{Q}{\int_{r_1}^{r_2} \frac{Q/l}{2\pi\varepsilon r} \, \mathrm{d}r} = \frac{2\pi\varepsilon l}{\ln \frac{r_2}{r_1}}, \qquad \text{with} \qquad \vec{E} = \frac{Q/l}{2\pi\varepsilon r} \vec{e}_r$$

2.1.10 Electrostatic Field at a Boundary

Figure 2.10 shows electrostatic field quantities at a boundary. At the boundary:

$$\vec{E}_{t2} = \vec{E}_{t1}, \qquad \text{and} \qquad \vec{E}_{n2} = \frac{\varepsilon_1}{\varepsilon_2} \cdot \vec{E}_{n1} \tag{2.16}$$

$$\vec{D}_{n2} = \vec{D}_{n1}, \qquad \text{and} \qquad \vec{D}_{t2} = \frac{\varepsilon_2}{\varepsilon_1} \cdot \vec{D}_{t1} \tag{2.17}$$

$$\tan \alpha_2 = \frac{\varepsilon_2}{\varepsilon_1} \cdot \tan \alpha_1 \tag{2.18}$$

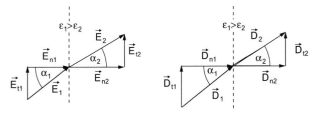

Fig. 2.10. Field quantities at a boundary

- The tangential component of the electric field strength is constant.
- The normal component of the electric field strength is inversely proportional to the dielectric constant.
- The tangential component of the electric displacement is proportional to the dielectric constant.
- The normal component of the electric displacement is constant.

2.1.11 Overview: Fields and Capacitances of Different Geometric Configurations

Table 2.1. Overview of fields and capacitances

Parallel-plate capacitor	$C = \varepsilon \cdot \dfrac{A}{d}$	$E = \dfrac{V}{d}$
Parallel-plate capacitor	$C = \dfrac{A}{\dfrac{d_1}{\varepsilon_1} + \dfrac{d_2}{\varepsilon_2}}$	$E_{1(2)} = \dfrac{V}{\varepsilon_{1(2)} \left(\dfrac{d_1}{\varepsilon_1} + \dfrac{d_2}{\varepsilon_2} \right)}$
Cylindrical capacitor	$C = \dfrac{2\pi\varepsilon \cdot l}{\ln \dfrac{r_2}{r_1}}$	$E = \dfrac{V}{r \cdot \ln \dfrac{r_2}{r_1}}$
Cylindrical capacitor	$C = \dfrac{2\pi l}{\dfrac{1}{\varepsilon_1} \ln \dfrac{r_2}{r_1} + \dfrac{1}{\varepsilon_2} \ln \dfrac{r_3}{r_2}}$	$E_{1(2)} = \dfrac{V}{\varepsilon_{1(2)} \cdot r} \times \dfrac{1}{\dfrac{1}{\varepsilon_1} \ln \dfrac{r_2}{r_1} + \dfrac{1}{\varepsilon_2} \ln \dfrac{r_3}{r_2}}$
Spherical capacitor	$C = 4\pi\varepsilon \cdot \dfrac{r_1 r_2}{r_2 - r_1}$	$E = \dfrac{V}{r^2} \cdot \dfrac{r_1 r_2}{r_2 - r_1}$
Parallel conductor	$C = \dfrac{\pi \varepsilon l}{\ln \left[\dfrac{a}{r_1} + \sqrt{\left(\dfrac{a}{r_1}\right)^2 - 1} \right]}$ $\approx \dfrac{\pi \varepsilon l}{\ln \dfrac{2a}{r_1}}$, for $a \gg r_1$	$E = \dfrac{V \dfrac{\sqrt{a^2 - r_1^2}}{a^2 - r_1^2 - x^2}}{\ln \left[\dfrac{a}{r_1} + \sqrt{\left(\dfrac{a}{r_1}\right)^2 - 1} \right]}$

Table 2.1. (cont.)

Single conductor-earth	$C = \dfrac{2\pi\varepsilon l}{\ln\left[\dfrac{a}{r_1} + \sqrt{\left(\dfrac{a}{r_1}\right)^2 - 1}\right]}$ $\approx \dfrac{2\pi\varepsilon l}{\ln\dfrac{2a}{r_1}}$ for $a \gg r_1$	$E = \dfrac{2V\dfrac{\sqrt{a^2 - r_1^2}}{a^2 - r_1^2 - x^2}}{\ln\left[\dfrac{a}{r_1} + \sqrt{\left(\dfrac{a}{r_1}\right)^2 - 1}\right]}$
Sphere-sphere	$C \approx \dfrac{2\pi\varepsilon}{\dfrac{1}{r_1} - \dfrac{1}{2a}}$	$E \approx \dfrac{V\left(\dfrac{1}{x^2} + \dfrac{1}{(2a-x)^2}\right)}{\dfrac{2}{r_1} - \dfrac{1}{a}}$
Sphere-infinity	$C = 4\pi\varepsilon r_1$	$E = V \cdot \dfrac{r_1}{r^2}$

2.1.12 Energy in an Electrostatic Field

Energy is required to create an electric field, as positive and negative charges must be separated. A charging current will flow if a voltage is applied to a capacitor in order to charge it. The energy supplied to the capacitor is stored in the electric field and not, as for a resistance, transformed into heat. The energy is given by:

$$W = \int_0^{t_1} v(t) \cdot \underbrace{i(t)\,\mathrm{d}t}_{\mathrm{d}Q} = \int_0^{Q_1} v(t)\,\underbrace{\mathrm{d}Q}_{C\,\mathrm{d}V} = C\int_0^{V_C} v\,\mathrm{d}v = \frac{1}{2}C \cdot V_C^2$$

In general:

$$W = \frac{1}{2}CV^2 = \frac{1}{2}QV = \frac{1}{2}\frac{Q^2}{C} \qquad (2.19)$$

If the integral quantities Q and V are replaced by the vector quantities $\vec{D}$ and $\vec{E}$, this becomes:

$$W = \frac{1}{2}\oint_A \vec{D}\,\mathrm{d}\vec{A} \cdot \int_s \vec{E}\,\mathrm{d}\vec{s} = \frac{1}{2}\int_V \vec{D}\cdot\vec{E}\,\mathrm{d}V, \quad \text{with}\quad \vec{e}_s \| \vec{e}_A \qquad (2.20)$$

The unity vector $\vec{e}_s$ points therefore in the same direction as the unity vector for the area $\vec{e}_A$, so that the integral $\int \mathrm{d}s \cdot \mathrm{d}A$ yields the volume element $\mathrm{d}V$. The **energy density** of an electrostatic field is:

$$\frac{\mathrm{d}W}{\mathrm{d}V} = \frac{1}{2}\vec{D}\cdot\vec{E} \qquad (2.21)$$

2.1.13 Forces in an Electrostatic Field

2.1.13.1 Force on a Charge

The force on a point charge in an electric field is given by:

$$\vec{F} = Q \cdot \vec{E} \tag{2.22}$$

EXAMPLE: The deflection of an electron after passing through an electric field is given by:

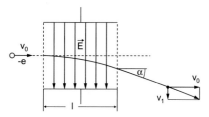

Fig. 2.11. Deflection of an electron passing through an electric field

A force is exerted on an electron during its passage through an electric field (Fig. 2.11). Because of the negative charge on the electron, the direction of the force opposes that of the electric field. The speed $\vec{v}_0$ of the electron perpendicular to the direction of the field remains unchanged during passage through the field. The time the electron requires to traverse the field is $t = l/v_0$.

$F = m \cdot a$ (force = mass × acceleration)

$v = \int a \, dt$ (velocity = time integral of the acceleration)

$$v_1 = \int_0^{l/v_0} \frac{e \cdot E}{m} \, dt = \frac{e}{mv_0} \cdot E \cdot l, \qquad \tan \alpha = \frac{e}{m} \cdot E \cdot l$$

2.1.13.2 Force at the Boundary

Both the boundary between dielectric and the conductive surface of electrodes and also the boundary between different dielectrics are subject to forces. These forces can be most easily derived by considering the energy balance between the mechanical, electrical and field energy. To do this, the boundary is assumed to have shifted infinitesimally and then the resulting change in the potential energy is calculated. The sum of the energy changes must be zero:

$$\boxed{d W_{\text{mech}} + d W_{\text{field}} + d W_{\text{electr}} = 0} \tag{2.23}$$

In order to carry out this energy balance it is necessary to know which of the changes is positive and which negative, i.e., which energy increases and which decreases. Consider the following thought experiment: A voltage source is applied to a parallel-plate capacitor. The plates of the capacitor are drawn to each other, as they are charged with opposite charges. Mechanical energy is applied if the plates are pulled apart. The capacitance will

2.1 Electrostatic Fields

simultaneously decrease, i.e. the stored field energy $\frac{1}{2}CV^2$ will decrease. The mechanical energy and the change in the stored field energy are balanced by the energy supplied by the voltage source. This energy balance can be more concretely stated as:

$$F \cdot \mathrm{d}s + \mathrm{d}\left(\frac{1}{2}CV^2\right) = \mathrm{d}Q \cdot V$$

With $Q = C \cdot V$, it follows therefore for the force:

$$F = \frac{1}{2}V^2 \frac{\mathrm{d}C}{\mathrm{d}s} \tag{2.24}$$

- The force on the plates of a capacitor is proportional to the change in the capacitance relative to the imaginary shift of the plates.

For boundaries this means, in general, that:

- The force at a boundary is proportional to the change in the capacitance relative to an imaginary shift of the boundary.
- The force at a boundary always tries to increase the capacitance.

NOTE: If a voltage source is not applied to the capacitor plates in the above thought experiment, but rather they hold the fixed charge Q, then the energy balance for the virtual shift is different. It then becomes:

$$F \cdot \mathrm{d}s + \mathrm{d}\left(\frac{1}{2}\frac{Q^2}{C}\right) = 0$$

The force is thus:

$$F = -\frac{1}{2}Q^2 \frac{\mathrm{d}}{\mathrm{d}s}\left(\frac{1}{C}\right) \tag{2.25}$$

The applied mechanical energy increases the field energy in this case. (In the first case with the voltage source the field energy was reduced!) Of course, both equations for the calculation of the force, $F = \frac{1}{2}V^2 \frac{\mathrm{d}C}{\mathrm{d}s}$ and $F = -\frac{1}{2}Q^2 \frac{\mathrm{d}}{\mathrm{d}s}\left(\frac{1}{C}\right)$, arrive at the same result.

EXAMPLE: Calculation of the force at the electrodes of a parallel plate capacitor (Fig. 2.12a):

$$F = \frac{1}{2}V^2 \frac{\mathrm{d}C}{\mathrm{d}s}, \quad C = \frac{\varepsilon A}{s}, \quad \frac{\mathrm{d}C}{\mathrm{d}s} = -\frac{\varepsilon A}{s^2} \quad \Rightarrow \quad F = -\frac{1}{2}V^2 \cdot \frac{\varepsilon A}{s^2}$$

Another approach is:

$$F = -\frac{1}{2}Q^2 \frac{\mathrm{d}}{\mathrm{d}s}\left(\frac{1}{C}\right), \quad \frac{1}{C} = \frac{s}{\varepsilon A}, \quad \frac{\mathrm{d}}{\mathrm{d}s}\left(\frac{1}{C}\right) = \frac{1}{\varepsilon A}$$

$$\Rightarrow F = -\frac{1}{2}Q^2 \frac{1}{\varepsilon A} = -\frac{1}{2}V^2 \cdot \frac{\varepsilon A}{s^2}$$

As expected, both results are equal. The minus sign in the result shows that the force resists against the plates being drawn apart.

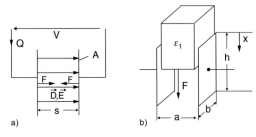

Fig. 2.12. Forces at boundaries: **a)** parallel to the flux lines; **b)** perpendicular to the flux lines

EXAMPLE: Calculation of the force drawing a dielectric between two capacitor plates (Fig. 2.12b):

$$F = \frac{V^2}{2} \cdot \frac{dC}{dx}, \quad C = \frac{\varepsilon_1 \varepsilon_0 b}{a} x + \frac{\varepsilon_0 b}{a}(h-x), \quad \frac{dC}{dx} = (\varepsilon_1 - 1)\frac{\varepsilon_0 b}{a}$$

$$\Rightarrow F = \frac{V^2}{2} \cdot (\varepsilon_1 - 1)\frac{\varepsilon_0 b}{a}$$

2.1.14 Overview: Characteristics of an Electrostatic Field

- Conducting media are field-free.
- The electric displacement and the electric field strength point in the same direction in isotropic materials.

$$\boxed{\vec{D} = \varepsilon \cdot \vec{E}}$$

- The electric field is a **charge field**. Electric flux lines always begin and end on electric charges. The positive direction is defined from negative to positive charges.
- The surface integral of the electric displacement over a closed surface is equal to the charge enclosed. This is obviously zero in a charge-free space.

$$\oint_A \vec{D} \, d\vec{A} = Q$$

- A space is charge-free if the divergence of the observed field in this space is zero.

$$\boxed{\nabla \cdot \vec{E} = 0} \quad \text{in a charge-free space.}$$

- If the observed space contains the charge density ϱ, then the divergence of $\vec{E}$ (Maxwell's third equation) is:

$$\boxed{\nabla \cdot \vec{E} = \frac{\varrho}{\varepsilon}}$$

- The curl of the electric field is zero at all points. If the integral of the electric field strength is calculated over a closed loop then the result is always zero, independent of the integration path chosen.

$$\boxed{\oint \vec{E} \, d\vec{s} = 0, \quad \nabla \times \vec{E} = 0}$$

- The electrostatic field is a **conservative field**. The line integral of the electric field strength is equal to the voltage (potential difference) between the beginning and end points of the path. It is therefore *not* relevant over *which* path the integration occurs.

$$\int_1^2 \vec{E} \cdot d\vec{s} = V_{12} = \varphi_1 - \varphi_2$$

- Flux lines pass perpendicularly through the equipotential surfaces. The electric field strength points in the direction of the greatest voltage change. The field vector points from higher to lower voltage levels (in the direction of the lower potential).

$$\vec{E} = -\nabla \cdot V$$

- The electrostatic field holds energy:

$$W = \frac{1}{2} \int_V \vec{D} \cdot \vec{E} \, dV$$

2.1.15 Relationship between the Electrostatic Field Quantities

$$
\begin{array}{ccc}
Q & \Leftarrow Q = \oint_A \vec{D} \, d\vec{A} \Rightarrow & \vec{D} \\
\Uparrow & & \Uparrow \\
Q = C \cdot V & & \vec{D} = \varepsilon \cdot \vec{E} \\
\Downarrow & & \Downarrow \\
V & \Leftarrow V = \int_s \vec{E} \, d\vec{s} \Rightarrow & \vec{E}
\end{array}
$$

2.2 Static Steady-State Current Flow

The **static steady-state current flow** describes the motion of charges in an electrical conductor and its effects when the electrical quantities do *not* change with time. The following section makes the assumption that $\dfrac{di}{dt} = 0$. There are therefore *no* induced voltages!

2.2.1 Voltage and Potential

Voltage causes directed charge motion in an electrical conductor. If a voltage is placed on an electrical conductor then a current I will flow. A voltage can be measured at any point on the surface of an electrical conductor when a current is flowing in the conductor. A voltage is also present at each internal point in the conductor, although this is not so easy to measure.

- The voltage decreases *uniformly* along the electrical conductor.

A point can be designated where the potential $\varphi = 0$. It is then possible to define potential φ at any point in the conductor. The voltage between two points is equal to the potential difference between the points:

$$V_{12} = \varphi_1 - \varphi_2 \tag{2.26}$$

Areas with the same voltage or with the same potential are known as **equipotential surfaces**, as in electrostatic fields.

2.2.2 Current

The **electric current** is the sum of charges that flow through a defined cross section per unit time:

$$I = \frac{dQ}{dt} \tag{2.27}$$

- The **direction of positive current** is the direction of motion of the positive charges. This is opposite to the direction of motion of the negative charges.

- Current always flows in a closed loop.

(See also Sect. 1.1)

2.2.3 Electric Field Strength

The **electric field strength** describes the change in the electric voltage over a given path. It is a vector that points in the direction of the *greatest* change. As the greatest change in the voltage is perpendicular to the equipotential surfaces, the electric field strength vector lies perpendicular to the equipotential surfaces (Fig. 2.13).

$$\vec{E} = \frac{dV}{ds_\perp} \cdot \vec{e}_{A_\perp} \tag{2.28}$$

The unity vector $\vec{e}_{A_\perp}$ is perpendicular to the equipotential surface, and the path element $ds_\perp$ passes perpendicularly through the equipotential surface. This may also be written as:

$$\vec{E} = -\nabla \cdot V \tag{2.29}$$

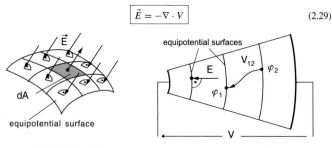

Fig. 2.13. Electric field strength, equipotential surfaces and electric voltage

- The **direction** of the electric field strength is the direction of the greatest change in voltage.
- The **magnitude** of the electric field strength yields the change in the voltage over the path.

The unit of electric field strength is volt per meter, $\frac{V}{m}$. The electric voltage V is the line integral of the electric field strength:

$$V_{12} = \int_{1}^{2} \vec{E} \cdot d\vec{s} \tag{2.30}$$

- The **electric voltage** between two points is equal to the line integral of the electric field strength between the two points. Therefore, it does not matter over *which* path the integration occurs.

A field is said to be **homogeneous** if the same field strength prevails at all points in that field, i.e. the magnitude and direction at all points are the same. The voltage in a homogeneous field is:

$$V_{12} = \vec{E} \cdot \vec{s}_{12} \tag{2.31}$$

2.2.4 Current Density

The current I is spread out over a conductor. This leads to the definition of **current density** $\vec{J}$:

$$\vec{J} = \frac{dI}{dA_{\perp}} \cdot \vec{e}_{A_{\perp}} \tag{2.32}$$

Thus $dA_{\perp}$ is the surface element of an equipotential surface. The unity vector $\vec{e}_{A_{\perp}}$ lies perpendicular to the equipotential surface.

- The **direction** of the current density points in the direction of the greatest voltage change. The current density vector lies perpendicular to the equipotential surface (Fig. 2.14a).
- The **magnitude** of the current density yields the amount of charge per cross section and per unit of time that passes through an equipotential surface.
- The current density points in the same direction as the electric field strength.

The unit of current density is amperes per square meter, $\frac{A}{m^2}$. The current I is the integral of the scalar product of the current density and an arbitrary area through which the current passes.

$$I = \int_{A} \vec{J} \cdot d\vec{A} \tag{2.33}$$

In a homogeneous field this integral becomes (Fig. 2.14b):

$$I = J \cdot A \cdot \cos\alpha \tag{2.34}$$

where α is the angle between the normal to the area and the current density.

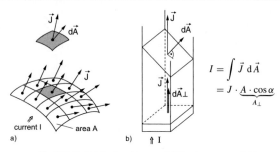

Fig. 2.14. Current density: **a)** in general; **b)** in a homogeneous field

2.2.5 Resistivity and Conductivity

The electric field strength is related to the current density by the **resistivity** ϱ and the **conductivity** σ.

$$\vec{E} = \varrho \cdot \vec{J}, \quad \text{and} \quad \vec{J} = \sigma \cdot \vec{E} \tag{2.35}$$

$$\sigma = \frac{1}{\varrho} \tag{2.36}$$

- The conductivity is the inverse of the resistivity.

- The unit of resistivity is ohm meters, Ωm.

NOTE: The unit of resistivity is often given as $\dfrac{\Omega \text{mm}^2}{\text{m}}$, since the lengths of electrical conductors are often given in meters and the cross-sectional area in millimeter^2. The electrical resistance of a homogeneous conductor may be calculated thus:

$$R = \varrho \cdot \frac{\text{length}}{\text{cross-sectional area}}$$

The unit of conductivity is siemens or mho per meter, $\dfrac{\text{S}}{\text{m}}$, or $\dfrac{\text{mho}}{\text{m}}$.

- The resistivity and the conductivity are material characteristics of the electrical conductor (Table 2.2).

- The resistivity and the conductivity are **temperature dependent**.

NOTE: The resistivity temperature dependence is given by the temperature coefficient α. The change in the resistivity with temperature may be calculated from:

$$\varrho\left(\vartheta_{2}\right) = \varrho\left(\vartheta_{1}\right) \cdot \left[1 + \alpha \cdot \left(\vartheta_{2} - \vartheta_{1}\right)\right]$$

The temperature coefficients of copper and aluminium are $\alpha = 0.004$ K^{-1}. Therefore, for a temperature change of 100 K, their resistivity changes by 40%.

Table 2.2. Resistivity of electrical conductors

Material	Conductivity σ (S/m $\cdot 10^{-6}$)	Resistivity ϱ ($\Omega \cdot$ mm^2/m)
Aluminium	37	0.027
Brass	14.3–12.5	0.07–0.08
Copper	59	0.017
Gold	45.5	0.022
Iron	10–2.5	0.1–0.4
Silver	62.5	0.016

2.2.6 Resistance and Conductance

Voltage is related to current by the **electrical resistance** R and the **electrical conductance** G.

$$V = R \cdot I, \quad \text{and} \quad I = G \cdot U \tag{2.37}$$

The SI unit of resistance R is the ohm, $1\,\Omega = \dfrac{1\,\text{V}}{1\,\text{A}}$, while the unit of conductance G is the mho or Siemens, $1 = \dfrac{1\,\text{A}}{1\,\text{V}}$. If the integral quantities V and I are represented as vector quantities, then R and G are calculated as:

$$R = \frac{V}{I} = \frac{\int_s \vec{E}\,\mathrm{d}\vec{s}}{\int_A \vec{J}\,\mathrm{d}\vec{A}}, \quad \text{and} \quad G = \frac{I}{V} = \frac{\int_A \vec{J}\,\mathrm{d}\vec{A}}{\int_s \vec{E}\,\mathrm{d}\vec{s}}, \quad \text{with} \quad \vec{e}_A = \vec{e}_s \tag{2.38}$$

Equation (2.38) is valid for the case where the path element $\mathrm{d}s$ lies perpendicular to the area element $\mathrm{d}A$. To evaluate this integral the field characteristics must be known qualitatively, i.e. the directions of the field strength and the current density must be known. For an isotropic conductor material of length l and cross-sectional area A with a homogeneous field distribution, $\boxed{R = \varrho \cdot \dfrac{l}{A}}$ and $\boxed{G = \sigma \cdot \dfrac{A}{l}}$

EXAMPLE: Calculation of the resistance of a quarter-ring: The contacts are perfect conductors, and the resistance material is isotropic. **a) Figure 2.15a:** the current is tangentially injected. The current density lines are tangential to the ring. The current is equally divided over the cross section. The individual current density lines may be considered as "current threads" $\mathrm{d}I$. The integration (addition) of these current threads thus yields the total current $I = \int \mathrm{d}I$. Each current thread $\mathrm{d}I$ is defined by the differential conductance $\mathrm{d}G$:

$$\mathrm{d}I = V \cdot \mathrm{d}G = V \cdot \sigma \frac{\mathrm{d}A}{l} = V \cdot \sigma \frac{b}{\pi/2} \cdot \frac{\mathrm{d}r}{r}$$

$$G = \frac{I}{V} = \int \frac{\mathrm{d}I}{V} = \int \frac{V}{V} \cdot \mathrm{d}G = \int \sigma \frac{\mathrm{d}A}{l} = \sigma \frac{b}{\pi/2} \cdot \int_{r_1}^{r_2} \frac{\mathrm{d}r}{r} = \sigma \cdot \frac{b}{\pi/2} \cdot \ln \frac{r_2}{r_1}$$

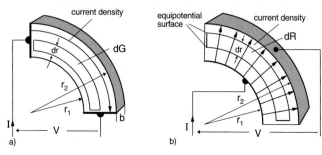

Fig. 2.15. Quarter-ring resistance: **a)** with tangential and **b)** with radial current injection

The resistance R is:

$$R = \frac{1}{G} = \frac{1}{\sigma} \cdot \frac{\pi/2}{b} \cdot \frac{1}{\ln(r_2/r_1)} \qquad (2.39)$$

An alternative method comes from analysing the geometry of the configuration: The total resistance can be considered as a combination of parallel resistors of length $\frac{\pi}{2}r$ and cross-sectional area $b \cdot dr$. The partial conductance dG is given by $\sigma \cdot \frac{b}{\pi/2} \cdot \frac{dr}{r}$. The total conductance G is therefore the integration (summation) of the partial conductances:

$$G = \int_{r_1}^{r_2} \sigma \cdot \frac{b}{\pi/2} \cdot \frac{dr}{r} = \sigma \cdot \frac{b}{\pi/2} \cdot \ln\frac{r_2}{r_1}$$

NOTE: The direction of the field must also be known for this method.

b) Figure 2.15b: the current is injected radially. The current is divided radially from the inside to the outside of the arc. The current density decreases from the inside to the outside of the ring, flowing away from the inner contact in a star-like pattern. The total current must pass through "resistance discs" dR of cross-sectional area $\frac{\pi}{2}r \cdot b$ and of length dr. The total resistance R can be considered as a series combination of resistance discs dR. The total resistance is therefore the integration (summation) of the partial resistances dR:

$$R = \int dR = \int \varrho \cdot \frac{dl}{A} = \int_{r_1}^{r_2} \varrho \cdot \frac{dr}{b \cdot (\pi/2) \cdot r} = \varrho \cdot \frac{\ln(r_2/r_1)}{b \cdot (\pi/2)}$$

2.2.7 Kirchhoff's Laws

2.2.7.1 Kirchhoff's First Law (Current Law)

The electric current always flows in a closed loop. For the current density this means that the flux lines of the current density always form a closed path. Kirchhoff's first law states that for a static steady-state current flow (Fig. 2.16):

$$\oint_A \vec{J}\, d\vec{A} = 0 \qquad (2.40)$$

This can also be written as:

$$\nabla \cdot \vec{J} = 0 \qquad (2.41)$$

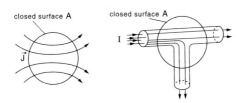

Fig. 2.16. Illustration of Kirchhoff's first law (current law)

- The surface integral of the current density over a closed surface is always zero.
- The current density $\vec{J}$ is **source-free**.

2.2.7.2 Kirchhoff's Second Law (Mesh Law)

The line integral of the electric field strength between two points is equal to the voltage between those points, independent of the path over which the integration is made. If the start and end points are the same then the result is obviously zero.

$$\oint_s \vec{E}\, d\vec{s} = 0 \qquad (2.42)$$

This can also be written as:

$$\nabla \times \vec{E} = 0 \qquad (2.43)$$

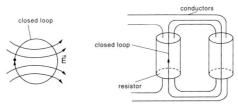

Fig. 2.17. Illustration of Kirchhoff's second law (mesh law)

- The line integral of the electric field strength over a closed loop is always zero (Fig. 2.17).
- The static steady-state current flow is **solenoidal**.

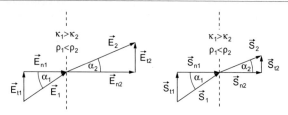

Fig. 2.18. Field quantities at a boundary

2.2.8 Static Steady-State Current Flow at Boundaries

A static steady-state current flow that passes a material boundary changes its direction depending on the conductivity of the materials (Fig. 2.18).

$$\boxed{\vec{J}_{n2} = \vec{J}_{n1}, \quad \text{and} \quad \vec{J}_{t2} = \frac{\varrho_1}{\varrho_2} \cdot \vec{J}_{t1} = \frac{\sigma_2}{\sigma_1} \cdot \vec{J}_{t1}} \tag{2.44}$$

$$\boxed{\vec{E}_{t2} = \vec{E}_{t1}, \quad \text{and} \quad \vec{E}_{n2} = \frac{\varrho_2}{\varrho_1} \cdot \vec{E}_{n1} = \frac{\sigma_1}{\sigma_2} \cdot \vec{E}_{n1}} \tag{2.45}$$

$$\boxed{\tan \alpha_2 = \frac{\varrho_1}{\varrho_2} \tan \alpha_1 = \frac{\sigma_2}{\sigma_1} \tan \alpha_1} \tag{2.46}$$

At the boundary:

- The normal component of the current density does not change.
- The change in the tangential component of the current density is proportional to the conductivity.
- The tangential component of the electric field strength does not change.
- The normal component of the electric field strength is proportional to the resistivity.

2.2.9 Overview: Fields and Resistances of Different Geometric Configurations

Table 2.3. Overview of fields and resistances

Parallel Plates		
	$R = \varrho \cdot \dfrac{d}{A}$	$E = \dfrac{V}{d}$ $S = \dfrac{I}{A}$
Parallel plates		
	$R = \dfrac{1}{A} \cdot (\varrho_1 d_1 + \varrho_2 d_2)$	$E_{1(2)} = \dfrac{\varrho_{1(2)} V}{\varrho_1 d_1 + \varrho_2 d_2}$ $S = \dfrac{I}{A}$
Coaxial cylinders		
	$R = \varrho \cdot \dfrac{\ln \dfrac{r_2}{r_1}}{2\pi l}$	$E = \dfrac{V}{r \cdot \ln \dfrac{r_2}{r_1}}$ $S = \dfrac{I}{2\pi r l}$
90° arc		
	$R = \varrho \cdot \dfrac{\pi}{2b} \dfrac{1}{\ln (r_2/r_1)}$	$E = \dfrac{V}{(\pi/2)r}$ $S = \dfrac{I}{br \ln (r_2/r_1)}$
Concentric spheres		
	$R = \varrho \cdot \dfrac{1}{4\pi} \cdot \left(\dfrac{1}{r_1} - \dfrac{1}{r_2}\right)$	$E = \dfrac{V}{r^2} \cdot \dfrac{r_1 r_2}{r_2 - r_1}$ $S = \dfrac{I}{4\pi r^2}$
Sphere-sphere		
	$R = \varrho \cdot \dfrac{1}{4\pi} \left(\dfrac{1}{r_1} - \dfrac{1}{2a}\right)$	$E \approx \dfrac{V \left(\dfrac{1}{x^2} + \dfrac{1}{(2a-x)^2}\right)}{\dfrac{2}{r_1} - \dfrac{1}{a}}$ $S \approx \dfrac{I \left(\dfrac{1}{x^2} + \dfrac{1}{(2a-x)^2}\right)}{4\pi}$

Table 2.3. cont.

Sphere-infinity		
(sphere with r_1)	$R = \dfrac{1}{4\pi\sigma r_1}$	$E = V \cdot \dfrac{r_1}{r^2}$ $S = \dfrac{I}{4\pi r^2}$
Parallel conductors		
(two parallel conductors r_1, separation $2a$, with x)	$R = \varrho \, \dfrac{\ln\left[\dfrac{a}{r_1} + \sqrt{\left(\dfrac{a}{r_1}\right)^2 - 1}\right]}{\pi\sigma l}$ $\approx \varrho \cdot \dfrac{\ln \dfrac{2a}{r_1}}{\pi l}$ for $a \gg r_1$	$E = \dfrac{V \dfrac{\sqrt{a^2 - r_1^2}}{a^2 - r_1^2 - x^2}}{\ln\left[\dfrac{a}{r_1} + \sqrt{\left(\dfrac{a}{r_1}\right)^2 - 1}\right]}$ $S = \dfrac{I\sqrt{a^2 - r_1^2}}{\pi l \left(a^2 - r_1^2 - x^2\right)}$

2.2.10 Power and Energy in Static Steady-State Current Flow

In a static steady-state current flow electrical energy is converted into heat. The electrical power is:

$$P = V \cdot I \tag{2.47}$$

For a resistance R or a conductance G this becomes:

$$P = V \cdot I = \frac{V^2}{R} = I^2 \cdot R, \quad \text{or} \quad P = V \cdot I = \frac{I^2}{G} = V^2 \cdot G \tag{2.48}$$

This also holds for each infinitesimal volume element:

$$dP = \underbrace{\frac{dV}{ds_\perp}}_{\vec{E}} ds_\perp \cdot \underbrace{\frac{dI}{dA_\perp}}_{\vec{J}} dA_\perp = \vec{E} \cdot \vec{J} \cdot dV \tag{2.49}$$

Thus $ds_\perp$ and $dA_\perp$ lie perpendicular to the equipotential surfaces. From Eq. (2.49) the **power density** of a static steady state current flow is defined as:

$$\frac{dP}{dV} = \vec{J} \cdot \vec{E} \tag{2.50}$$

The power P is then given by:

$$P = \int_V \vec{J} \cdot \vec{E} \, dV \tag{2.51}$$

The **energy** is the integral of the power over time:

$$W = \int_{t_1}^{t_2} P(t) \, dt \tag{2.52}$$

NOTE: The static steady-state current flow makes assumptions for *time-invariant* quantities. The use of $P(t)$ in the above equations seems to contradict this. However, for slow time variations of the electric quantities, the magnetic effects can be ignored (for example, current pinching, skin effect).

NOTE: If the power is a time-varying periodic progression, then the average power $\bar{P}$ is usually used to refer to the *power loss*. This is calculated as the arithmetic average of the work done divided by the period:

$$\bar{P} = \frac{1}{T} \cdot W(T) = \frac{1}{T} \int_0^T P(t)\, dt \qquad (2.53)$$

2.2.11 Overview: Characteristics of Static Steady-State Current Flow

- The static steady-state current flow is a **conservative field**. The line integral of the electric field strength is equal to the voltage (potential difference) between the beginning and end points of the path. The path over which the integration is made is therefore *not* relevant. The flux lines pass perpendicularly through the equipotential surfaces.

$$\int_1^2 \vec{E} \cdot d\vec{s} = V_{12} = \varphi_1 - \varphi_2, \quad \text{or} \quad \vec{E} = -\nabla \cdot V \qquad (2.54)$$

- The static steady-state current flow is **solenoidal**. The line integral of the electric field strength over a closed loop is always zero (Kirchhoff's second law, $\sum V = 0$).

$$\oint \vec{E} \cdot d\vec{s} = 0, \quad \text{or} \quad \nabla \times \vec{E} = 0 \qquad (2.55)$$

- The static steady-state current flow is **source-free**. The current always flows in a closed loop. The surface integral over a closed surface is always zero (Kirchhoff's first law, $\sum I = 0$).

$$\oint_A \vec{J} \cdot d\vec{A} = 0, \quad \text{or} \quad \nabla \cdot \vec{J} = 0 \qquad (2.56)$$

- In a static steady-state current flow electrical power is converted into heat:

$$P = \int_V \vec{J} \cdot \vec{E}\, dV \qquad (2.57)$$

2.2.12 Relationship Between Quantities in Static Steady-State Current Flow

$$
\begin{array}{ccc}
I & \Leftarrow I = \int_A \vec{J} \, \mathrm{d}\vec{A} \Rightarrow & \vec{S} \\
\Uparrow & & \Uparrow \\
V = R \cdot I & & \vec{J} = \sigma \cdot \vec{E} \\
\Downarrow & & \Downarrow \\
V & \Leftarrow V = \int_S \vec{E} \, \mathrm{d}\vec{s} \Rightarrow & \vec{E}
\end{array}
$$

2.3 Magnetic Fields

The **magnetic field** describes the effects of stationary and time-varying currents inside and outside an electrical conductor. Electric charges in motion are subject to Coulomb forces and also to other forces caused by the magnetic field. The unit of electrical current can be defined in terms of the forces in a magnetic field:

- If, for two straight, parallel, infinitely long conductors with negligibly small diameter separated by a distance of $r = 1$ m and with the same time-invariant current I flowing through them, each 1-m conductor length exerts a force $F = 2 \cdot 10^{-7}$ N on the other, then the current I has a value of 1 A.

The effect of *time-varying* currents and *time-varying* magnetic fields is described in Sect. 2.3.13.2 by Faraday's law. Faraday's law is the basis of many technical applications, such as, for example, the electric motor, transformers, relays and the electrical energy supply by rotating generators.

Direction-Pointing Convention

In this book, three-dimensional physical relationships are illustrated when considering magnetic fields. To do this, the following representation is normally used for directions:

⊗: Direction pointer or vector pointing into the page;
⊙: Direction pointer or vector pointing out of the page towards the observer.

Direction pointer: Direction convention for scalar quantities, such as current, voltage or magnetic flux. **Vector**: Direction convention for field quantities, such as flux density and magnetic field strength. **Cross product, vector product**: For example: $\vec{F} = (\vec{v} \times \vec{B})$, or: "F is equal to v cross B". The **magnitude** of $\vec{F}$ is

$$|\vec{F}| = |\vec{v}| \cdot |\vec{B}| \cdot \sin \alpha$$

where α is the angle between vectors $\vec{v}$ and $\vec{B}$. The vector $\vec{F}$ lies *perpendicular* to the plane formed by the vectors $\vec{v}$ and $\vec{B}$. The **direction** of $\vec{F}$ can be given by the **corkscrew rule**. If a corkscrew is turned over the smallest angle from $\vec{v}$ to $\vec{B}$, then the direction of the corkscrew gives the direction of $\vec{F}$ (Fig. 2.19). **Source pointer system**: If the direction pointers for V and I in a basic circuit element point in the same directions, then it may be assumed that the basic element is a source and thus supplies energy. This does not mean

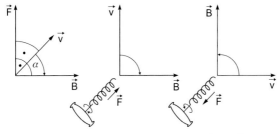

Fig. 2.19. Corkscrew rule for a cross product, here: $\vec{F} = \vec{v} \times \vec{B}$

that it definitely is a source, which can only be ascertained after some network calculations. Only if the application of Kirchhoff's law yields a positive result for current and voltage, can it be confirmed that the element was indeed a source. If the result is negative, then the assumption that the element was a source was wrong, and the element is an energy consumer. **Consumer pointer system**: If the direction pointers for V and I in a basic circuit element point in opposite directions, then it may be presumed that the particular element is a consumer and thus absorbs energy.

2.3.1 Force on a Moving Charge

Moving electric charges exert forces on one another that *cannot* be explained by Coulomb's law. The magnetic force on two point charges in *uniform motion* on parallel paths, while at the same height, is given by:

$$F = \frac{\mu}{4\pi} \cdot \frac{(Q_1 v_1) \cdot (Q_2 v_2)}{r^2}, \quad \text{with} \quad \mu = \mu_0 \cdot \mu_r \quad (2.58)$$

μ: permeability;

μ_0: free-space permeability, $\mu_0 = 4\pi \cdot 10^{-7} \frac{\text{Vs}}{\text{Am}} = 1.257 \cdot 10^{-6} \frac{\text{Vs}}{\text{Am}}$;

μ_r: relative permeability;

r: distance between the paths.

The permeability is a constant that depends on the medium in which the charges are moving. In a vacuum and in air the relative permeability is $\mu_r = 1$.

- For the same sign on the products $(Q_1 \cdot v_1)$ and $(Q_2 \cdot v_2)$ the charges are drawn closer, but for opposite signs the charges are drawn apart (Fig. 2.20).

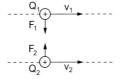

Fig. 2.20. Force on moving point charges

2.3.2 Definition of Magnetic Flux Density

The magnetic flux density $\vec{B}$ is derived from the force on charges in motion. By definition:

$$F = (Q_1 v_1) \cdot \underbrace{\frac{\mu}{4\pi} \frac{(Q_2 v_2)}{r^2}}_{B} = Q_1 v_1 \cdot B$$

B is the **magnetic flux density**. The magnetic flux density is a field quantity with direction, that is, a vector. The SI unit of magnetic flux density is the **tesla**, $1\,\text{T} = \dfrac{1\,\text{Vs}}{1\,\text{m}^2}$. The **direction** of the magnetic flux density can be found using a magnetic dipole (e.g. a compass needle). The positive direction of magnetic flux density is the direction to which the north pole of the magnetic dipole points.

NOTE: The compass needle is a magnetic dipole. The north pole of the compass needle points north. This means that the geographic north pole actually is the magnetic south pole of the earth.

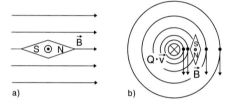

Fig. 2.21. **a)** Compass needle in a magnetic field; **b)** magnetic flux density around a moving charge

Magnetic fields are illustrated by means of **flux lines**. To show the field, lines are drawn whose tangents correspond at each point to the direction in which an infinitesimally small magnetic dipole would point. The density of flux lines is thus a measure of the magnitude of the field strength.

- It can be shown experimentally that the magnetic force lines (flux lines) are tangential in a **clockwise** sense about the direction of motion of the charge (Fig. 2.21).

- The flux lines in magnetic induction have no beginning or end since they form closed loops.

Right-hand rule: The direction of the magnetic field around a moving charge or around an electric current can be found with the right-hand rule. If the right-hand thumb points in the direction of the moving charge (or in the direction of the current), then the curved fingers of the right hand show the rotation direction of the field. **Corkscrew rule**: The direction of the magnetic field around a moving charge or around an electric current can also be found with the corkscrew rule. If a corkscrew were turned in the direction of the moving charge (in the direction of the current), then the rotation direction of the corkscrew is the direction of the field. The force on a moving charge can be found from the field direction:

$$\boxed{\vec{F} = Q \cdot \left(\vec{v} \times \vec{B} \right)} \tag{2.59}$$

- The force on a moving charge is known as the **Lorentz force**.

2.3 Magnetic Fields

Lorentz Force on a Current Carrying Conductor

The current I can be considered as a directed motion of charges. If a charge quantity ΔQ moves in an electrical conductor, then it can be described by $\Delta Q = I \cdot \Delta t$. The velocity of the charges can be written as $v = \dfrac{\Delta l}{\Delta t}$. Therefore:

$$\Delta Q \cdot v = \Delta Q \cdot \frac{\Delta l}{\Delta t} = \frac{\Delta Q}{\Delta t} \cdot \Delta l = I \cdot \Delta l$$

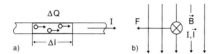

Fig. 2.22. Lorentz force: **a)** moving charge in a conductor; **b)** force on the conductor

The Lorentz force on a straight current carrying conductor therefore amounts to:

$$\boxed{\vec{F} = I \cdot \left(\vec{l} \times \vec{B}\right)} \quad (2.60)$$

The vector $\vec{l}$ points thus in the direction of the current I (Fig. 2.22).

EXAMPLE: Calculate the rotation direction and the torque of an electric motor: In a permanent magnet electric motor the flux density in the air gap is $B = 0.5$ T (permanent magnet: the magnetic field is created by a permanent magnet). The rotor of the motor has an active length of $l = 10$ cm (the conductor length is 10 cm in the magnetic field) and a diameter of $d = 10$ cm. There are four current-carrying conductors in the magnetic field on either side (Fig. 2.23a). Each conductor carries a current of 1 A.

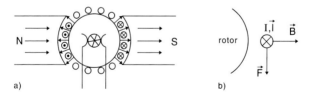

Fig. 2.23. a) Simplified representation of an electric motor; **b)** system to find the rotation direction

Solution:
The motor turns clockwise (Fig. 2.23b). The direction of force for each of the current-carrying conductors is clockwise. The torque is:

$$M = F_{\text{total}} \cdot \frac{d}{2} = 8 \cdot I \cdot s \cdot B \cdot \frac{d}{2} = 8 \cdot 1\,\text{A} \cdot 0.1\,\text{m} \cdot 0.5 \frac{\text{Vs}}{\text{m}^2} \cdot \frac{0.1\,\text{m}}{2} = 20 \cdot 10^{-3}\,\text{VAs} = 20 \cdot 10^{-3}\,\text{Nm}$$

NOTE: The conversion from electrical into mechanical units can be done by considering the energy: 1 VAs = 1 Ws = 1 J = 1 Nm.

2.3.3 Biot–Savart's Law

Biot–Savart's law gives the magnetic flux density at any given point, in magnitude and direction, caused by a moving point charge (Fig. 2.24).

$$\vec{B} = \frac{\mu}{4\pi} \cdot \frac{Q}{r^2} (\vec{v} \times \vec{e}_r) \tag{2.61}$$

where $\vec{e}_r$ is the unity vector in direction $\vec{r}$.

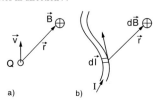

Fig. 2.24. Biot–Savart's law: **a)** for a moving charge; **b)** for a current-carrying conductor

In order to calculate the magnetic flux density caused by a very thin arbitrarily shaped current-carrying conductor, each infinitesimal current-carrying conductor element $\mathrm{d}I \cdot \vec{l}$ is considered as a moving charge $Q \cdot \vec{v}$ (see also Sect. 2.3.2). In this case, each conductor element creates an infinitesimal flux density $\mathrm{d}\vec{B}$ in the space under consideration. Biot–Savart's law states for the current-carrying conductor that:

$$\mathrm{d}\vec{B} = \frac{\mu}{4\pi} \cdot \frac{I}{r^2} \cdot \left(\mathrm{d}\vec{l} \times \vec{e}_r\right) \tag{2.62}$$

The magnetic flux density $\vec{B}$ can be found by integrating according to the superposition theorem, $\vec{B} = \int \mathrm{d}\vec{B}$.

EXAMPLE: Calculation of the magnetic field of an infinitely long current-carrying conductor with a negligibly small diameter: Given that the field must be radially symmetric around the conductor, the evaluation using Biot–Savart's law can be treated as a planar problem (Fig. 2.25). In this case, Biot–Savart's law can be

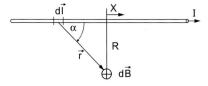

Fig. 2.25. Field of a straight current-carrying conductor

simplified. Note that from vector algebra $|\vec{a} \times \vec{b}| = |\vec{a}| \cdot |\vec{b}| \cdot \sin \angle \vec{a}, \vec{b}$:

$$\mathrm{d}B = \frac{\mu}{4\pi} \cdot \frac{I}{r^2} \cdot \mathrm{d}l \cdot \sin \alpha$$

with $r = \sqrt{x^2 + R^2}$, and $\sin \alpha = \dfrac{R}{\sqrt{x^2 + R^2}}$, then

$$B = \int_{-\infty}^{+\infty} \frac{\mu}{4\pi} \cdot \underbrace{\frac{I}{x^2+R^2}}_{I/r^2} \cdot \underbrace{\frac{R}{\sqrt{x^2+R^2}}}_{\sin\alpha} \, dx = \frac{\mu \cdot I \cdot R}{2\pi} \cdot \int_{0}^{+\infty} (x^2+R^2)^{-3/2} \, dx = \frac{\mu \cdot I}{2\pi R}$$

The magnetic flux density lies tangentially around the conductor and decreases in proportion to the distance from the conductor.

NOTE: Biot–Savart's law is also valid for the magnetic field strength $\vec{H}$. Since $\vec{B} = \mu \cdot \vec{H}$, then:

$$\boxed{\vec{H} = \frac{1}{4\pi} \cdot \frac{Q}{r^2} (\vec{v} \times \vec{e}_r)} \tag{2.63}$$

2.3.4 Magnetic Field Strength

The **magnetic field strength** $\vec{H}$ is, after the magnetic flux density $\vec{B}$, the second field quantity of the magnetic field. The magnetic flux density $\vec{B}$ was defined in Sect. 2.3.2 in terms of the force on moving charges. It is dependent on the surrounding medium through the permeability μ. The magnetic field strength $\vec{H}$ is defined *independently* of the surrounding medium:

$$F = (Q_1 v_1) \cdot \underbrace{\mu \cdot \frac{(Q_2 v_2)}{4\pi r^2}}_{B} = (Q_1 v_1) \cdot \mu \cdot \underbrace{\frac{(Q_2 v_2)}{4\pi r^2}}_{H}$$

It follows that:

$$\boxed{\vec{H} = \frac{1}{\mu} \cdot \vec{B}, \quad \text{or} \quad \vec{B} = \mu \cdot \vec{H}} \tag{2.64}$$

The SI unit of magnetic field strength $\vec{H}$ is amperes per meter, $\frac{A}{m}$. For isotropic media:

- The cause of the magnetic field strength $\vec{H}$ is the moving charge or the electric current I.
- Vectors $\vec{B}$ and $\vec{H}$ point in the same direction.
- The **permeability** μ is the proportionality constant between the magnetic flux density $\vec{B}$ and the magnetic field strength $\vec{H}$.

The permeability is formed by the permeability of free space μ_0 and the relative permeability μ_r:

$$\boxed{\mu = \mu_0 \cdot \mu_r} \tag{2.65}$$

The value of the **permeability of free space** is:

$$\boxed{\mu_0 = 4\pi \cdot 10^{-7} \, \frac{Vs}{Am} = 1.257 \cdot 10^{-6} \, \frac{Vs}{Am}} \tag{2.66}$$

The **relative permeability** can have values less than one (**diamagnetism**), and also greater than one (**paramagnetism**).

- The relative permeability of a vacuum is $\mu_r = 1$. This is also the value for air and for most gases.

2.3.5 Magnetic Flux

The **magnetic flux** Φ is the surface integral of the magnetic flux density.

$$\Phi = \int_A \vec{B} \, d\vec{A} \tag{2.67}$$

The SI unit of magnetic flux is weber, $1 \text{ Wb} = 1 \text{ Vs}$.

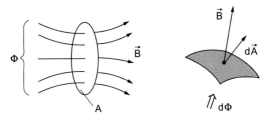

Fig. 2.26. Magnetic flux and magnetic flux density

NOTE: The magnetic flux density defines the magnetic flux per unit area.

Magnetic flux always forms a closed loop. The surface integral of the magnetic flux density is always zero (Fig. 2.26).

$$\oint_A \vec{B} \, d\vec{A} = 0 \tag{2.68}$$

This can also be written as:

$$\nabla \cdot \vec{B} = 0 \tag{2.69}$$

- The magnetic flux density is a **solenoidal field**.
- Magnetic flux always forms a closed loop.

Flux Linkage

The **flux linkage** Ψ is the flux encircled by a conductor winding that can consist of more than one turn. Of particular interest is the special case where N windings of a conductor are linked N times by the *same* flux. Then:

$$\Psi = N \cdot \Phi \tag{2.70}$$

The unit of flux linkage is volt seconds, Vs. The flux linkage is the effective flux for the application of Faraday's laws. It is used specifically in the calculations for transformers and electric machines.

2.3.6 Magnetic Voltage and Ampere's Law

The **magnetic voltage** V is the line integral of the magnetic field strength.

$$V_{12} = \int_1^2 \vec{H} \, d\vec{s} \tag{2.71}$$

- The magnetic voltage between two points is equal to the line integral of the magnetic field strength between the points. It is therefore *not* independent of the path over which the integration occurs. Depending on how often the current I loops, the result can vary by $n \cdot I$, $n = \pm 1, 2, \ldots, i$ (Fig. 2.27).

The SI unit of the magnetic voltage is the ampere, A.

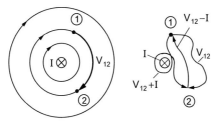

Fig. 2.27. Magnetic voltage V

Ampere's Law

If the magnetic voltage is calculated over a *closed* loop, then the result for the magnetic voltage is equal to the circulating current. **Ampere's law** states:

$$\oint \vec{H} \, d\vec{s} = \sum I = \Theta = \text{MMF} \tag{2.72}$$

- The magnetic field strength $\vec{H}$ is directly related to the current I.

- The circular integral of the magnetic field strength is equal to the circulated current. If a current I is circulated n times, then the result of the circular integral is $\sum I = n \cdot I$ (Fig. 2.28a).

- If the current flows through a coil with N windings and if the line integral is calculated for all of the windings, then the result of the circular integral is $I \cdot N$ (Fig. 2.28b).

- The sum of the circulating currents is known as the **magnetomotive force (MMF)**, or also as **ampere turns** Θ.

If the current $\sum I$ is described by the current density and the electric flux density, then Ampere's law states:

$$\oint_s \vec{H} \cdot d\vec{s} = \int_A \left(\vec{J} + \frac{d\vec{D}}{dt} \right) d\vec{A} \tag{2.73}$$

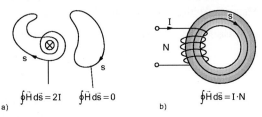

$$\oint \vec{H}\,d\vec{s} = 2I \qquad \oint \vec{H}\,d\vec{s} = 0 \qquad\qquad \oint \vec{H}\,d\vec{s} = I \cdot N$$

a)   b)

Fig. 2.28. Application of Ampere's law: **a)** circular integral enlosing the current I; **b)** flux through a coil with N windings

In this form Ampere's law is Maxwell's first equation. This is also written as:

$$\nabla \times \vec{H} = \vec{J} + \frac{d\vec{D}}{dt} \tag{2.74}$$

- The curl of the magnetic field is not zero.

Ampere's law permits the calculation of the magnetic field strength in simple geometric configurations, namely when the flux lines and their direction are known.

EXAMPLE: Calculation of the magnetic field strength of an infinitely long, straight current-carrying conductor with a circular cross section (Fig. 2.29): The current density in the conductor is homogeneous. For symmetry reasons the magnetic field strength lies tangentially (circularly) around the centre of the conductor. The calculation can be carried out in two sections: a) inside, and b) outside the conductor:

$$\text{a)} \oint \vec{H}\,d\vec{s} = \int \vec{J}\,d\vec{A}, \quad J = \frac{I}{\pi r_1^2} \Rightarrow H \cdot 2\pi r = \frac{I}{\pi r_1^2}\pi r^2$$

$$\Rightarrow H(r) = \frac{I}{2\pi r_1^2}\cdot r, \quad \text{for } r \leq r_1$$

$$\text{b)} \oint \vec{H}\,d\vec{s} = I \qquad\qquad\qquad \Rightarrow H \cdot 2\pi r = I$$

$$\Rightarrow H(r) = \frac{I}{2\pi r}, \qquad \text{for } r \geq r_1$$

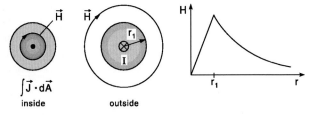

Fig. 2.29. Magnetic field of a round conductor

2.3.7 Magnetic Resistance, Magnetic Conductance, Inductance

The **magnetic resistance** R_m is defined as:

$$R_m = \frac{V}{\Phi} \tag{2.75}$$

- It depends only on the geometric configuration and the permeability.

The unit of magnetic resistance is the inverse henry, $\frac{1}{H} = \frac{A}{Vs}$. If the quantities V and Φ are replaced by vector field quantities, then R_m may be calculated as:

$$R_m = \frac{V}{\Phi} = \frac{\int \vec{H}\, d\vec{s}}{\int \vec{B}\, d\vec{A}}, \quad \text{with} \quad \vec{e}_A \parallel \vec{e}_s \tag{2.76}$$

Equation (2.76) is valid for the case where the path element $d\vec{s}$ lies perpendicular to the area element $d\vec{A}$ ($\vec{e}_A = \vec{e}_s$). To evaluate this integral, the field must be known qualitatively, i.e. the direction of the magnetic flux density and the magnetic field strength must be known. For a homogeneous magnetic material with a homogeneous field distribution, the magnetic resistance for a length l and a cross-sectional area A is:

$$R_m = \frac{1}{\mu} \cdot \frac{l}{A} = \frac{1}{\mu_0 \mu_r} \cdot \frac{l}{A} \tag{2.77}$$

- The magnetic resistance is proportional to the magnetic path length and inversely proportional to the cross section of the magnetic resistance.

The **magnetic conductance** G_m is the reciprocal of the magnetic resistance:

$$G_m = \frac{1}{R_m} \tag{2.78}$$

The **total magnetic resistance** of a *closed* loop is (see also following section):

$$R_{total} = \frac{I \cdot N}{\Phi} \tag{2.79}$$

The reciprocal of the resistance R_{total} is the magnetic conductance A_L:

$$A_L = \frac{\Phi}{I \cdot N} \tag{2.80}$$

The unit of A_L is the henry, $1\,H = 1\,\frac{Vs}{A}$.

- The value of $\mathbf{A_L}$ depends only on the geometric dimensions and the material characteristics of the configuration (of the magnetic loop).

- For the saturation of ferromagnetic materials, the value of A_L decreases with increasing saturation (the magnetic resistance increases).

NOTE: The value of A_L is given in data books for cores for the construction of chokes with its dependence on the air gap, mostly in nH. The bigger the air gap is, the smaller is the value of A_L. It should be mentioned here that most of the energy in an inductance is stored in the air gap, i.e. the air gap is necessary in a choke (see Sect. 2.3.16).

Inductance

The **inductance** L defines the relationship between current and voltage (see Sect. 1.1.7): With Faraday's law $v = N \cdot \dfrac{d\Phi}{dt}$ (see Sect. 2.3.13.2), the relationship between current and voltage $v = L \cdot \dfrac{di}{dt}$ and the value of $A_L = \dfrac{\Phi}{i \cdot N}$ follows:

$$L = N^2 \cdot A_L \tag{2.81}$$

The unit of inductance is the henry, $1\,\text{H} = 1\,\dfrac{\text{Vs}}{\text{A}}$.

- The inductance L is the product of the value of A_L and the square of the number of windings.

- The inductance depends only on the geometric dimensions and the material characteristics of the core as well as the number of windings.

The relationship between the inductance L, the current I and the magnetic flux Φ is given by the value of A_L: with $A_L = \dfrac{\Phi}{I \cdot N}$ and $L = N^2 \cdot A_L$, it follows that $L = \dfrac{N \cdot \Phi}{I}$, or:

$$L \cdot I = N \cdot \Phi \tag{2.82}$$

NOTE: The value of inductance for ferromagnetic materials decreases with increasing saturation.

2.3.8 Materials in a Magnetic Field

The **permeability** μ is determined by the permeability of free-space μ_0 and the relative permeability μ_r.

$$\mu = \mu_0 \cdot \mu_r \tag{2.83}$$

$$\mu_0 = 4\pi \cdot 10^{-7}\,\dfrac{\text{Vs}}{\text{Am}} = 1.257 \cdot 10^{-6}\,\dfrac{\text{Vs}}{\text{Am}} \tag{2.84}$$

The **relative permeability** μ_r can have values less than one (**diamagnetism**) or greater than one (**paramagnetism**).

- The relative permeability of a vacuum is $\mu_r = 1$. This is also approximately the value for air and for most gases.

- Paramagnetic materials concentrate the magnetic flux, while diamagnetic materials spread the magnetic flux out.

Materials with $\mu_r \gg 1$, known as **ferromagnetic** materials, are of particular importance. The characteristics of materials with very high μ_r are:

- They concentrate the present magnetic flux. This is used, for example, in shielding (Fig. 2.30).
- Large inductance values can be realised with them.
- For an applied voltage at the winding (and the consequent magnetic flux), only a small winding current is required (Fig. 2.31).
- For an applied current to the winding (and the consequent magnetic field strength), large magnetic fluxes or high flux densities can be obtained (Fig. 2.31).

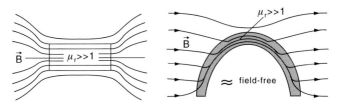

Fig. 2.30. Concentration of a magnetic field by materials with $\mu_r \gg 1$, that is, ferromagnetics

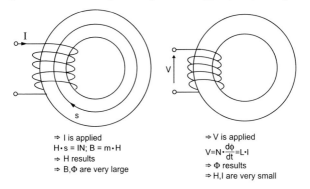

⇒ I is applied
$H \cdot s = IN$; $B = m \cdot H$
⇒ H results
⇒ B, Φ are very large

⇒ V is applied
$V = N \cdot \dfrac{d\phi}{dt} = L \cdot I$
⇒ Φ results
⇒ H, I are very small

Fig. 2.31. Effect of materials with $\mu_r \gg 1$ for an applied current and voltage

2.3.8.1 Ferromagnetic Materials

Ferromagnetics: Ferromagnetics are of particular technical importance. Their relative permeability depends on the magnetic field strength. Values typically used are much bigger than one, amd usually lie between 1000 and 100 000. **Weiss domains**: Ferromagnetic materials contain small crystalline magnetic dipoles, whose directions are statistically distributed in the unmagnetised state. These are known as Weiss domains. With increasing magnetic field strength the Weiss domains become orientated, i.e. the magnetic path is shortened, and the magnetic flux density is high. With further increases in field strength saturation occurs, when practically all Weiss domains are orientated. At this point the flux density no longer increases in the same manner, but increases approximately with μ_0.

Hysteresis loop: The relationship between the magnetic flux density B and the electric field strength H for ferromagnetics is represented by the hysteresis loop (Fig. 2.32).

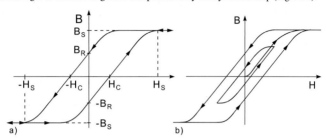

Fig. 2.32. Hysteresis loop: **a)** hysteresis quantities; **b)** hysteresis loss

Magnetic saturation:

The saturation flux density B_S is the flux density that is present after the orientation of all the Weiss domains. Any further increase in the magnetic field strength only yields a minor increase in the the flux density. The saturation flux density for iron usually lies in the range between 1 and 2 Tesla, for ferrites between 0.3 and 0.4 Tesla. **Remanent flux density**: The remanent flux density B_R is the flux density, which remains after the magnetisation up to the saturation flux density and the subsequent return to $H = 0$. In practice it occurs when a closed ring of constant cross section is magnetised up to the saturation flux density and then the coil current is switched off. If an air gap is inserted in the magnetic ring, then the remaining flux density will be smaller (see Sect. 2.3.11). **Coercivity**: The coercive magnetic field strength H_C is the field strength required to bring the flux density back to zero after being magnetised up to the saturation flux density. **Hysteresis loss, iron loss**: Work must be done to magnetise ferromagnetic materials. The enclosed area of the hysteresis loop is a measure of the work required. The enclosed area has the dimension work per unit volume $\frac{dW}{dV}$. It is dissipated as heat during every full passage through the hysteresis loop. The corresponding work for a given core volume is thus:

$$W = \frac{dW}{dV} \cdot V \tag{2.85}$$

and the power loss at a frequency f:

$$P = \frac{dW}{dV} \cdot V \cdot f \tag{2.86}$$

The hysteresis losses can be found by calculation of the enclosed area of the hysteresis loop (if they are not known from other sources, e.g. data sheets). At high-frequency magnetisation the hysteresis losses can lead to excessive overheating of the material. In this case the magnetisation level must be decreased. The losses decreases quadratically with the magnetisation level, i.e. by halving the maximum field strength, the losses are quartered (Fig. 2.32b).

NOTE: The expression **iron losses** also includes the eddy current losses with the hysteresis losses.

Soft iron: Soft iron is the name given to a ferromagnetic material that has a narrow hysteresis curve, i.e. a small enclosed area, and therefore with small hysteresis losses. Soft iron materials are used, for example, in transformers and electric machines. **Hard iron**: Hard iron is the name given to a ferromagnetic material that has a large hysteresis curve. They have a high remanence flux density and coercivity. They are suitable for permanent magnets.

Demagnetisation: There are several ways to demagnetise a core:

- Surpass the Curie temperature. The Curie temperature is the temperature at which the molecular thermal motion in the material is so great that the fixed orientation of the Weiss domains cannot be maintained, and they return to a random orientation.
- Demagnetisation through high-frequency withdrawal of the magnetisation level. The core magnetisation level is controlled at high frequency, while the level is slowly decreased. The magnetisation returns slowly to zero.
- By mechanical vibration: mechanical motion can destroy the orientation of the Weiss domains. Strong blows remove the magnetisation. Modern hard magnetic materials are occasionally very sensitive to mechanical forces.

2.3.9 Magnetic Fields at Boundaries

A magnetic field that passes a material boundary changes its direction depending on the permeability of the materials (Fig. 2.33).

$$\vec{H}_{t2} = \vec{H}_{t1}, \quad \text{and} \quad \vec{H}_{n2} = \frac{\mu_1}{\mu_2} \cdot \vec{H}_{n1} \tag{2.87}$$

$$\vec{B}_{n2} = \vec{B}_{n1}, \quad \text{and} \quad \vec{B}_{t2} = \frac{\mu_2}{\mu_1} \cdot \vec{B}_{t1} \tag{2.88}$$

$$\tan \alpha_2 = \frac{\mu_2}{\mu_1} \cdot \tan \alpha_1 \tag{2.89}$$

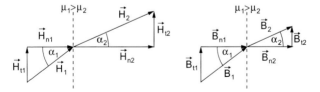

Fig. 2.33. Field quantities at boundaries

At the boundary:

- The tangential component of the magnetic field strength is constant.
- The normal component of the magnetic field strength is inversely proportional to the permeability.
- The tangential component of the magnetic flux density is proportional to the permeability.
- The normal component of the magnetic flux density is constant.

2.3.10 The Magnetic Circuit

In analogy to the current circuits in electrical networks, the magnetic circuit is defined for the magnetic field. The magnetic circuit can, as in electrical networks, be represented by an equivalent circuit diagram. The magnetic voltage V is put in instead of the electric voltage V, the magnetic flux Φ is put in instead of the electric current I and the magnetic resistance R_m is put in instead of the ohmic resistance R (Fig. 2.34).

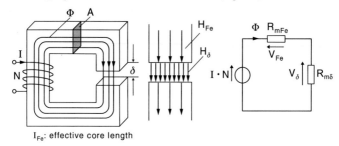

l_{Fe}: effective core length

Fig. 2.34. Magnetic circuit

For the magnetic circuit with the air gap shown in Fig. 2.34 the following relations may be written:

- The **magnetic flux** Φ forms closed loops.
- For a constant cross section A the **magnetic flux density** $\vec{B}$ is equal in all places to: $B = \dfrac{\Phi}{A}$. This is true both in the iron and also in the air gap: $B_{Fe} = B_\delta$ (field widening in the air gap and field narrowing in the corners of the core are negligible).
- The **magnetic field strength** $\vec{H}$ is bigger in the air gap than in the iron by a factor μ_r.

$$\boxed{B = \mu \cdot H \quad \Rightarrow \quad B = \mu_0 \mu_r \cdot H_{Fe} = \mu_0 \cdot H_\delta \quad \Rightarrow \quad \mu_r \cdot H_{Fe} = H_\delta} \tag{2.90}$$

- The **magnetic voltage drop** V is equal to the magnetic flux multiplied by the magnetic resistance:

$$\boxed{V = \Phi \cdot R_m} \tag{2.91}$$

- The **magnetomotive force** $I \cdot N$ is equal to the sum of the magnetic voltage drops:

$$\boxed{I \cdot N \approx H_{Fe} \cdot l_{Fe} + H_\delta \cdot \delta} \tag{2.92}$$

- The magnitudes of the **magnetic resistances** R_m are:

$$\boxed{R_{m\,Fe} \approx \dfrac{1}{\mu_0 \mu_r} \cdot \dfrac{l_{Fe}}{A}, \quad \text{and} \quad R_{m\delta} \approx \dfrac{1}{\mu_0} \cdot \dfrac{\delta}{A}} \tag{2.93}$$

- The magnitude of the **value of** A_L is:

$$\boxed{A_L = \dfrac{\mu_0 \cdot A}{\left(\dfrac{l_{Fe}}{\mu_r} + \delta\right)}} \tag{2.94}$$

2.3 Magnetic Fields

- The magnitude of the **inductance** L of the circuit is:

$$L = N^2 \cdot A_L = N^2 \cdot \frac{\mu_0 \cdot A}{\left(\dfrac{l_{Fe}}{\mu_r} + \delta\right)} \tag{2.95}$$

- The magnitude of the **linked magnetic flux** Ψ is:

$$\Psi = N \cdot \Phi = L \cdot I \tag{2.96}$$

NOTE: The iron path length l_{Fe} is divided only by μ_r in the calculation of the inductance. Considering that usual values for $\mu_r \approx 1000\text{–}10\,000$, it can be seen that the inductance mainly depends on the length of the air gap.

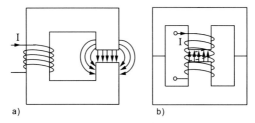

Fig. 2.35. a) Air gap outside the winding: large field broadening; b) air gap inside the winding: small field broadening

NOTE: The field broadening in the air gap is not insignificant and in practical calculations should *not* be ignored. Field broadening causes a much smaller value of inductance compared to a homogeneous field distribution. This is especially true when the air gap lies *outside* the winding (Fig. 2.35a). For this reason, the air gap lies *inside* the winding for practical cores (Fig. 2.35b). It is better in any case to calculate with the measured value of A_L from the manufacturer's data sheets and not to try to derive this from the core geometry.

EXAMPLE: The following data are from the datasheet of a core: Value of A_L: $A_L = 250$ nH, minimum cross-sectional area: $A_{min} = 280$ mm², maximum flux density: $B_{max} = 0.3$ T Question: What is the maximum inductance L that can be achieved with this core for a current of $I = 2$ A, and how many windings are required? With $L = N^2 \cdot A_L$ and $L \cdot I = N \cdot \Phi$ then:

$$N^2 \cdot A_L \cdot I = N \cdot \Phi$$

The magnitude of the maximum allowable flux is:

$$\Phi_{max} = B_{max} \cdot A_{min}$$

Thus the maximum number of windings can be calculated:

$$N = \frac{\Phi_{max}}{A_L \cdot I} = \frac{B_{max} \cdot A_{min}}{A_L \cdot I} = 168$$

The magnitude of the maximum inductance is:

$$L = N^2 \cdot A_L = 7 \text{ mH}$$

2.3.11 Magnetic Circuit with a Permanent Magnet

The purpose of a magnetic circuit with a permanent magnet is usually to produce a magnetic field in an air gap, for example, in a permanently magnetised electric motor, an electric measurement device or a loudspeaker. The configuration in Fig. 2.36 is used in the analysis of such a magnetic circuit. A permanent magnet has the air gap as a load, and the magnetic resistance of the iron is negligible. The application of Ampere's law yields:

$$\oint \vec{H} \, d\vec{s} = H_M \cdot l_M + H_\delta \cdot \delta = 0$$

It follows that:

$$H_M = -H_\delta \cdot \frac{\delta}{l_M} \qquad (2.97)$$

- The magnetomotive force is zero! Thus the magnetic voltage drops over the permanent magnet and the air gap are equally large and opposite to one another.

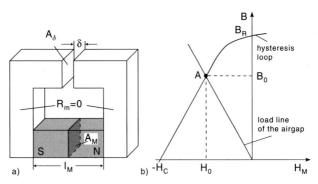

Fig. 2.36. Magnetic circuit with a permanent magnet

The magnetic flux Φ forms a closed loop. The flux Φ in the permanent magnet is the same as in the air gap. It follows that:

$$\Phi = \text{const.} = B_M \cdot A_M = B_\delta \cdot A_\delta$$

and:

$$B_M = B_\delta \cdot \frac{A_\delta}{A_M} \qquad (2.98)$$

It further holds for the air gap that:

$$B_L = \mu_0 \cdot H_\delta \qquad (2.99)$$

From Eqs. (2.97)–(2.99) it follows that the **load line** $B_M(H_M)$ of the air gap:

$$B_M(H_M) = B_\delta \cdot \frac{A_\delta}{A_M} = H_\delta \cdot \mu_0 \frac{A_\delta}{A_M} = -H_M \cdot \mu_0 \cdot \frac{A_\delta}{A_M} \cdot \frac{l_M}{\delta} \qquad (2.100)$$

$$B_M = -H_M \cdot \mu_0 \cdot \frac{A_\delta}{A_M} \cdot \frac{l_M}{\delta} \qquad (2.101)$$

The connection of an active element (the permanent magnet) and a passive element (the air gap) leads to an intersection of the load lines in the graphical solution. This intersection is the **operating point** (H_0, B_0).

- The permanent magnet is *loaded* by the air gap. The higher the magnetic resistance of the air gap, the smaller the air gap flux density.

- For small magnetic resistance of the air gap, the operating point moves on the hysteresis loop of the magnetic material near the remanence flux density, and for large magnetic resistance it moves near the coercivity.

The flux density and the magnetic field strength in the air gap can be calculated from the operating point with Eqs. (2.97) and (2.98):

$$B_\delta = B_0 \cdot \frac{A_M}{A_\delta} \quad \text{and} \quad H_\delta = -H_0 \cdot \frac{l_M}{\delta} \qquad (2.102)$$

NOTE: The minus sign on H_0 highlights the fact that in the magnet the magnetic flux density and the magnetic field strength have opposite directions.

Designing a Permanent Magnet

Question: What size of a permanent magnet is needed for certain air gap dimensions A_δ and l_δ and a defined air gap energy W_δ? The permanent magnet gives maximum energy when the product of B_0 and H_0 in the operating point is a maximum. (For the straight-line hysteresis curve of a permanent magnet, this maximum lies at $B_0 = B_R/2$ and $H_0 = H_C/2$). With the chosen operating point (B_0, H_0) the required volume V_M of the magnet can be found:

$$W_\delta = \frac{1}{2} \cdot \underbrace{H_\delta}_{B_0 \cdot \frac{A_M}{A_\delta}} \cdot \underbrace{B_\delta}_{H_0 \cdot \frac{l_M}{\delta}} = \frac{1}{2} B_0 H_0 V_M \quad \Rightarrow \quad V_M = \frac{2W_\delta}{B_0 H_0}$$

Equation (2.101) yields the ratio of the magnet's cross-sectional area A_M to the length l_M:

$$\frac{A_M}{l_M} = \frac{H_0}{B_0} \cdot \mu_0 \frac{A_\delta}{\delta}$$

It follows for the magnet measurements that:

$$A_M = \frac{1}{B_0} \cdot \sqrt{2W_\delta \cdot \mu_0 \frac{A_\delta}{\delta}}, \quad \text{and} \quad l_M = \frac{1}{H_0} \cdot \sqrt{2W_\delta \cdot \frac{\delta}{\mu_0 \cdot A_\delta}} \qquad (2.103)$$

NOTE: Besides maximising the air gap energy and minimising the volume of the magnet, there are further (if necessary, more important) reasons for the choice of the operating point. For example, the question of the demagnetisation of a permanent magnet by the operating current (or also by a short-circuit current) in permanently magnetised electric motors.

2.3.12 Overview: Inductances of Different Geometric Configurations

Table 2.4. Overview of inductances

Parallel round conductors	$L = \dfrac{\mu}{\pi} \cdot l \left(\ln \dfrac{2a}{r_1} + \dfrac{1}{4} \right)$
Parallel rectangular conductors	$L = \dfrac{2\mu}{\pi} \cdot l \cdot \ln \left(1 + \dfrac{b}{b+h} \right)$, for $\begin{cases} a \ll b \\ a \ll h \end{cases}$ $L = \dfrac{\mu}{\pi} \cdot l \cdot \dfrac{2b}{h+b}$, for $\begin{cases} a \ll b \\ a \ll h \\ b \ll h \end{cases}$
Coaxial conductor	$L = \dfrac{\mu_0}{2\pi} \cdot l \cdot \ln \dfrac{r_2}{r_1}$ Without internal conductor and coating
Toroid	$L = \mu_0 R \left(\ln \dfrac{R}{d/2} + \dfrac{1}{4} \right)$
Coil around toroidal core	$L = N^2 \cdot \mu_r \mu_0 \cdot \dfrac{b}{2\pi} \ln \dfrac{r_2}{r_1}$
Solenoid	$L \approx N^2 \cdot \mu_0 \cdot \dfrac{\pi R^2}{l}$

2.3.13 Induction

2.3.13.1 Induction in a Moving Electrical Conductor

The induction on a moving charge in a magnetic field, known as the **magnetic induction**, can be explained as follows: an electrical conductor is moved in a magnetic field (Fig. 2.37). A force $\vec{F}_M$, which in Fig. 2.37 points backwards, is exerted on the positive charge carriers in the electrical conductor. A force that points forwards is exerted on the negative charges. The positive and negative charge carriers are separated by the Lorentz force. At the same time, a Coulomb force $\vec{F}_E$ builds up between the positive and negative charge carriers, which works against the separation of the charge carriers. In the steady-state case, these forces balance each other.

$$\vec{F}_M = Q \cdot \left(\vec{v} \times \vec{B} \right) = -\vec{F}_E = -Q \cdot \vec{E}$$

It follows that:

$$\boxed{-\left(\vec{v} \times \vec{B} \right) = \vec{E}} \qquad (2.104)$$

- The electric field strength in a moving conductor is equal to the cross-product of the velocity and the magnetic induction.

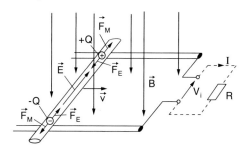

Fig. 2.37. Moving conductor in a magnetic field

The voltage v_i is the line integral of the electrical field strength over the length l of the moving conductor:

$$\boxed{v_i = \int_l \vec{E} \, d\vec{l} = \int_l -\left(\vec{v} \times \vec{B} \right) d\vec{l}} \qquad (2.105)$$

If the electrical conductor is straight and if the vectors $\vec{v}$, $\vec{B}$ and $\vec{s}$ are each perpendicular to one another, the calculation is simplified:

$$\boxed{v_i = l \cdot v \cdot B} \qquad (2.106)$$

If a resistance R is connected across the voltage v_i, it can be seen that a charge equalisation in the moving conductor is possible: a current I flows in direction shown in Fig. 2.37. The conductor loop with the moving conductor becomes a generator.

2.3.13.2 Faraday's Law of Induction

Around a *time-varying* magnetic field there is an electric field.

- The line integral of the electric field strength along a closed loop is equal to the negative change in the magnetic flux enclosed by the integration path (Fig. 2.38a).

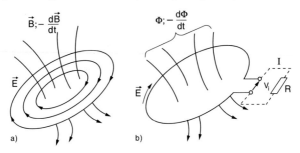

Fig. 2.38. Faraday's law of induction: **a)** ; **b)**

$$\oint_s \vec{E} \, d\vec{s} = -\frac{d\Phi}{dt} = -\frac{d}{dt} \int_A \vec{B} \, d\vec{A} \qquad (2.107)$$

This is also written as:

$$\nabla \times \vec{E} = -\frac{d\vec{B}}{dt} \qquad (2.108)$$

If a conductor loop is placed around the time-varying magnetic flux Φ, then **Faraday's law** states that the terminal voltage V_i (Fig. 2.38b) is:

$$v_i = -\frac{d\Phi}{dt} \qquad (2.109)$$

- The induced voltage at the terminals of a conductor loop is proportional to the time variation in the magnetic flux, which cuts through the conductor loop.

- In Fig. 2.38b the induced voltage v_i is in the direction shown, if the change in the magnetic flux is negative, i.e. if the magnetic flux is decreasing. If the magnetic field is increasing, then the sign of the induced voltage changes.

If a resistor terminates the conductor loop, then the current flows in the direction shown in Fig. 2.38b. **Lenz's law**: The current always flows in a direction such that its magnetic field opposes the flux responsible for inducing the voltage.

NOTE: Lenz's law is most useful in determining the direction of the induced voltage. The current that flows when a conductive load is present creates a magnetic field that opposes the original magnetic field. The direction of the current can be determined therefore from the right-hand rule. If the direction of the current is known, then the voltage on the load resistance and and the direction of v_i are also known.

2.3 Magnetic Fields

- If the conductor loop is shorted and if it is an ideal conductor (superconductor), then the size of the current flowing in it is always large enough to cancel out any change in the flux in the loop.

If the flux Φ links the conductor N times, then the magnitude of the induced voltage is (Fig. 2.38b):

$$v_i = -N \cdot \frac{d\Phi}{dt} \quad (2.110)$$

A conductor loop with N windings is known as a **coil**. The converse of the Faraday's law is also true. If a voltage is applied to an ideal conductor loop (with N windings), then the flux change in it is:

$$N \cdot \frac{d\Phi}{dt} = v(t) \quad (2.111)$$

It can be seen that v_i and v have opposite signs, since v_i produces energy and v comsumes energy. The flux Φ is the integral of the voltage over the time:

$$N \cdot \Phi(t_1) = \int_0^{t_1} v(t) \, dt + \Phi(0) \quad (2.112)$$

- The magnetic flux in a conductor loop (in a coil, in an inductance) depends *only* on the integral of the applied voltage over time $\int v \, dt$.

EXAMPLE: The application of the Faraday's law to coupled coils that are linked by the same flux: At time t_0 the voltage V_0 is applied to winding 1. At time t_1 switch B is closed, and at time t_2 switch A is opened. Figure 2.39a shows the configuration and Fig. 2.39b shows the related voltages, currents and magnetic flux. The solution to the problem could equally be provided by the equivalent circuit in Fig. 2.39c.

Calculation:

$t_0 < t < t_1$ v_1 is given by V_0	$t_1 < t < t_2$ v_1 is given by V_0	$t > t_2$ Φ is continuous, $i_{2(t2)} = -\dfrac{N_2 \cdot \Phi(t_2)}{L_2}$
$v_1 = V_0$	$v_1 = V_0$	$i_2 = i_{2(t2)} \cdot e^{-\frac{t}{L_2/R}}$
$i_1 = \dfrac{1}{L_1} \cdot \displaystyle\int_{t_0}^{t_1} v_1 \, dt$	$i_1 = \dfrac{1}{L_1} \cdot \displaystyle\int_{t_1}^{t_2} v_1 \, dt + i_{1(t1)} + \dfrac{V_2}{R}\dfrac{N_2}{N_1}$	$v_2 = i_2 \cdot R$
$\Phi = \dfrac{1}{N_1} \displaystyle\int_{t_0}^{t_1} v_1 \, dt$	$\Phi = \dfrac{1}{N_1} \displaystyle\int_{t_1}^{t_2} v_1 \, dt + \Phi_{(t1)}$	$\Phi = \dfrac{1}{N_2} \displaystyle\int_{t_2}^{t} v_2 \, dt + \Phi_{(t2)}$
$v_2 = -N_2 \dfrac{d\Phi}{dt} = v_1 \dfrac{N_2}{N_1}$	$v_2 = -N_2 \dfrac{d\Phi}{dt} = v_1 \dfrac{N_2}{N_1}$	$i_1 = 0$
$i_2 = 0$	$i_2 = \dfrac{V_2}{R}$	$v_1 = v_2 \dfrac{N_1}{N_2}$

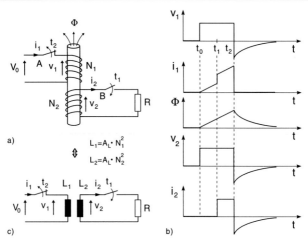

Fig. 2.39. Voltages, currents and magnetic flux for coupled coils

Calculation of the Induced Voltage

A change in the magnetic flux linking a conductor loop can come about in two ways: first if the flux density is time-varying, and second if the enclosed area is changing:

$$v_i = -\frac{d\Phi}{dt} = -\frac{d}{dt}\left(\vec{B} \cdot \vec{A}\right) = -\left(\vec{B} \cdot \frac{\partial \vec{A}}{\partial t} + \vec{A} \cdot \frac{\partial \vec{B}}{\partial t}\right) \tag{2.113}$$

Induction caused by motion as in Fig. 2.37 is also covered by Eq. (2.113). The conductor motion causes the enclosed area to decrease, so that:

$$-\vec{B} \cdot \frac{d\vec{A}}{dt} = \int_l \left(\frac{d\vec{s}}{dt} \times \vec{B}\right) d\vec{l}$$

In this case Faraday's law states:

$$v_i = -\left[\int_l \left(\vec{v} \times \vec{B}\right) d\vec{l} + \vec{A} \cdot \frac{\partial \vec{B}}{\partial t}\right] \tag{2.114}$$

EXAMPLE: A conductor loop with $N = 200$ windings is rotating with an angular speed of $\omega = 314\ \text{s}^{-1}$ (radians per second) in a homogeneous magnetic field with a magnetic flux density of $B = 50$ mT. The conductor loop dimensions are 10×10 cm (Fig. 2.40a). What voltage v_i can be measured at its terminals?

a) Application of Eq. (2.113):

$$v_i = -N \cdot \left(\vec{B} \cdot \frac{\partial \vec{A}}{\partial t} + \underbrace{\vec{A} \cdot \frac{\partial \vec{B}}{\partial t}}_{=0}\right) = -N \cdot B \cdot \frac{d[A \cdot \cos \omega t]}{dt}$$

$$= -\underbrace{N \cdot B \cdot A \cdot \omega}_{31.4\ \text{V}} \cdot \underbrace{\sin \omega t}_{50\ \text{Hz}} = -31.4\ \text{V} \cdot \sin \omega t$$

b) Application of Eq. (2.114) (motion induction):

$$v_i = -N \cdot \left(\int_l \left(\vec{v} \times \vec{B} \right) d\vec{l} + \underbrace{\vec{A} \cdot \frac{\partial \vec{B}}{\partial t}}_{=0} \right)$$

$$= -N \cdot \underbrace{\omega \cdot \frac{b}{2}}_{v} \cdot B \cdot \sin(\omega t) \cdot 2 \cdot l = -31.4 \text{ V} \cdot \sin \omega t$$

The direction of the induced voltage in a moving conductor (Fig. 2.40) can be best determined with Eq. (2.104): $\vec{v} \times \vec{B} = -\vec{E}$. Since $\vec{v} \times \vec{B}$ is a vector product, the direction of $\vec{E}$ and therefore also v_i can be determined by the corkscrew rule.

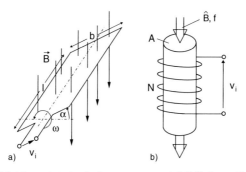

Fig. 2.40. a) Rotating conductor loop in a homogeneous magnetic field; **b)** change of flux in a coil

EXAMPLE: A coil with $N = 5$ windings is linked by $\hat{B} = 1.5$ T, $f = 50$ Hz. Its cross-sectional area is 200 mm² (Fig. 2.40b). What is the peak value of the voltage v_i?

$$v_i = -N \cdot \left(\underbrace{\vec{B} \cdot \frac{\partial \vec{A}}{\partial t}}_{=0} + \vec{A} \cdot \frac{\partial \vec{B}}{\partial t} \right) = -N \cdot A \cdot \frac{d(\hat{B} \cdot \sin \omega t)}{dt}$$

$$= -N \cdot \hat{B} \cdot A \cdot \omega \cos \omega t \Rightarrow \hat{V}_i = 0.47 \text{ V}$$

2.3.13.3 Self-Induction

If a current I flows in a conductor loop, then this causes a magnetic flux Φ. If the current is switched off, then the magnetic flux is simultaneously removed. The instant that the current is switched off there is a large rate of change of flux. This flux change creates an induced voltage at the terminals of the conductor. This process is known as **self-induction**.

- According to Lenz's law, the induced voltage opposes the original voltage.

With $v_i = -N \cdot \dfrac{d\Phi}{dt}$, and $L \cdot I = N \cdot \Phi$ then:

$$\boxed{v_i = -N \cdot \frac{d\Phi}{dt} = -L \cdot \frac{di}{dt}} \qquad (2.115)$$

Voltage v_i is given in Eq. (2.115). In the instant the current I is turned off, $\frac{di}{dt}$ is negative. The induced voltage in Fig. 2.41a lies across the inductance L.

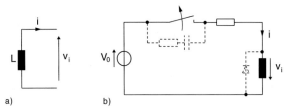

Fig. 2.41. The induced voltage on an inductance: **a)** induced voltage polarity; **b)** switch protection in inductive circuits

NOTE: Switching off an inductive current can cause large induced voltages. By examining the circuit shown in Fig. 2.41b, it can be seen that in the instant the current is removed that practically all of the induced voltage falls across the switch. This can cause arcing in the switch and can destroy it. The destruction of the switch can also be explained by performing an energy analysis: the magnetic field stores energy. This energy is forced out of the inductor when the switch is opened. This energy is transformed into heat in the switch and thus can lead to its destruction. A diode parallel to the inductance or an RC circuit across the switch gives protection.

2.3.14 Mutual Induction

When a coil's magnetic flux, or a part thereof, links another coil, this is referred to as **magnetic coupling**. In Fig. 2.42 the portion Φ_{21} of the magnetic flux Φ_1 links coil 2. The coupling coefficient is defined to describe this condition: **Coupling coefficient k_1**:

$$\boxed{\Phi_{21} = k_1 \cdot \Phi_1} \tag{2.116}$$

For coil 1 there is a corresponding definition:

$$\boxed{\Phi_{12} = k_2 \cdot \Phi_2} \tag{2.117}$$

The coupling coefficients can be determined from the geometry of the configuration with the aid of the equivalent magnetic circuit:

$$\frac{\Phi_{21}}{\Phi_1} = k_1 = \frac{R_{m3}}{R_{m2} + R_{m3}}, \quad \text{and} \quad k_2 = \frac{R_{m3}}{R_{m1} + R_{m3}}$$

Alternatively, a determination based on measurement is also possible: an alternating voltage V_1' is applied to N_1, then the voltage V_2' is measured, or V_2'' is applied to N_2 and V_1'' is measured. With $V_2 = N_2 \cdot \frac{d\Phi_{21}}{dt} = N_2 \cdot \frac{k_1 d\Phi_1}{dt}$, and $V_1 = N_1 \cdot \frac{d\Phi_1}{dt}$, k_1 and the corresponding k_2 are given by:

$$k_1 = \frac{V_2'}{V_1'} \cdot \frac{N_1}{N_2}, \quad \text{and} \quad k_2 = \frac{V_1''}{V_2''} \cdot \frac{N_2}{N_1}$$

- The coupling coefficients can have a maximum value of 1. For smaller coupling coefficients the expression 'loose coupling' is used.

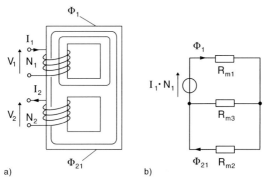

Fig. 2.42. Magnetically coupled coils: **a)** geometric configuration; **b)** equivalent circuit

Mutual Inductance

The **mutual inductances** M_{21} and M_{12} can be defined as:

$$M_{12} = \frac{N_1 \cdot \Phi_{12}}{I_2}, \quad \text{and} \quad M_{21} = \frac{N_2 \cdot \Phi_{21}}{I_1}$$

For isotropic magnetic materials it is always true that:

$$\boxed{M_{21} = M_{12} = M}$$

With the coupling coefficients k_1 and k_2 and the individual inductances L_1 and L_2, the **mutual inductance** M can be calculated:

$$\boxed{M = \sqrt{k_1 \cdot k_2 \cdot L_1 \cdot L_2} = k \cdot \sqrt{L_1 \cdot L_2}, \quad \text{with} \quad k = \sqrt{k_1 \cdot k_2}} \quad (2.118)$$

where k is the **total coupling coefficient**.

The mutual inductance is required to describe a magnetically coupled system by Kirchhoff's laws. For two magnetically coupled resistor-terminated coils as given in Fig. 2.42:

$$\boxed{\begin{aligned} v_1 &= +L_1 \cdot \frac{di_1}{dt} + i_1 \cdot R_1 - M \cdot \frac{di_2}{dt}, \\ v_2 &= -L_2 \cdot \frac{di_2}{dt} - i_2 \cdot R_2 + M \cdot \frac{di_1}{dt} \end{aligned}} \quad (2.119)$$

For sinusoidal operation in the complex domain:

$$\boxed{\begin{aligned} v_1 &= +i_1 \cdot (j\omega L_1 + R_1) - i_2 \cdot j\omega M, \\ v_2 &= -i_2 \cdot (j\omega L_2 + R_2) + i_1 \cdot j\omega M \end{aligned}} \quad (2.120)$$

2.3.15 Transformer Principle

The **transformer** is made out of magnetically coupled coils, whose coupling coefficient $k \approx 1$. In Fig. 2.43a a voltage V_1 is applied to winding 1. The magnetic flux change in winding 1 is thus $\dfrac{d\Phi}{dt} = \dfrac{V_1}{N_1}$. The flux Φ passes through the iron core and therefore also links winding 2. The induced voltage in winding 2 is thus $V_2 = -N_2 \cdot \dfrac{d\Phi}{dt}$.

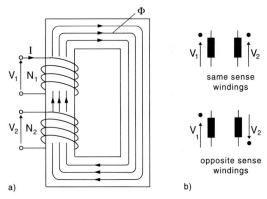

Fig. 2.43. Principle of the transformer: **a)** flux linking and winding sense; **b)** circuit representation of the winding sense

For correct orientation of the flux direction through winding 2, the voltage V_2 for the winding sense shown in Fig. 2.43a is given by:

$$V_2 = V_1 \cdot \frac{N_2}{N_1} \tag{2.121}$$

- The voltage V_2 only depends on the voltage V_1 and the number of windings.
- The voltages on a transformer are proportional to the number of windings.

The circuit representation of the winding sense is shown in Fig. 2.43b). The winding sense is given by the dots. For windings with the same sense with respect to the magnetic flux, the dots are drawn on the same side of the terminals of the inductance. For windings with an opposite sense, the dots are drawn on different sides of the terminals.

2.3.16 Energy in a Magnetic Field

Like the electrostatic field, the magnetic field stores energy. An inductance's energy is stored in its magnetic field. Its magnitude is:

$$W = \frac{1}{2} L \cdot I^2 = \frac{1}{2} \cdot N \cdot I \cdot \Phi \tag{2.122}$$

2.3 Magnetic Fields

This can also be represented by the field quantities $\vec{B}$ and $\vec{H}$:

$$W = \frac{1}{2} \oint_S \vec{H} \, d\vec{s} \cdot \int_A \vec{B} \, d\vec{A} = \frac{1}{2} \int_V \vec{H} \cdot \vec{B} \, dV, \quad \text{with} \quad \vec{e}_s \parallel \vec{e}_A \quad (2.123)$$

The unity vector $\vec{e}_s$ points in the same direction as the unity vector of the normal to the area $\vec{e}_A$, so that the integral $\int ds \, dA$ yields the volume element dV. The magnitude of the **energy density** of the magnetic field is:

$$\frac{dW}{dV} = \frac{1}{2} \cdot \vec{H} \cdot \vec{B} \quad (2.124)$$

2.3.16.1 Energy in a Magnetic Circuit with an Air Gap

The magnetic circuit with an air gap is shown in Fig. 2.44. Magnetic circuits, which store energy, are known as **choking coils** or **chokes**. The magnitude of the energy stored by a choke is:

$$W = \frac{1}{2} \cdot L \cdot I^2 \quad (2.125)$$

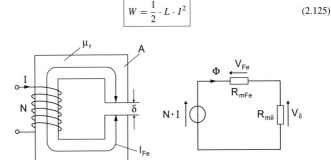

Fig. 2.44. Magnetic circuit with an air gap

This energy is stored in the form of magnetic field energy, both in the iron and also in the air gap.

$$W = W_{Fe} + W_\delta = \frac{1}{2} B_{Fe} \cdot H_{Fe} \cdot V_{Fe} + \frac{1}{2} B_\delta \cdot H_\delta \cdot V_\delta \quad (2.126)$$

For the same cross-sectional area over the entire magnetic path $B_\delta = B_{Fe}$, the magnetic field strength in the air gap is higher than in the iron by a factor of μ_r. Recalling that the relative permeability is usually in the region of $\mu_r = 1000$–$10\,000$ then it can be said to a good approximation that the magnetic energy is predominately concentrated in the air gap. The iron path is required in order to concentrate the magnetic flux and thus to create high magnetic field strength in the air gap. Large amounts of energy for small dimensions can be achieved using this construction.

- The energy stored in chokes is predominately concentrated in the air gap.

- Actual chokes always have an air gap.

92 2 Electric Fields

NOTE: The air gap in actual chokes is not always realised in the form of a 'real' air gap. In so-called **powder cores** a loosely glued union of iron powder is 'distributed' inside the core.

- A large air gap is required to store a lot of magnetic energy.

If Eq. (2.123) is evaluated, then:

$$W = \frac{1}{2}\underbrace{\oint_s \vec{H}\,\mathrm{d}\vec{s}}_{V_{Fe}+V_\delta} \cdot \underbrace{\int_A \vec{B}\,\mathrm{d}\vec{A}}_{\Phi} = \frac{1}{2}\cdot(\underbrace{V_{Fe}}_{\Phi\cdot R_{mFe}} + \underbrace{V_L}_{\Phi\cdot R_{m\delta}})\cdot\Phi$$

It follows that:

- The magnetic energy is divided in proportion to the magnetic resistances.

2.3.17 Forces in a Magnetic Field

For all forces in a magnetic field:

- Forces in a magnetic field always point in the direction that flux lines would seek to shorten their path (Fig. 2.45).

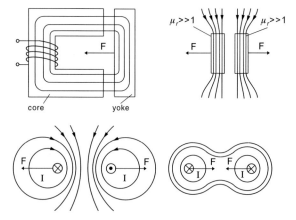

Fig. 2.45. Forces in a magnetic field

2.3.17.1 Force on a Current-Carrying Conductor

See also Sect. 2.3.2. The magnitude of the force on a straight current-carrying conductor in a homogeneous magnetic field is:

$$\boxed{\vec{F} = I \cdot \left(\vec{l} \times \vec{B}\right)} \qquad (2.127)$$

The vector $\vec{l}$ thus points thus in the direction of the current I (Fig. 2.46).

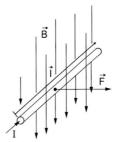

Fig. 2.46. Force on a current-carrying conductor

2.3.17.2 Force at the Boundaries

Forces are present at the boundaries between magnetically linked materials of different permeabilities. The calculation of the forces is most easily carried out by an energy balance between the mechanical, the electrical and the field energy. The boundary is assumed to have shifted infinitesimally, and the resulting change in potential energy is calculated. The sum of the energy changes must be zero:

$$\boxed{\mathrm{d}W_{\mathrm{mech}} + \mathrm{d}W_{\mathrm{field}} + \mathrm{d}W_{\mathrm{electr}} = 0} \tag{2.128}$$

In order to be able to evaluate the energy balance, the sign of these energy changes must be known, i.e. which energy increases and which decreases. Consider the following thought experiment: The magnetic circuit in the upper left of Fig. 2.45 is supplied by a constant current, i.e. is connected to a constant current source I_0. The yoke is drawn towards the core. If the yoke is pulled away from the core, then mechanical energy is supplied. The inductance will simultaneously decreases (the magnetic conductance decreases), i.e. the field energy stored $\frac{1}{2}LI^2$ decreases. Both energies, the mechanical and the change in the field energy, are absorbed by the current source. The energy balance above states therefore with $N \cdot \Phi = \int V\,\mathrm{d}t$, and $L \cdot I = N \cdot \Phi$:

$$\underbrace{F \cdot \mathrm{d}\underline{s}}_{\mathrm{d}W_{\mathrm{mech}}} + \underbrace{\mathrm{d}\left(\frac{1}{2}LI_0^2\right)}_{\mathrm{d}W_{\mathrm{field}}} = \underbrace{\mathrm{d}(I_0 \cdot \underbrace{V \cdot t}_{N \cdot \Phi})}_{\mathrm{d}W_{\mathrm{electr}}} = \mathrm{d}(I_0 \cdot \underbrace{N \cdot \Phi}_{L \cdot I_0}) = \mathrm{d}\left(L \cdot I_0^2\right)$$

It follows that:

$$\boxed{F = \frac{1}{2}I^2 \cdot \frac{\mathrm{d}L}{\mathrm{d}s}} \tag{2.129}$$

- The force on the boundaries in a magnetic configuration is proportional to the change in the inductance relative to the shift in the boundaries.

- The force on a boundary points in the direction that increases the inductance.

2.3.18 Overview: Characteristics of a Magnetic Field

- The magnetic flux density $\vec{B}$ and the magnetic field strength $\vec{H}$ point in the same direction in isotropic materials.

$$\boxed{\vec{B} = \mu \cdot \vec{H}} \tag{2.130}$$

- The magnetic flux always forms a closed loop.
- At boundaries of different permeability μ, the normal component of the flux density is constant and the normal component of the magnetic field strength is variable (it increases if μ is smaller).
- The magnetic field is a **solenoidal field**. The surface integral of the magnetic flux density over a closed surface area is always zero:

$$\boxed{\oint_A \vec{B} \, d\vec{A} = 0, \quad \text{or} \quad \nabla \cdot \vec{B} = 0} \tag{2.131}$$

- The magnetic flux is directly related to the integral of the applied voltage over time:

$$\boxed{N \cdot \Phi = \int_0^{t_1} v \, dt + \Phi(0)} \tag{2.132}$$

- The magnetic field strength is directly related to the electric current. Ampere's law yields the relationship:

$$\boxed{\oint \vec{H} \, d\vec{s} = i \cdot N} \tag{2.133}$$

- The complete form of Ampere's law is Maxwell's first equation and states:

$$\boxed{\oint_s \vec{H} \, d\vec{s} = \int_A \left(\vec{J} + \frac{d\vec{D}}{dt} \right) d\vec{A}, \quad \text{or} \quad \nabla \times \vec{H} = \vec{J} + \frac{d\vec{D}}{dt}} \tag{2.134}$$

- A time-varying magnetic flux induces a voltage in a conductor linked by the flux. If the conductor loop is linked N times by the flux, then the induced voltage is N times higher.

$$\boxed{v_i = -N \cdot \frac{d\Phi}{dt}} \tag{2.135}$$

- The relationship between the electrical quantity I and the flux Φ is:

$$\boxed{L \cdot I = N \cdot \Phi} \tag{2.136}$$

- The magnetic field stores energy:

$$\boxed{W = \int_V \vec{B} \cdot \vec{H} \, dV} \tag{2.137}$$

2.3.19 Relationship between the Magnetic Field Quantities

Formulae of a magnetic circuit:

$$
\begin{array}{ccc}
\Phi & \Leftarrow \quad \Phi = \int_A \vec{B} \, d\vec{A} \quad \Rightarrow & \vec{B} \\
\Uparrow & & \Uparrow \\
I \cdot N = R_\mathrm{m} \cdot \Phi & & \vec{B} = \mu \cdot \vec{H} \\
\Downarrow & & \Downarrow \\
I \cdot N & \Leftarrow \quad I \cdot N = \oint_s \vec{H} \, d\vec{s} \quad \Rightarrow & \vec{H}
\end{array}
$$

Faraday's law:

$$v_\mathrm{i}(t) = -N \cdot \frac{d\Phi}{dt}, \quad \text{or} \quad \Phi(t_1) = \frac{1}{N} \int_0^{t_1} v(t) \, dt + \Phi(0) \tag{2.138}$$

2.4 Maxwell's Equations

The numerous physical phenomena described in the sections on electrostatic fields, static steady-state current flow and magnetic fields, can be expressed together in four equations, Maxwell's equations. **Maxwell's first equation (Ampere's law):**

$$\oint_s \vec{H} \, d\vec{s} = \int_A \left(\vec{J} + \frac{d\vec{D}}{dt} \right) d\vec{A}, \quad \text{or} \quad \nabla \times \vec{H} = \vec{J} + \frac{d\vec{D}}{dt} \tag{2.139}$$

The electric current creates the magnetic field strength. Maxwell's first equation states that the circular integral of the magnetic field strength is equal to the enclosed current (Fig. 2.47). This is independent of whether the current is due to charge carriers or due to a time-varying alternating electric displacement.

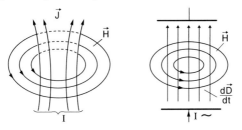

Fig. 2.47. Maxwell's first equation

Maxwell's second equation (Faraday's law):

$$\oint_s \vec{E} \, d\vec{s} = -\frac{d}{dt} \int_A \vec{B} \, d\vec{A}, \quad \text{or} \quad \nabla \times \vec{E} = -\frac{d\vec{B}}{dt} \tag{2.140}$$

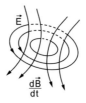

Fig. 2.48. Maxwell's second equation

A time-varying magnetic flux density creates an electric field strength. Maxwell's second equation states that the circular integral of the electric field strength is equal to the negative change in the enclosed magnetic flux (Fig. 2.48). **Maxwell's third equation** :

$$\oint_A \vec{B} \, d\vec{A} = 0, \quad \text{or} \quad \nabla \cdot \vec{B} = 0 \tag{2.141}$$

The magnetic flux density is a solenoidal field. Maxwell's third equation states that the surface integral of the magnetic flux density over an enclosed area is always zero. **Maxwell's fourth equation (Gauss's Law)**:

$$\oint_A \vec{D} \, d\vec{A} = \int_V \varrho \, dV, \quad \text{or} \quad \nabla \cdot \vec{D} = \varrho \tag{2.142}$$

where ϱ: volume charge density. The electric displacement is a charge field. Maxwell's fourth equation states that the integral of the electric displacement over a closed surface is equal to the enclosed charge.

2.5 Notation Index

a	acceleration (m/s²)
A	area (m²)
A_L	magnetic conductance of the total magnetic circuit (H = Vs/A), normally given in (nH)
$A_\perp$	area element of an equipotential surface (m²)
$\vec{A}$	area vector perpendicular to the area (m²)
$\vec{B}$	magnetic flux density (T = Vs/m²)
B_R	remanence density (T = Vs/m²)
C	capacitance (F = As/V)
$\vec{D}$	electric displacement (As/m²)
d	separation, distance (m)
e	elemental charge, $e = \pm 1.602 \cdot 10^{-19}$ As
$\vec{e}$	unity vector (index shows the respective quantities, e.g. $\vec{e}_r = \vec{r}/|\vec{r}|$
$\vec{E}$	electric field strength (V/m)
f	frequency (Hz)
F	force (N)

2.5 Notation Index

G	conductance (S = A/V)
G_m	magnetic conductance (H = Vs/A)
$\vec{H}$	magnetic field strength (A/m)
H_C	coercivity strength (A/m)
i	time-varying current (A)
I	DC current (A)
$\vec{J}$	current density (A/m^2)
k	coupling coefficient
l	length (m)
L	inductance (H = Vs/A)
m	mass (kg)
M	mutual inductance (H = Vs/A)
M	momentum (Nm)
M	subscript: magnet
N	number of windings
N	magnetic north pole
P	power (W = VA)
Q	charge (C = As)
r, R	radius, distance for polar coordinates (m)
R	resistance (Ω = V/A)
R_m	magnetic resistance (1/H = A/Vs)
s	path (m)
S	magnetic south pole
t	time (s)
T	period (s)
v	time varying voltage (V)
V	voltage (V)
V_i	induced voltage (V)
v	velocity (m/s)
V	volume (m^3)
V	magnetic voltage (A)
W	energy (Ws = VAs)
δ	air gap length (m)
δ	subscript: air gap
ε	dielectric constant (As/Vm)
ε_0	free-space permittivity, absolute dielectric constant, $8.85 \cdot 10^{-12}$ As/Vm
ε_r	relative permittivity, relative dielectric constant
ϑ	temperature (K, °C)
η	charge carrier concentration (As/m^3)
σ	specific conductance, conductivity (S/m)
λ	line charge density (As/m)
μ	permeability (Vs/Am)
μ_0	permeability of free space, $1.257 \cdot 10^{-6}$ Vs/Am

μ_r	relative permeability
ϱ	volume charge density (As/m^3)
ϱ	specific resistance, resistivity (Ωm), (Ωmm^2/m)
σ	surface charge (As/m^2)
φ	potential (V)
Φ	magnetic flux (Wb = Vs)
Ψ	linked magnetic flux (Vs)
ω	angular speed or frequency (s^{-1})
Θ	magnetomotive force, MMF

2.6 Further Reading

DUFFIN, W. J.: *Electricity and Magnetism, 4th Edition*
McGraw-Hill (1990)

FLOYD, T. L.: *Electric Circuits Fundamentals, 5th Edition*
Prentice Hall (2001)

FLOYD, T. L.: *Electronics Fundamentals: Circuits, Devices, and Applications, 5th Edition*
Prentice Hall (2000)

FLOYD, T. L.: *Electronic Devices, 5th Edition*
Prentice Hall (1998)

GIANCOLI, D. C.: *Physics for Scientists and Engineers, Volume 1, 3rd Edition*
Prentice Hall (2000)

GROB, B.: *Basic Electronics, 8th Edition*
McGraw-Hill (1996)

MUNCASTER, R.: *A-Level Physics*
Stanley Thornes Ltd. (1997)

NELKON, M.; PARKER, P.: *Advanced Level Physics*
Heinemann (1995)

RAO, N. N.: *Elements of Engineering Electromagnetics, 5th Edition*
Prentice Hall (1999)

SOMEDA, C. G.: *Elecromagnetic Waves, 1st Edition*
Chapman-Hall (1997)

3 AC Systems

AC quantities are described by trigonometric functions, complex numbers and complex functions. For clarity in graphical representation phasors are used. The mathematical basis and relations are explained in the following section.

3.1 Mathematical Basics of AC

3.1.1 Sine and Cosine Functions

A sine function is given by

$$v = \hat{v} \sin \varphi$$

where $\hat{v}$ is the **peak magnitude** or **amplitude**. The **phase** φ often varies with time

$$v(t) = \hat{v} \sin(\omega t + \varphi_0)$$

here $v(t)$ is called the **instantaneous value** or **transient** or **actual value** of the function. ω is the **angular frequency** and φ_0 the **phase shift**. The sine function is periodic with a period of 2π (Fig. 3.1).

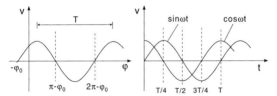

Fig. 3.1. Period and phase shift of the sine function; sine and cosine functions

The interval of time between two identical values of the function is called the **period** T. The **frequency** f of the sine function is the inverse of the period.

$$T = \frac{2\pi}{\omega} \quad f = \frac{1}{T} \quad \omega = 2\pi f \tag{3.1}$$

The cosine is a similar function

$$v = \hat{v} \cos \varphi$$

Both functions are related thus

$$\sin \varphi = \cos(\pi/2 - \varphi) \tag{3.2}$$

$$\cos \varphi = \sin(\pi/2 + \varphi) \tag{3.3}$$

Sine and cosine functions together with the exponential function with imaginary exponents are known as **harmonic functions**.

3.1.1.1 Addition of Sinusoidal Waveforms

- The sum (difference) of two sinusoidal waveforms of the *same frequency* results in a sinusoidal waveform of the same frequency (Fig. 3.2).

The addition of **cosine functions**

$$v_1(t) = \hat{v}_1 \cdot \cos(\omega t + \varphi_1), \qquad v_2(t) = \hat{v}_2 \cdot \cos(\omega t + \varphi_2)$$

results in the sum signal $v_s = v_1 + v_2$

$$v_s(t) = \hat{v}_s \cdot \cos(\omega t + \varphi_s)$$

with the parameters

$$\hat{v}_s = \sqrt{\hat{v}_1^2 + \hat{v}_2^2 + 2\hat{v}_1\hat{v}_2\cos(\varphi_1 - \varphi_2)}, \qquad \tan\varphi_s = \frac{\hat{v}_1\sin\varphi_1 + \hat{v}_2\sin\varphi_2}{\hat{v}_1\cos\varphi_1 + \hat{v}_2\cos\varphi_2} \quad (3.4)$$

The addition of **sine functions**

$$v_1 = \hat{v}_1 \sin(\omega t + \varphi_1), \qquad v_2 = \hat{v}_2 \sin(\omega t + \varphi_2)$$

results in the sum signal $v_s = v_1 + v_2$

$$v_s = \hat{v}_s \sin(\omega t + \varphi_s), \tag{3.5}$$

with the parameters $\hat{v}_s$ and φ_s as in Eq. (3.4).

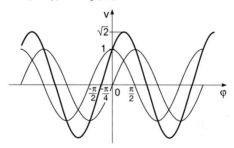

Fig. 3.2. The addition of sinusoidal waveforms results in a sinusoidal waveform

EXAMPLE: Calculation of the sum of the sinusoidal functions $v_1(t) = \sin(\omega t)$ and $v_2(t) = \sin\left(\omega t + \frac{\pi}{2}\right)$.

According to Eq. (3.4) the amplitude of the sine waveform is

$$v_s = \sqrt{1 + 1 + 2 \cdot \cos(0 - \pi/2)} = \sqrt{2} \approx 1.41$$

For the phase shift of the sum signal

$$\tan\varphi_s = \frac{\hat{v}_2}{\hat{v}_1} = 1 \quad \Rightarrow \quad \varphi_s = \frac{\pi}{4} \qquad \text{(or 45°)}$$

The sum signal is therefore

$$v_s(t) = \sqrt{2} \cdot \sin\left(\omega t + \frac{\pi}{4}\right)$$

These parameters are shown in Fig. 3.2.

NOTE: The calculation becomes considerably easier when the time functions are represented as phasors. This method is explained in Sect. 3.1.7.

NOTE: In general, the sum of harmonic functions of *different frequencies* is not a harmonic function. It cannot be represented by stationary phasors.

3.1.2 Complex Numbers

The real numbers $\mathbb{R}$ are extended to the complex numbers $\mathbb{C}$ by joining with the imaginary numbers. **Imaginary unit**

$$j = \sqrt{-1}, \qquad j^2 = -1$$

NOTE: In mathematical literature the imaginary unit is named i. However, in electrical engineering the letter j is commonly used to avoid confusion with the symbol for current.

Powers of j

$$j^1 = j \qquad j^{-1} = \frac{1}{j} = -j$$

$$j^2 = -1 \qquad j^{-2} = \frac{1}{j^2} = -1$$

$$j^3 = -j \qquad j^{-3} = \frac{1}{j^3} = j$$

$$j^4 = 1 \qquad j^{-4} = \frac{1}{j^4} = 1$$

$$j^5 = j \qquad j^{-5} = \frac{1}{j^5} = -j$$

Imaginary Numbers

An imaginary number is the product of a real number with the imaginary unit. Examples: $5j$, $2\pi j$, jb.

- The product of two imaginary numbers is real (since $j \cdot j = -1$).

Complex numbers can be represented as the sum of a real number x and an imaginary number jy.

$$\underline{z} = x + jy$$

NOTATION: To emphasise that the number z is complex, it is represented in this chapter with an underscore ($\underline{z}$).

x is known as the **real part** of the complex number $\underline{z}$: $x = \text{Re}(\underline{z})$,
y is known as the **imaginary part** of the complex number $\underline{z}$: $y = \text{Im}(\underline{z})$.

- The imaginary part is a real number.
- Two complex numbers are equal if both their real parts as well as their imaginary parts are equal.
- Every real number is also complex (with an imaginary part of zero).

For the number $\underline{z} = x + jy$ the number $\underline{z}^* = x - jy$ is called the **complex conjugate**.

$$(\underline{z}^*)^* = \underline{z}$$

For a real number $w \in \mathbb{R}$, it follows that $w^* = w$.

The product of a complex number with its conjugate is called the **absolute value squared**.

$$z \cdot z^* = |z|^2$$

It is $\quad z \cdot z^* = x^2 + y^2 = (\mathrm{Re}(z))^2 + (\mathrm{Im}(z))^2$

$\sqrt{z \cdot z^*} = \sqrt{x^2 + y^2} = \sqrt{|z|^2} = |z|$ is called the **absolute value** or **magnitude** of the complex number z.

- The absolute value is a non-negative real number (positive or zero).

3.1.2.1 Complex Arithmetic

The **addition** and **subtraction** of complex numbers is done by adding or subtracting the relevant components.

$$\text{If } z_1 = x_1 + jy_1, \quad z_2 = x_2 + jy_2$$
$$z_1 + z_2 = (x_1 + x_2) + j \cdot (y_1 + y_2)$$
$$z_1 - z_2 = (x_1 - x_2) + j \cdot (y_1 - y_2)$$

The **multiplication** is done like the multiplication of two binomial expressions given that $j \cdot j = -1$.

$$z_1 \cdot z_2 = (x_1 + jy_1) \cdot (x_2 + jy_2) = (x_1 x_2 - y_1 y_2) + j \cdot (x_1 y_2 + x_2 \cdot y_1)$$

The **division** is done as follows

$$\frac{z_1}{z_2} = \frac{x_1 x_2 + y_1 y_2}{x_2^2 + y_2^2} + j \cdot \frac{x_2 y_1 - x_1 y_2}{x_2^2 + y_2^2}$$

Division by a complex number can be transformed into a division by a real number. This can be done by multiplying the numerator and denominator with the complex conjugate of the denominator.

$$\frac{z_1}{z_2} = \frac{z_1}{z_2} \cdot \frac{z_2^*}{z_2^*} = \frac{z_1 z_2^*}{|z_2|^2}$$

The basic rules for the addition and multiplication of real numbers are also valid for complex numbers:

$$z + 0 = z$$
$$z \cdot 1 = z$$
$$z \cdot 0 = 0$$
$$z_1 + z_2 = z_2 + z_1 \quad\quad\quad\quad\quad\quad\quad \text{Commutative laws}$$
$$z_1 \cdot z_2 = z_2 \cdot z_1$$
$$z_1 + z_2 + z_3 = (z_1 + z_2) + z_3 = z_1 + (z_2 + z_3) \quad \text{Associative laws}$$
$$z_1 \cdot z_2 \cdot z_3 = (z_1 \cdot z_2) \cdot z_3 = z_1 \cdot (z_2 \cdot z_3)$$
$$z_1 \cdot (z_2 + z_3) = z_1 \cdot z_2 + z_1 \cdot z_3 \quad\quad\quad\quad \text{Distributive law}$$

Division by zero is also not defined for complex numbers.

3.1.2.2 Representation of Complex Numbers

Cartesian Form

The real and the imaginary parts of the complex number z are interpreted as coordinates of a point in a plane. This plane is called the **complex plane** or **Argand diagram** (Fig. 3.3). The coordinates $z = (x, y)$ define a phasor.

- In this representation **complex conjugate** numbers are positioned symmetrically relative to the real axis.

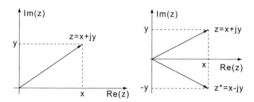

Fig. 3.3. Complex plane and complex conjugate numbers

Trigonometric or Polar Form

In polar form a complex number is represented by the length r of its phasor and the angle φ between the phasor and the real axis.

Polar coordinate system: $(x, y) \rightarrow (r, \varphi)$
The **absolute value** or **magnitude** of the complex number $r = |z|$; the **phase**, angle or **argument** is φ.

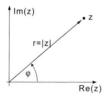

Fig. 3.4. Trigonometric representation of a complex number

- φ is ambiguous. Every rotation of 2π (360°) leads to the same point. The **principal value** of the argument is the angle measured anticlockwise between the phasor and the positive real axis (Fig. 3.4).

Trigonometric Form

$$x = r \cdot \cos\varphi, \qquad y = r \cdot \sin\varphi$$
$$z = r(\cos\varphi + j \cdot \sin\varphi)$$
$$= |z|(\cos\varphi + j \cdot \sin\varphi)$$
$$\mathrm{Re}(z) = |z| \cdot \cos\varphi, \qquad \mathrm{Im}(z) = |z| \cdot \sin\varphi$$

Because of the symmetrical position of the **complex conjugate** numbers, they differ only in the sign of the argument (Fig. 3.3).

$$z^* = |z| \cdot (\cos \varphi - j \cdot \sin \varphi) = |z| \cdot (\cos(-\varphi) + j \cdot \sin(-\varphi))$$

The absolute values of both numbers are equal

$$|z| = |z^*|$$

Exponential Form

The **Euler formula**

$$e^{j\varphi} = \cos \varphi + j \sin \varphi$$

leads to a compact representation of complex numbers

$$z = r \cdot e^{j\varphi} = |z| \cdot e^{j\varphi}$$

The **absolute value** or **magnitude** of the complex number $r = |z|$; The **phase**, angle or **argument** is φ.

- r is a real number.
- $e^{j\varphi}$ is a complex number with an absolute value of one.

$$|z| = |z \cdot e^{j\varphi}| = |z| \cdot |e^{j\varphi}| = r \cdot |e^{j\varphi}| = r \cdot 1 = r$$

A **complex exponential function** with an imaginary exponent is periodic with a period of 2π. The values are located on a unity circle in the complex plane.

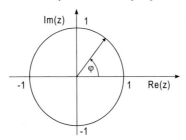

Fig. 3.5. Function values of the exponential function with imaginary exponent

Reading the values from Fig. 3.5 leads to:

$$e^{j \cdot 0} = e^{j \cdot 0°} = 1, \qquad e^{j\pi} = e^{j \cdot 180°} = -1, \qquad e^{j\frac{\pi}{2}} = e^{j \cdot 90°} = j, \qquad e^{j\frac{3\pi}{2}} = e^{j \cdot 270°} = -j$$

NOTE: In some literature the **versor** representation is also used.

$$|z| \cdot e^{j\varphi} = |z| \cdot \underline{/\varphi}$$

This reads as: z magnitude *versor* φ. Therefore: $5 \, \Omega \cdot e^{j\frac{\pi}{2}}$ is written as $5 \, \Omega \underline{/\frac{\pi}{2}}$ or $5 \, \Omega \underline{/90°}$.

3.1.2.3 Changing Between Different Representations of Complex Numbers

Changing from (x, y) to (r, φ)

$$r = \sqrt{x^2 + y^2}, \qquad \varphi = \arctan\left(\frac{y}{x}\right) \quad \text{for } x \neq 0$$

NOTE: On calculators the arctan function is labelled $\tan^{-1}$.

Table 3.1. Special Cases, conversion from (x, y) to (r, φ)

Real part x	Imaginary part y	Argument φ	z
$x = 0$	$y = 0$	undefined	$z = 0$
$x = 0$	$y > 0$	$\varphi = \dfrac{\pi}{2}$	z positive imaginary
$x = 0$	$y < 0$	$\varphi = \dfrac{3\pi}{2}$	z negative imaginary
$x > 0$	$y = 0$	$\varphi = 0$	z positive real
$x < 0$	$y = 0$	$\varphi = \pi$	z negative real

Changing from (r, φ) to (x, y)

$$x = r \cdot \cos \varphi, \qquad y = r \cdot \sin \varphi$$

NOTE: Scientific calculators often provide a function to change coordinates from the polar notation to the Cartesian notation and vice versa.

3.1.3 Complex Calculus

3.1.3.1 Complex Addition and Subtraction

Complex numbers are added by summing their real components and their imaginary components, respectively. Therefore the addition of complex phasors can be performed as an **addition of vectors** (Fig. 3.6). The **subtraction** can be done geometrically by the addition of an inverted orientation of the phasor.

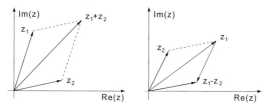

Fig. 3.6. Geometrical addition and subtraction of complex numbers

NOTE: There is a limitation for the analogy between vectors and phasors. The product of two complex numbers is neither similar to the scalar product nor the vector product of two vectors. The absolute value squared is, however, similar to the scalar product of a vector with itself.

In order to solve problems in electrical engineering, it is often necessary to determine the **absolute values of sums** of complex variables (Fig. 3.7). The cosine formula yields

$$|z_1 + z_2| = \sqrt{z_1^2 + z_2^2 + 2 \cos(\varphi_1 - \varphi_2)}$$

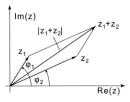

Fig. 3.7. Absolute value of the sum of two complex numbers

3.1.3.2 Multiplication of Complex Numbers

The **multiplication** of a complex number with a (positive) real number α increases its magnitude by the factor α. The orientation of the phasor is not affected. If $\alpha < 1$ the magnitude of the phasor is decreased, and if $\alpha < 0$ the orientation of the phasor is inverted. **Multiplication** of two **complex numbers** in trigonometric and exponential representation is given by

$$z = z_1 \cdot z_2 = |z_1||z_2|e^{j(\varphi_1 + \varphi_2)}$$

- The absolute value of the product of two complex numbers is identical to the product of their absolute values. The argument of the product is the sum of their arguments.

$$r = r_1 \cdot r_2, \quad \varphi = \varphi_1 + \varphi_2$$

The multiplication of a complex number with an **absolute value of one** $|z| = 1$ is a special case. Any complex number with an absolute value of 1 can be represented as

$$z = e^{j\varphi}$$

For the multiplication of this number with a complex number z_1 it follows that

$$z_1 \cdot z = |z_1| \cdot \underbrace{|z|}_{=1} \cdot e^{j(\varphi_1 + \varphi)} = |z_1|e^{j(\varphi_1 + \varphi)}$$

The absolute value of the product remains unchanged, only the argument is changed. The phasor is **rotated** by an angle φ. If the argument of the number e^{φ} is a function of time, especially a linear function, then the representation is

$$z_1 = e^{j\omega t}$$

The parameter ω is called the **angular frequency**. The product of a complex number z and a phasor $e^{j\omega t}$ is equivalent to a **rotation** of the complex phasor z with the angular frequency ω, and $|z|e^{j\omega t}$ is a **rotating phasor**.

$$
\begin{aligned}
\alpha \cdot z &= \alpha x + j\alpha y & &\text{Cartesian form} \\
&= \alpha |z| \cdot (\cos \varphi + j \sin \varphi) & &\text{trigonometric form} \\
&= \alpha \cdot |z| \cdot e^{j\varphi} & &\text{exponential form}
\end{aligned}
$$

3.1.4 Overview: Complex Number Arithmetic

Table 3.2. Overview of complex number arithmetic

$z_1, z_2 \neq 0$	Cartesian
z	$z = x + jy$
Re(z)	x
Im(z)	y
z^*	$z = x - jy$
$z_1 + z_2$	$(x_1 + x_2) + j(y_1 + y_2)$
$z_1 - z_2$	$(x_1 - x_2) + j(y_1 - y_2)$
$\|z_1 + z_2\|$	$\sqrt{(x_1 + x_2)^2 + (y_1 + y_2)^2}$
$z_1 \cdot z_2$	$(x_1 x_2 - y_1 y_2) + j(x_1 y_2 + x_2 y_1)$
$\dfrac{z_1}{z_2}$	$\dfrac{z_1 z_2^*}{\|z_2\|^2}$
$1/z$	$\dfrac{z^*}{\|z\|^2}$
z^n	Change to exponential repres.
$\sqrt[n]{z}$	Change to exponential repres.

$z_1, z_2 \neq 0$	Trigonometric	Exponential
z	$z = \|z\| \cdot (\cos\varphi + j\sin\varphi)$	$z = \|z\| \cdot e^{j\varphi}$
Re(z)	$\|z\| \cdot \cos\varphi$	$\|z\| \cdot \cos\varphi$
Im(z)	$\|z\| \cdot \sin\varphi$	$\|z\| \cdot \sin\varphi$
z^*	$z = \|z\| \cdot (\cos\varphi - j\sin\varphi)$	$z = \|z\| \cdot e^{-j\varphi}$
$z_1 + z_2$	Change to Cartesian representation	
$z_1 - z_2$	Change to Cartesian representation	
$\|z_1 + z_2\|$	$\sqrt{r_1^2 + r_2^2 + 2r_1 r_2 \cos(\varphi_1 - \varphi_2)}$	
$z_1 \cdot z_2$	$\|z_1\|\|z_2\|\left[\cos(\varphi_1 + \varphi_2) + j\sin(\varphi_1 + \varphi_2)\right]$	$\|z_1\|\|z_2\|e^{j(\varphi_1 + \varphi_2)}$
$\dfrac{z_1}{z_2}$	$\dfrac{\|z_1\|}{\|z_2\|}\left[\cos(\varphi_1 - \varphi_2) + j\sin(\varphi_1 - \varphi_2)\right]$	$\dfrac{\|z_1\|}{\|z_2\|}e^{j(\varphi_1 - \varphi_2)}$
$1/z$	$\dfrac{1}{\|z\|}(\cos\varphi - j\sin\varphi)$	$\dfrac{1}{\|z\|}e^{-j\varphi}$
z^n	$\|z\|^n(\cos n\varphi + j\sin n\varphi)$	$\|z\|^n e^{jn\varphi}$
$\sqrt[n]{z}$	$\sqrt[n]{\|z\|}\left[\cos\left(\dfrac{\varphi}{n}\right) + \sin\left(\dfrac{\varphi}{n}\right)\right]$	$\sqrt[n]{\|z\|}e^{j\frac{\varphi}{n}}$

3.1.5 The Complex Exponential Function

For real numbers the exponential function can be defined by a power series

$$e^x = \sum_{n=0}^{\infty} \frac{x^n}{n!} = 1 + \frac{x}{1!} + \frac{x^2}{2!} + \frac{x^3}{3!} + \ldots$$

This power series converges for any real number and therefore defines a function $f : \mathbb{R} \to \mathbb{R}$. This can be extended to complex numbers. The power series converges for any complex number. In general, the function value will be complex, which means $f : \mathbb{C} \to \mathbb{C}$.

NOTE: For the exponential function e^z the representation as $\exp(z)$ can also be found. The latter is preferred when the exponent is a lengthy term.

3.1.5.1 Exponential Function with Imaginary Exponents

The exponential function with purely imaginary exponents has a special significance. Because of the relation

$$e^{j\omega t} = \cos \omega t + j \sin \omega t$$

the real and the imaginary parts of the function value are defined. The function is periodic with a period of 2π. The following holds for the **derivatives**

$$\frac{d}{dt} e^{j\omega t} = j\omega \cdot e^{j\omega t}, \qquad \frac{d^2}{dt^2} e^{j\omega t} = -\omega^2 \cdot e^{j\omega t}$$

Because of the second equation the exponential function with imaginary exponents is a **harmonic function** like the sine and the cosine functions. It follows for the **integrals** that

$$\int e^{j\omega t} \, dt = \frac{1}{j\omega} \cdot e^{j\omega t}, \qquad \int \left(\int e^{j\omega t} \, dt \right) dt = \frac{-1}{\omega^2} \cdot e^{j\omega t}$$

The **multiplication** of a phasor z with the term $e^{j\varphi}$ results in a **rotation** of the phasor by an angle of φ (Fig. 3.8).

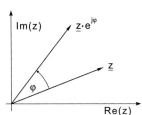

Fig. 3.8. Rotation of a phasor as a result of a multiplication with $e^{j\varphi}$

3.1.5.2 Exponential Function with Complex Exponents

The exponential function with complex exponents $s = \sigma + j\omega$ can be separated into a real exponent and an imaginary exponent.

$$e^s = e^{\sigma + j\omega} = e^\sigma \cdot e^{j\omega}$$

The term $e^{j\omega}$ is a harmonic function, and the term e^σ can be regarded as an amplitude factor. This becomes clear when examining the function e^{st}

$$f(t) = e^{st} = e^{\sigma t} \cdot e^{j\omega t}$$

For $\sigma = 0$ the term $e^{\sigma t} = 1$. The function is thus a harmonic function of time. For $\sigma < 0$ the term $e^{\sigma t}$ leads to a **damped oscillation**. For $\sigma > 0$ an exponentially **increasing oscillation** is the result (Fig. 3.9).

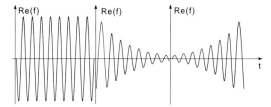

Fig. 3.9. Real part of the function $e^{\sigma t} \cdot e^{j\omega t}$ for $\sigma = 0, \sigma < 0$ and $\sigma > 0$

3.1.6 Trigonometric Functions with Complex Arguments

Sine and **cosine functions** can be extended to **complex arguments** like the exponential function. Their relationships to the exponential function are as follows

$$\cos z = \frac{1}{2}\left(e^{jz} + e^{-jz}\right) \tag{3.6}$$

$$\sin z = \frac{1}{2j}\left(e^{jz} - e^{-jz}\right) \tag{3.7}$$

Adding Eq. (3.6) to Eq. (3.7) multiplied by j leads to the **Euler formula**

$$e^{jz} = \cos z + j \sin z$$

Furthermore

$$\cos^2 z + \sin^2 z = 1$$

as for real numbers.

NOTE: **Addition theorems** are found in Appendix A. Calculations with trigonometric functions are often simplified by using the transformations given in Eqs. (3.6) and (3.7). This eliminates the necessity to use the addition theorems. This particularly simplifies the calculation of integrals as only products of exponential functions will occur.

3.1.7 From Sinusoidal Waveforms to Phasors

The trigonometric and the exponential representations of complex numbers lead to a geometric analogy, which can be used to explain many aspects in science.

3.1.7.1 Complex Magnitude

A real harmonic function $v(t) = \hat{v} \cdot \cos(\omega t + \varphi)$ can be written as the real component of a complex function.

$$v(t) = \hat{v} \cdot \text{Re}\{e^{j(\omega t + \varphi)}\} = \text{Re}\underbrace{\{\hat{v} \cdot e^{j(\omega t + \varphi)}\}}_{\text{complex function of time}}$$

Formally, the latter term in brackets is regarded as a **complex function of time**, denoted in this chapter as $\underline{v}$.

$$\underline{v}(t) = \hat{v} \cdot e^{j(\omega t + \varphi)} = \underbrace{\hat{v} \cdot e^{j\varphi}}_{\text{complex amplitude}} \cdot e^{j\omega t} = \underline{\hat{v}} \cdot e^{j\omega t}$$

The **complex amplitude** $\underline{\hat{v}}$ is a product of the amplitude $\hat{v}$ and the phase factor $e^{j\varphi}$.

- The absolute value of a complex amplitude equals the real amplitude $|\underline{\hat{v}}| = \hat{v}$.

The **complex RMS value** $\underline{V}$ may be defined in analogy to the root mean square (RMS) value V of sinusoidal waveforms. The amplitude and the complex RMS value of a sinusoidal waveform are represented as **phasors** in the complex plane.

NOTE: Similarly it is possible to look at a time function as the imaginary part of an exponential oscillation. Both models have equal qualities, but they must not be used simultaneously.

Time function as real part	
Sinusoidal	$\mapsto$ Phasor
$\hat{v}\cos(\omega t + \varphi)$	$\mapsto \quad \hat{v} \cdot e^{j\varphi}$
$\hat{v}\sin(\omega t + \varphi)$	$\mapsto \quad \hat{v} \cdot e^{j(\varphi - \pi/2)}$
Phasor	$\mapsto$ Sinusoidal
$\hat{v} \cdot e^{j\varphi}$	$\mapsto \quad \hat{v}\cos(\omega t + \varphi)$

Time function as imaginary part	
Sinusoidal	$\mapsto$ Phasor
$\hat{v}\sin(\omega t + \varphi)$	$\mapsto \quad \hat{v} \cdot e^{j\varphi}$
$\hat{v}\cos(\omega t + \varphi)$	$\mapsto \quad \hat{v} \cdot e^{j(\varphi + \pi/2)}$
Phasor	$\mapsto$ Sinusoidal
$\hat{v} \cdot e^{j\varphi}$	$\mapsto \quad \hat{v}\sin(\omega t + \varphi)$

The amplitude $\hat{v}$ can be replaced by the RMS value V.

Time invariant complex phasors are called **operators** (e.g. complex impedance).

Table 3.3. Functions and their complex counterparts

Symbol	Example	Notation
$v(t) =$	$\hat{v} \cdot \cos(\omega t + \varphi)$	Time-varying voltage
$\hat{v} =$		Amplitude
$\underline{v}(t) =$	$\hat{v} \cdot e^{j(\omega t + \varphi)}$	Complex time-varying voltage
$\underline{\hat{v}} =$	$\hat{v} \cdot e^{j\varphi}$	Complex amplitude
$V =$	$\dfrac{\hat{v}}{\sqrt{2}}$	RMS value
$\underline{V} =$	$\dfrac{\underline{\hat{v}}}{\sqrt{2}}$	Complex RMS value

3.1.7.2 Relationship Between Sinusoidal Waveforms and Phasors

The sine function can be regarded as the vertical projection of the **rotating phasor**.

In the **phasor diagram** the phasor rotates with a constant angular frequency ω in a mathematically positive sense, i.e. anticlockwise. The vertical magnitudes of the phasors are drawn onto the time axis (Fig. 3.10). A phasor is described by four characteristic values:

- The physical quantity represented by the phasor. Often these are the voltage $\underline{v}$ or the current $\underline{i}$, but it can also be the flux $\underline{\Phi}$ and other quantities. The symbol is written next to the phasor.

- The absolute value of the phasor is represented by the length of the phasor, where either amplitude or RMS are chosen.

- The phase shift φ_0 is represented by the orientation of the phasor with respect to the zero line (which is usually horizontal).

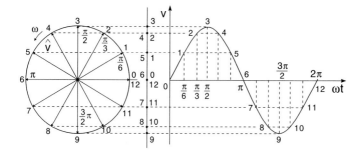

Fig. 3.10. Phasor diagram and time diagram of the sine function

- The angular frequency of the phasor is equal to the angular frequency of the represented quantity. In most cases it will be clearly defined by the problem and is not explicitly noted.

When the angular speed of all concerned phasors is the same (which means that the frequencies of the sine waves are equal) then only the relative phases of the phasors need to be considered. This leads to the representation with stationary (i.e. nonrotating) phasors.

NOTE: The cosine function can be similarly regarded as the horizontal projection of the phasor.

3.1.7.3 Addition and Subtraction of Phasors

- The sum (difference) of sinusoidal waveforms of the *same frequency* results in a sinusoidal waveform of the same frequency.

Sums and differences of sinusoidal waveforms can be obtained from the phasor diagram (Fig. 3.11).

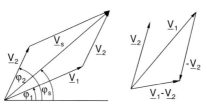

Fig. 3.11. Sum and difference of sinusoidal waveforms in the phasor diagram

The sum of the cosine voltages

$$v_1 = \hat{v}_1 \cos(\omega t + \varphi_1), \qquad v_2 = \hat{v}_2 \cos(\omega t + \varphi_2)$$

results in the sum signal $v_s = v_1 + v_2$

$$v_s = \hat{v}_s \cos(\omega t + \varphi_s) \qquad (3.8)$$

For the calculation the phasor representation is used.

$$v_1 \mapsto \underline{v}_1 = \hat{v}_1 \cdot e^{j(\omega t + \varphi_1)} \mapsto \underline{\hat{v}}_1 \cdot e^{j\varphi_1}, \qquad v_2 \mapsto \underline{v}_2 = \hat{v}_2 \cdot e^{j(\omega t + \varphi_2)} \mapsto \underline{\hat{v}}_2 \cdot e^{j\varphi_2}$$

The addition is done using complex notation $\underline{v}_s = \underline{v}_1 + \underline{v}_2$. The amplitude of the sum signal $v_s(t)$ is obtained from the complex sum amplitude.

$$\hat{v}_s = |\underline{v}_1 + \underline{v}_2| = \sqrt{\text{Re}^2\{\underline{v}_s\} + \text{Im}^2\{\underline{v}_s\}}, \qquad \tan\varphi_s = \frac{\text{Im}\{\underline{v}_1 + \underline{v}_2\}}{\text{Re}\{\underline{v}_1 + \underline{v}_2\}}$$

This leads to the following:

$$\hat{v}_s = \sqrt{\hat{v}_1^2 + \hat{v}_2^2 + 2\hat{v}_1\hat{v}_2\cos(\varphi_1 - \varphi_2)}, \qquad \tan\varphi_s = \frac{\hat{v}_1\sin\varphi_1 + \hat{v}_2\sin\varphi_2}{\hat{v}_1\cos\varphi_1 + \hat{v}_2\cos\varphi_2} \quad (3.9)$$

NOTE: In general, the sum of harmonic functions of *different frequencies* is not a harmonic function. It cannot be represented by stationary phasors.

3.2 Sinusoidal Waveforms

When considering voltages and currents the following distinctions are made (Fig. 3.12):

Constant quantity: a quantity that is constant over time, $v(t) = \text{const}$.

    EXAMPLE: DC current, DC voltage, magnetic flux of a permanent magnet.

Pulsating quantity: a quantity with changing instantaneous value, but with a constant sign.

    EXAMPLE: Chopped DC voltage, 'humming' DC voltage.

Alternating quantity: a time-varying quantity with an average mean (over a longer period) of zero.

    EXAMPLE: Telephone signals, AC from 230 V mains.

Mixed quantity: a waveform with an alternating instantaneous value and magnitude. The RMS is not necessarily zero. It is also known as a **general alternating quantity**.

Periodic quantity: a waveform with a repeating progression after an interval of T.
    DEFINITION: A waveform of time is periodic, if there is a T with $s(t) = s(t + T)$ for all t.
    T is called the **period** of the signal $s(t)$.

Sinusoidal quantity: an alternating waveform with a sinusoidal (i.e. **harmonic**) progression. Sinusoidal waveforms are elementary signals in AC. All periodic alternating signals (and for certain assumptions also nonperiodic signals) can be represented by sinusoidal signals (Fourier analysis and synthesis).

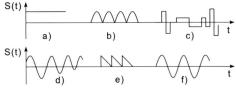

Fig. 3.12. Comparison of different waveforms: **a)** DC quantity, **b)** pulsating DC quantity; **c)** alternating quantity (non periodic); **d)** mixed quantity; **e)** periodic quantity (nonharmonic); **f)** sinusoidal quantity

3.2.1 Characteristics of Sinusoidal Waveforms

Sinusoidal currents and voltages can be represented by

$$v(t) = \hat{v} \cdot \sin \omega t, \quad \text{or} \quad v(t) = \hat{v} \cdot \cos \omega t$$

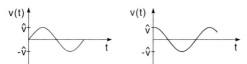

Fig. 3.13. Harmonic signals

Both signals appear identical. The instantaneous value at time $t = 0$ is zero in one case, and the maximum value in the other case (Fig. 3.13). The **instantaneous value** of the time signal $v(t)$ varies between the two values $\hat{v}$ and $-\hat{v}$. The most positive value is called the **amplitude** or **peak value**. The parameter ω is called the **angular frequency** or **radian frequency**.

The **frequency** of the signal is

$$f = \frac{\omega}{2\pi}, \quad \omega = 2\pi f$$

The unit of frequency is Hz (hertz), and the unit of the angular frequency is s^{-1} or rad/s. The **period** of the signal is

$$T = \frac{1}{f} = \frac{2\pi}{\omega}$$

and is the distance between two consecutive maxima (minima) of the signal.

EXAMPLE: A sinusoidal AC voltage with an amplitude of 300 V and a frequency of 50 Hz is measured with an oscilloscope. What is the instantaneous value of the signal 12 ms after the zero-crossing?

$$v(t) = \hat{v} \cdot \sin \omega t, \quad \hat{v} = 300 \text{ V}, \quad \omega = 2\pi \cdot 50 \text{ s}^{-1} \approx 314.16 \text{ s}^{-1}$$

The signal has a zero-crossing at the time $t = 0$.

$$v(12 \text{ ms}) = \hat{v} \cdot \sin(\omega \cdot 12 \cdot 10^{-3} \text{ s}) = 300 \text{ V} \cdot \sin(3.770) = -176 \text{ V}$$

For a single signal the positioning of the zero-crossing, $t = 0$, can be made by choice. However, for interrelations between harmonic signals, the **phase shift** must be known.

$$v(t) = \hat{v} \cdot \sin(\omega t + \varphi_0)$$

At $t = 0$ the **phase shift** φ_0 is present.

The relative **phase position** is regarded as **leading** for a positive phase shift φ_0, otherwise it is regarded as **lagging**.

NOTE: Usually the phase of the voltage is given with respect to the current, i.e. $\varphi = \varphi_V - \varphi_I$. The phase of the complex impedance and the complex power is defined likewise. An exception is the complex admittance; its phase is expressed with respect to the voltage.

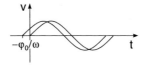

Fig. 3.14. Phase-shifted harmonic signals

EXAMPLE: Two sinusoidal currents i_1 and i_2 with the same amplitude have a phase shift of 30°. Therefore, i_2 is leading i_1. What is the instantaneous value of i_2 at the zero-crossing of i_1?

$$i_1 = \hat{\imath} \cdot \sin \omega t, \qquad i_2 = \hat{\imath} \cdot \sin(\omega t + 30°), \qquad i_2(t=0) = \hat{\imath} \cdot \sin(30°) = 0.5\,\hat{\imath}$$

For the description of alternating functions further quantities are used. The **average** or **arithmetic mean** is defined as

$$\bar{v} = \frac{1}{T} \int\limits_{t_0}^{t_0+T} v(\tau)\, d\tau = \frac{1}{T} \int\limits_{0}^{T} v(\tau)\, d\tau \qquad (3.10)$$

This value represents the area underneath the time function over *one period*. Because of the periodicity of the function, the value $\bar{v}$ is independent of the starting point t_0. For sinusoidal functions this value is zero.

Fig. 3.15. Visualisation of average, average rectified value and RMS value of sinusoidal functions

The average rectified value is the average of the magnitude of the signal (Fig. 3.15)

$$\overline{|v|} = \frac{1}{T} \int\limits_{0}^{T} |v(\tau)|\, d\tau \qquad (3.11)$$

NOTE: The average rectified value has to be regarded when calculating the charge of capacitors after rectification, or for electrolytical processes. The dimensioning of rectifier diodes can also be based on the average rectified value of the current, as the voltage drop across the diode is nearly constant.

Specifically for sinusoidal voltages (and currents)

$$\overline{|v|} = \frac{1}{T} \int\limits_{0}^{T} \hat{v} \cdot |\sin \omega t|\, dt = \frac{2}{T} \hat{v} \cdot \int\limits_{0}^{T/2} \sin \omega t\, dt = \frac{1}{\pi}\hat{v}\Big[-\cos \omega t \Big]_{\omega t=0}^{\omega t = \pi} = \frac{2}{\pi}\hat{v} \approx 0.637\hat{v}$$

The **RMS value** of an AC voltage is related to the power. In Fig. 3.16 a DC voltage source with a 1 V terminal voltage causes a power dissipation of P in the resistor R. An AC voltage source, which causes the same average power dissipation in the same resistor, i.e. it causes the same temperature increase, has a terminal voltage with an RMS of 1 V. This definition is independent of the actual shape of the AC voltage.

The definition of the RMS of a function is

$$V_{\text{RMS}} = \sqrt{\frac{1}{T} \int_0^T v^2(t) \, dt} \qquad (3.12)$$

For the special case of sinusoidal voltages and currents the result is

$$V_{\text{RMS}} = \sqrt{\frac{1}{T} \int_0^T (\hat{v} \sin \omega t)^2 \, dt} = \sqrt{\frac{1}{T} \hat{v}^2 \int_0^T \sin^2 \omega t \, dt} = \frac{\hat{v}}{\sqrt{2}} \approx 0.707 \hat{v}$$

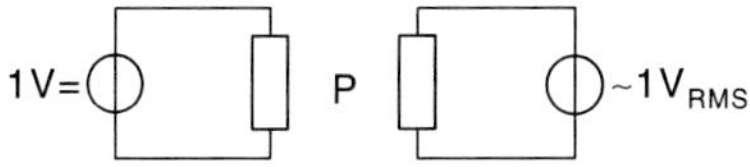

Fig. 3.16. Concept of the RMS value

The square of the RMS is

$$V_{\text{RMS}}^2 = \frac{\hat{v}^2}{2} \quad \Rightarrow \quad \frac{V_{\text{RMS}}^2}{R} = \frac{1}{2} \frac{\hat{v}^2}{R} \qquad \text{for sinusoidal waveforms}$$

That is, the average dissipated power of an AC voltage source is only half the value of the dissipation of a DC voltage source whose terminal voltage is the same value as the peak value of the AC voltage.

For general forms of alternating functions it is always true that:

- The RMS value is always smaller than or equal to the peak value.

NOTE: The **RMS value** of the voltage or the current has to be considered for the correct thermal dimensioning of resistive components. The **peak value** must be considered to choose the breakdown voltage of a capacitor or the reverse voltage of semiconductors.

3.2.2 Characteristics of Nonsinusoidal Waveforms

The **crest factor** is the ratio of the peak value $\hat{v}$ to the RMS value V_{RMS} of an alternating function of any shape.

$$k_c = \frac{\hat{v}}{V_{\text{RMS}}}$$

The **form factor** is the relationship between the RMS value and the average rectified value.

$$k_f = \frac{V_{\text{RMS}}}{|\bar{v}|}$$

The crest factor and the form factor are characteristic values providing a rough description of the shape of an alternating function. The 'flatter' the shape of the curve, the more the form factor approaches a value of 1 (from above). Table 3.4 shows crest and form factors for some selected waveforms.

Table 3.4. Crest and form factors for selected waveforms

Waveform	k_c	k_f
Sine	1.414	1.111
Triangle	1.732	1.155
Square-wave, DC-free	1.000	1.000
Sawtooth	1.732	1.155
Half-wave rectified sine	2.000	1.571
Full-wave rectified sine	1.414	1.111
Three-phase rectified sine	1.190	1.017

APPLICATION: The deflection of the pointer of a **moving-coil** meter with rectifier bridges is proportional to the average rectified value of the AC. On the contrary, **moving-iron** meters display the RMS value. However, the scales of both instruments are calibrated to the RMS value of a sinusoidal current. Therefore errors occur when a nonsinusoidal current is measured with a moving coil meter. This error can be rectified if the form factor of the measured AC is known.

EXAMPLE: A square-wave voltage with a peak value of ±1 V is measured with a moving-coil meter. The RMS and the average rectified value are 1 V for this waveform. The deflection of the pointer is proportional to the average rectified value.

A sinusoidal AC voltage with a average rectified value of 1 V has a RMS of $k_f \cdot |\bar{v}| \approx 1.11$ V. For the rectangular voltage a moving-coil meter would show 1.11 V, which results in an error of 11%.

NOTE: In telecommunications additional quantities are commonly used to characterise the deviation from the sinusoidal waveform, particularly the nonlinear distortion factor (total harmonic distortion, THD).

3.3 Complex Impedance and Admittance

3.3.1 Impedance

The **complex impedance** is defined analogously to the definition of DC resistance as

$$\underline{Z} = \frac{\underline{v}}{\underline{i}} = \frac{\hat{v} \cdot e^{j\varphi_V}}{\hat{i} \cdot e^{j\varphi_I}} = \frac{\hat{v}}{\hat{i}} \cdot e^{j(\varphi_V - \varphi_I)} \tag{3.13}$$

Since $\underline{Z}$ is a complex value, it can be represented in exponential form

$$\boxed{\underline{Z} = Z \cdot e^{j\varphi_Z}} \tag{3.14}$$

The phase angle φ_Z represents the **phase shift** of the **voltage** with respect to the **current** flowing through the AC impedance. In Cartesian form it is represented as follows:

$$\boxed{\underline{Z} = R + jX} \tag{3.15}$$

with

$$Z = \sqrt{R^2 + X^2}, \qquad \varphi_Z = \arctan\left[\frac{\mathrm{Im}(\underline{Z})}{\mathrm{Re}(\underline{Z})}\right] = \arctan\left(\frac{X}{R}\right) \tag{3.16}$$

R is referred to as **resistance**;
X is referred to as **reactive impedance** or **reactance**;
Z is referred to as **impedance**.

The unit of impedance is the ohm, Ω.

- The **complex resistance** is the ratio of the voltage amplitude to the current amplitude (or their RMS values) *and* the phase shift of the voltage relative to the current flowing through the impedance.

- The **impedance** is the ratio of the voltage amplitude to the current amplitude (or of their RMS values), *without* considering the relative phase value.

The following relationships hold

$$Z = \frac{V}{I} \tag{3.17}$$

$$R = Z \cdot \cos\varphi_Z, \qquad X = Z \cdot \sin\varphi_Z \tag{3.18}$$

The representation of the complex impedance as in Eq. (3.14) leads to a phasor analogy.

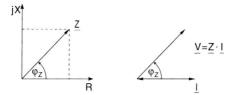

Fig. 3.17. Phasor of the impedance and of the voltage and current for an AC impedance

The phasor of the impedance is represented in the **complex impedance plane** (Fig. 3.17). According to the expression $\underline{V} = \underline{Z} \cdot \underline{I}$, the relationship between voltage and current can be represented by a phasor diagram. The impedance rotates the voltage phasor by an angle of φ_Z with respect to the current phasor. The ratio of the absolute values of voltage to current is Z.

EXAMPLE: A current $i(t) = \hat{\imath}\cos(\omega t + \varphi_I)$ flows through a component with an impedance $\underline{Z}$. How does the time function of the voltage behave?

Change to the phasor representation $i(t) \mapsto \underline{I}$

$$\underline{V} = \underline{Z} \cdot \underline{I} = Z \cdot e^{j\varphi_Z} \cdot \underline{I} = Z \cdot I \cdot e^{j\varphi_Z}$$
$$\implies \underline{v}(t) = Z \cdot \hat{\imath} \cdot e^{j(\varphi_Z + \varphi_I)}$$
$$\implies v(t) = \mathrm{Re}(\underline{v}) = Z \cdot \hat{\imath} \cdot \cos(\omega t + \varphi_Z + \varphi_I)$$

EXAMPLE: A sinusoidal voltage with an amplitude of 1 V is applied to an impedance of $(4+j3)\ \Omega$ (Fig. 3.18). What are the absolute value and the phase of the current?

The impedance is $Z = \sqrt{4^2 + 3^2}\ \Omega = 5\ \Omega$. The current flowing through the impedance has an amplitude of $\hat{\imath} = \hat{v}/Z = 200$ mA. The RMS value is $I = \hat{\imath}/\sqrt{2} = 141$ mA. The phase shift of the voltage with respect to the current is $\varphi_Z = \arctan(3/4) = 0.64$ (37°). The current lags the voltage by 37° ($\varphi_I = -37°$).

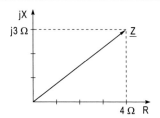

Fig. 3.18. Impedance phasor

3.3.2 Complex Impedance of Passive Components

This topic is treated (focusing on time-varying signals) more extensively in Sect. 3.4.

3.3.2.1 Resistor

For the voltage and the current in a resistor R it follows from Ohm's law

$$v(t) = R \cdot i(t) \qquad \text{across the resistor}$$

If the current $\underline{i}$ is complex then

$$\boxed{\underline{Z} = \frac{\underline{v}}{\underline{i}} = \frac{R \cdot \underline{i}}{\underline{i}} = R} \tag{3.19}$$

- The complex impedance of the resistor is real and equals R.

3.3.2.2 Inductor

In an inductor the induced voltage is proportional to the change of the current di/dt.

$$v(t) = L \cdot \frac{di}{dt}$$

For $\underline{i}(t) = \hat{i} \cdot e^{j\omega t}$ is then

$$\underline{v}(t) = L \cdot \hat{i} \tfrac{d}{dt} e^{j\omega t} = j\omega L \cdot \hat{i} \cdot e^{j\omega t}$$

$$\boxed{\underline{Z} = \frac{\underline{v}}{\underline{i}} = j\omega L} \tag{3.20}$$

- The complex impedance of an inductor is imaginary positive. It is proportional to the inductance and to the angular frequency.

3.3.2.3 Capacitor

The voltage across a capacitor is proportional to the integral of the current flowing through the capacitor.

$$v(t) = \frac{1}{C} \int i(t)\, dt$$

For $\underline{i}(t) = \hat{\imath} \cdot e^{j\omega t}$ is then

$$\underline{v}(t) = \frac{1}{C} \int \hat{\imath} \cdot e^{j\omega t} \, dt = \frac{1}{C} \frac{1}{j\omega} \cdot \hat{\imath} \cdot e^{j\omega t}$$

$$\boxed{\underline{Z} = \frac{\underline{v}}{\underline{i}} = -j\frac{1}{\omega C} = \frac{1}{j\omega C}} \tag{3.21}$$

- The complex impedance operator of the capacitor is imaginary negative. It is inversely proportional to the capacitance and to the angular frequency.

3.3.3 Admittance

The **complex conductance** is defined analogously to the definition of the DC admittance.

$$\underline{Y} = \frac{\underline{i}}{\underline{v}} = \frac{\hat{\imath} \cdot e^{j(\omega t + \varphi_{\mathrm{I}})}}{\hat{v} \cdot e^{j(\omega t + \varphi_{\mathrm{V}})}} = \frac{\hat{\imath}}{\hat{v}} \cdot e^{j(\varphi_{\mathrm{I}} - \varphi_{\mathrm{V}})} \tag{3.22}$$

Because $\underline{Y}$ is a complex quantity it can be represented in exponential form

$$\boxed{\underline{Y} = Y \cdot e^{j\varphi_{\mathrm{Y}}}} \tag{3.23}$$

For the complex conductance the phase angle φ_Y represents the **phase shift** of the **current** relative to the **voltage** in the AC impedance. In Cartesian form it is

$$\boxed{\underline{Y} = G + jB} \tag{3.24}$$

With

$$Y = \sqrt{G^2 + B^2}, \qquad \varphi_Y = \arctan\left[\frac{\mathrm{Im}(\underline{Y})}{\mathrm{Re}(\underline{Y})}\right] = \arctan\left(\frac{B}{G}\right) \tag{3.25}$$

G is referred to as **conductance**;
B is referred to as **susceptance**;
Y is referred to as **admittance**.

Sometimes the complex conductance is also called admittance. The unit for the complex conductance is siemens (S) or mho (℧).

- The **complex conductance** is the ratio of the current amplitude to the voltage amplitude (or their RMS values) *and* the phase shift of the current relative to the voltage at the component.

- The **admittance** is the ratio of the current amplitude to the voltage amplitude (or their RMS values) *without* considering the relative phase value.

The following relationships hold

$$Y = \frac{I}{V} \tag{3.26}$$

$$G = Y \cdot \cos\varphi_Y, \qquad B = Y \cdot \sin\varphi_Y \tag{3.27}$$

The representation of the complex admittance as in Eq. 3.23 leads to a phasor analogy.

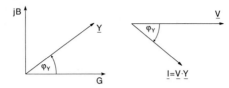

Fig. 3.19. Phasor diagram of the admittance and of the current and voltage at the AC impedance

The phasor of the admittance is represented in the **complex admittance plane**. According to the expression $\underline{I} = \underline{Y} \cdot \underline{V}$ the relationship of voltage and current at the resistor can be represented by a phasor diagram (Fig. 3.19). The admittance rotates the current phasor by an angle of φ_Y relative to the voltage phasor. The ratio of the absolute values of current to voltage is Y.

The following relationship between impedance and admittance holds

$$\boxed{\underline{Y} = \frac{1}{\underline{Z}}} \qquad (3.28)$$

Therefore

$$\underline{Y} = \frac{1}{R + jX} = \frac{R}{R^2 + X^2} - j\frac{X}{R^2 + X^2} = \underbrace{\frac{R}{Z^2}}_{=G} \underbrace{-j\frac{X}{Z^2}}_{=jB} \qquad (3.29)$$

For the conductance and susceptance it follows directly that

$$G = \frac{R}{Z^2}, \qquad B = -\frac{X}{Z^2} \qquad (3.30)$$

- A positive susceptance is equivalent to a negative impedance and vice versa.

Furthermore Eq. (3.28) leads to

$$Y = \frac{1}{Z}, \qquad \text{and} \qquad \varphi_Y = -\varphi_Z \qquad (3.31)$$

- The phase of the complex admittance equals the phase of the negative impedance.

3.3.4 Complex Admittance of Passive Components

From Eq. (3.28) the complex admittances of the resistor, inductor and capacitor are

Resistor: $\quad \underline{Y} = G = 1/R$

Inductor: $\quad \underline{Y} = -j\dfrac{1}{\omega L}$

Capacitor: $\quad \underline{Y} = j\omega C$

3.3.5 Overview: Complex Impedance

Terminology

Symbol	Terminology
$\underline{Z} = R + jX$	(Complex) impedance
Z	Impedance
X	Reactance
R	Resistance
$\underline{Y} = G + jB$	(Complex) admittance
Y	Admittance
B	Susceptance
G	Conductance

Impedance and Admittance of Passive Components

Table 3.5. Impedance and admittance of passive components

General expression	Resistor R	Inductor L	Capacitor C
$\underline{Z} = R + jX$	R	$j\omega L$	$-j\dfrac{1}{\omega C}$
R	R	0	0
X	0	ωL	$-\dfrac{1}{\omega C}$
$Z = \sqrt{R^2 + X^2}$	R	ωL	$\dfrac{1}{\omega C}$
$\varphi_Z = \arctan(X/R)$	0	$+\pi/2$	$-\pi/2$
$\underline{Y} = G + jB$	$1/R$	$-j\dfrac{1}{\omega L}$	$j\omega C$
G	$1/R$	0	0
B	0	$-\dfrac{1}{\omega L}$	ωC
$Y = \sqrt{G^2 + B^2}$	$1/R$	$\dfrac{1}{\omega L}$	ωC
$\varphi_Y = \arctan(B/G)$	0	$-\pi/2$	$+\pi/2$

$$\underline{Y} = \frac{1}{\underline{Z}}, \qquad Y = \frac{1}{Z}, \qquad \varphi_Y = -\varphi_Z$$

$$\underline{Y} = G + jB, \qquad G = \frac{R}{Z^2}, \qquad B = \frac{-X}{Z^2}$$

3.4 Impedance of Passive Components

Passive linear electrical networks are composed of resistors, inductors and capacitors. This section examines the behaviour of these passive components for sinusoidal voltages and currents. See Table 3.5 for a summary.

For a **resistor** R current and voltage are in phase. The resistance is

$$Z = |\underline{Z}| = \left|\frac{v}{i}\right| = R$$

For the **conductance** therefore

$$Y = \frac{1}{Z} = \frac{1}{R} = G$$

For an **inductor** L the induced voltage is proportional to the rate of change of current di/dt

$$v(t) = L \cdot \frac{di}{dt}$$

For a sinusoidal **current**

$$v(t) = L \cdot \frac{d}{dt}(\hat{i} \sin \omega t) = \hat{i} \, \omega L \cos \omega t = \hat{i} \, \omega L \sin\left(\omega t + \frac{\pi}{2}\right)$$

- The voltage across an inductor leads the current by 90° or $\pi/2$, see Fig. 3.20.

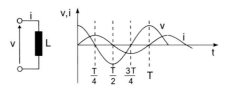

Fig. 3.20. Voltage and current in an inductor

The **inductive reactance** X_L is

$$X_L = \omega L$$

For DC voltage the impedance of an ideal inductor is zero. It increases linearly with the frequency. The complex impedance is

$$\underline{Z} = jX_L = j\omega L \tag{3.32}$$

For the admittance

$$\underline{Y} = \frac{1}{\underline{Z}} = \frac{1}{j\omega L} = -j\frac{1}{\omega L} \tag{3.33}$$

For a **capacitor** C the voltage is the integral of the current flowing through the capacitor

$$v(t) = \frac{1}{C}\int i(t)\, dt$$

Differentiation of both sides of the equation leads to

$$\frac{dv}{dt} = \frac{1}{C} \cdot i(t) \quad \Rightarrow \quad i(t) = C \cdot \frac{dv}{dt}$$

For a sinusoidal **voltage**

$$i(t) = C \cdot \frac{d}{dt}(\hat{v} \sin \omega t) = \hat{v}\omega C \cos \omega t = \hat{v}\omega C \sin\left(\omega t + \frac{\pi}{2}\right)$$

- The voltage across a capacitor lags the current by 90° or $\pi/2$ ($\varphi_V = -90°$), see Fig. 3.21.

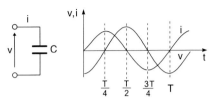

Fig. 3.21. Voltage and current in a capacitor

The **capacitive reactance** X_C is

$$X_C = -\frac{1}{\omega C}$$

For DC voltage the impedance of an ideal capacitor is infinite. It decreases in inverse proportion to the frequency. For the complex impedance it follows that

$$\underline{Z} = jX_C = -j\frac{1}{\omega C} \tag{3.34}$$

For the admittance

$$\underline{Y} = \frac{1}{\underline{Z}} = jB_C = j\omega C \tag{3.35}$$

3.5 Combinations of Passive Components

3.5.1 Series Combinations

3.5.1.1 General Case

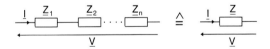

Fig. 3.22. Series combination of passive components

Figure 3.22 shows the general case of a series combination of passive components. For AC currents and voltages (in analogy to DC)

$$\underline{V} = \underline{I} \cdot \underline{Z}$$

The notation shows clearly that these are *complex* values. The same current flows through all components. For the total impedance $\underline{Z}$ it follows that

$$\underline{Z} = \underline{Z}_1 + \underline{Z}_2 + \cdots + \underline{Z}_n$$

As complex numbers are added by summing up the real parts and the imaginary parts, the resistive part and the reactive part of the impedance can be added separately

$$\underline{Z} = \sum_{i=1}^{n} \text{Re}(\underline{Z}_i) + j \sum_{i=1}^{n} \text{Im}(\underline{Z}_i) = \sum_{i=1}^{n} R_i + j \sum_{i=1}^{n} X_i$$

3.5.1.2 Resistor and Inductor in Series

The current flowing through both components is identical. For the resistor, the current and voltage are in phase; for the inductor the voltage leads the current by 90° or $\pi/2$. The terminal voltage of the combination is the sum of the partial voltages (Fig. 3.23).

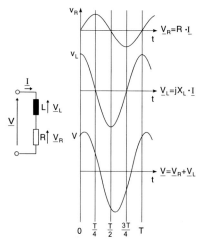

Fig. 3.23. Series combination of resistor and inductor

The phasor diagram yields (using the Pythagorean theorem)

$$V = \sqrt{V_R^2 + V_L^2} = \sqrt{I^2 R^2 + I^2 X_L^2} = I \cdot \sqrt{R^2 + X_L^2}$$

X_L is the inductive reactance.

$$\frac{V}{I} = \sqrt{R^2 + X_L^2}$$

The ratio V/I is called the **impedance** Z

$$Z = |\underline{Z}| = \sqrt{R^2 + X_L^2}$$

This result can be obtained directly by examining the impedance phasors in the complex impedance plane (Fig. 3.24b).

- The phase of the voltage relative to the current lies between 0° and 90° ($\pi/2$). The larger the resistive part in the series combination, the closer the phase value is to zero.

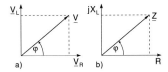

Fig. 3.24. a RMS phasors of the voltages; **b** phasor in the complex impedance plane

The phase can be obtained from the phasor triangle

$$\tan \varphi = \frac{V_L}{V_R} = \frac{X_L}{R} = \frac{\omega L}{R}$$

In complex notation the complex impedances can be simply added

$$\underline{Z} = R + jX_L = R + j\omega L \tag{3.36}$$

with

$$\boxed{\begin{aligned} \underline{Z} &= |\underline{Z}|e^{j\varphi} \\ Z &= |\underline{Z}| = \sqrt{R^2 + X_L^2} = \sqrt{R^2 + (\omega L)^2} \\ \varphi &= \arctan\left(\frac{X_L}{R}\right) = \arctan\left(\frac{\omega L}{R}\right) \end{aligned}} \tag{3.37}$$

3.5.1.3 Resistor and Capacitor in Series

The current flowing through both components is identical. For the resistor, the current and voltage are in phase, for the capacitor the voltage lags the current by 90° or π/2. The terminal voltage of the combination is the sum of the partial voltages (Fig. 3.25).

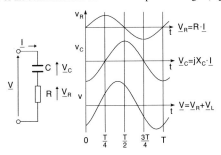

Fig. 3.25. Series combination of resistor and capacitor

The phasor diagram yields (using the Pythagorean theorem)

$$V = \sqrt{V_R^2 + V_C^2} = \sqrt{I^2 R^2 + I^2 X_C^2} = I \cdot \sqrt{R^2 + X_C^2}$$

where X_C is the capacitive reactance. The impedance Z follows as

$$Z = |\underline{Z}| = \sqrt{R^2 + X_C^2}$$

This result can be obtained directly by examining the impedance phasors in the complex impedance plane (Fig. 3.26b).

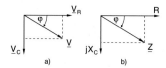

Fig. 3.26. a) RMS phasors of the voltages; b) phasor in the complex impedance plane

- The phase of the voltage relative to the current lies between 0° and −90° (−π/2). The larger the resistive part in the series combination, the smaller is the phase value.

The phase can be obtained from the phasor triangle

$$\tan\varphi = \frac{V_C}{V_R} = \frac{X_C}{R} = -\frac{1}{\omega RC}$$

In complex notation the complex impedances can be simply added

$$\underline{Z} = R + jX_C = R - j\frac{1}{\omega C} \tag{3.38}$$

with

$$\boxed{\begin{aligned} \underline{Z} &= |\underline{Z}|e^{j\varphi} \\ Z &= |\underline{Z}| = \sqrt{R^2 + X_C^2} = \sqrt{R^2 + \left(\frac{1}{\omega C}\right)^2} \\ \varphi &= \arctan\left(\frac{X_C}{R}\right) = -\arctan\left(\frac{1}{\omega RC}\right) \end{aligned}} \tag{3.39}$$

3.5.1.4 Resistor, Inductor and Capacitor in Series

NOTE: In practice a series combination of a pure capacitance and a pure inductance does not occur. This is because real components such as coils and capacitors always exhibit losses, which can be modelled as a resistor in series.

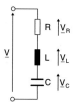

Fig. 3.27. Series combination of a resistor, an inductor and a capacitor

The arrangement shown in Fig. 3.27 is called a **series-resonant circuit**. The current flowing through all three components is identical. For the resistor current and voltage are in phase. For the inductor the voltage leads the current by +90° π/2, while for the capacitor the voltage lags the current by $-90°$ $-\pi/2$. Consequently, the voltages across L and C have opposite signs. The terminal voltage of the combination is the sum of the partial voltages.

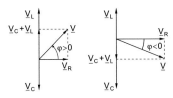

Fig. 3.28. RMS phasors of the voltages in the series-resonant circuit

The phasor diagram yields (Fig. 3.28)

$$V = \sqrt{V_R^2 + (V_L + V_C)^2} = \sqrt{I^2 R^2 + I^2 (X_L + X_C)^2}$$

where X_L is the inductive reactance, and X_C is the capacitive reactance. The impedance Z follows as

$$Z = |\underline{Z}| = \sqrt{R^2 + (X_L + X_C)^2}, \qquad X_L = \omega L, \qquad X_C = -\frac{1}{\omega C}$$

The reactances X_L and X_C have opposite signs. Depending on the values of the capacitor and the inductor, either the capacitive or the inductive reactance dominates, as shown in Fig. 3.29.

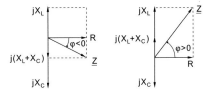

Fig. 3.29. Phasor diagrams for different reactance values of X_L, X_C

- The phase of the voltage relative to the current in the series-resonant circuit lies between $-90°$ and $+90°$ ($\pm \pi/2$). If the capacitive reactance dominates, the circuit behaves like an RC combination; if the inductive reactance dominates, the circuit behaves like an RL combination.

The phase φ can be obtained from the phasor diagram

$$\tan \varphi = \frac{V_L + V_C}{V_R} = \frac{X_L + X_C}{R} = \frac{\omega L - \dfrac{1}{\omega C}}{R} = \frac{\omega^2 LC - 1}{\omega RC}$$

In complex notation

$$\underline{Z} = \underline{Z}_R + \underline{Z}_L + \underline{Z}_C = R + j(X_L + X_C) = R + j\left(\omega L - \frac{1}{\omega C}\right) \tag{3.40}$$

with

$$\underline{Z} = |\underline{Z}| \cdot e^{j\varphi}$$
$$Z = \sqrt{R^2 + (X_L + X_C)^2} = \sqrt{R^2 + \left(\omega L - \frac{1}{\omega C}\right)^2}$$
$$\varphi = \arctan\left(\frac{X_L + X_C}{R}\right) = \arctan\left(\frac{\omega L - \frac{1}{\omega C}}{R}\right)$$
(3.41)

The reactances vary with the frequency. The inductive reactance increases proportionally with the frequency, while the capacitive reactance decreases inversely with the frequency. At the **resonant frequency** both reactances are equal in magnitude, but have an opposite sign. As they cancel each other out at this frequency, only the resistance appears at the terminals of the combination. At **resonance** the voltages across L and C have the same magnitude.

$$|V_L| = |V_C|, \quad |X_L| = |X_C|, \quad X_L + X_C = 0 \quad \Rightarrow \quad \omega_r L = \frac{1}{\omega_r C}$$

The equality of the magnitudes of the reactances at the resonant frequency leads to

$$\omega_r = \frac{1}{\sqrt{LC}} \quad \Rightarrow \quad f_r = \frac{1}{2\pi}\frac{1}{\sqrt{LC}}$$

- Below the resonant frequency the circuit behaves like a resistor–capacitor combination, and above the resonant frequency it behaves like a resistor–inductor combination (Fig. 3.30).

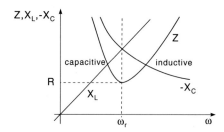

Fig. 3.30. Reactance and impedance of the series-resonant circuit with frequency

3.5.2 Parallel Combinations

3.5.2.1 General Case

Figure 3.31 shows the general AC case of parallel combinations of passive components. Similar to the DC case the alternating current and voltage are related by using the admittance

$$\underline{I} = \underline{V} \cdot \underline{Y}$$

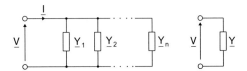

Fig. 3.31. Parallel combination of passive components

The notation clearly shows that these are complex quantities. Each circuit element has the same voltage drop across its terminals. This leads to a general expression for the total admittance $\underline{Y}$

$$\underline{Y} = \underline{Y}_1 + \underline{Y}_2 + \cdots + \underline{Y}_n$$

As complex numbers are summed by adding the individual real and imaginary components, the conductive component G_S and the susceptive component B_S of the admittance can be added separately

$$\underline{Y} = \sum_{i=1}^{n} G_i + j \cdot \sum_{i=1}^{n} B_i$$

The impedance $\underline{Z}$ of the parallel combination is given by

$$\underline{Z} = \frac{1}{\underline{Y}} = \frac{1}{\underline{Z}_1} + \frac{1}{\underline{Z}_2} + \cdots \frac{1}{\underline{Z}_n}$$

In the special case of only two components in parallel, the resulting expression for the impedance is analogous to the case of two resistances in parallel

$$\underline{Z} = \frac{\underline{Z}_1 \cdot \underline{Z}_2}{\underline{Z}_1 + \underline{Z}_2} \tag{3.42}$$

and if the impedance is separated into its conductive and susceptive components $\underline{Z}_i = R_i + j \cdot X_i$

$$\underline{Z} = \frac{R_1(R_2^2 + X_2^2) + R_2(R_1^2 + X_1^2)}{(R_1 + R_2)^2 + (X_1 + X_2)^2} + j \cdot \frac{X_1(R_2^2 + X_2^2) + X_2(R_1^2 + X_1^2)}{(R_1 + R_2)^2 + (X_1 + X_2)^2}$$

3.5.2.2 Resistor and Inductor in Parallel

Each circuit element has the same voltage drop across its terminals. For the resistor, the current and voltage are in phase, while for the inductor the current lags the voltage by 90°

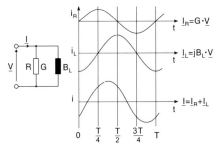

Fig. 3.32. Parallel combination of a resistor and an inductor

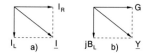

Fig. 3.33. a) RMS current phasor; **b)** phasor in the complex admittance plane

or $\pi/2$ (Fig. 3.32). The total current through the combination is the sum of the current in the individual branches.

The phasor diagram yields (using the Pythagorean theorem)

$$I = \sqrt{I_R^2 + I_L^2} = \sqrt{V^2 G^2 + V^2 B_L^2} = V \cdot \sqrt{G^2 + B_L^2}$$

where B_L is the inductive susceptance.

$$\frac{I}{V} = \sqrt{G^2 + B_L^2}$$

The ratio I/V is the **admittance Y**

$$Y = |\underline{Y}| = \sqrt{G^2 + B_L^2} \tag{3.43}$$

This result can be obtained directly by using phasors in the complex admittance plane (Fig. 3.33b).

- The phase of the current relative to the voltage lies between $0°$ and $-90°$ ($-\pi/2$). This decreases in magnitude as the inductance increases.

The phase difference can be obtained from the phasor diagram

$$\tan \varphi_Y = \frac{I_L}{I_R} = \frac{B_L}{G} = -\frac{R}{\omega L}$$

In complex notation the complex admittances can be simply added.

$$\underline{Y} = G + jB_L = \frac{1}{R} - j\frac{1}{\omega L} \tag{3.44}$$

with

$$\boxed{\begin{aligned} \underline{Y} &= |\underline{Y}| e^{j\varphi_Y} \\ Y &= |\underline{Y}| = \sqrt{G^2 + B_L^2} = \sqrt{\left(\frac{1}{R}\right)^2 + \left(\frac{1}{\omega L}\right)^2} \\ \varphi_Y &= \arctan\left(\frac{B_L}{G}\right) = \arctan\left(\frac{R}{\omega L}\right) \end{aligned}} \tag{3.45}$$

3.5.2.3 Resistor and Capacitor in Parallel

Each circuit element has the same voltage drop across its terminals. For the resistor, the current and voltage are in phase, while for the capacitor the current leads the voltage by $90°$ or $\pi/2$ (Fig. 3.34). The total current through the combination is the sum of the currents in the individual branches.

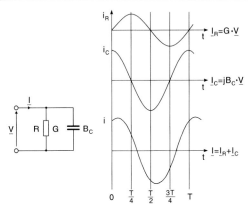

Fig. 3.34. Parallel combination of resistor and capacitor

The phasor diagram yields (using the Pythagorean theorem)

$$I = \sqrt{I_R^2 + I_C^2} = \sqrt{V^2 G^2 + V^2 B_C^2} = V \cdot \sqrt{G^2 + B_C^2}$$

where B_C is the capacitive susceptance.

$$\frac{I}{V} = \sqrt{G^2 + B_C^2}$$

The ratio I/V is the **admittance** Y

$$Y = |\underline{Y}| = \sqrt{G^2 + B_C^2}$$

This result can be obtained directly by using phasors in the complex admittance plane (Fig. 3.35 b)).

- The phase of the current relative to the voltage lies between $0°$ and $-90°$ ($-\pi/2$). This decreases in magnitude as the capacitance decreases.

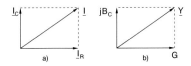

Fig. 3.35. a) RMS phasor; b) phasor in the complex admittance plane

The phase difference can be obtained from the phasor diagram

$$\tan \varphi_Y = \frac{I_C}{I_R} = \frac{B_C}{G} = \omega RC$$

In complex notation the complex admittances can be simply added.

$$\underline{Y} = G + jB_C = \frac{1}{R} + j\omega C \tag{3.46}$$

with

$$\boxed{\begin{aligned}\underline{Y} &= |\underline{Y}|e^{j\varphi_Y} \\ Y &= |\underline{Y}| = \sqrt{G^2 + B_C^2} = \sqrt{\left(\frac{1}{R}\right)^2 + (\omega C)^2} \\ \varphi_Y &= \arctan\left(\frac{B_C}{G}\right) = \arctan(\omega RC)\end{aligned}}$$
(3.47)

NOTE: It can be seen that the phase difference φ_Y is measured with respect to the voltage, and $\varphi_Y = -\varphi_v$.

3.5.2.4 Resistor, Inductor and Capacitor in Parallel

NOTE: A parallel combination of a pure capacitance and a pure inductance is not realistic. Real capacitors and inductors suffer losses, which can be modelled by a resistor in parallel.

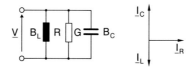

Fig. 3.36. A parallel combination of a resistor, a capacitor and an inductor

The arrangement shown in Fig. 3.5.2.4 is called a parallel-resonant circuit. Each circuit element has the same voltage drop across its terminals. For the resistor the current and voltage are in phase. For the inductor, the current lags the voltage by 90° or $\pi/2$; for the capacitor, the current leads the voltage by 90° or $\pi/2$. The currents through L and C have opposite signs. The total current through the combination is the sum of the current in the individual branches.

The phasor diagram yields

$$I = \sqrt{I_R^2 + (I_L + I_C)^2} = \sqrt{V^2 G^2 + V^2(B_L + B_C)^2} = V\sqrt{G^2 + (B_L + B_C)^2}$$

where B_L is the inductor susceptance, and B_C is the capacitive susceptance. The admittance Y follows as

$$Y = |\underline{Y}| = \sqrt{G^2 + (B_L + B_C)^2}, \qquad B_L = -\frac{1}{\omega L}, \qquad B_C = \omega C$$

The susceptances B_L and B_C have opposite signs. Depending on the size of the inductance or capacitance, either the inductive or capacitive susceptance will dominate, as shown in Fig. 3.37.

- The phase φ_Y of the voltage with respect to the current varies between $-90°$ and $+90°$ ($\pm\pi/2$). If the capacitive susceptance dominates, the circuit behaves like an RC combination; if the inductive susceptance dominates, the circuit behaves like an RL combination.

The phase angle φ_Y of the susceptance may be derived from the phasor diagram

$$\tan\varphi_Y = \frac{I_L + I_C}{I_R} = \frac{B_L + B_C}{G} = \frac{\omega C - \dfrac{1}{\omega L}}{G} = R\left(\omega C - \frac{1}{\omega L}\right)$$

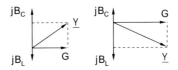

Fig. 3.37. Phasor diagram for different susceptance values B_L, B_C

In complex notation

$$\underline{Y} = \underline{Y}_R + \underline{Y}_L + \underline{Y}_C = G + j(B_L + B_C) = \frac{1}{R} + j\left(\omega C - \frac{1}{\omega L}\right) \quad (3.48)$$

with

$$\begin{aligned} \underline{Y} &= |\underline{Y}| \cdot e^{j\varphi_Y} \\ Y &= \sqrt{G^2 + (B_L + B_C)^2} = \sqrt{\left(\frac{1}{R^2}\right) + \left(\omega C - \frac{1}{\omega L}\right)^2} \\ \varphi_Y &= \arctan\left(\frac{B_L + B_C}{G}\right) = \arctan\left[R\left(\omega C - \frac{1}{\omega L}\right)\right] \end{aligned} \quad (3.49)$$

The susceptances are frequency dependent. The inductive susceptance decreases with frequency, while the capacitive susceptance increases with frequency (Fig. 3.38). At resonance both susceptances are equal in magnitude, but have opposite signs. As they cancel each other out at this frequency, only the resistance appears at the terminals of the combination. At resonance the currents through L and C have the same absolute value.

$$|I_L| = |I_C|, \qquad |B_L| = |B_C|, \qquad B_L + B_C = 0 \quad \Rightarrow \quad \frac{1}{\omega_r L} = \omega_r C$$

It can be seen by equating the susceptances at the resonant frequency that

$$\omega_r = \frac{1}{\sqrt{LC}} \quad \Rightarrow \quad f_r = \frac{1}{2\pi} \frac{1}{\sqrt{LC}}$$

- Below the resonant frequency, the circuit behaves like a parallel resistor–inductor combination, while above the resonant frequency the circuit behaves like a parallel resistor–capacitor combination.

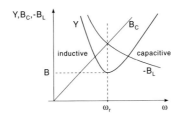

Fig. 3.38. Susceptance and admittance characteristic curves for a parallel-resonant circuit

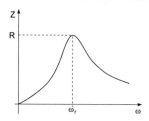

Fig. 3.39. Impedance characteristic curve for a parallel-resonant circuit

3.5.3 Overview of Series and Parallel Circuits

Table 3.6. Series circuit

	Phasor diagram	$\underline{Z}$	$\tan \varphi_Z = \dfrac{X}{R}$
RL		$\sqrt{R^2 + (\omega L)^2}$	$\dfrac{\omega L}{R}$
RC		$\sqrt{R^2 + \left(\dfrac{1}{\omega C}\right)^2}$	$-\dfrac{1}{\omega RC}$
LC*		$\left|\omega L - \dfrac{1}{\omega C}\right|$	$\pm \infty$
RLC	See series-resonant circuits	$\sqrt{R^2 + \left(\omega L - \dfrac{1}{\omega C}\right)^2}$	$\dfrac{\omega^2 LC - 1}{\omega RC}$
Phase values are given with respect to total current			

*The ideal case is with $R = 0$. For frequencies below the resonant frequency the phase shift is $\varphi = -90°$; above the phase shift is $\varphi = 90°$.

Table 3.7. Series resonant circuit

Frequency	Phasor diagram	$\underline{Z}$	φ_Z
$f < f_r$		Resistive–capacitive	$-90°$ to $0°$
$f = f_r$		Purely–resistive	$0°$
$f > f_r$		Resistive–inductive	$0°$ to $90°$
Resonant frequency $f_r = \dfrac{1}{2\pi}\sqrt{\dfrac{1}{LC}}$			

Table 3.8. Parallel circuit

	Phasor diagram	Y	$\tan\varphi_Y = \dfrac{B}{G}$
RL	G V; $B_L = -\dfrac{1}{\omega L}$; $\underline{Y}$	$\dfrac{\sqrt{R^2 + (\omega L)^2}}{\omega RL}$	$\dfrac{-R}{\omega L}$
RC	$\underline{Y}$; $B_C = \omega C$; G V	$\dfrac{\sqrt{(\omega RC)^2 + 1}}{R}$	ωRC
LC^*	B_C; B_L	$\left|\dfrac{\omega^2 LC - 1}{\omega L}\right|$	$\pm\infty^*$
RLC	See parallel resonant circuit	$\dfrac{\sqrt{R^2(\omega^2 LC - 1)^2 + (\omega L)^2}}{\omega RL}$	$\dfrac{R(\omega^2 LC - 1)}{\omega L}$

Phase values are given with respect to total voltage

NOTE: $Z = 1/Y$, $R = 1/G$, $\varphi_Z = -\varphi_Y$. *Ideally $R = \infty$.

Table 3.9. Parallel resonant circuit

Frequency	Phasor diagram	$\underline{Z}$	φ_Z
$f < f_r$	G, $\underline{Y}$, B_C, B_L	Resistive–inductive	90° to 0°
$f = f_r$	B_C, $\underline{Y}$, G, B_L	Purely resistive	0°
$f > f_r$	B_L, B_C, $\underline{Y}$, G	Resistive–capacitive	0° to −90°
Resonant frequency $f_r = \dfrac{1}{2\pi}\sqrt{\dfrac{1}{LC}}$			

3.6 Network Transformations

3.6.1 Transformation from Parallel to Series Circuits and Vice Versa

Any circuit consisting of the series combination of a resistive and a reactive component can be transformed into a parallel circuit consisting of a conductive and a susceptive component (Fig. 3.40). If an identical AC voltage causes identical AC currents to flow through two such circuits, they are known as **equivalent circuits** (see also Sect. 1.3.6.1).

The equivalence of the circuits implies that their impedances are equal.

$$R_s + jX_s = \underline{Z} = \frac{1}{G_p + jB_p}$$

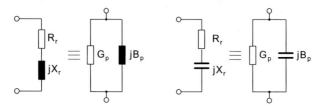

Fig. 3.40. Transformation of a series circuit into an equivalent parallel circuit and vice versa

Expanding by multiplying the numerator and the denominator by the complex conjugate of the denominator yields

$$R_s + jX_s = \underline{Z} = \frac{G_p - jB_p}{G_p^2 + B_p^2} = \frac{G_p}{G_p^2 + B_p^2} - j\frac{B_p}{G_p^2 + B_p^2}$$

Thus the transformation of a **parallel circuit to the equivalent series circuit** yields the following for the resistive and reactive components:

$$R_s = \frac{G_p}{G_p^2 + B_p^2}, \qquad X_s = -\frac{B_p}{G_p^2 + B_p^2} \tag{3.50}$$

The transformation of a series circuit into the equivalent parallel circuit therefore implies the equality of the complex admittances.

$$G_p + jB_p = \underline{Y} = \frac{1}{R_s + jX_s}$$

Expanding by multiplying the numerator and the denominator by the complex conjugate of the denominator yields

$$G_p + jB_p = \underline{Y} = \frac{R_s - jX_s}{R_s^2 + X_s^2} = \frac{R_s}{G_s^2 + X_s^2} - j\frac{X_s}{R_s^2 + X_s^2}$$

Thus the transformation of a **series circuit to the equivalent parallel circuit** yields the following for the resistive and reactive components:

$$G_p = \frac{R_s}{R_s^2 + X_s^2}, \qquad B_p = -\frac{X_s}{R_s^2 + X_s^2} \tag{3.51}$$

or, if expressed in terms of resistive and reactive values,

$$R_p = \frac{R_s^2 + X_s^2}{R_s} = \frac{Z^2}{R_s}, \qquad X_p = \frac{Z^2}{X_s}$$

- These transformations are valid **only for a fixed angular frequency** ω. All impedances are frequency dependent and thus experience different values in the transformed circuit as the frequency changes.

- The equivalence of the circuits only holds for sinusoidal voltages and currents.

NOTE: For network analysis series–parallel transformations are unnecessary, if the basic rules for combining impedances in series and parallel are applied.

3.6.2 Star–Delta (Wye–Delta) and Delta–Star (Delta–Wye) Transformations

For very complicated circuits, mesh or nodal analysis is frequently used. Depending on the type of analysis a star–delta or a delta–star transformation is required.*

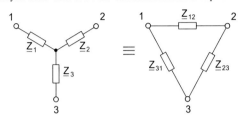

Fig. 3.41. Star–delta and delta–star transformations

For a **delta–star transformation**, using the notation of Fig. 3.41

$$
\begin{aligned}
\underline{Z}_1 &= \frac{\underline{Z}_{12} \cdot \underline{Z}_{31}}{\underline{Z}_{12} + \underline{Z}_{23} + \underline{Z}_{31}}, \\
\underline{Z}_2 &= \frac{\underline{Z}_{12} \cdot \underline{Z}_{23}}{\underline{Z}_{12} + \underline{Z}_{23} + \underline{Z}_{31}}, \\
\underline{Z}_3 &= \frac{\underline{Z}_{23} \cdot \underline{Z}_{31}}{\underline{Z}_{12} + \underline{Z}_{23} + \underline{Z}_{31}}
\end{aligned}
\qquad (3.52)
$$

Using the notation for impedance, resistance and reactance that $\underline{Z}_i = R_i + \mathrm{j} \cdot X_i$

$$R_1 = \frac{(R_{12}R_{31} - X_{12}X_{31})R + (R_{12}X_{31} + R_{31}X_{12})X}{R^2 + X^2}$$

$$X_1 = \frac{(R_{12}X_{31} + R_{31}X_{12})R - (R_{12}R_{31} - X_{12}X_{31})X}{R^2 + X^2}$$

$$R_2 = \frac{(R_{12}R_{23} - X_{12}X_{23})R + (R_{23}X_{12} + R_{12}X_{23})X}{R^2 + X^2}$$

$$X_2 = \frac{(R_{23}X_{12} + R_{12}X_{23})R - (R_{12}R_{23} - X_{12}X_{23})X}{R^2 + X^2}$$

$$R_3 = \frac{(R_{23}R_{31} - X_{23}X_{31})R + (R_{31}X_{23} + R_{23}X_{31})X}{R^2 + X^2}$$

$$X_3 = \frac{(R_{31}X_{23} + R_{23}X_{31})R - (R_{23}R_{31} - X_{23}X_{31})X}{R^2 + X^2}$$

with

$$R = R_{12} + R_{23} + R_{31}, \quad \text{and} \quad X = X_{12} + X_{23} + X_{31}$$

---
* In American literature the term *Wye* is used instead of *Star*.

For a **star–delta transformation**, using the notation of Fig. 3.41

$$\boxed{\begin{aligned}
\underline{Z}_{12} &= \underline{Z}_1 + \underline{Z}_2 + \frac{\underline{Z}_1 \cdot \underline{Z}_2}{\underline{Z}_3}, \\
\underline{Z}_{23} &= \underline{Z}_2 + \underline{Z}_3 + \frac{\underline{Z}_2 \cdot \underline{Z}_3}{\underline{Z}_1}, \\
\underline{Z}_{31} &= \underline{Z}_1 + \underline{Z}_3 + \frac{\underline{Z}_3 \cdot \underline{Z}_1}{\underline{Z}_2}
\end{aligned}} \qquad (3.53)$$

Using the notation for impedance, resistance and reactance that $\underline{Z}_i = R_i + \mathrm{j} \cdot X_i$

$$R_{12} = R_1 + R_2 + \frac{(R_1 R_2 - X_1 X_2) R_3 + (R_1 X_2 + R_2 X_1) X_3}{R_3^2 + X_3^2}$$

$$X_{12} = X_1 + X_2 + \frac{(R_1 X_2 + R_2 X_1) R_3 - (R_1 R_2 - X_1 X_2) X_3}{R_3^2 + X_3^2}$$

$$R_{23} = R_2 + R_3 + \frac{(R_2 R_3 - X_2 X_3) R_1 + (R_2 X_3 + R_3 X_2) X_1}{R_1^2 + X_1^2}$$

$$X_{23} = X_2 + X_3 + \frac{(R_2 X_3 + R_3 X_2) R_1 - (R_2 R_3 - X_2 X_3) X_1}{R_1^2 + X_1^2}$$

$$R_{31} = R_1 + R_3 + \frac{(R_1 R_3 - X_1 X_3) R_2 + (R_3 X_1 + R_1 X_3) X_2}{R_2^2 + X_2^2}$$

$$X_{31} = X_1 + X_3 + \frac{(R_1 X_3 + R_1 X_3) R_2 - (R_1 R_3 - X_1 X_3) X_2}{R_2^2 + X_2^2}$$

EXAMPLE: The bridged T-configuration found in filters can be analysed by referring to the star–delta transformation (Fig. 3.42).

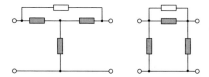

Fig. 3.42. Application of the star–delta transformation

- These transformations are valid **only for a fixed angular frequency** ω. All impedances are frequency dependent and thus experience different values in the transformed circuit as the frequency changes.

- The equivalence of the circuits only holds for sinusoidal voltages and currents, unless the circuit is purely resistive.

3.6.3 Circuit Duality

Two passive elements are called **dual** if the impedance of one of the circuits is proportional to the admittance of the other for *all frequencies*. This can be expressed as follows:

$$\underline{Z}_2 = R_D^2 \cdot \underline{Y}_2 \quad \Longleftrightarrow \quad \underline{Y}_2 = G_D^2 \cdot \underline{Z}_2 \tag{3.54}$$

R_D^2 and G_D^2 are real constants and are known as the **duality constants**. For elementary impedances the duality relationships in Table 3.10 hold.

Table 3.10. Duality relationships for circuit elements

Passive element	Dual element
Resistance R	Resistance R_D^2/R
Inductor L	Capacitor $C = L/R_D^2$
Capacitor C	Inductor $L = R_D^2 C$
Voltage source V_S, R_S	Current source $I_S = V_S/R_S$, $G_S = 1/R_S$
Current source I_S, G_S	Voltage source $V_S = I_S G_S$, $R_S = 1/G_S$
short circuit	Open circuit

There are also duality relationships for active elements. A voltage source with a voltage of V_S and an internal resistance of R_S is dual to a current source with a current of $I_S = V_S/R_S$. The current source has an admittance in parallel with the value $G_S = 1/R_S$.

The following quantities are dual pairs: voltage to current, resistance to admittance. Where the same current flows through two elements in a circuit, then in the dual circuit the voltage across the two elements will be the same and vice versa. The circuits given in Table 3.11 are dual.

Table 3.11. Duality relationships for circuits

Circuit	Dual circuit
Series circuit	Parallel circuit
Series-resonant circuit	Parallel-resonant circuit
Longitudinal resistor	Transverse resistor
Longitudinal inductor	Transverse capacitor
Longitudinal capacitor	Transverse inductor
T-configuration	Π-configuration
Mesh	node
Delta circuit	Star circuit
Voltage driven	Current driven
Current source	Voltage source

- When a circuit consists of a voltage source, with an internal resistance R_S and a load resistance R_L, the duality constant is given by $R_D = R_S \cdot R_L$. Voltage or current sources can be converted into their dual elements. For circuits that have been designed for open- or short-circuit operation, R_D can be arbitrarily chosen.

NOTE: The duality constants should be chosen such that resulting quantities can easily be realised by available components.

EXAMPLE: In the circuit shown in Fig. 3.43 a 1 MHz source with a 50 Ω internal resistance supplies a 50 Ω load. The circuit should be transformed into an equivalent circuit with fewer inductors.

In the dual circuit the two series inductors can be replaced by two parallel transverse capacitors. The capacitor can be translated into a series inductor. The duality constant is $R_D^2 = R_S \cdot R_L = 2500 \, \Omega^2$. The dual quantities are

$$C_D = \frac{L}{R_D^2} = \frac{8.2 \, \mu H}{2500 \, \Omega^2} = 3.3 \, nF, \quad L_D = C \cdot R_D^2 = 2.2 \, nF \cdot 2500 \, \Omega^2 = 5.5 \, \mu H$$

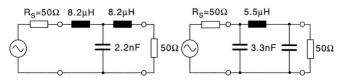

Fig. 3.43. Circuit and dual circuit

3.7 Simple Networks

3.7.1 Complex Voltage and Current Division

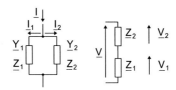

Fig. 3.44. Current and voltage division with complex impedances

With **current division** both impedances have the same alternating voltage across their terminals (Fig. 3.44). Therefore

$$\boxed{\frac{\underline{I}_1}{\underline{I}_2} = \frac{\underline{Y}_1}{\underline{Y}_2} = \frac{\underline{Z}_2}{\underline{Z}_1}} \tag{3.55}$$

- The total current is divided proportionally to the values of the admittances.

With **voltage division** both impedances have the same alternating current flowing through them. Therefore

$$\boxed{\frac{\underline{V}_1}{\underline{V}_2} = \frac{\underline{Z}_1}{\underline{Z}_2}} \tag{3.56}$$

- The total voltage is divided proportionally to the values of the impedances.

For a voltage $\underline{V}$ applied to a voltage divider, the output voltage $\underline{V}_2$ is given by

$$\underline{V}_2 = \underline{V} \cdot \frac{\underline{Z}_2}{\underline{Z}_1 + \underline{Z}_2} \tag{3.57}$$

NOTE: For the special case where all impedances are purely resistive, the rules for DC voltage and current division apply.

NOTE: In general, the **voltage-divider ratio** is **frequency dependent**, since the reactive components of the impedance are frequency dependent. Circuits providing frequency-dependent ratios between the input and output voltages are known as **filters**.

For measurement purposes, it is preferable to have a voltage divider with a division ratio that is independent of frequency. Equation (3.56) yields

$$\frac{V_1}{V_2} = \frac{\underline{Z}_1}{\underline{Z}_2} = \frac{Z_1 \cdot e^{j\varphi_1}}{Z_2 \cdot e^{j\varphi_2}} = \frac{Z_1}{Z_2} \cdot e^{j(\varphi_1 - \varphi_2)}$$

This ratio is thus frequency independent only when it is real, which means that the exponential expression must be real. For angles φ between $-90°$ and $+90°$, this means that $\varphi_1 = \varphi_2$. This also means that

$$\frac{X_1}{R_1} = \frac{X_2}{R_2} \quad \Rightarrow \quad \frac{R_1}{R_2} = \frac{X_1}{X_2} \tag{3.58}$$

- The voltage ratio is thus frequency independent if the resistive and reactive components of the voltage divider are in the same ratio. Put another way, the respective time constants $\tau = R \cdot C$ or L/R of the impedances must be equal.

APPLICATION: In oscilloscope measurements, the truest possible representation of the signal is required.

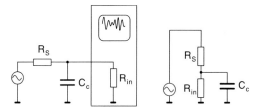

Fig. 3.45. Oscilloscope on a voltage source with an internal resistance, and the equivalent circuit

The AC voltage source with the internal resistance R_S is loaded by the input resistance R_{in} of the oscilloscope. The unavoidable capacitance of the cable lies in parallel. Thus for sources with high source resistances the voltage divider ratio falls off at higher frequencies. This can be seen in the circuit diagram in Fig. 3.45. This can be remedied by use of a **probe** with a **compensated voltage divider** (Fig. 3.46).

By equalising the capacitance in the probe, equal time constants (phase angles) $R_1 C_1 = R_2 C_2$ of the two RC parallel circuits can be achieved. This means a frequency-independent voltage-divider ratio. The increased voltage-divider ratio (usually 10 : 1) with a probe can be compensated in the oscilloscope by higher amplification. Simultaneously, the probe raises the input resistance of the measurement system by the division ratio.

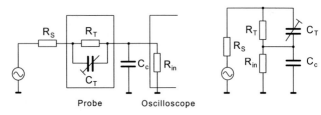

Fig. 3.46. An oscilloscope probe and the equivalent circuit

3.7.2 Loaded Complex Voltage Divider

The relationship defined in Eq. (3.56) is valid for unloaded voltage division, that is, open circuit.

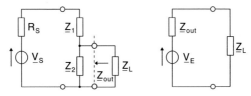

Fig. 3.47. Loaded voltage division on a voltage source with an internal resistance, and its equivalent circuit

In a real application a voltage divider is supplied by a voltage source with an internal resistance R_S and loaded by Z_L (Fig. 3.47). The voltage source is loaded by the **input impedance** of the voltage divider and load circuit. The load sees a voltage source with a complex internal impedance, which corresponds to the **output impedance** of the voltage divider.

The input impedance is

$$\underline{Z}_{in} = \underline{Z}_1 + \underline{Z}_2 || \underline{Z}_L = \underline{Z}_1 + \frac{\underline{Z}_2 \cdot \underline{Z}_L}{\underline{Z}_2 + \underline{Z}_L}$$

The output impedance $\underline{Z}_{out}$ of the voltage divider is given by

$$\underline{Z}_{out} = \underline{Z}_2 || (\underline{Z}_1 + R_S) = \frac{\underline{Z}_2 \cdot (\underline{Z}_1 + R_S)}{\underline{Z}_1 + \underline{Z}_2 + R_S}$$

In the unloaded configuration, the open-circuit voltage of the voltage divider for a voltage source $\underline{V}_S$, is defined as

$$\underline{V}_\infty = \underline{V}_S \cdot \frac{\underline{Z}_2}{\underline{Z}_1 + \underline{Z}_2 + R_S}$$

The short-circuit current is

$$\underline{I}_0 = \frac{V_S}{\underline{Z}_1 + R_S}$$

The output impedance $\underline{Z}_{out}$ is therefore given by

$$\underline{Z}_{out} = \frac{\text{Open-circuit voltage}}{\text{Short-circuit current}} = \frac{\underline{V}_\infty}{\underline{I}_0} \tag{3.59}$$

The voltage source of the equivalent circuit 'seen' by the load is defined as

$$\underline{V}_E = \underline{V}_S \cdot \frac{\underline{Z}_2 || \underline{Z}_L}{\underline{Z}_2 || \underline{Z}_L + \underline{Z}_1 + R_S}$$

3.7.3 Impedance Matching

In RF communications it is important that the signal source and the load have the same impedance in order to avoid reflections over the connecting medium.

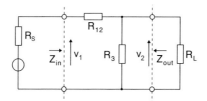

Fig. 3.48. Impedance matching for $R_S > R_L$

To that end, the circuit in Fig. 3.48 can be used. It is suitable when the **internal impedance** of the source is **higher than** the **load impedance**. The impedances must be so chosen so that the input impedance of the circuit with the load connected is the same as the internal impedance of the voltage source. On the other hand, the output impedance of the circuit must be the same as that of the load. For the standard situation where loads and source impedances are real

$$Z_{in} = R_{12} + R_3 || R_L = R_{12} + \frac{R_3 \cdot R_L}{R_3 + R_L}, \qquad Z_{out} = \frac{R_3(R_{12} + R_S)}{R_3 + R_{12} + R_S}$$

To match the impedances the following must hold (Fig. 3.49):

$$Z_{in} = R_S, \quad \text{and} \quad Z_{out} = R_L$$

From these conditions, the resistances may be derived

$$\boxed{R_{12} = R_S \cdot \sqrt{1 - \frac{R_L}{R_S}}, \quad \text{and} \quad R_3 = \frac{R_L}{\sqrt{1 - \frac{R_L}{R_S}}}, \quad \text{for } R_S > R_L} \qquad (3.60)$$

The voltage ratio

$$\boxed{\frac{V_1}{V_2} = \frac{R_S}{R_L}\left(1 + \sqrt{1 - \frac{R_L}{R_S}}\right) = \frac{1}{1 - \sqrt{1 - \frac{R_L}{R_S}}}} \qquad (3.61)$$

is always greater than 1 for the condition that $R_S > R_L$, so that the circuit attenuates the signal. Without changing the functionality of the circuit, the resistance R_{12} can be divided between the two symmetrically placed resistors R_1 and R_2. Fig. 3.50 shows an example.

For the case where the **load resistance is higher than** the **source resistance** of the voltage source, the circuit in Fig. 3.49 is suitable. Here the transverse resistance is in parallel with the input terminals of the circuit.

The input and output impedance of the circuit are given by

$$Z_{in} = R_3 ||(R_{12} + R_L) = \frac{R_3 \cdot (R_{12} + R_L)}{R_3 + R_{12} + R_L}, \qquad Z_{out} = R_{12} + (R_3 || R_S) = R_{12} + \frac{R_3 \cdot R_S}{R_3 + R_S}$$

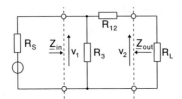

Fig. 3.49. Resistance matching for $R_L > R_S$

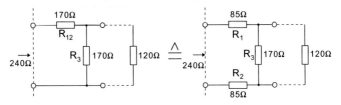

Fig. 3.50. Resistive circuit to match a source with an internal resistance of 240 Ω to a load of 120 Ω

From the conditions for impedance matching where $Z_{in} = R_S$ and $Z_{out} = R_L$, the resistances are given by

$$R_{12} = R_L \cdot \sqrt{1 - \frac{R_S}{R_L}}, \quad \text{and} \quad R_3 = \frac{R_S}{\sqrt{1 - \frac{R_S}{R_L}}} \quad \text{for } R_L > R_S \tag{3.62}$$

The voltage ratio

$$\frac{V_1}{V_2} = 1 + \frac{R_{12}}{R_L} = 1 + \sqrt{1 - \frac{R_S}{R_L}} \tag{3.63}$$

is always greater than 1, so the network attenuates the signal. Without changing the functioning of the circuit, the resistance R_{12} can be divided between the two symmetrically placed resistors R_1 and R_2. An example is given in Fig. 3.51.

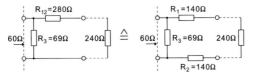

Fig. 3.51. Resistive circuit to match a source with an internal resistance of 60 Ω to a load of 240 Ω

NOTE: To reduce the losses in the resistive components, transformers are frequently used for impedance matching.

3.7.4 Voltage Divider with Defined Input and Output Resistances

Voltage dividers can be realised using **T-configurations** or **Π-configurations**, which load the source with an input impedance that is equal to the load resistance R_L, while showing the load a source with a given internal resistance (Fig. 3.52).

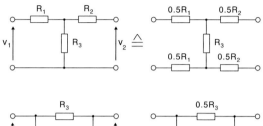

Fig. 3.52. T-configurations and Π-configurations and their symmetrical variations

For a predetermined voltage ratio of $a = \dfrac{V_1}{V_2} > 1$ (attenuation), the values of the **T-configuration** are given by

$$R_1 = R_2 = R_L \cdot \frac{a-1}{a+1}, \quad \text{for } a > 1 \quad \text{and } R_S = R_L$$

$$R_3 = R_L \cdot \frac{2a}{a^2 - 1}$$

For a **Π-configuration**

$$R_1 = R_2 = R_L \cdot \frac{a+1}{a-1}, \quad \text{for } a > 1 \quad \text{and } R_S = R_L$$

$$R_3 = R_L \cdot \frac{a^2 - 1}{2a}$$

NOTE: Both circuit types are dual to each other or form a dual pair with the duality constant $R_S R_L$.

EXAMPLE: A T-configuration and a Π-configuration are required that divide the terminal voltage by a factor of $a = 5$ for a source with an internal resistance of 600 Ω and a load of 600 Ω (Fig. 3.53).

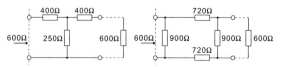

Fig. 3.53. T- and Π-configurations as a 1 : 5 voltage divider for a 600 Ω source and a 600 Ω load

3.7.5 Phase-Shifting Circuits

The phase shift between two sinusoidal waveforms is

$$\tan \varphi = \frac{\text{imaginary part of AC quantity}}{\text{real part of AC quantity}}$$

The conditions for a phase shift of 45°, 90° or 180° are relatively easy to formulate.

Table 3.12. Phase-shifting conditions

Phase shift	Condition
$45° = +\pi/4$	$\text{Re}(\underline{V}_2) = \text{Im}(\underline{V}_2)$
$-45° = -\pi/4$	$\text{Re}(\underline{V}_2) = -\text{Im}(\underline{V}_2)$
$90° = +\pi/2$	$\underline{V}_2 = j\underline{V}_1 \cdot k$
$-90° = -\pi/2$	$\underline{V}_2 = -j\underline{V}_1 \cdot k$
$180° = +\pi$	$\underline{V}_2 = -\underline{V}_1 \cdot k$

In Table 3.12 k is thus a positive real constant, dependent on the R, L, C components of the circuit.

3.7.5.1 RC Phase Shifter

To achieve a phase shift of 45° between the input and output voltage waveforms, an RC configuration can be used (Fig. 3.54).

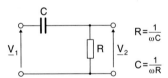

Fig. 3.54. RC phase shifter for 45°

The output voltage $\underline{V}_2$ is given by

$$\underline{V}_2 = \underline{V}_1 \cdot \frac{R}{R - j\frac{1}{\omega C}} = \underline{V}_1 \cdot \frac{\omega RC}{\omega RC - j}$$

Separating into real and imaginary components

$$\underline{V}_2 = \underline{V}_1 \cdot \frac{\omega RC(\omega RC + j)}{(\omega RC)^2 + 1} = \underline{V}_1 \cdot \frac{\omega^2 R^2 C^2 + j\omega RC}{\omega^2 R^2 C^2 + 1}$$

For a phase shift of 45° $\text{Re}(\underline{V}_2) = \text{Im}(\underline{V}_2)$ must apply.

$$\omega^2 R_{45}^2 C^2 = \omega R_{45} C \quad \Rightarrow \quad R_{45} = \frac{1}{\omega C} \tag{3.64}$$

The voltage ratio in this situation is given by

$$\frac{|\underline{V}_2|}{|\underline{V}_1|} = \left| \frac{R_{45}}{R_{45} - j\frac{1}{\omega C}} \right| = \left| \frac{1}{1-j} \right| = \frac{1}{\sqrt{2}} \approx 0.707$$

NOTE: The value $\omega = \dfrac{1}{RC}$ is known as the **corner** or **critical frequency**, and $R \cdot C$ is known as the **time constant** of the RC configuration.

A phase shift of 90° cannot be achieved with a simple RC configuration, as the resistance would have to be zero ($R = 0$). Two cascaded RC configurations can resolve this problem (Fig. 3.55). If the circuit is considered a voltage divider, then after some calculation with the capacitor impedance $\underline{Z}_C$

$$\underline{V}_2 = \underline{V}_1 \cdot \frac{R^2}{\underline{Z}_C^2 + 3R\underline{Z}_C + R^2} \tag{3.65}$$

The condition that $\underline{V}_2$ should be shifted by 90° with respect to $\underline{V}_1$ means that $\underline{V}_1 = jk\underline{V}_2$. By expanding Eq. (3.65) in j, it can be reduced to the following:

$$\underline{V}_2 = j\underline{V}_1 \frac{R^2}{j\underbrace{(\underline{Z}_C^2 + R^2)}_{\text{real}} + j2\underbrace{R\underline{Z}_C}_{\text{imaginary}}}$$

The expression in brackets in the denominator must disappear if the fraction is to be real.

$$\underline{Z}_C^2 + R^2 = 0 \quad \Rightarrow \quad \text{with } \underline{Z}_C = -j\frac{1}{\omega C} : R_{90} = \frac{1}{\omega C} \tag{3.66}$$

The voltage ratio in this situation is given by

$$\boxed{\frac{|\underline{V}_2|}{|\underline{V}_1|} = \frac{\omega RC}{2} = \frac{1}{2}} \tag{3.67}$$

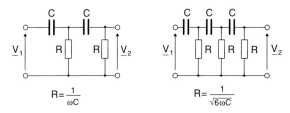

Fig. 3.55. RC phase shifter for 90° (*left*) and 180° (*right*)

A phase shift of 180° can be achieved with a three-stage RC configuration (Fig. 3.55). The fairly complicated analysis of the circuit under the condition that $\underline{V}_2 = -k\underline{V}_1$ leads to the following result:

$$R_{180} = \frac{1}{\sqrt{6}\,\omega C}$$

The voltage ratio in this situation is given by

$$\boxed{\frac{|\underline{V}_2|}{|\underline{V}_1|} = \frac{1}{29}} \tag{3.68}$$

NOTE: In order to achieve defined phase shifts, all-pass or AC bridges are employed.

3.7.5.2 Alternative Phase-Shifting Circuits

For suitable values, the circuit shown in Fig. 3.56 causes a phase shift of the current I_2 by 90° with respect to the voltage across the circuit.

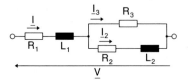

Fig. 3.56. Circuit for a phase shift of the current I_2

The total voltage across the circuit is

$$\underline{V} = \underline{V}_1 + \underline{V}_2 = (\underline{I}_2 + \underline{I}_3)(R_1 + j\omega L_1) + \underline{I}_2(R_2 + j\omega L_2)$$

The same voltage appears across the resistor R_3 as appears across $R_2 L_2$, so $\underline{I}_3$ can be equated as follows:

$$\underline{I}_3 \cdot R_3 = \underline{I}_2(R_2 + j\omega L_2) \quad \Rightarrow \quad \underline{I}_3 = \underline{I}_2 \frac{R_2 + j\omega L_2}{R_3}$$

$$\underline{V} = \underline{I}_2 \left(R_1 + \frac{R_1 R_2}{R_3} + j\frac{\omega R_1 L_2}{R_3} + j\omega L_1 + j\frac{\omega R_2 L_1}{R_3} - \frac{\omega^2 L_1 L_2}{R_3} + R_2 + j\omega L_2 \right)$$

The expression in parentheses must be purely imaginary if the current is to lag by 90°. Therefore all real components must disappear.

$$R_1 R_3 + R_1 R_2 - \omega^2 L_1 L_2 + R_2 R_3 = 0$$

For

$$R_3 = \frac{\omega^2 L_1 L_2 - R_1 R_2}{R_1 + R_2} \tag{3.69}$$

a current lag of $\pi/2$ in the branch will be achieved.

The circuit shown in Fig. 3.57 also achieves a phase shift of the current through L_2, in that the parallel resistor in Fig. 3.56 is replaced by a low-loss capacitor. A 90° phase shift of the inductor current with respect to the total voltage is achieved for

$$C = \frac{R_1 + R_2}{\omega^2 (L_1 R_2 + L_2 R_1)} \tag{3.70}$$

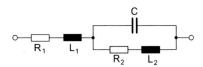

Fig. 3.57. Circuit for a 90° phase shift of the inductor current with respect to the total voltage

A curiosity of AC analysis is the circuit shown in Fig. 3.58. A sinusoidal voltage is applied to the circuit. The resistance R_2 should be chosen such that the ammeter deflection does not change while throwing the switch.

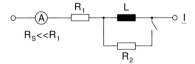

Fig. 3.58. An AC paradox

A constant deflection of the meter means that the *magnitude* of AC current is equal in both cases. The impedance of the circuit when the switch is open is given by

$$\underline{Z}_\infty = R_1 + j\omega L \quad \Rightarrow \quad Z_\infty^2 = R_1^2 + (\omega L)^2$$

When the switch is closed

$$\underline{Z}_0 = R_1 + \frac{j\omega L \cdot R_2}{j\omega L + R_2} = \frac{R_1 R_2 + j\omega L(R_1 + R_2)}{j\omega L + R_2} \quad \Rightarrow \quad Z_0^2 = \frac{R_1^2 R_2^2 + \omega^2 L^2 (R_1 + R_2)^2}{\omega^2 L^2 + R_2^2}$$

By equating both quadratic equations for the impedance, the requirement for the closed-switch circuit resistance R_2 is given by

$$R_2 = \frac{(\omega L)^2}{2R_1} \tag{3.71}$$

Closing the switch changes the phase of the current, but not its magnitude.

3.7.6 AC Bridges

3.7.6.1 Balancing Condition

Several representations of bridge circuits are shown in Fig. 3.59.

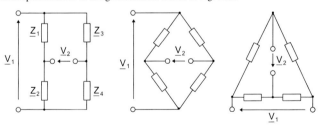

Fig. 3.59. Various representations of bridge circuits

The same current flows through the impedances $\underline{Z}_1$ and $\underline{Z}_2$ for an unloaded bridge. Both impedances function as a voltage divider. The same is true for $\underline{Z}_3$ and $\underline{Z}_4$. The output voltage $\underline{V}_2$ is the difference between the voltages at the terminals of the voltage divider. It follows that

$$\underline{V}_2 = 0 \quad \Longleftrightarrow \quad \boxed{\frac{\underline{Z}_1}{\underline{Z}_2} = \frac{\underline{Z}_3}{\underline{Z}_4}} \tag{3.72}$$

Under these conditions, this is known as a **balanced bridge** (Fig. 3.59).

The impedance ratio in Eq. (3.72) is complex. Therefore both of the following conditions must hold

$$\frac{Z_1}{Z_2} = \frac{Z_3}{Z_4}, \quad \text{and} \quad \varphi_1 - \varphi_2 = \varphi_3 - \varphi_4$$

- A balanced AC bridge must fulfil two conditions, one for the magnitudes and the other for the phases of the complex impedances.

NOTE: The balancing condition for AC bridges only holds for a single frequency. Measurement bridges are therefore used only with sinusoidal waveforms. Frequency independence can be achieved only with special bridge circuits.

3.7.6.2 Application: Measurement Technique

The balanced condition from Eq. (3.72) can be used to determine the value of an unknown impedance on one of the bridge branches.

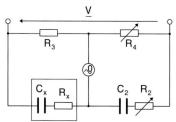

Fig. 3.60. Wien bridge to measure a capacitance C_x and its loss resistance R_x

Figure 3.60 shows a **Wien bridge** for measuring the equivalent circuit of a capacitor. A null instrument is placed in the diagonal branch (earphones could also be used), so that the minimum condition can be found. Initially the imaginary part of the unknown impedance is found by varying R_4 to find a minimum in the detector. Then

$$\frac{X_{C_x}}{X_{C_2}} = \frac{\frac{1}{\omega C_x}}{\frac{1}{\omega C_2}} = \frac{C_2}{C_x} = \frac{R_3}{R_4} \quad \text{1. Balance}$$

Then the real part of the impedance can be balanced using R_2.

$$\frac{R_x}{R_2} = \frac{R_3}{R_4} \quad \text{2. Balance}$$

The magnitude on the right side of the equation does not change during this adjustment. The equivalent circuit values of the capacitor are thus

$$C_x = C_2 \cdot \frac{R_4}{R_3}, \qquad R_x = R_2 \cdot \frac{R_3}{R_4}$$

3.8 Power in AC Circuits

3.8.1 Instantaneous Power

The **instantaneous power** of an AC waveform is given by

$$p(t) = v(t) \cdot i(t)$$

3.8.1.1 Power in a Resistance

The current and voltage are in phase on the resistor (Fig. 3.61). For a sinusoidal voltage instantaneous power is

$$p(t) = \hat{v} \sin \omega t \cdot \hat{i} \sin \omega t = \hat{v}\,\hat{i}\, \sin^2 \omega t = VI\,(1 - \cos 2\omega t)$$

Here V and I are the RMS values of the voltage and current. The instantaneous power is a periodic value and is always positive in a resistor. The power consumption oscillates at twice the frequency of the voltage.

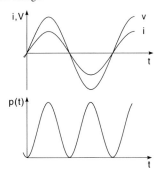

Fig. 3.61. Current, voltage and instantaneous power for a resistance

3.8.1.2 Power in a Reactive Element

The voltage on a capacitor leads the current by 90° ($\varphi = -\pi/2$), where φ is the phase angle of the voltage, relative to the current. The product $p(t) = v(t)i(t)$ has thus both positive and negative values.

The positive and negative parts of the power curve are equal in magnitude (Fig. 3.62, left). The capacitor temporarily stores energy and gives it back to the source in the following quarter period. A mixed result exists in the case of a capacitive–resistive load (Fig. 3.62, right). One part of the power is consumed in the resistive component of the load, while the other part flows back to the source. The power in an inductive and an inductive–resistive load produces an analogous result.

For sinusoidal current the instantaneous power can be written as

$$p(t) = \underbrace{VI \cdot \cos \varphi}_{\text{constant}} - \underbrace{VI \cdot \cos(2\omega t - \varphi)}_{\text{varying}}$$

Where φ is the phase difference of the voltage with respect to the current.

- The instantaneous power has a constant component and a component which varies at twice the frequency of the current or voltage waveforms.

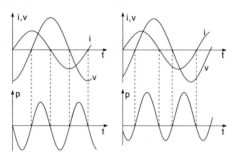

Fig. 3.62. Current, voltage and instantaneous power in a capacitor and in a resistive–capacitive load

Alternatively, the instantaneous power can be represented by

$$p(t) = \underbrace{VI\cos\varphi[1 - \cos 2\omega t]}_{\text{resistive component}} - \underbrace{VI\sin\varphi \sin 2\omega t}_{\text{reactive component}} \tag{3.73}$$

The first term, known as the **resistive component**, is always positive. The second term alternates between positive and negative values and is known as the **reactive component** (Fig. 3.63).

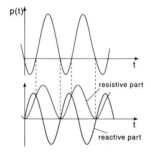

Fig. 3.63. Separation of the instantaneous power into a resistive and a reactive component

3.8.2 Average Power

The **average power** is defined as

$$P = \bar{p} = \frac{1}{T} \int\limits_{t}^{t+T} p(t) \, dt \tag{3.74}$$

When reference is made to *power* in circuit theory, usually the average power is meant.

3.8.2.1 Real Power

The **real power** for sinusoidal current and voltage waveforms is given by

$$P = V \cdot I \cdot \cos \varphi \tag{3.75}$$

where V and I are the RMS values of the voltage and current. The expression $\cos \varphi$ is known as the **power factor**. The unit of real power is the **watt** (W).

- For purely resistive circuits ($\varphi = 0$) the power factor $\cos \varphi = 1$, and the real power is $P = VI$.

- For purely reactive circuits ($\varphi = \pm 90°$) the power factor $\cos \varphi = 0$, and the power is thus zero.

- For resistor–capacitor and resistor–inductor loads ($-90° < \varphi < 90°$) the real power is positive.

- The real power can be converted into other forms of power (heat, mechanical power, etc.).

When a complex circuit is represented by its parallel equivalent circuit of a resistance and a reactance the power factor $\cos \varphi$ can be modelled in relation to the current. This is known as **in-phase** or **real current** (Fig. 3.64).

$$I_{\text{real}} = I \cdot \cos \varphi \tag{3.76}$$

$$P = I_{\text{real}} \cdot V, \qquad P = \frac{V^2}{R_p} \tag{3.77}$$

- The real power is given by the product of the in-phase current and the RMS voltage. This approach can only be used on parallel combinations, where all of the components experience the same voltage across their terminals.

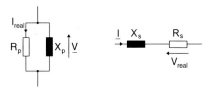

Fig. 3.64. In-phase currents and voltages for the equivalent circuit of a complex circuit

When a complex circuit is represented by its series equivalent circuit of a resistance and a reactance the power factor $\cos \varphi$ can be modelled in relation to the voltage. This is known as **in-phase** or **real voltage**.

$$V_{\text{real}} = V \cdot \cos \varphi \tag{3.78}$$

$$P = V_{\text{real}} \cdot I, \qquad P = I^2 \cdot R_r \tag{3.79}$$

- The real power is given by the product of the in-phase voltage and the RMS current. This approach can only be used for series combinations, where the same current flows through all of the components.

NOTE: The real power is *not* the product of the in-phase current and the in-phase voltage. These quantities are derived from and applied to different equivalent circuits.

3.8.2.2 Reactive Power

The **reactive power** is defined as

$$Q = V \cdot I \cdot \sin \varphi \tag{3.80}$$

V and I are the RMS values of the voltage and current, and φ is the phase difference of the voltage with respect to the current. The factor $\sin \varphi$ is known as the **leading** or **lagging power factor**.[†] The unit for reactive power is the **volt–ampere reactive** (VAR).

- For purely resistive impedances ($\varphi = 0$) the reactive power is zero.
- The reactive power in an inductive–resistive load is positive, while in a capacitive–resistive load it is negative.
- Reactive power *cannot* be converted into other forms of power.

When a complex circuit is represented by its parallel equivalent circuit of a resistance and a reactance the power factor $\sin \varphi$ can be modelled in relation to the current. This is known as **reactive** or **out-of-phase current** (Fig. 3.65).

$$I_{\text{react}} = -I \cdot \sin \varphi_Y \tag{3.81}$$

$$Q = -I_{\text{react}} \cdot V, \qquad Q = \frac{V^2}{X_p} \tag{3.82}$$

- The reactive power is given by the product of the negative reactive current and the RMS voltage. This approach can only be used on parallel combinations, where all of the components experience the same voltage across their terminals.

NOTE: The negative sign on the reactive current comes from the fact that in the parallel equivalent circuit the phase is expressed with respect to the voltage ($\varphi_Y = -\varphi_Z$).

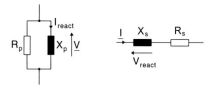

Fig. 3.65. Out-of-phase currents and voltages for the equivalent circuit of a complex circuit element

When a complex circuit is represented by its series equivalent circuit of a resistance and a reactance the power factor $\sin \varphi$ can be modelled in relation to the current. This is known as **reactive** or **out-of-phase voltage**.

$$V_{\text{react}} = V \cdot \sin \varphi \tag{3.83}$$

$$Q = V_{\text{react}} \cdot I, \qquad Q = I^2 \cdot X_r \tag{3.84}$$

- The reactive power is given by the product of the out-of-phase voltage and the RMS current. This approach can only be used for series combinations, where the same current flows through all of the components.

NOTE: The reactive power is *not* the product of the out-of-phase current and the out-of-phase voltage. These quantities are derived from different equivalent circuits.

[†] For inductive loads this is the *lagging power factor*, whereas for capacitive loads this the *leading power factor*. Lagging and leading describe the lagging or leading of the current with respect to the voltage.

3.8.2.3 Apparent Power

The **apparent power** is defined as

$$S = V \cdot I \tag{3.85}$$

V and I are the RMS values of the voltage and the current, and φ is the phase difference of the voltage with respect to the current. The unit of apparent power is the **volt–ampere** (VA). Thus

$$P = S \cdot \cos\varphi, \qquad Q = S \cdot \sin\varphi \tag{3.86}$$

This relation can be best represented geometrically (Fig. 3.66).

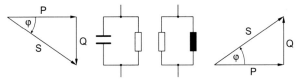

Fig. 3.66. Phasor triangle of the resistive, reactive and apparent power for a resistor–capacitor load and a resistor–inductor load

It can be seen from the phasor diagram that

$$S = \sqrt{P^2 + Q^2} \tag{3.87}$$

- Apparent powers from elements with **different power factors** cannot be added. On the contrary, resistive and reactive powers must be added independently. This yields the overall apparent power.

3.8.3 Complex Power

Complex power is defined as

$$\underline{S} = \underline{V} \cdot \underline{I}^* \tag{3.88}$$

- The complex power is the product of the voltage and complex conjugate of the current.

$$\underline{S} = V e^{j\varphi_V} \cdot I e^{-j\varphi_I} = VI e^{j(\varphi_V - \varphi_I)} = VI e^{j\varphi}$$

where φ represents the phase shift of the voltage with respect to the current. It follows that

$$\underline{S} = \underbrace{VI \cos\varphi}_{P} + j \underbrace{VI \sin\varphi}_{Q}$$

and thus

$$\underline{S} = P + jQ, \qquad S = |\underline{S}| = \sqrt{P^2 + Q^2} \tag{3.89}$$

- The **real power** is the **real part** of the complex power.
- The **reactive power** is the **imaginary part** of the complex power.
- The **apparent power** is the **magnitude** of the complex power.

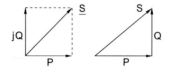

Fig. 3.67. The complex power in a phasor diagram

As for other complex quantities, the complex power can be represented by a phasor diagram (Fig. 3.67).

3.8.4 Overview: AC Power

Table 3.13. Summary for AC power

Load	$P = S \cos \varphi$	$Q = S \sin \varphi$	S	$\cos \varphi$
Purely inductive	0	Positive	Q	0
Resistor–inductor	Positive	Positive	$\sqrt{P^2 + Q^2}$	0 to 1
Purely resistive	Positive	0	P	1
Resistor–capacitor	Positive	Negative	$\sqrt{P^2 + Q^2}$	0 to 1
Purely capacitive	0	Negative	Q	0

$$S = \sqrt{P^2 + Q^2} \tag{3.90}$$

$$P = S \cdot \cos \varphi \tag{3.91}$$

$$Q = S \cdot \sin \varphi \tag{3.92}$$

$$Q = P \cdot \tan \varphi \tag{3.93}$$

$$P = Q \cdot \cot \varphi \tag{3.94}$$

$$\tan \varphi = \frac{Q}{P} \tag{3.95}$$

3.8.5 Reactive Current Compensation

The **power factor** specifies the fractional contribution of real power P to the apparent power S.

Although the reactive current does not contribute to transferable power, it must nonetheless be transported from the7 supply to the load. In order to get useable power from the power

3.8 Power in AC Circuits

Table 3.14. Summary for equivalent circuits

	Parallel equivalent circuit	Series equivalent circuit
Configuration	(circuit with I_{react}, B, I_{real}, G, V)	(circuit with I, X, R, V_{react}, V_{real})
Complex impedance Complex admittance	$\underline{Y} = G + jB$	$\underline{Z} = R + jX$
Impedance magnitude Admittance magnitude	$Y = \sqrt{G^2 + B^2}$	$Z = \sqrt{R^2 + X^2}$
Real impedance Real admittance	$G = Y \cos \varphi_Y = I_{real}/V$	$R = Z \cos \varphi = V_{real}/I$
Reactance Susceptance	$B = -Y \sin \varphi_Y = -I_{react}/V$	$X = Z \sin \varphi = V_{react}/I$
Complex power	$\underline{S} = \underline{Y}^* V^2 = (G - jB)V^2$	$\underline{S} = \underline{Z}I^2 = (R + jX)I^2$
Real power	$P = I_{real} V = I_{real}^2/G = V^2 G$	$P = V_{real} I = V_{real}^2/R = I^2 R$
Reactive power	$Q = -I_{react} V = -I_{react}^2/B = -V^2 B$	$Q = V_{react} I = V_{react}^2/X = I^2 X$
Apparent power	$S = VI = V\sqrt{I_{real}^2 + I_{react}^2}$	$S = VI = I\sqrt{V_{real}^2 + V_{react}^2}$
Power factor	$\cos \varphi = G/Y$	$\cos \varphi = R/Z$
In-phase current In-phase voltage	$I_{real} = I \cos \varphi_Y = GV$	$V_{real} = V \cos \varphi = RI$
Reactive current Reactive voltage	$I_{react} = -I \sin \varphi_Y = -BV$	$V_{react} = V \sin \varphi = XI$

source efforts must be made to minimise the reactive current. These actions are known as **power factor correction** (Fig. 3.68).

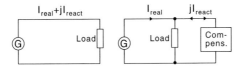

Fig. 3.68. Principle of power factor correction

A reactance is placed in parallel with the load and absorbs all of the reactive current. A capacitor is used for the resistive–inductive loads most often encountered. The reactive current of the capacitor must be equal in magnitude to that of the load. The effect is that the reactive current is diverted from the load to the compensation element, and thus the supply is no longer loaded (Fig. 3.69). If the reactive current of the compensation element exceeds that of the load, then this is known as **overcompensation**. In practice, compensation is designed for a power factor of about $\cos \varphi = 0.9$.

EXAMPLE: A motor with 230 V/16 A/$\cos \varphi = 0.8$ should be compensated by a capacitor.

A real current of $16 \text{ A} \cdot \cos \varphi = 12.8$ A flows through the motor. The reactive current is $I_{react} = \sqrt{I^2 - I_{real}^2} = \sqrt{16^2 - 12.8^2} = 9.6$ A. The reactive current

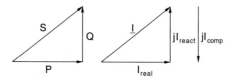

Fig. 3.69. Power and current phasor diagram for reactive current compensation

must be absorbed by the capacitor. Its reactance is $X_C = 230\text{ V}/9.6\text{ A} = 24\text{ }\Omega$. It follows therefore for a frequency of 50 Hz a capacitor of $C = 1/X_C \cdot \omega = (24\text{ }\Omega \cdot 2\pi \cdot 50\text{ s}^{-1})^{-1} = 133\text{ }\mu\text{F}$. For compensation of a power factor of $\cos\varphi = 0.9$ a reactive current of only 6.2 A must be compensated, for which a capacitor of 86 µF suffices.

NOTE: The power factors of transformers and motors decline when they are unloaded. The reactive currents are caused by the buildup and reduction of the magnetic fields.

3.9 Three-Phase Supplies

3.9.1 Polyphase Systems

Figure 3.70 shows a circular arrangement of several coils, in whose centre a permanent magnet is rotating at a constant angular velocity. An alternating voltage is induced in each of the coils with the same frequency and a constant phase shift with respect to each other.

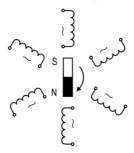

Fig. 3.70. Basic arrangement to produce constant-frequency alternating voltage in a polyphase system

Such arrangements of alternating voltage generators, conductors and loads are known as **polyphase systems**. For n voltages there is an n-**phase system**. If the voltages have the same magnitude and frequency and a constant phase shift between them, then the polyphase system is said to be **balanced**.

The presence of such balanced voltages applies for generators with shifted windings (the rotating magnet is replaced by a DC excited rotor). Such polyphase voltage systems are produced by generators with shifted windings, where the voltage is induced by a rotating magnetic field (e.g. synchronous generators). More recent systems (e.g. uninterruptable power supplies, UPS) produce a three-phase system by using static inverters, using switching semiconductors. On the other hand, a rotating field can be created by applying

3.9 Three-Phase Supplies

a polyphase voltage to symmetrically positioned coils, as shown in Fig. 3.70 (this is used in asynchronous and synchronous motors).

The **three-phase system** is of particular importance in the distribution of electricity. The advantages of the three-phase systems are

- fewer power lines compared to three single phase lines (three, four or five power lines instead of six);
- constant power delivery from the generator for symmetric loads (see Sect. 3.10.1);
- several voltage options available to the user;
- simple motor construction.

3.9.2 Three-Phase Systems

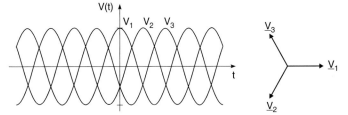

Fig. 3.71. Time variation of the voltages in a symmetrical three-phase system and their RMS values in a phasor diagram

In a three-phase circuit, only three or four wires must be brought to the user rather than six. Figure 3.72 shows the representation of the voltage sources and their transmission lines.

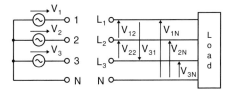

Fig. 3.72. Representation of the voltage sources and their transmission lines in a three-phase circuit

The terminal voltages V_{12} etc. between the **line conductors** L_1, L_2 and L_3 are known as **line-to-line voltages**. The voltages with respect to the **neutral conductor** N are known as the **phase voltages**. The generator voltages V_1, V_2, V_3 of the phase windings are also **phase voltages**. **Line currents** are the currents in the lines, and **phase currents** are the currents flowing in the generator windings.

NOTE: The most commonly used three-phase system in Europe employs voltages of 230 V/400 V.

In symmetrical three-phase circuits the instantaneous values of the phase voltages may be represented as

$$v_1(t) = \hat{v} \cdot \cos \omega t,$$
$$v_2(t) = \hat{v} \cdot \cos\left(\omega t - \frac{2\pi}{3}\right),$$
$$v_3(t) = \hat{v} \cdot \cos\left(\omega t + \frac{2\pi}{3}\right) \tag{3.96}$$

They are shifted 120° ($2\pi/3$) with respect to each other (Fig. 3.73). The complex RMS values of the voltages are given by

$$\underline{V}_1 = \frac{\hat{v}}{\sqrt{2}},$$
$$\underline{V}_2 = \frac{\hat{v}}{\sqrt{2}} \cdot e^{-j2\pi/3},$$
$$\underline{V}_3 = \frac{\hat{v}}{\sqrt{2}} \cdot e^{+j2\pi/3} \tag{3.97}$$

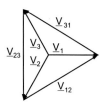

Fig. 3.73. Phasor diagram of the phase voltages and line voltages for the arrangement in Fig. 3.72

3.9.2.1 Properties of the Complex Operator $\underline{a}$

For convenience in this chapter the complex operator $e^{j2\pi/3}$ is more compactly represented as $\underline{a}$ (Fig. 3.74). When multiplied by a phasor it causes a rotation of $2\pi/3$ (120°) in the

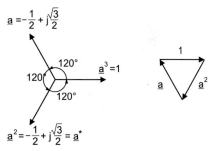

Fig. 3.74. The complex operators $\underline{a}$, $\underline{a}^2$ and $\underline{a}^3$

complex plane. Some properties of $\underline{a}$ are as follows:

$$\underline{a} = e^{j2\pi/3} = \frac{1}{2}\left(-1 + j\sqrt{3}\right) = -\frac{1}{2} + j\frac{\sqrt{3}}{2}$$
$$\underline{a}^2 = e^{j4\pi/3} = e^{-j2\pi/3} = \frac{1}{2} \cdot \left(-1 - j\sqrt{3}\right) = \underline{a}^* \qquad (3.98)$$
$$\underline{a}^3 = e^{j2\pi} = 1$$

By use of the complex operator, the complex three-phase phasor can be written as

$$\underline{V}_1 = \underline{V}_1, \qquad \underline{V}_2 = \underline{V}_1 \cdot \underline{a}^2, \qquad \underline{V}_3 = \underline{V}_1 \cdot \underline{a} \qquad (3.99)$$

- A single application of the complex operator $\underline{a}$ rotates by $2\pi/3$ (120°), two applications rotate by $4\pi/3$ (240°) and three applications rotate by 2π (360°).

$$1 + \underline{a} + \underline{a}^2 = 0 \qquad (3.100)$$

More rules applying to $\underline{a}$ can be derived from Fig. 3.75.

$$1 - \underline{a}^2 = \frac{3}{2} + j\frac{\sqrt{3}}{2} = -j\sqrt{3} \cdot \underline{a} \qquad (3.101)$$
$$\underline{a}^2 - \underline{a} = -j\sqrt{3}$$
$$\underline{a} - 1 = -\frac{3}{2} + j\frac{\sqrt{3}}{2} = -j\sqrt{3} \cdot \underline{a}^2$$

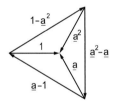

Fig. 3.75. Sums and differences of the complex operators

3.9.3 Delta-Connected Generators

In the **delta-connected generator** the three-phase windings are connected one after the other in a daisy chain, and the series combination circuit is shorted (Fig. 3.76). The terminals

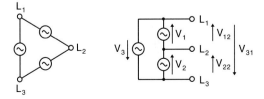

Fig. 3.76. Two representations of delta-connected generators

of the phase windings are mounted on a terminal board with standard markings. The phase windings and the terminal board of a three-phase generator in a delta configuration are shown in Fig. 3.77.

$$\underline{V}_{12} = \underline{V}_1, \qquad \underline{V}_{23} = \underline{V}_2, \qquad \underline{V}_{31} = \underline{V}_3$$

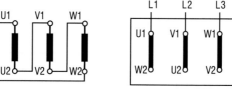

Fig. 3.77. Phase windings and the terminal board configuration in a delta-connected generator

- For **delta-connected generators** the line voltages are equal to the **phase voltages** (generator terminal voltages).

The sum of the phase voltages is given by (Fig. 3.78)

$$\underline{V}_s = \underline{V}_1 + \underline{V}_2 + \underline{V}_3 = \underline{V}_1 \cdot \underbrace{(1 + \underline{a}^2 + \underline{a})}_{=0} = 0$$

By inserting Eq. (3.100) the sum of the complex operators in the parentheses disappears.

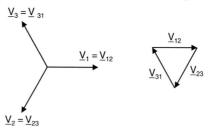

Fig. 3.78. Phasor diagram of the complex RMS voltages in the symmetrical delta-connected generators

- For ideal symmetrical generator voltages the sum of the voltages is zero. Therefore, for a short circuit (delta circuit) of the series combination of the phase windings no current will flow.

3.9.4 Star-Connected Generators

In the **star-connected generator** one side of each of the phase windings is connected to the **generator star point** (Fig. 3.79 and 3.80).

The following holds for the **line voltages** $\underline{V}_{12}$, etc. (Fig. 3.81)

$$\begin{aligned}
\underline{V}_{12} &= \underline{V}_1 - \underline{V}_2 = \underline{V}_1 \cdot (1 - \underline{a}^2) = -j\sqrt{3}\underline{a}\,\underline{V}_1 = -j\sqrt{3}\underline{V}_3 \\
\underline{V}_{23} &= \underline{V}_2 - \underline{V}_3 = \underline{V}_1 \cdot (\underline{a}^2 - \underline{a}) = -j\sqrt{3}\underline{V}_1 \\
\underline{V}_{31} &= \underline{V}_3 - \underline{V}_1 = \underline{V}_1 \cdot (\underline{a} - 1) = -j\sqrt{3}\underline{a}^2\underline{V}_1 = -j\sqrt{3}\underline{V}_2
\end{aligned} \qquad (3.102)$$

Equations (3.98), (3.99) and (3.100) are instrumental in solving the above equations.

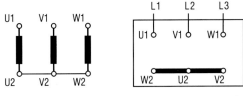

Fig. 3.79. Phase windings and terminal board configuration for a star-connected generator

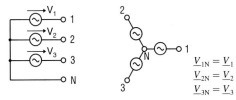

Fig. 3.80. Two representations of star-connected generators

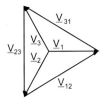

Fig. 3.81. Phasor diagram of line and phase voltages for a symmetrical star-connected generator

- The **line voltages** exceed the phase voltages by a factor of $\sqrt{3}$ in the symmetrical **star-connected generator**.

$$V_{1N} = V_{2N} = V_{3N}$$
$$V_{12} = V_{21} = V_{31} = \sqrt{3} \cdot V_{1N} = \sqrt{3} \cdot V_{2N} = \sqrt{3} \cdot V_{3N}$$

- The line voltages are, like the phase voltages, **phase-shifted** by $2\pi/3$ (120°) **with respect to each other**.

$$\underline{V}_{23} = \underline{a}^2 \cdot \underline{V}_{12}, \qquad \underline{V}_{31} = \underline{a} \cdot \underline{V}_{12}$$

- The line voltages are **phase-shifted** by $\pi/2$ (90°) with respect to the **opposite** phase voltages.

$$\underline{V}_{12} = -j\sqrt{3}\underline{V}_3, \qquad \underline{V}_{23} = -j\sqrt{3}\underline{V}_1, \qquad \underline{V}_{31} = -j\sqrt{3}\underline{V}_2$$

NOTE: This property is used in the measurement of the reactive power in a three-phase system. A 90° phase-shifted voltage can thus be measured without employing a phase-shifting circuit.

3.10 Overview: Symmetrical Three-Phase Systems

Three-phase systems with symmetrical generators and loads are shown in Fig. 3.82 and summarised in Table 3.15. There are four different combinations of star and delta circuits. In all cases the line voltage is V, and all load impedances have the same value $\underline{Z}$.

Table 3.15. Symmetrical three-phase systems, see Fig. 3.82

generator-load combination	star–star	star–delta	delta–star	delta–delta
phase voltages	$\dfrac{V}{\sqrt{3}}$	$\dfrac{V}{\sqrt{3}}$	V	V
voltage across the load $\underline{Z}$	$V_{1N},\ V_{2N},\ V_{3N}$ $\dfrac{V}{\sqrt{3}}$	$V_{12},\ V_{23},\ V_{31}$ V	$V_{1N},\ V_{2N},\ V_{3N}$ $\dfrac{V}{\sqrt{3}}$	$V_{12},\ V_{23},\ V_{31}$ V
currents through the load $\underline{Z}$	$I_{1N},\ I_{2N},\ I_{3N}$ $\dfrac{1}{\sqrt{3}}\cdot\dfrac{V}{Z}$	$I_{12},\ I_{23},\ I_{31}$ $\dfrac{V}{Z}$	$I_{1N},\ I_{2N},\ I_{3N}$ $\dfrac{1}{\sqrt{3}}\cdot\dfrac{V}{Z}$	$I_{12},\ I_{23},\ I_{31}$ $\dfrac{V}{Z}$
line currents	$I_1,\ I_2,\ I_3$ $\dfrac{1}{\sqrt{3}}\cdot\dfrac{V}{Z}$	$I_1,\ I_2,\ I_3$ $\sqrt{3}\cdot\dfrac{V}{Z}$	$I_1,\ I_2,\ I_3$ $\dfrac{1}{\sqrt{3}}\cdot\dfrac{V}{Z}$	$I_1,\ I_2,\ I_3$ $\sqrt{3}\cdot\dfrac{V}{Z}$
total real power	$\dfrac{V^2}{Z}\cdot\cos\varphi$	$3\cdot\dfrac{V^2}{Z}\cdot\cos\varphi$	$\dfrac{V^2}{Z}\cdot\cos\varphi$	$3\cdot\dfrac{V^2}{Z}\cdot\cos\varphi$
line voltages $V_{12}=V_{23}=V_{31}=V$			

- Three times more power is transferred to the resistive load in the delta circuit than in the star circuit.

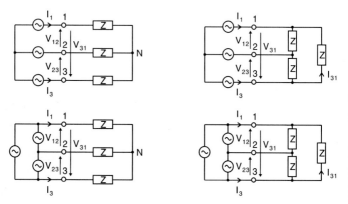

Fig. 3.82. Symmetrical three-phase systems

NOTE: This property is used in three-phase motors for the so-called **star–delta start**. The motor is started in the star configuration and then switched over to the delta configuration. In this manner, unnecessarily high transient currents are avoided.

- When using the line currents I_1, I_2, I_3 and voltages V_{12}, V_{23}, V_{31}, the configuration of the generator circuit is irrelevant for the power delivered.

3.10.1 Power in a Three-Phase System

See Sect. 4.4.3.1 on power measurement in three-phase systems.

The average real power delivered by a symmetrical three-phase system is

$$P = V \cdot I \cdot \sqrt{3} \tag{3.103}$$

The **instantaneous** real power is

$$p(t) = \frac{v_1^2(t)}{R_1} + \frac{v_2^2(t)}{R_2} + \frac{v_3^2(t)}{R_3}$$

where R is the real component of the load impedance. For a symmetrical load $R_1 = R_2 = R_3 = R$, the instantaneous power is given by

$$\begin{aligned} p(t) &= \frac{\hat{V}^2}{R} \cdot \left[\cos^2 \omega t + \cos^2 \left(\omega t - \frac{2}{3}\pi \right) + \cos^2 \left(\omega t + \frac{2}{3}\pi \right) \right] \\ &= \frac{\hat{V}^2}{2R} \cdot \left[1 + \cos 2\omega t + 1 + \cos \left(2\omega t - \frac{4}{3}\pi \right) + 1 + \cos \left(2\omega t + \frac{4}{3}\pi \right) \right] \\ &= \frac{3\hat{V}^2}{2R} \end{aligned} \tag{3.104}$$

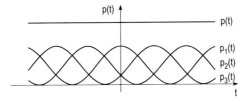

Fig. 3.83. Instantaneous power delivered by each individual winding of the three-phase system $p_i(t)$ and the total power $p(t)$

- The **total real power** delivered by the generator is **constant**, although the power varies in each individual winding (Fig. 3.83).

This property has great advantages in the construction of electrical machines, because this also means that the mechanical torque is constant over a rotation, thus considerably reducing vibration.

Polyphase systems with constant power delivery are said to be **balanced**; otherwise they are said to be unbalanced.

NOTE: The property of constant power delivery can also be achieved in n-phase systems.

3.11 Notation Index

a	voltage ratio
$\underline{a}$	complex operator $e^{j2\pi/3}$
B	susceptance (S)
B	bandwidth (Hz)
C	capacitor (F)
f	frequency (Hz)
f_r	resonant frequency
G	conductance
G_S	source conductance
i	time-varying current
$\hat{\imath}$	peak value of the current
I	RMS value of the current
I_c	compensation current
I_{react}	reactive current
I_{real}	real current
Im()	imaginary part
k_f	form factor
k_c	crest factor
L_1, L_2, L_3	line
N	neutral conductor
p	subscript: parallel combination
p	instantaneous power (W)
P	average power (W)
Q	reactive power (VAR)
r	subscript: resonant
r	magnitude of complex number in polar coordinates
R	resistor, resistance
R_D^2	duality constant (Ω^2)
R_L	load resistance
R_s	series resistor (in Sect. 3.6.1)
R_S	source resistance
R_{45}, R_{90}	resistance for phase shift of 45° or 90°
$R||C$	R in parallel to C
Re()	real part
s	subscript: series combination
S	apparent power (VA)
$\underline{S}$	complex power
T	period, periodic time
v	time-varying voltage
$\underline{v}$	complex time-varying voltage
$\hat{v}$	peak value of the voltage
$\underline{\hat{v}}$	complex amplitude
$\bar{v}$	average value
$\overline{|v|}$	average rectified value

V	RMS value of the voltage
$\underline{V}$	complex RMS value of the voltage
V_1	input voltage
V_2	output voltage
V_{12}, V_{23}, V_{31}	line voltages
V_{1N}, V_{2N}, V_{3N}	star voltages
V_C	voltage across a capacitor
V_L	voltage across an inductor
V_R	voltage across a resistor
V_{react}	reactive voltage
V_{real}	real voltage
V_S	source voltage
X	reactance
X_C	capacitive reactance
X_L	inductive reactance
Y	admittance
$\underline{Y}$	complex conductance, admittance
z^*	complex conjugate
Z	impedance
$\underline{Z}$	complex resistance, impedance
Z_{in}	input impedance
Z_{out}	output impedance
φ	phase difference (rad)
φ_0	phase shift
φ_I	phase of the current
φ_S	phase of the sum signal
φ_V	phase of the voltage
φ_Y	phase difference of the admittance
φ_Z	phase difference of the impedance
ω	angular frequency (s^{-1})
ω_r	resonant frequency (s^{-1})

3.12 Further Reading

BIRD, J. O.: *Electrical Circuit Theory and Technology*
Butterworth/Heinemann (1999)

BOYLESTAD, R. L.: *Introductory Circuit Analysis, 9th Edition*
Prentice Hall (1999)

CHAPRA, S. C.; CANALE, R. P.: *Numerical Methods for Engineers, 3rd Edition*
McGraw-Hill (1998)

DE WOLF, D. A.: *Essentials of Electromagnetics for Engineering*
Cambridge University Press (2000)

DORF, R. C.: *The Electrical Engineering Handbook, Section I*
CRC press (1993)

FLOYD, T. L.: *Electric Circuits Fundamentals, 5th Edition*
Prentice Hall (2001)

FLOYD, T. L.: *Electronics Fundamentals: Circuits, Devices, and Applications, 5th Edition*
Prentice Hall (2000)

FLOYD, T. L.: *Electronic Devices, 5th Edition*
Prentice Hall (1998)

GROB, B.: *Basic Electronics, 8th Edition*
McGraw-Hill (1996)

HUGHES, E.: *Electrical Technology, 7th Edition*
Longman (1995)

JONES, G. R.; LAUGHTON, M. A.; SAY, M. G.: *Electrical Engineers Reference Book, 14th Edition*
Butterworth (1993)

KOVETZ, A.: *Electromagnetic Theory, 1st Edition*
Oxford University Press (2000)

MUNCASTER, R.: *A-Level Physics*
Stanley Thornes Ltd. (1997)

NELKON, M.; PARKER, P.: *Advanced Level Physics*
Heinemann (1995)

O'NEIL, P. V.: *Advanced Engineering Mathematics, 4th Edition*
Brooks/Cole Publishing Company (1997)

4 Current, Voltage and Power Measurement

This chapter focuses on the most basic measurement methods for electrical quantities using electrical measuring instruments.

4.1 Electrical Measuring Instruments

Electrical measuring instruments measure an electrical quantity by deflecting a pointer using magnetic or mechanical principles.

4.1.1 Moving-Coil Instrument

In a **moving-coil instrument** a coil turns in the field of a permanent magnet. The current flowing through the coil creates a torque, which is compensated by a reset spring. The rotation of the coil is displayed by a pointer. See also Sect. 2.3.17.1 on the force on a current-carrying conductor in a magnetic field.

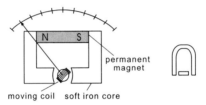

Fig. 4.1. Principle of the moving-coil instrument and its circuit symbol

- The scale of a moving-coil instrument is linear for DC.

- The moving-coil instrument displays the **arithmetic average value** of the current. For purely AC current the pointer stays at zero.

- Moving-coil instruments with a rectifier display the **rectified value**.

- The moving-coil instrument is the most sensitive analogue instrument.

NOTE: **Galvanometers** are particularly sensitive moving-coil instruments.

4.1.2 Ratiometer Moving-Coil Instrument

The **ratiometer moving-coil instrument** works on the moving-coil principle, using two crossed coils mounted on the same iron core at 30° to 60° with respect to each other. The coils are configured such that the currents flowing through them exert opposing torques (Fig. 4.2). The position in which both torques are equal depends on the *ratio* of the currents in both coils. For this reason the instrument is known as a **ratio instrument**.

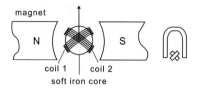

Fig. 4.2. Principle of the ratiometer-type moving-coil instrument and its circuit symbol

- The ratiometer-type moving-coil instrument displays the **quotients** of two coil currents.

- The scale is not linear, but has a wide linear range around the centre of the scale.

4.1.3 Electrodynamic Instrument

The **electrodynamic instrument** is similar in principle to the moving-coil instrument, except that the instrument's field is produced by a second current flowing in a measurement coil (Fig. 4.3). This was previously known as a **dynamometer**.

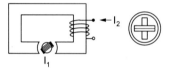

Fig. 4.3. Principle of the electrodynamic instrument and its circuit symbol for the iron-screened realisation

- The deflection of an electrodynamic instrument is proportional to the **product** of the currents in both coils.

- If both coils are excited by sinusoidal currents (of the same frequency), then the display is proportional to the product of the currents and depends on the relative phase shift. Maximum deflection occurs for $\Delta\varphi = 0°$, whereas there is no deflection for $\Delta\varphi = 90°$.

If both measurement coils are connected in series, then the same current flows through each coil.

- The electrodynamic instrument displays the root mean square (**RMS**) of the measurement current. The display is then to a large degree independent of the shape of the waveform. In this mode the scale is quadratic.

The main use for electrodynamic instrumenta is in **power measurement**. One of the coils is excited by the measurement current, while the other coil is excited by a current that is proportional to the voltage.

- The electrodynamic instrument serves as a power meter for both direct and alternating current and is to a large degree independent of the shape of the waveform.

4.1.4 Moving-Iron Instrument

The **moving-iron instrument** (**soft-iron instrument**) uses the opposing forces on equally polarised, magnetised soft-iron vanes in the magnetic field of a coil with a measurement current (Fig. 4.4). By suitably shaping the air gap the scale can cover a wide range.

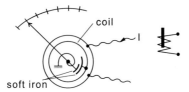

Fig. 4.4. Principle of the moving-iron instrument and its circuit symbol

NOTE: Graduations are often extended in the upper region for accurate reading (operational instruments), or compressed in order to be able to quantitatively determine overloads.

- The deflection of the moving-iron instrument is independent of the current direction. It is thus equally suitable for DC and AC (for low frequencies such as the mains frequency).

- The moving-iron instrument is an **RMS** meter.

- The moving-iron instrument has a **high internal power consumption**.

- The moving-iron instrument can by its nature withstand **high overloads**.

NOTE: In the application of moving-iron instruments as current meters, the display is independent of the shape of the current waveform. Voltage measurements of nonsinusoidal waveforms require caution (Sect. 3.2.2). The large inductance of the meter attenuates the higher frequencies. This is why shunt resistors are rarely used for range extension. Instead, the current coil may be designed with several terminals, or, alternatively, a current transformer can be used.

4.1.5 Other Instruments

Rotary magnet instrument: In this case a small permanent magnet rotates in the field created by a coil carrying the measurement current. The reflex torque is provided by an additional magnet. The rotary magnet meter is very robust. Unlike the moving-coil instrument, no current leads are required to the moving parts.

Electrostatic movement: This technique uses the electrostatic force of two capacitor electrodes. It can only measure voltages, but with very small internal power consumption. Its application is for DC and AC voltage measurement (up to the RF range). The electrostatic instruments measure the RMS values of the voltage.

Thermal instruments: These instruments use the thermal expansion of current-carrying conductors and are implemented as **hot-wire measuring systems** or as **bimetallic instruments**. Their characteristics are high internal power consumption and long settling times. Thermal instruments are RMS meters.

Induction instruments: Two coils shifted 90° with respect to each other have AC currents of the same frequency passing through them. They induce eddy currents in an aluminium cylinder, which produces a torque on a spring. Induction measurement devices are instruments that use the product of the two currents (only for AC currents). The domestic electricity meter uses an aluminium disc that continuously rotates in the field of a permanent magnet (Fig. 4.5).

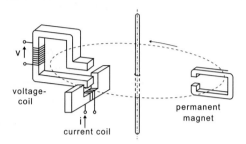

Fig. 4.5. Construction of an induction instrument to measure the electrical work performed

Electrodynamic ratio meter: This instrument is derived from the ratiometer-type moving-coil instrument seen earlier, but the outer magnetic field is generated by a second current coil. The pointer deflection depends on the **quotients** of the moving coil currents and on the **phase** of the measurement current with respect to the induction current. Electrodynamic ratio instruments are mostly used as power factor meters. The **cross-coil instrument** uses two right-angled inductor coils instead of the usual cross-coil. The inductor coils can rotate freely in its rotation field, thus enabling the use of a 360° scale.

Vibration instrument: Several tuned steel reeds are spring-mounted in the alternating magnetic field of a current-carrying coil. The reed, whose resonant frequency corresponds to the actual current frequency, oscillates with the largest amplitude (Fig. 4.6).

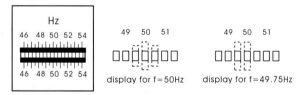

Fig. 4.6. Scale of a reed frequency meter

4.1.6 Overview: Electrical Instruments

A summary of the types of electrical instruments is given in Table 4.1.

Table 4.1. Summary of electrical instruments

Circuit symbol	Instrument	Measured quantity		Scale function
	Moving-coil	I, V	—	$\alpha = c \cdot \bar{i}$ Average value
	Moving-coil with rectifier	I, V	$\simeq$	$\alpha = c \cdot \overline{\lvert i \rvert}$ Rectified value
	Moving-coil with thermoconverter	I	$\simeq$	$\alpha = c \cdot I^2$ RMS
	Moving-iron	$I, (V)$	$\simeq$	$\alpha = f(I^2)$ RMS
	Moving magnet	I, V	—	$\alpha = c \cdot \bar{I}$ Arithmetic average value
	Ratio moving-coil	R	—	$\alpha = f\left(\dfrac{I_1}{I_2}\right)$
	Electrodynamic	P	$\simeq$	$\alpha = f(I_1 \cdot I_2 \cdot \cos\varphi_{12})$
	Electrostatic	V	$\simeq$	$\alpha = f(V^2)$ RMS
	Hot-wire bimetallic	I	$\simeq$	$\alpha = f(I^2)$ RMS
	Induction	W	$\sim$	$\sigma = c \cdot \displaystyle\int I_1 \cdot I_2 \cdot \cos\varphi_{12}\, dt$
	Electrodynamic ratio	$\cos\varphi$	$\sim$	$\alpha = f\left(\dfrac{I_1}{I_2}, \varphi_{13}, \varphi_{23}\right)$

The scale function α represents the relationship between the measured quantities and the pointer deflection, and c is the respective device constant.

4.2 Measurement of DC Current and Voltage

4.2.1 Moving-Coil Instrument

The moving-coil instrument is the most used DC measurement instrument because of its comparatively small internal power consumption and the high accuracy it achieves. The measured current flows through the instrument coil. Usual measured currents, for which the instrument displays full-scale deflection, lie between 10 µA and 10 mA. The internal resistance of an unconnected moving-coil instrument is relatively high. Due to this quality the moving-coil instrument can be used as voltmeter as well. The current through the measurement coil is proportional to the applied voltage. The scale is calibrated in volts.

$$V_M = I_M \cdot R_M \tag{4.1}$$

V_M: voltage on the instrument for full-scale deflection;
I_M: measurement current for full-scale deflection;
R_M: internal resistance of the instrument.

4.2.2 Range Extension for Current Measurements

To extend the measurement range, the measured current is split between the instrument coil and a parallel **shunt resistor** R_{Sh}. By varying the value of the shunt resistor, different measurement ranges can be obtained (Fig. 4.7).

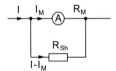

Fig. 4.7. Measurement range extension using a shunt resistor

EXAMPLE: An instrument with $I_M = 50$ µA full-scale deflection and internal resistance $R_M = 2$ kΩ is to be extended to a measurement range of 10 mA.

The magnitude of the voltage drop on the instrument for full-scale deflection is 100 mV. Therefore the value of the shunt resistor is $R_{Sh} = 100\,\text{mV}/9950\,\text{µA} = 10.05$ Ω.

In general,

$$R_{Sh} = \frac{I_M \cdot R_M}{I - I_M} \tag{4.2}$$

therefore I is the current for full-scale deflection in the desired measurement range.

If several measurement ranges were desired, the switch would lie in series with the shunt resistor. In this case, the resistance of the switch would not be negligible because of the very low impedance of the shunt resistor R_{Sh}. This is avoided in the circuit in Fig. 4.8.

Depending on the switch position, the resistors R_{Sh1}, $R_{Sh1} + R_{Sh2}$ or $R_{Sh1} + R_{Sh2} + R_{Sh3}$ act as the shunt resistor. The instrument, with some resistors in series, lies in parallel. The sum of all of the measurement resistances and the internal resistance of the instrument is

$$R_{Sum} = R_{Sh1} + R_{Sh2} + R_{Sh3} + R_M$$

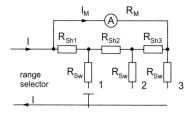

Fig. 4.8. Current measurement range extension, where the conductivity resistance R_C of the switch does not affect the measurement

In order to determine the individual resistor values, R_{Sum} is calculated as follows (Fig. 4.8):

Switch in position 3: The voltage drop on the instrument is equal to the drop over all the shunt resistors. Therefore, I_3 is the current for full-scale deflection in measurement range 3.

$$I_M \cdot R_M = V_M = (I_3 - I_M) \cdot (R_{Sh1} + R_{Sh2} + R_{Sh3})$$
$$\Rightarrow \quad R_{Sum} = \frac{I_M \cdot R_M}{I_3 - I_M} + R_M$$

Switch in position 1: Equating the voltage drop as before yields

$$(I_1 - I_M) \cdot R_{Sh1} = I_M \cdot (R_{Sh2} + R_{Sh3} + R_M) = I_M \cdot (R_{Sum} - R_{Sh1})$$
$$\Rightarrow \quad R_{Sh1} = \frac{I_M}{I_1} \cdot R_{Sum}$$

Switch in position 2:
$$R_{Sh2} = \frac{I_M}{I_2} \cdot R_{Sum} - R_{Sh1}$$

Switch in position 3:
$$R_{Sh3} = \frac{I_M}{I_3} \cdot R_{Sum} - (R_{Sh1} + R_{Sh2})$$

EXAMPLE: An instrument with $I_M = 500\,\mu A$ and an internal resistance $R_M = 1\,k\Omega$ is to be extended to an ammeter with a measurement range $I_1 = 100$ mA, $I_2 = 30$ mA and $I_3 = 10$ mA. The magnitude of the sum resistance is $R_{Sum} = 1052.63\,\Omega$. The values of the shunt resistors are $R_{Sh1} = 5.26\,\Omega$, $R_{Sh2} = 12.28\,\Omega$ and $R_{Sh3} = 35.09\,\Omega$. The sum resistance value is given to several decimal digits as the equations require very similar resistance values to be subtracted from each other.

4.2.3 Range Extension for Voltage Measurements

To measure larger voltages, resistors are used in series with the moving-coil instrument. For a full-scale deflection at the voltage V, the series resistor R_1 is given by

$$R_1 = \frac{V}{I_M} - R_M \tag{4.3}$$

where I_M is the current through the instrument at full-scale deflection, and R_M is the internal resistance of the instrument. The internal resistance of the voltmeter (instrument and series resistance) is often related to the voltage at full-scale deflection.

- The **voltage-related internal resistance** is the reciprocal of the instrument current at full-scale deflection (expressed in Ω/V).

EXAMPLE: A voltmeter is to be realised for the measurement range 10 V, 30 V and 100 V using a moving-coil instrument with $I_M = 50\ \mu\text{A}$ and $R_M = 1\ \text{k}\Omega$. The magnitude of the instrument's voltage-related internal resistance is $20\ \text{k}\Omega/\text{V}$. Thus the total resistance in the measurement range of 10 V is $200\ \text{k}\Omega$, of 30 V is $600\ \text{k}\Omega$ and of 100 V is $2\ \text{M}\Omega$. The actual values of the measurement resistors are given in Fig. 4.9.

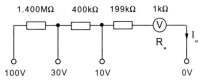

Fig. 4.9. Voltmeter with series resistors

4.2.4 Overload Protection

In order to avoid an overload in the moving-coil instrument, it is bridged by two parallel opposite-sense diodes (Fig. 4.10). If the voltage on the instrument exceeds about 0.7 V, the diodes shunt the excess current away. A fast-blowing microfuse in the current arm handles longer-lasting overloads.

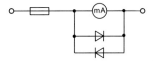

Fig. 4.10. Overload protection in a moving-coil instrument

4.2.5 Systematic Measurement Errors in Current and Voltage Measurement

Current Measurement

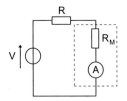

Fig. 4.11. Systematic measurement error in current measurement

Without the measurement instrument, the current flowing is $I = V/R$. By inserting the ammeter with the internal resistance R_M the current reduces to $I = V/(R + R_M)$, see Fig. 4.11.

- In **current measurement** the current is usually measured too low. The absolute measurement error decreases with decreasing ammeter internal resistance.

The magnitude of the **systematic** relative **measurement error** is

$$\frac{\Delta I}{I} = -\frac{R_M}{R_M + R} \approx -\frac{R_M}{R}, \qquad R_M \ll R \tag{4.4}$$

EXAMPLE: A systematic measurement error smaller than 1% is achieved if the ammeter internal resistance is at least 100 times smaller than the resistance in the current loop.

Voltage Measurement

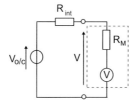

Fig. 4.12. Systematic measurement error in voltage measurement

By measuring a voltage $V_{o/c}$ and R_{int} the voltmeter resistance R_M loads the voltage source, and therefore the terminal voltage decreases a little. The measured voltage is V (Fig. 4.12).

- In **voltage measurement** the voltage is generally measured too low. The measurement error decreases with increasing voltmeter internal resistance R_M.

The magnitude of the **systematic** relative **measurement error** is

$$\frac{V - V_{o/c}}{V_{o/c}} = \frac{R_M}{R_{int} + R_M} - 1 = \frac{-R_{int}}{R_{int} + R_M} \approx -\frac{R_{int}}{R_M}, \qquad R_M \gg R_{int} \tag{4.5}$$

EXAMPLE: A systematic measurement error smaller than 1% is achieved if the voltmeter internal resistance is at least 100 times higher than the internal resistance of the voltage source.

4.3 Measurement of AC Voltage and AC Current

4.3.1 Moving-Coil Instrument with Rectifier

The configuration most frequently employed in measuring AC voltages is a moving-coil instrument equipped with a rectifier (Fig. 4.13).

The diodes rectify the measured current. For small measured voltages the threshold voltages of the diodes are noticeable. This effect is less apparent in the circuit on the right in Fig. 4.13.

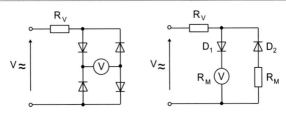

Fig. 4.13. Moving-coil instrument with rectifier; *left* with bridge, *right* with one-way rectifier

Here a single diode lies in series with the instrument. The replication of the measurement arm through the resistor R_M and the diode D_2 ensures that AC current flows through the configuration.

- Moving-coil instruments with an average-value rectifier display the **rectified value**.

NOTE: The scale on instruments are usually calibrated to display the RMS values for sinusoidal voltages. Measurements of nonsinusoidal waveforms must be corrected by a **form factor** (see Sect. 3.2.2).

EXAMPLE: A rectangular voltage with a peak value ± 1 V is measured with a moving-coil instrument. The RMS and the rectified values are 1 V for this waveform. The instrument deflection is proportional to the rectified value.

On the other hand, a sinusoidal voltage with a rectified value of 1 V has an RMS value of $k_f \cdot \overline{|v|} \approx 1.11$ V. For the rectangular voltage a moving-coil instrument displays thus an RMS voltage of 1.11 V. Consequently, there is a systematic measurement error of 11%.

- For small AC voltages the scale is clearly nonlinear.

NOTE: The AC voltage to be measured can be increased by transformers. The influence of the diode's characteristic curve then decreases. The transformer, however, limits the frequency range, where the lower limit is approx. 30 Hz and the upper limit is approx. 10 kHz.

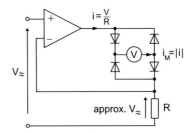

Fig. 4.14. Voltage–current transformation to measure small AC voltages

The circuit in Fig. 4.14 converts the input AC voltage into a proportional AC current using the resistance R. The output current of the operational amplifier changes in a manner such that the voltages on the inverting and the noninverting inputs are always equal. This conversion to current means that the display is independent of the nonlinearities of the diodes. A suitable choice of R allows very small AC voltages (1 mV) to be measured.

$$i_M = \frac{|V_{AC}|}{R} \quad \Rightarrow \quad \text{Selection:} \quad R = \frac{V_{AC}}{I_M} \tag{4.6}$$

where V_{AC} is the AC voltage for full-scale deflection, and I_M is the the instrument current for full-scale deflection. Measurement range extension can be achieved by inserting a voltage divider before the converter circuit.

AC currents flowing into the instrument are to be avoided when measuring high-frequency AC voltages. A peak-value rectifier is set up away from the instrument in a RF probe and passes only DC voltage to the instrument (Fig. 4.15). The display is suitably calibrated to be proportional to the peak-to-peak value.

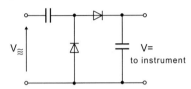

Fig. 4.15. Peak-value rectifier for measuring high-frequency voltages

4.3.2 Moving-Iron Instruments

The simplest instrument for measuring AC voltages and currents is the moving-iron instrument. This is an RMS meter, which therefore also displays the correct values for non-sinusoidal currents. Care must be taken when using it as a voltmeter as its large inductance attenuates harmonics. Therefore, the instrument must be frequency-compensated.

Moving-iron instruments were frequently used for monitoring machinery, usually in conjunction with current or voltage transformers.

4.3.3 Measurement Range Extension Using an Instrument Transformer

Apart from measurement range extension through series and shunt resistors, as outlined previously, current and voltage transformers also offer the possibility to measure extensive AC quantities. Instrument transformers have high-tolerance conversion ratios. They also offer the advantage that the measurement is electrically isolated from the mains.

Voltage transformers step down/up the measured voltage according to the winding ratio. Common secondary-side voltages are 100 V or $100/\sqrt{3}$ V for three-phase applications. The specification is given on the identification plate, e.g. 380 V/100 V. The terminals of the voltage transformer on the primary side are denoted by U and V, and on the secondary side by u and v. Unused voltage transformers that are connected on the primary side are left open-circuit on the secondary side. The primary winding usually consists of many windings of thin copper wire (high voltage, small current).

Current transformers step down the measurement current in inverse proportion to the winding ratio. Common secondary side currents are 5 A, and occasionally 1 A. The specification is given on the identification plate, e.g. 25 A/5 A. The terminals of the current transformer are denoted on the primary side by K and L, and on the secondary side by k

and l. Unused current transformers that are connected on the primary side have to be short-circuited on the secondary side. The primary winding usually consists of few windings of thick copper wire, which surrounds one or more times the core linking the secondary coil. A special construction is a hinged version of the current transformer, which can be looped around the conductor. When used with an ammeter this is known as a clip-on ammeter.

NOTE: When using instrument transformers, it is possible to sum up the currents (voltages) of several sections on the display. The secondary side of the instrument transformer must be connected with the correct poles in parallel for current transformation, and in series for voltage transformation.

For an instrument transformer a **current** or **voltage error** is specified. This is the guaranteed upper limit of the error of the secondary side current (voltage) from the correct value. Because of the (tiny) losses in the transformer, there is a small phase shift of some angular minutes between the input sinusoidal quantity and the output quantities. This is specified as the **phase error**. The phase error is important when two measurement quantities are related, for example, in power measurement. Instrument transformers must be loaded with their **nominal load** to stay within their specified error limit. **Current transformers** act at their secondary side like a current source. Therefore the nominal load has a very low resistance (nearly short-circuit). **Voltage transformers** act at their secondary side like a voltage source. Therefore the nominal load has a very high resistance (nearly infinity).

An instrument transformer's **load** is often also given as the maximum deliverable **apparent power** in units of VA. For current transformers values are in the range 1–60 VA, and for voltage transformers 10–300 VA.

4.3.4 RMS Measurement

The scales of most measurement instruments for AC quantities are calibrated for RMS. But if the measurement device is not a true RMS meter, then the displayed value is correct only for the waveform the instrument has been calibrated for (normally sinusoidal).

For the measurement of RMS values (so-called true RMS measurement) there are different options.

RMS meters are measurement instruments that measure the RMS value because of their operating principle. These include:

- moving-iron instruments,
- thermal instruments,
- electrodynamic instruments with both coils in series,
- electrostatic instruments (for voltages).

Moving-coil instrument with thermal transformer: The current to be measured heats up a resistor, whose temperature is measured by a thermal element. A moving-coil instrument that is calibrated to give RMS values of the current is connected to this thermal element (Fig. 4.16).

Instrument with analogue RMS calculator: This circuit is the electronic (analogue) realisation of the defining equation for the RMS value (Fig. 4.17).

Fig. 4.16. Symbol for a moving-coil instrument with thermal transformer for RMS measurement. In the configuration on the *right* the thermal element is isolated from the measurement loop

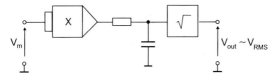

Fig. 4.17. Principle of the circuit for RMS generation of the measurement quantity. Actual application circuits are realised in a sightly different manner

Digital measurement devices: The quantity to be measured is sampled, and the sampled values inserted into the RMS defining equation and calculated by a microprocessor, before being displayed.

NOTE: The RMS voltage measurement is often problematic. Voltmeters are used in parallel and represent a frequency-dependent load. Different frequencies are weighted in different ways. For this reason the instrument should be frequency-compensated.

4.4 Power Measurement

4.4.1 Power Measurement in a DC Circuit

The power dissipated in a load can be determined through the measurement of the current through and the voltage drop across the load.

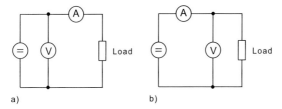

Fig. 4.18. Determination of the power in a DC circuit: circuit for **a)** correct current measurement; **b)** correct voltage measurement

The **voltage error circuit** or **correct current measurement** is shown in Fig. 4.18a. The voltage actually measured as being across the load is higher by the voltage drop on the ammeter.

The **current error circuit** or **correct voltage measurement** is shown in Fig. 4.18b. The current actually measured as flowing through the load is higher by the amount flowing through the voltmeter.

NOTE: If the delivered source power is to be precisely measured, rather than the power dissipated in the load, the circuits in Figs. 4.18a and 4.18b swap roles as summarised in Table 4.2.

Table 4.2. Measurement of DC quantities

Measurement quantity	Load	Suitable circuit
Load power	High ohmic	Correct current (a)
	Low ohmic	Correct voltage (b)
Source power	High ohmic	Correct current (b)
	Low ohmic	Correct voltage (a)

To **directly display the measurement**, electrodynamic measurement instruments are used as power meters. One of the coils is used as a **current path**, and the other as a **voltage path**. The basic circuit and its systematic error are analogous to the measurement with two instruments.

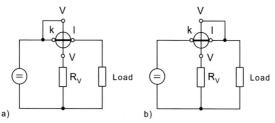

Fig. 4.19. Power determination in a DC circuit with a wattmeter: **a)** configuration for correct current; **b)** configuration for correct voltage

Figure 4.19a shows the configuration for **correct current**, while that in Fig. 4.19b shows the configuration for **correct voltage**. The comments on measurement with two instruments apply here as well.

NOTE: The display on wattmeters shows the **product** of the **currents** in the current and voltage paths. An **overload** of an individual path is possibly not visible on the display. For this reason, it must be made sure that neither the current nor the voltage exceed the permitted values.

NOTE: In applications where the current flow can reverse (e.g. in rechargeble batteries with charging circuitry), wattmeters with centred null positions or those with toggle switches are placed in the path.

4.4.2 Power Measurement in an AC Circuit

In an AC circuit with sinusoidal currents and voltages, the power measurement (Fig. 4.20) must differentiate between:

- apparent power $S = V \cdot I$ given in VA,
- real power $P = V \cdot I \cdot \cos \varphi$ given in W,
- reactive power $Q = V \cdot I \cdot \sin \varphi$ given in var,

Here, I and V are the RMS values of current or voltage (sinusoidal quantities!). The **apparent power**, like DC power, is determined with a voltmeter and an ammeter, as shown in Fig. 4.18 and the related comments.

Real power is measured with electrodynamic or induction instruments, that takes into account the phase shift of current and voltage. As for the DC measurement there is a correct current and correct voltage measurement configuration (Fig. 4.19).

Reactive power is measured with a wattmeter, in whose voltage path the current is shifted by 90° by a phase-shifting circuit.

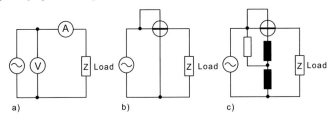

Fig. 4.20. Measurement configuration to determine **a)** the apparent power; **b)** the real power; and **c)** the reactive power

NOTE: In current loops with a high proportion of reactive power, a wattmeter can be overloaded without giving any indication on the display. In these cases the current in the current path must be controlled.

NOTE: Care must be taken with nonsinusoidal voltages and/or currents! The power measurement can be in error because:

- any phase-shifting circuit present may only be designed for *one* frequency,
- in the voltage path the harmonics are very strongly attenuated

This is regularly the case for loads that use SCRs*, current rectifiers or similar components. Any power factor measurement is then usually error-prone or even pointless.

4.4.2.1 Three-Voltmeter Method

A known resistance is inserted before a complex load.

The **real power** can be calculated from the three measured voltages (Fig. 4.21):

$$P = \frac{V_{\text{total}}^2 - V_R^2 - V_Z^2}{2R} \tag{4.7}$$

The **three-ammeter method** works in a similar fashion. A known resistance is inserted parallel to the complex load.

The real power can be calculated from the three measured currents (Fig. 4.22):

$$P = \frac{R}{2} \cdot \left(I_{\text{total}}^2 - I_R^2 - I_Z^2\right) \tag{4.8}$$

---

* SCR: silicon controlled rectifier such as thyristor, triac

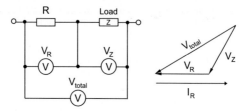

Fig. 4.21. Three-voltmeter method for power determination and the related vector diagram

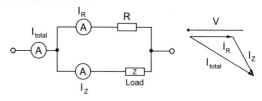

Fig. 4.22. Three-ammeter method for power determination and the related vector diagram

The **reactive power** Q can also be calculated from the readings in the three-voltmeter method, by applying the following relationship

$$\cos \varphi = \frac{V_{total}^2 - V_R^2 - V_Z^2}{2 \cdot V_R \cdot V_Z}, \qquad \sin \varphi = \sqrt{1 - \cos^2 \varphi}, \qquad S = V_Z \cdot \frac{V_R}{R}$$

which yields

$$Q = \sqrt{1 - \cos^2 \varphi} \cdot V_Z \cdot \frac{V_R}{R}$$

4.4.2.2 Power Factor Measurement

The power factor (PF) can be calculated from the apparent power (measured with a voltmeter and an ammeter) and the real power (measured with a wattmeter).

A **directly displayed measurement** is carried out by the electrodynamic quotient instrument, as shown in Fig. 4.23.

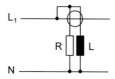

Fig. 4.23. Power factor measurement with electrodynamic quotient instrument

The application is limited to a narrow frequency range due to the requirement of a phase-shifting inductance (typically 49.5–50.5 Hz).

The display is approximately proportional to $\tan \varphi$, where φ is the phase angle. The scale is mostly calibrated, however, in values of $\cos \varphi$, e.g. (+0.4 capacitive to +0.4 inductive). The instrument does not have reset/return-to-zero capability, so the display is not defined for currentless situationa. Some special constructions have a 360°-scale.

4.4.3 Power Measurement in a Multiphase System

This section describes the measurement of the

- **apparent power** with three voltmeters and three ammeters,
- **real power** with one, two or three wattmeters,
- **reactive power** with a suitable phase shift of the voltage. No phase shift is required in a multiphase network, because the 90°-shifted voltage is available.

4.4.3.1 Measurement of the Real Power in a Multiphase System

Three wattmeters are required for **asymmetrical loading** in a **four-conductor system** (Fig. 4.24). The total real power is the sum of the powers measured on each of the outer conductors.

$$P = P_1 + P_2 + P_3 = V_{1N} \cdot I_1 \cdot \cos \varphi_1 + V_{2N} \cdot I_2 \cdot \cos \varphi_2 + V_{3N} \cdot I_3 \cdot \cos \varphi_3$$

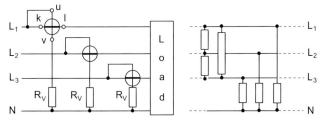

Fig. 4.24. Real power measurement in a four-conductor system with asymmetrical loading

NOTE: The powers P_1 to P_3 could also be measured step by step with a single wattmeter.

For symmetrical loading $V_{1N} = V_{2N} = V_{3N}$, $I_1 = I_2 = I_3$ and $\cos \varphi_1 = \cos \varphi_2 = \cos \varphi_3$. For **symmetrical loading** in a **four-conductor system**, *one* wattmeter, whose scale is suitably calibrated, is sufficient (Fig. 4.25). The power measured by the instrument is P_M. The total power is therefore

$$P = 3 \cdot P_M$$

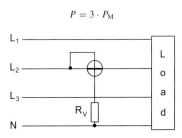

Fig. 4.25. Real power measurement in a symmetrically loaded four-conductor system

In a **three-conductor system** the neutral conductor is missing. An **artificial zero-point** can be created for power measurement. The resistance $R_V + R_M$ is the total resistance of the wattmeter's voltage path (Fig. 4.26).

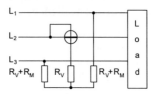

Fig. 4.26. Real power measurement in a symmetrically loaded three-conductor system with an artificial zero-point

For **asymmetrical loading** in a **three-conductor system** the circuit in Fig. 4.24 is extended by adding an artificial zero-point. The **two-wattmeter method** is used for the same purpose with less handling required (Fig. 4.27, an Aron-circuit).

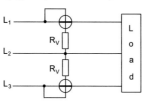

Fig. 4.27. Real power measurement in an asymmetrically loaded three-conductor system using the two-wattmeter method

The voltage in the instrument's voltage path is higher by a factor of $\sqrt{3}$ with respect to the measurement with the neutral conductor. The real power in the load is equal to the sum of the displayed values on both wattmeters.

$$P = P_{M1} + P_{M3}$$

NOTE: A display can be negative for large phase shifts and the signs must therefore be taken into account. Therefore the correct polarity must be carefully considered. Wattmeters with central zero-points or toggle switches are used.

4.4.3.2 Measurement of the Reactive Power in a Multiphase System

The measurement of **reactive power** in a multiphase system is possible without using a phase-shifting circuit, since the 90°-shifted voltage is available on other outer conductors.

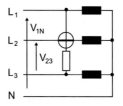

Fig. 4.28. Reactive power measurement in a symmetrically loaded four-conductor system

The measurement of the **reactive power** for **symmetrical loading** is shown in Fig. 4.28 as an example. The voltage V_{23} between the outer conductors L_2 and L_3 is shifted by 90° with respect to the voltage V_{1N}. However, it is greater by a factor of $\sqrt{3}$. The total (symmetrical) reactive power is thus

$$Q_{\text{total}} = 3 \cdot \frac{Q}{\sqrt{3}} = \sqrt{3} \cdot Q$$

The correct reading can be obtained by suitable choice of the series resistors or by using an instrument transformer with a suitable conversion ratio.

A **three-wattmeter circuit** permits the measurement of the **reactive power** for **asymmetrical loading** in which the voltage paths are fed respectively with voltages shifted by 90° with respect to the real power measurement (Fig. 4.29). This only produces correct results provided that the voltage vectors are not shifted by the load.

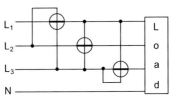

Fig. 4.29. Reactive power measurement in an asymmetrically loaded three-conductor system (series resistors are not shown)

The magnitudes of the voltages must be corrected by a factor of $\sqrt{3}$:

$$Q = \frac{1}{\sqrt{3}} \cdot (Q_1 + Q_2 + Q_3)$$

4.5 Measurement Errors

4.5.1 Systematic and Random Errors

Every measurement process is subject to error. Error sources can be in the measurement device or in the measurement method, in external quantities such as temperature or stray fields, as well as in the reading of the device. Errors are classified as systematic or random errors.

Systematic errors are caused by inadequacies of the measurement devices or an inappropriate measurement technique. Such errors are reproducible and can be compensated.

Random errors do not have definite causes. They are normally not the same if the measurement is repeated and therefore cannot be corrected.

- The **display error** (expressed in %) usually is expressed with respect to the full-scale value. For scales whose zero-point does not lie at the scales' boundaries, the sum of both scales' end values is taken together as the reference value.

NOTE: Instruments with highly nonlinear scales, without a zero-point or reed frequency meters are not covered by this definition. The reference value then is the true value or is shown on the instrument.

4.5.2 Guaranteed Error Limits

Manufacturers of measurement devices guarantee that their instrument's display error will not exceed certain limits, under defined environmental and operating conditions. For measurement instruments **classes of precision** have been defined (Table 4.3).

Table 4.3. Classes of precision for measurement instruments

Classes of Precision (VDE 0410)
Specification (%)
0.1　0.2　0.5　1　1.5　2.5　5

EXAMPLE: An instrument of precision class 1.5 with a maximum value of 300 V displays a measured value of 100 V. How large can the relative error of the display be? The absolute measurement error can be up to 4.5 V. For a display value of 100 V this yields $\frac{4.5 \text{ V}}{100 \text{ V}} = 4.5\%$

- The measurement range of an instrument should always be chosen so that the meaured value is in the upper third of the scale.

4.6 Overview: Symbols on Measurement Instruments

A summary of the symbols found on measurement instruments is given in Table 4.4. See also instrument symbols in Sect. 4.1.6.

Table 4.4. Symbols on measurement instruments

Symbol	Description
—	DC instrument
∿	AC instrument
≂	DC and AC
≈	Multiphase instrument with one movement
≈∥	Multiphase instrument with two movements
≈∥∥	Multiphase instrument with three movements
☆	Isolation-testing voltage 500 V
☆(2)	Isolation-testing voltage higher than 500 V, here 2 kV
☆(0)	No voltage testing

Table 4.4. (cont.)

Symbol	Description
⊥	Perpendicular operation position
⊓	Horizontal operation position
∠60	Diagonal operation position, specification of the angle of inclination
1.5	Class signs for display error, with respect to the measurement range end value
⟨1.5⟩	Class signs for display error, with respect to the scale length
(1.5)	Class signs for display error, with respect to the true value
→▶—	Rectifier in a device (in addition to the instrument symbol)
⊕	Electronic circuit in a device
⊣□⊢	Indication of separate shunt resistance
—⟦R⟧—	Indication of separate series resistance
(⋯)	Electrostatic shielding
◯	Magnetic shielding
ast	Astatic instrument
⎵5⎵	Maximum allowable quantity of an interfering field, here 5 mT
⏚	Protective conductor termination
◠	Pointer zero-position adjustment
⚠	Warning! Follow operating instructions
⚡	Testing voltage does not comply with VDE
(V)⚡	Danger, high-voltage exposure on instrument

4.7 Overview: Measurement Methods

Check Table 4.5 where to find information about the measurement of the electrical quantity given in the first column.

Table 4.5. Cross-reference for measuring electrical quantities

Measurement quantity	Section
DC current	4.2.2
DC voltage	4.2.3
AC current	4.3.2
AC voltage	4.3.1
RMS	4.3.4
Power in a DC circuit	4.4.1
Real power	4.4.2
Reactive power	4.4.2
Power factor	4.4.2.2
Power in a multiphase circuit	4.4.3.1
Reactive power	4.4.3.1
Impedance	3.7.6.2

4.8 Notation Index

c	device constant
$\cos \varphi$	power factor
$\bar{I}$	arithmetic average value of current
ΔI	systematic current-measurement error
I_1, I_2, I_3	outer conductor currents
I_M	current through instrument for full-scale deflection
I_R	current through resistance
I_Z	current through unknown impedance $\underline{Z}$
k_f	form factor
P	real power
P_{M1}, P_{M2}	displayed power
Q	reactive power
R_A	internal resistance of ammeter
R_i	internal resistance of voltage source
R_M	instrument's internal resistance
R_{Sh}	shunt resistor
R_{Sum}	total resistance
R_V	series resistance
S	apparent power
$\overline{|v|}$	rectified value
V_{1N}, V_{2N}, V_{3N}	star voltages
$V_{o/c}$	open circuit voltage, terminal voltage
V_M	voltage across instrument for full-scale deflection

V_R	voltage across resistor
V_Z	voltage across unknown impedance $\underline{Z}$
φ_{13}	phase angle between I_1 and I_3
σ	scale function

4.9 Further Reading

BENTLY, J. P.: *Principles of Measurement, 3rd Edition*
Longman (1997)

FLOYD, T. L.: *Electric Circuits Fundamentals, 5th Edition*
Prentice Hall (2001)

5 Networks at Variable Frequency

Often in communications the internal structure of a system is of no particular interest. It is more interesting to consider the behaviour of the input and output signals, which in most cases are voltages (Fig. 5.1). The system is described by a function that represents the transformation of an input signal to an output signal. Such systems are often called **black-box** systems.

$$v_{\text{out}} = T(v_{\text{in}})$$

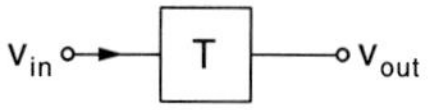

$$V_{\text{in}} \circ\!\!-\!\!\boxed{T}\!\!-\!\!\circ V_{\text{out}}$$

Fig. 5.1. A system with input and output signals

5.1 Linear Systems

Many systems can be considered **linear** to a good approximation. It then holds that

$$T(\alpha \cdot v_{\text{in}}) = \alpha \cdot T(v_{\text{in}}) \tag{5.1}$$

- The output signal is proportional to the input signal.

$$T(v_1 + v_2) = T(v_1) + T(v_2) \tag{5.2}$$

- Either input signal is treated independently of any other input signal (Fig. 5.2).

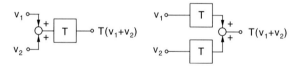

Fig. 5.2. Principle of superposition in linear systems

The approach shown in Eq. (5.2) and Fig. 5.2 is called the **principle of superposition**.

- When linear systems are fed with a harmonic input signal they produce a harmonic output signal of the **same frequency**, while amplitude and phase usually change.

NOTE: Systems that react to harmonic input signals with nonharmonic output signals are called **nonlinear systems**. The output signal of such systems contains components of frequencies different from the input signal.

5.1.1 Transfer Function, Amplitude and Phase Response

The behaviour of linear systems in response to harmonic input signals of different frequencies is described by the **transfer function** $G(\omega)$.

$$\text{transfer function} = \frac{\text{output value}}{\text{input value}}$$

An independent variable of the transfer function is the (angular) frequency of the harmonic input signals.

$$G(\omega) = \frac{v_{\text{out}}}{v_{\text{in}}}, \qquad \text{only for harmonic signals} \tag{5.3}$$

This equation causes trouble for signals at zero-crossings. The following equation is therefore more suitable:

$$\boxed{v_{\text{out}}(\omega) = G(\omega) \cdot v_{\text{in}}(\omega)} \tag{5.4}$$

In general, the transfer function is complex valued. This implies that the amplitude as well as the phase of the input signal is affected.

EXAMPLE: Figure 5.3 shows a low-pass filter.

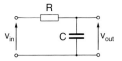

Fig. 5.3. Low-pass filter as voltage divider

The transfer function is

$$G(\omega) = \frac{v_{\text{out}}}{v_{\text{in}}} = \frac{1/j\omega C}{1/j\omega C + R} = \frac{1}{1 + j\omega RC}$$

The transfer function is also known as the (complex) **frequency response**. It can be split up into the magnitude and the phase components.

$$G(\omega) = |G(\omega)| \cdot e^{j\varphi(\omega)} \tag{5.5}$$

$|G(\omega)|$ or $|G(f)|$ are called the **magnitude** or **gain frequency characteristic**, or just the **magnitude** or **gain response** of a system. $\varphi(\omega)$ is called the **phase frequency characteristic** or the **phase response**. Often $G(\omega)$ is represented in a logarithmic form. The **magnitude response** can be written as

$$A(\omega) = 20 \log_{10} |G(\omega)| \qquad \text{(dB)} \tag{5.6}$$

This is a ratio of two values expressed in **decibels** (dB).

Table 5.1. Typical values of amplification and magnitude response

Typical values	
Amplification $v = \|G(\omega)\|$	Magnitude response $A(\omega)$
1	0 dB
$\sqrt{2}$	≈ 3 dB
$1/\sqrt{2}$	≈ -3 dB
2	≈ 6 dB
4	≈ 12 dB
10	20 dB
0.1	−20 dB

EXAMPLE: What is the gain of a system with an amplification of 14 dB?

$A(\omega) = 20 \log_{10} |G(\omega)|$. The table yields

$$14 \text{ dB} = 20 \text{ dB} - 6 \text{ dB} \Rightarrow v = \frac{10}{2} = 5, \qquad G(\omega) = 10^{\frac{A(\omega)}{20}} = 10^{\frac{14}{20}} = 5$$

NOTE: The following representation is also used in communications:

$$G(\omega) = e^{-(\tilde{A}(\omega) + jB(\omega))} = e^{-\tilde{A}(\omega)} \cdot e^{-jB(\omega)} \qquad (5.7)$$

In this case $\tilde{A}(\omega)$ is the **attenuation factor**, and $B(\omega)$ is the **phase factor** of a system. Table 5.1 gives some typical values for amplification and magnitude response.

The transfer function is often represented in a Bode plot, where the gain response is drawn against the logarithm of the frequency (Fig. 5.4). The phase is represented separately.

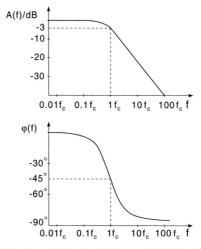

Fig. 5.4. Bode plot of the transfer function of the low-pass filter in the previous example

5.2 Filters

Filter circuits are circuits with transfer functions that enable the magnitude and the phase of the individual frequency components of the input signal to be modified by different amounts, for example,

- low-pass filters (LPF),
- high-pass filters (HPF),
- bandpass filters (BPF),
- band-stop or notch filters,
- all-pass filters (APF).

Ideally, signals in the **pass-band** should pass through the filter without being changed. Signals in the **stop-band** should be attenuated as much as possible.

5.2.1 Low-Pass Filter

Figures 5.5 and 5.6 show the schematic symbols and characteristic plots of a low-pass filter, respectively.

Fig. 5.5. Schematic symbols for low-pass filters

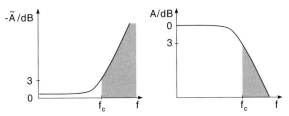

Fig. 5.6. Characteristic plot of the attenuation and the magnitude response of the low-pass filter. The stop-band is shaded in grey

- At the **cutoff frequency** f_c, the amplitude of the signal is $1/\sqrt{2} = 0.707$ times smaller than for a DC signal. This means that the gain response has decreased to -3 dB, or the attenuation has a value of 3 dB.
- The **pass-band** reaches from DC up to the cutoff frequency.
- The **stop-band** commences for frequencies above the cutoff frequency.

5.2.2 High-Pass Filter

Figures 5.7 and 5.8 show the schematic symbols and characteristic plots of a high-pass filter, respectively.

Fig. 5.7. Schematic symbols for high-pass filters

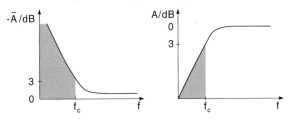

Fig. 5.8. Characteristic plot of the attenuation and the magnitude response of the high-pass filter. The stop-band is shaded in grey

- At the **cutoff frequency** f_c, the amplitude of the signal is $1/\sqrt{2} = 0.707$ times smaller than for very high frequencies. This means that the gain response has decreased to -3 dB, or the attenuation has a value of 3 dB.
- The **pass-band** commences for frequencies above the cutoff frequency.
- The **stop-band** reaches from DC up to the cutoff frequency.

5.2.3 Bandpass Filter

Figures 5.9 and 5.10 show the schematic symbols and characteristic plots of a bandpass filter, respectively.

Fig. 5.9. Schematic symbols for bandpass filters

- The bandpass filter has a **lower cutoff frequency** f_{cl} and an **upper cutoff frequency** f_{cu}.
- The **centre frequency** f_0 is the arithmetic mean of both cutoff frequencies.

$$f_0 = \frac{f_{cl} + f_{cu}}{2}$$

- The **bandwidth** B is the difference between the two cutoff frequencies.
- The **relative bandwidth** is the ratio of the bandwidth to the centre frequency expressed in percent.

$$B_{\text{rel}} = \frac{B}{f_0} \cdot 100\%$$

- The **quality factor** Q, or **Q-factor** is the ratio of the centre frequency to the bandwidth.

$$Q = \frac{f_0}{B}$$

- The **shape factor** F is a measure of the steepness of the bandpass filter slopes. It is the ratio of the 3 dB and the 20 dB bandwidths.

$$F = \frac{B_{3\,\text{dB}}}{B_{20\,\text{dB}}}$$

The closer this value is to 1 the steeper is the roll-off of the filter.

NOTE: The harmonic mean of both cutoff frequencies is also referred to as the centre frequency.

$$f_0 = \sqrt{f_{cl} \cdot f_{cu}}$$

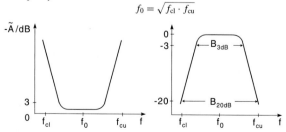

Fig. 5.10. Characteristic plot of the attenuation and the magnitude response of the bandpass filter

5.2.4 Stop-Band Filter

Stop-band filters are the complement to bandpass filters (Fig. 5.11). Stop-band filters are used to suppress a specific frequency range. A **notch filter** is used to suppress a specific single frequency.

Fig. 5.11. Schematic symbols for stop-band filters

5.2.5 All-Pass Filter

All-pass filters have a constant magnitude response over the frequency, that is, the attenuation is identical for all frequencies. However, the phase is changed depending on the frequency.

5.3 Simple Filters

5.3.1 Low-Pass Filter

Figure 5.12 shows a first-order low-pass filter. Its (complex) **transfer function** is given by

$$G(\omega) = \frac{1/j\omega C}{1/j\omega C + R} = \frac{1}{1 + j\omega RC} \tag{5.8}$$

NOTE: The circuit is regarded as a voltage divider to determine the transfer function.

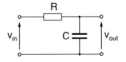

Fig. 5.12. First-order low-pass filter

The **magnitude response** is the magnitude of the transfer function:

$$|G(\omega)| = \frac{1}{\sqrt{1 + (\omega RC)^2}} \tag{5.9}$$

The **phase response** is the phase difference between the output voltage and tha input voltage $\varphi(\omega) = \varphi_{v\,out} - \varphi_{v\,in}$:

$$\varphi(\omega) = \arctan\left[\frac{\mathrm{Im}(G(\omega))}{\mathrm{Re}(G(\omega))}\right] = -\arctan(\omega RC) \tag{5.10}$$

The gain response and the phase response are represented in Fig. 5.13 in a Bode plot.

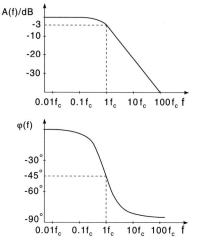

Fig. 5.13. Bode plot of a low-pass filter

For the special angular frequency $\omega_c = 1/RC$, it holds that

$$|G(\omega_c)| = \frac{1}{\sqrt{2}} \triangleq -3 \text{ dB}$$

$f_c = \omega_c/2\pi$ is the **cutoff frequency** or **corner frequency** of the low-pass filter. The phase at the cutoff frequency is given by

$$\varphi(\omega_c) = \arctan(-1) = -\frac{\pi}{4}, \quad \text{or} \quad (-45°)$$

- At the cutoff frequency ω_c the gain of the low-pass filter is 3 dB lower than the DC gain. The phase shift between the input signal and the output signal is then $\frac{\pi}{4}$, or 45°.

5.3.1.1 Rise Time

The step response of a low-pass filter can be estimated in the time domain from its cutoff frequency f_c.

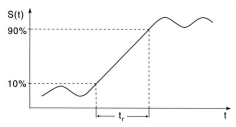

Fig. 5.14. Definition of rise time

The **rise time** is the time interval required by the signal to rise from 10% to 90% of the steady-state value Fig. 5.14. Between the rise time t_r and the critical frequency f_c the following relationship holds:

$$t_r \approx \frac{1}{3 f_c} \approx \frac{2}{\omega_c} \tag{5.11}$$

EXAMPLE: An oscilloscope with a critical frequency of 30 MHz has a rise time t_r of approximately $1/(3 \cdot 30 \cdot 10^6)$ s ≈ 10 ns.

5.3.2 Frequency Normalisation

All low-pass filters with a structure as in Fig. 5.12 have similar transfer functions except for the parameter ω_c. In order to describe all low-pass filters with this structure uniformly, a **frequency normalisation** relative to the cutoff frequency is done:

$$\textbf{Normalisation:} \quad \Omega := \frac{\omega}{\omega_c} = \frac{f}{f_c} \tag{5.12}$$

$$\textbf{De-normalisation:} \quad \omega = \Omega \cdot \omega_c, \qquad f = \Omega \cdot f_c \tag{5.13}$$

Ω is called the **normalised frequency** and has no unit. Therefore the normalised critical frequency of any low-pass filter is $\Omega = 1$.

It follows that the normalised transfer function of a low-pass filter is

$$G(\Omega) = \frac{1}{1 + j\Omega}$$

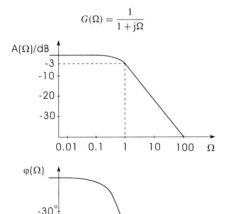

Fig. 5.15. Bode plot of a low-pass filter in frequency-normalised representation

The normalised magnitude response is

$$|G(\Omega)| = \frac{1}{\sqrt{1+\Omega^2}}$$

Figure 5.15 shows the normalised frequency response of a low-pass filter.

5.3.2.1 Approximation of the Magnitude Response

The magnitude response of a low-pass filter is represented in normalised form by

$$A(\Omega) = 20 \log_{10} \frac{1}{\sqrt{1+\Omega^2}} = 20 \log_{10} \frac{1}{\sqrt{1+\left(\dfrac{\omega}{\omega_c}\right)^2}}$$

For angular frequencies that are much larger than the cutoff frequency, Ω is much greater than 1. The following approximation can then be made

$$A(\Omega) \approx 20 \log_{10} \frac{1}{\Omega} = -20 \log_{10} \Omega, \quad \text{for } \Omega \gg 1$$

- Below the cutoff frequency the magnitude response is approximately constant.

- The magnitude response drops by 20 dB for a tenfold increase (decade) in frequency. This is described as a roll-off of -20 dB/decade or -6 dB/octave (Fig. 5.16).

- At the cutoff frequency $\Omega_c = 1$, the magnitude response is -3 dB.

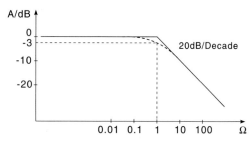

Fig. 5.16. Approximate magnitude response for a low-pass filter

5.3.3 High-Pass Filter

Figure 5.17 shows a first-order high-pass filter. Its (complex) **transfer function** is given by

$$G(\omega) = \frac{R}{1/j\omega C + R} = \frac{j\omega RC}{1 + j\omega RC} \tag{5.14}$$

NOTE: The circuit is regarded as a voltage divider to determine the transfer function.

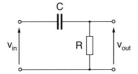

Fig. 5.17. First-order high-pass filter

The **magnitude response** is the magnitude of the transfer function

$$|G(\omega)| = \frac{(\omega RC)}{\sqrt{1 + (\omega RC)^2}} \tag{5.15}$$

In **normalised** representation the transfer function and the magnitude response are

$$G(\Omega) = \frac{j\Omega}{1 + j\Omega}, \qquad |G(\Omega)| = \left| \frac{\Omega}{\sqrt{1 + \Omega^2}} \right| \tag{5.16}$$

The **phase response** of the high-pass filter is

$$\varphi(\omega) = \arctan\left[\frac{\mathrm{Im}(G(\omega))}{\mathrm{Re}(G(\omega))}\right] = \arctan\left(\frac{1}{\omega RC}\right) = \arctan\left(\frac{\omega_c}{\omega}\right) \tag{5.17}$$

In normalised representation

$$\varphi(\Omega) = \arctan\left(\frac{1}{\Omega}\right) \tag{5.18}$$

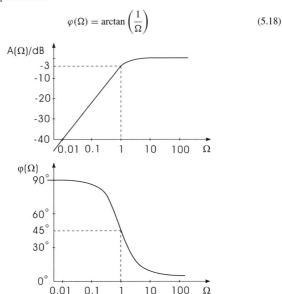

Fig. 5.18. Bode plot for a high-pass filter in frequency-normalised representation

The magnitude and phase characteristics are represented as Bode plot in Fig. 5.18. For the special angular frequency $\omega_c = 1/RC$, it holds that

$$|G(\omega_c)| = \frac{1}{\sqrt{2}} \,\widehat{=}\, -3 \text{ dB}$$

Here $f_c = \omega_c/2\pi$ is the **cutoff frequency** or **corner frequency** of the high-pass filter. The phase at the cutoff frequency is given by

$$\varphi(\omega_c) = \arctan(1) = \frac{\pi}{4} \quad \text{or} \quad 45°$$

- At the cutoff frequency ω_c the gain of the low-pass filter is 3 dB lower than the gain at very high frequencies ($\omega \gg \omega_c$). The phase shift between the input signal and the output signal is then $\dfrac{\pi}{4}$, or $45°$.

5.3.3.1 Approximation of the Magnitude Response

- The normalised critical frequency of the high-pass filter is $\Omega_c = 1$.

- The magnitude response increases by 20 dB for a tenfold increase (decade) in frequency (Fig. 5.19). Above the cutoff frequency, the magnitude is approximately constant.

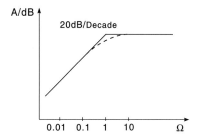

Fig. 5.19. Approximate magnitude response of a high-pass filter

5.3.4 Higher-Order Filters

Filters of higher-order are obtained when two filters are combined such that the output signal of the first filter is the input signal of the following filter (**cascade circuit**, Fig. 5.20). The **filter order** is dependent on the number of independent energy-storing elements (that is, capacitors or inductors). With higher-order filters sharper roll-offs can be obtained.

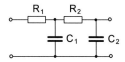

Fig. 5.20. Cascaded second-order low-pass filter

Fig. 5.21. Second-order RLC low-pass filter

The RLC filter in Fig. 5.21 is a second-order low-pass filter. Its transfer function is

$$G(\omega) = \frac{\dfrac{1}{j\omega C}}{\dfrac{1}{j\omega C} + R + j\omega L} = \frac{1}{1 + j\omega RC - \omega^2 LC} \qquad (5.19)$$

Similar to the series resonant circuit, a resonant frequency ω_r can be defined. The transfer function can be frequency normalised as follows:

$$\omega_r = \frac{1}{\sqrt{L \cdot C}}, \qquad \Omega = \frac{\omega}{\omega_r}$$

In normalised form the transfer function is

$$G(\Omega) = \frac{1}{1 + jR\sqrt{\dfrac{C}{L}}\Omega - \Omega^2} \qquad (5.20)$$

The quantity

$$D = \frac{R}{2}\sqrt{\frac{C}{L}}$$

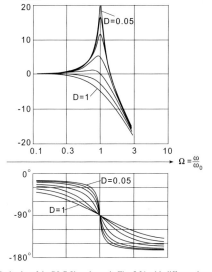

Fig. 5.22. Bode plot of the RLC filter shown in Fig. 5.21 with different damping ratios

is called the **damping ratio** (see also Sect. 1.2.6). Using this quantity the normalised transfer function is

$$G(\Omega) = \frac{1}{1 + 2jD\Omega - \Omega^2} \tag{5.21}$$

The shape of the amplitude/frequency characteristic and the phase response is essentially determined by the damping ratio D. Figure 5.22 shows the Bode plot of an RLC filter with the damping ratio as a parameter.

With low damping ratios the low-pass filter shows a pronounced resonant characteristic and behaves similarly to a bandpass filter. The characteristic plot of the phase becomes steeper as the damping ratio decreases.

5.3.5 Bandpass Filter

Figure 5.23 shows a series resonant circuit acting as a bandpass filter.

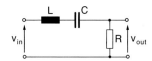

Fig. 5.23. Example of an RLC bandpass filter

Regarding this as a complex voltage divider, it follows that the transfer function is

$$G(\omega) = \frac{R}{R + j\omega L + \dfrac{1}{j\omega C}} = \frac{j\omega RC}{j\omega RC - \omega^2 LC + 1}$$

The frequency is normalised to the resonant frequency $\omega_0 = 1/\sqrt{LC}$ of the resonant circuit

$$G(\Omega) = \frac{j\Omega RC \dfrac{1}{\sqrt{LC}}}{j\Omega RC \dfrac{1}{\sqrt{LC}} - \Omega^2 + 1}, \quad \text{with } \Omega = \frac{\omega}{\omega_0}$$

Using $D = \dfrac{R}{2}\sqrt{\dfrac{C}{L}}$, the normalised **transfer function** becomes

$$\boxed{G(\Omega) = \frac{2jD\Omega}{2jD\Omega - \Omega^2 + 1}} \tag{5.22}$$

where D is the **damping ratio**. The normalised **magnitude response** is

$$\boxed{|G(\Omega)| = \frac{2D\Omega}{\sqrt{4D^2\Omega^2 + (1-\Omega^2)^2}}} \tag{5.23}$$

At the resonant frequency ω_0, which is also the centre frequency of the bandpass filter, the transfer function is

$$G(\Omega = 1) = 1 \quad \Rightarrow \quad |G(\omega = \omega_0)| = 1$$

The output signal at the lower and upper cutoff frequencies of the bandpass filter is 3 dB lower than at the centre frequency.

$$\frac{|G(\Omega_{3\,\mathrm{dB}})|}{|G(\Omega = 1)|} = \frac{1}{\sqrt{2}} \quad \Rightarrow \quad |G(\Omega_{3\,\mathrm{dB}})| = \frac{1}{\sqrt{2}}$$

The indices for the cutoff frequencies are omitted in the following analysis:

$$|G(\Omega)| = \frac{2D\Omega}{\sqrt{4D^2\Omega^2 + (1 - \Omega^2)^2}} = \frac{1}{\sqrt{2}}$$

This leads to the equation

$$4D^2\Omega^2 = (1 - \Omega^2)^2$$

This equation has four solutions, but only two of them yield positive frequencies

$$\omega_{\mathrm{lwr}} = \sqrt{D^2 + 1} - D, \qquad \omega_{\mathrm{upr}} = \sqrt{D^2 + 1} + D$$

where ω_{lwr} and ω_{upr} are the lower and upper cutoff frequencies, respectively. The **normalised bandwidth** of the filter is $2D$.

$$D = \frac{R}{2}\sqrt{\frac{C}{L}}, \qquad B = \frac{R}{2\pi L}, \qquad Q = \frac{1}{R}\sqrt{\frac{L}{C}} \qquad (5.24)$$

The bandwidth of the filter decreases with decreasing the resistance R. The normalised **phase response** is

$$\boxed{\varphi(\Omega) = \arctan\left[\frac{\mathrm{Im}(G(\Omega))}{\mathrm{Re}(G(\Omega))}\right] = \arctan\left(\frac{1 - \Omega^2}{2D\Omega}\right)} \qquad (5.25)$$

Figure 5.24 shows the Bode plot of the bandpass filter for different damping ratios.

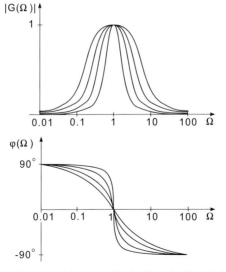

Fig. 5.24. Bode plot of the bandpass filter for different damping ratios D

NOTE: For this filter the centre frequency ω_0 is the harmonic mean of the lower and the upper cutoff frequencies ω_{lwr} and ω_{upr}. In normalised notation:

$$\sqrt{\Omega_{\text{lwr}} \cdot \Omega_{\text{upr}}} = \sqrt{(\sqrt{D^2+1} - D) \cdot (\sqrt{D^2+1} + D)} = 1$$

5.3.6 Filter Realisation

Electrical filters can be realised in a variety of ways. Some options are listed below.

RC filters consist only of resistors and capacitors. A disadvantage is the high attenuation.

LRC filters employ additional inductors to obtain a resonant network. These have sharper roll-offs than pure RC filters.

Reactance filters consist only of inductances and capacitances. Except for losses in inductors and capacitors no resistive components appear. As a consequence, they have high quality factors and steep slopes. They are mainly used in RF technology.

Active filters compensate for the losses of filters using operational amplifiers. With suitable circuits inductors can be completely avoided. Their use for high frequencies is limited by the critical frequency of the amplifiers (see Sect. 7.7 for details).

Switched capacitor filters (SC filters) are a type of active filters. Resistors are simulated by charging and discharging a capacitor at high frequency. The advantage is the possibility of varying the filter parameters with the frequency of the switching signal.

Quartz and ceramic filters are mechanical resonators with low losses. Both quality factor and stability are very high for quartz filters.

Mechanical filters were once the only possibility to obtain filters with steep slopes and were widely used in telephony.

Surface acoustic wave filters (SAW filters) convert electric signals into acoustic surface waves on a substrate. By suitable tapping of the crystal surface, the filter properties can be adjusted as required. These filters are suitable for high frequencies.

Digital filters work numerically on sampled signals. They have no inaccuracies caused by ageing, production tolerances or ambient temperature. Thanks to the progress in semiconductor manufacturing the usable frequency range is increasing while prices are decreasing.

5.4 Notation Index

A	voltage gain
$A(\omega)$	gain response (dB)
$\tilde{A}(\omega)$	attenuation (dB)
B	bandwidth (Hz)
$B_{3\,\text{dB}}$	3 dB bandwidth (Hz)
B_{rel}	relative bandwidth
$B(\omega)$	logarithmic phase response
D	damping ratio
F	shape factor (filter)
f_0	centre frequency, resonant frequency (Hz)
f_c	cutoff or corner frequency

f_{cl}	lower cutoff frequency
f_{cu}	upper cutoff frequency
$G(\omega)$	transfer function
$|G(\omega)|$	magnitude response
$G(\Omega)$	frequency-normalised transfer function
Im()	imaginary part
Q	quality factor, Q-factor
Re()	real part
T	transformation through a system
t_r	rise time
v_{in}	input voltage
v_{out}	output voltage
$\varphi(\omega)$	phase response
ω_0	resonant angular frequency (s^{-1})
ω_c	angular cutoff frequency
ω_{lwr}	lower cutoff frequency
ω_{upr}	upper cutoff frequency
Ω	normalised frequency
$\Omega_{3\ dB}$	normalised frequency, where the magnitude of the transfer function has decreased by 3 dB
Ω_{lwr}	lower normalised cutoff frequency
Ω_{upr}	upper normalised cutoff frequency

5.5 Further Reading

CHEN, C. T.: *Linear System Theory and Design, 3rd Edition*
Oxford University Press (1998)

DORF, R. C.: *The Electrical Engineering Handbook*
CRC Press (1999)

KENNEDY, G.; DAVIS, B.: *Electric Communication Systems*
McGraw-Hill (1992)

ZVEREV, A. I.: *Handbook of Filter Synthesis*
John Wiley & Sons (1967)

6 Signals and Systems

6.1 Signals

6.1.1 Definitions

In communications and electrical engineering signals are characterised in different classes.

Periodic signals are signals that repeat themselves after a definite time interval T (Fig. 6.1).
Definition: a value T exists, such that for all times t

$$f(t) = f(t + T)$$

T is the **period** of the signal $f(t)$.

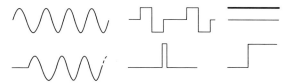

Fig. 6.1. Examples of periodic signals (*top*) and nonperiodic signals (*bottom*)

Nonperiodic signals are all signals that are not periodic according to the definition given above.
Causal signals are signals that have nonzero values only after time $t = 0$. The name is related to the definition of causal systems.

The normalised **power** of a signal is defined as

$$P = \lim_{T \to \infty} \frac{1}{2T} \int_{-T}^{T} |f(t)|^2 \, dt \qquad (6.1)$$

Analogously, the normalised **energy** of a signal is defined as

$$E = \lim_{T \to \infty} \int_{-T}^{T} |f(t)|^2 \, dt = \int_{-\infty}^{\infty} |f(t)|^2 \, dt \qquad (6.2)$$

Power signals have a finite normalised power P according to Eq. (6.1). For nonzero power signals $E = \infty$.
Energy signals have a finite normalised energy E. For energy signals $P = 0$.

- All periodic signals are power signals, but not all power signals are periodic.

Fig. 6.2. A power signal and two energy signals

EXAMPLE: The following signal is an energy signal (see Fig. 6.2, centre).

$$f(t) = \begin{cases} 0 & \text{for } t < 0 \\ e^{-t/\tau} & \text{for } t \geq 0 \end{cases}$$

$$E = \int_{-\infty}^{\infty} |f(t)|^2 \, dt = \int_0^{\infty} e^{-2t/\tau} \, dt = \left[-\frac{\tau}{2} e^{-2t/\tau} \right]_0^{\infty} = \frac{\tau}{2} < \infty$$

6.1.2 Symmetry Properties of Signals

A function is an **even function** if it holds for all t that

$$f(t) = f(-t)$$

These functions have an **axial symmetry** with respect to the ordinate (y-axis). They are also known as **symmetric functions** (Fig. 6.3).

A function is an **odd function** if it holds for all t that

$$f(t) = -f(-t)$$

Such functions have **point symmetry** with respect to the origin. They are also known as **antisymmetric functions** (Fig. 6.3).

Fig. 6.3. Examples of even (*left*) and odd functions (*right*)

EXAMPLE: The cosine is an even function, while the sine is an odd function.

NOTE: These properties are mutually exclusive. A function can either be *even* or *odd*, but not both (except the null function). However, there are functions that are neither even nor odd.

A signal has **full-wave symmetry** if it holds for all t

$$f\left(t + \frac{T}{2}\right) = f(t)$$

which means the signal effectively has a shorter period of $T/2$.

A signal has **half-wave symmetry** if it holds for all t

$$f\left(t + \frac{T}{2}\right) = -f(t)$$

This means that the half-waves would be axially symmetric about the time axis if they were shifted over each other (Fig. 6.4).

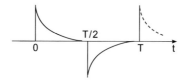

Fig. 6.4. Example of a signal with half-wave symmetry

EXAMPLE: A DC-free triangular signal has a half-wave symmetry.

6.2 Fourier Series

- Any periodic signal with a period T can be represented as a sum of harmonic signals. The lowest frequency is $1/T$. All other frequencies are integer multiples of this fundamental frequency. These signal components are called **harmonics**.

6.2.1 Trigonometric Form

If the signal $f(t)$ is periodic with a period T it can be represented by a Fourier series:

$$f(t) = \frac{a_0}{2} + \sum_{n=1}^{\infty} \left[a_n \cdot \cos(n\omega t) + b_n \cdot \sin(n\omega t) \right] \tag{6.3}$$

where ω is the **fundamental** (angular) **frequency** of the signal.

$$\omega = \frac{2\pi}{T} = 2\pi f$$

The **Fourier coefficients** a_n and b_n are

$$a_n = \frac{2}{T} \int_0^T f(t) \cdot \cos(n\omega t) \, dt$$

$$b_n = \frac{2}{T} \int_0^T f(t) \cdot \sin(n\omega t) \, dt$$

for $n = 0, 1, 2 \ldots$ \quad (6.4)

- $\dfrac{a_0}{2} = \dfrac{1}{T}\displaystyle\int_0^T f(t)\,\mathrm{d}t$ is the average value of the signal over one period, i.e. the **DC part** of the signal. Note that b_0 is always zero.

- The trigonometric representation of the Fourier series depends on the choice of the starting time $t = 0$ of the signal.

NOTE: Since the signal $f(t)$ is periodic, it is irrelevant whether the integration limits are 0 to T, or $-T/2$ to $+T/2$.

NOTE: The equivalent Fourier representation is also found in the literature.

$$f(t) = \sum_{n=0}^{\infty}\Big[a_n \cdot \cos(n\omega t) + b_n \cdot \sin(n\omega t)\Big]$$

In that case a_0 has to be defined separately:

$$a_0 = \frac{1}{T}\int_0^T f(t)\,\mathrm{d}t$$

NOTE: The mathematical conditions for convergence of the Fourier series in Eq. (6.3) are:

- The signal has a finite number of noncontinuous points;
- The average value over one period is finite;
- The signal has a finite number of maxima and minima.

These conditions always hold for signals that can be physically realised.

6.2.1.1 Symmetry Properties

- For pure alternating signals $a_0 = 0$.
- Even functions do not contain sine components. This means all $b_n = 0$.
- Odd functions do not contain cosine components. This means all $a_n = 0$.
- Waveforms with full-wave symmetry have only even harmonics with frequencies $0, 2\omega, 4\omega \ldots$
- Waveforms with half-wave symmetry have only odd harmonics with frequencies $\omega, 3\omega, 5\omega \ldots$

6.2.2 Amplitude–Phase Form

The addition of sine and cosine functions with the same frequency results in a harmonic function of this frequency.

$$a_n \cdot \cos(n\omega t) + b_n \cdot \sin(n\omega t) = A_n \cdot \cos(n\omega t + \varphi_n)$$

This leads to the **amplitude–phase form** of the Fourier series (Fig. 6.5).

$$\boxed{f(t) = \frac{a_0}{2} + \sum_{n=1}^{\infty} A_n \cdot \cos(n\omega t + \varphi_n)} \tag{6.5}$$

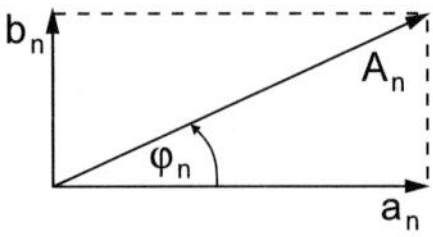

Fig. 6.5. Combination of the Fourier coefficients a_n and b_n for the amplitude–phase form

with

$$A_n = \sqrt{a_n^2 + b_n^2}, \qquad \varphi_n = -\arctan\left(\frac{b_n}{a_n}\right) \qquad \text{for } n = 1, 2, 3\ldots \tag{6.6}$$

where a_n and b_n are the Fourier coefficients according to Eq. (6.4). The set of all A_n is known as the **amplitude spectrum**, and the set of φ_n is the **phase spectrum**.

- The amplitude spectrum is independent of the choice of the starting point $t = 0$. This does not hold for the phase spectrum.

6.2.3 Exponential Form

Applying

$$\cos(n\omega t) = \frac{1}{2}\left(\mathrm{e}^{\mathrm{j}n\omega t} + \mathrm{e}^{-\mathrm{j}n\omega t}\right), \qquad \sin(n\omega t) = \frac{1}{2\mathrm{j}}\left(\mathrm{e}^{\mathrm{j}n\omega t} - \mathrm{e}^{-\mathrm{j}n\omega t}\right) \tag{6.7}$$

the trigonometric form of the Fourier series can be converted into the **complex normal form**, or the **exponential form**.

$$f(t) = \sum_{n=-\infty}^{\infty} c_n \cdot \mathrm{e}^{\mathrm{j}n\omega t} \tag{6.8}$$

The **complex Fourier coefficients** c_n are calculated as

$$c_n = \frac{1}{T}\int_0^T f(t) \cdot \mathrm{e}^{-\mathrm{j}n\omega t}\, \mathrm{d}t \tag{6.9}$$

The set of all c_n is called the **complex spectrum**. Positive and negative frequency parameters $n\omega$ and $-n\omega$ appear in this representation of the Fourier series. This leads to the concept of a **positive** and **negative frequency spectrum**.

- The coefficient c_0 represents the DC component. It therefore is equivalent to $a_0/2$.
- The spectral component of a harmonic with an angular frequency of $n\omega$ is

$$c_n \cdot \mathrm{e}^{\mathrm{j}n\omega t} + c_{-n} \cdot \mathrm{e}^{-\mathrm{j}n\omega t}$$

- The spectral coefficients c_n and c_{-n} are complex conjugates (for real-valued signals), i.e. $c_n^* = c_{-n}$.
- The complex Fourier coefficients have a magnitude that is half the value of the corresponding amplitude elements in the amplitude–phase form: $2|c_n| = A_n$.

EXAMPLE: Figure 6.6 shows the two-sided spectra of the cosine and the sine functions. Using Eq. (6.7), both functions can be expressed as

$$\cos \omega_0 t = \underbrace{+\frac{1}{2}\,\mathrm{e}^{\mathrm{j}\omega_0 t}}_{c_1} + \underbrace{\frac{1}{2}\,\mathrm{e}^{-\mathrm{j}\omega_0 t}}_{c_{-1}},$$

$$\sin \omega_0 t = \underbrace{-\frac{\mathrm{j}}{2}\,\mathrm{e}^{\mathrm{j}\omega_0 t}}_{c_1} + \underbrace{\frac{\mathrm{j}}{2}\,\mathrm{e}^{-\mathrm{j}\omega_0 t}}_{c_{-1}}$$

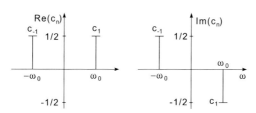

Fig. 6.6. Frequency spectrum of the cosine and sine functions with components of positive and negative frequencies

6.2.3.1 Symmetry Properties

- Even functions have purely real spectral coefficients c_n.
- Odd functions have purely imaginary spectral coefficients c_n.

6.2.4 Overview: Fourier Series Representations

Tables 6.1 and 6.2 present a summary of the Fourier series representations and coefficients as well as conversion between these representations.

Table 6.1. Summary of the Fourier series representations

Series representations	Coefficients
Real normal form $$f(t) = \frac{a_0}{2} + \sum_{n=1}^{\infty}\left[a_n \cdot \cos(n\omega t) + b_n \cdot \sin(n\omega t)\right]$$	$a_n = \dfrac{2}{T}\displaystyle\int_0^T f(t)\cdot\cos(n\omega t)\,\mathrm{d}t$ $b_n = \dfrac{2}{T}\displaystyle\int_0^T f(t)\cdot\sin(n\omega t)\,\mathrm{d}t$ for $n = 0, 1, 2\ldots$
Amplitude–phase form $$f(t) = \frac{a_0}{2} + \sum_{n=1}^{\infty} A_n \cdot \cos(n\omega t + \varphi_n)$$	$A_n = \sqrt{a_n^2 + b_n^2}$ $\varphi_n = -\arctan\left(\dfrac{b_n}{a_n}\right)$ for $n = 1, 2, 3\ldots$
Complex normal form $$f(t) = \sum_{n=-\infty}^{\infty} c_n \cdot \mathrm{e}^{\mathrm{j}n\omega t}$$	$c_n = \dfrac{1}{T}\displaystyle\int_0^T f(t)\cdot\mathrm{e}^{-\mathrm{j}n\omega t}\,\mathrm{d}t$

Table 6.2. Summary of conversion between the Fourier series represenations

	Conversion of representations		
	Fourier coefficients	Spectral coefficients	Complex Fourier coefficients
$a_n =$	a_n	$A_n \cdot \cos \varphi_n$	$c_n + c_n^* = 2 \cdot \mathrm{Re}(c_n)$
$b_n =$	b_n	$A_n \cdot \sin \varphi_n$	$\mathrm{j}(c_n - c_n^*) = 2 \cdot \mathrm{Im}(c_n)$
$A_n =$	$\sqrt{a_n^2 + b_n^2}$	A_n	$2 \cdot |c_n|$
$\varphi_n =$	$-\arctan\left(\dfrac{b_n}{a_n}\right)$	φ_n	$-\arg(c_n)$
$c_n =$	$\begin{cases} \dfrac{a_0}{2} & n = 0 \\ \dfrac{a_n}{2} - \mathrm{j}\dfrac{b_n}{2} & n > 0 \\ \dfrac{a_n}{2} + \mathrm{j}\dfrac{b_n}{2} & n < 0 \end{cases}$	$\dfrac{A_n}{2} \cdot \mathrm{e}^{-\mathrm{j}\varphi_n}$	c_n

6.2.5 Useful Integrals for the Calculation of Fourier Coefficients

The average value of the sine and cosine functions over one period is zero.

$$\int_0^T \cos n\omega t \, \mathrm{d}t = 0 \tag{6.10}$$

$$\int_0^T \sin n\omega t \, \mathrm{d}t = 0 \tag{6.11}$$

The sine and the cosine function are **orthogonal**:

$$\int_0^T \sin n\omega t \cdot \sin k\omega t \, \mathrm{d}t = 0, \quad \text{for } n \neq k \tag{6.12}$$

$$\int_0^T \sin n\omega t \cdot \cos k\omega t \, \mathrm{d}t = 0 \tag{6.13}$$

$$\int_0^T \cos n\omega t \cdot \cos k\omega t \, \mathrm{d}t = 0, \text{ for } n \neq k \tag{6.14}$$

For integrals over the product of sine and cosine functions of the same frequency $n\omega$, it holds respectively that

$$\int_0^T \sin^2 n\omega t \, \mathrm{d}t = \frac{T}{2} \tag{6.15}$$

$$\int_0^T \cos^2 n\omega t \, \mathrm{d}t = \frac{T}{2} \tag{6.16}$$

6.2 Fourier Series

The orthogonality conditions can be summarised by

$$\int_0^T \cos n\omega t \cdot \cos k\omega t \, dt = \delta_{nk} \cdot \frac{T}{2}$$

The same holds for sine functions. Here δ_{nk} is the **Kronecker symbol**. Its value is unity for $n = k$, otherwise zero.

6.2.6 Useful Fourier Series

The Fourier series of several functions are given in Table 6.3. Table 6.4 gives the amplitude spectra of the signals.

Table 6.3. Useful Fourier series

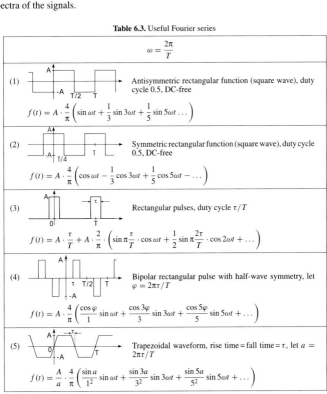

$$\omega = \frac{2\pi}{T}$$

(1) Antisymmetric rectangular function (square wave), duty cycle 0.5, DC-free

$$f(t) = A \cdot \frac{4}{\pi} \left(\sin \omega t + \frac{1}{3} \sin 3\omega t + \frac{1}{5} \sin 5\omega t \ldots \right)$$

(2) Symmetric rectangular function (square wave), duty cycle 0.5, DC-free

$$f(t) = A \cdot \frac{4}{\pi} \left(\cos \omega t - \frac{1}{3} \cos 3\omega t + \frac{1}{5} \cos 5\omega t - \ldots \right)$$

(3) Rectangular pulses, duty cycle τ/T

$$f(t) = A \cdot \frac{\tau}{T} + A \cdot \frac{2}{\pi} \cdot \left(\sin \pi \frac{\tau}{T} \cdot \cos \omega t + \frac{1}{2} \sin \pi \frac{2\tau}{T} \cdot \cos 2\omega t + \ldots \right)$$

(4) Bipolar rectangular pulse with half-wave symmetry, let $\varphi = 2\pi\tau/T$

$$f(t) = A \cdot \frac{4}{\pi} \left(\frac{\cos \varphi}{1} \sin \omega t + \frac{\cos 3\varphi}{3} \sin 3\omega t + \frac{\cos 5\varphi}{5} \sin 5\omega t + \ldots \right)$$

(5) Trapezoidal waveform, rise time = fall time = τ, let $a = 2\pi\tau/T$

$$f(t) = \frac{A}{a} \cdot \frac{4}{\pi} \left(\frac{\sin a}{1^2} \sin \omega t + \frac{\sin 3a}{3^2} \sin 3\omega t + \frac{\sin 5a}{5^2} \sin 5\omega t + \ldots \right)$$

Table 6.3. (cont.)

(6)	 $$f(t) = A \cdot \frac{8}{\pi^2} \left(\sin \omega t - \frac{1}{3^2} \sin 3\omega t + \frac{1}{5^2} \sin 5\omega t - \ldots \right)$$	Antisymmetric triangular waveform (half-wave symmetry), DC-free
(7)	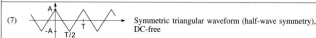 $$f(t) = A \cdot \frac{8}{\pi^2} \left(\cos \omega t + \frac{1}{3^2} \cos 3\omega t + \frac{1}{5^2} \cos 5\omega t + \ldots \right)$$	Symmetric triangular waveform (half-wave symmetry), DC-free
(8)	 $$f(t) = A \cdot \frac{2}{\pi} \left(\sin \omega t + \frac{1}{2} \sin 2\omega t + \frac{1}{3} \sin 3\omega t + \ldots \right)$$	Sawtooth waveform, DC-free, antisymmetric
(9)	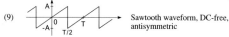 $$f(t) = A \cdot \frac{2}{\pi} \left(\sin \omega t - \frac{1}{2} \sin 2\omega t + \frac{1}{3} \sin 3\omega t - \ldots \right)$$	Sawtooth waveform, DC-free, antisymmetric
(10)	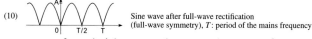 $$f(t) = A \cdot \frac{2}{\pi} - A \cdot \frac{4}{\pi} \cdot \left(\frac{1}{1 \cdot 3} \cos 2\omega t + \frac{1}{3 \cdot 5} \cos 4\omega t + \frac{1}{5 \cdot 7} \cos 6\omega t + \ldots \right)$$	Sine wave after full-wave rectification (full-wave symmetry), T: period of the mains frequency
(11)	 $$f(t) = A \cdot \frac{2}{\pi} + A \cdot \frac{4}{\pi} \cdot \left(\frac{1}{1 \cdot 3} \cos 2\omega t - \frac{1}{3 \cdot 5} \cos 4\omega t + \frac{1}{5 \cdot 7} \cos 6\omega t - \ldots \right)$$	Cosine wave after full-wave rectification (full-wave symmetry), T: period of the mains frequency
(12)	 $$f(t) = A \cdot \frac{1}{\pi} + A \cdot \frac{2}{\pi} \cdot \left(\pi \cos \omega t + \frac{1}{1 \cdot 3} \cos 2\omega t + \frac{1}{3 \cdot 5} \cos 4\omega t \ldots + \right)$$	Cosine wave after half-wave rectification
(13)	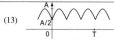 $$f(t) = A \cdot \frac{3\sqrt{3}}{\pi} \cdot \left(\frac{1}{2} - \frac{1}{2 \cdot 4} \cos 3\omega t - \frac{1}{5 \cdot 7} \cos 6\omega t - \frac{1}{8 \cdot 10} \cos 9\omega t - \ldots \right)$$	Rectified three-phase current, T: period of the mains frequency
(14)	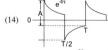 $$f(t) = A \cdot \frac{2}{\pi} \cdot \sum_{n=0}^{\infty} \frac{\gamma \cos \left[(2n+1)\omega t\right] + (2n+1) \sin \left[(2n+1)\omega t\right]}{\gamma^2 + (2n+1)^2}$$	Rectangular waveform passing through RC circuit, time constant τ, let $\gamma = T/2\pi\tau$

Table 6.4. Amplitude spectra A_n of the signals

Signal	Factor	Harmonic								
number	A	1	2	3	4	5	6	7	8	9
1 and 2	$4/\pi$	1	0	1/3	0	1/5	0	1/7	0	1/9
3($\tau/T = 1/3$)	$2/\pi$	.87	.43	0	.22	.17	0	.12	.11	0
3($\tau/T = 1/5$)	$2/\pi$	.59	.48	.32	.15	0	.098	.14	.12	.065
6 and 7	$8/\pi^2$	1	0	1/9	0	1/25	0	1/49	0	1/81
8 and 9	$2/\pi$	1	1/2	1/3	1/4	1/5	1/6	1/7	1/8	1/9
10 and 11	$4/\pi$	0	1/3	0	1/15	0	1/35	0	1/63	0
12	$2/\pi$	π	1/3	0	1/15	0	1/35	0	1/63	0
13	$3\sqrt{3}/\pi$	0	0	1/8	0	0	1/35	0	0	1/80

6.2.7 Application of the Fourier Series

6.2.7.1 Spectrum of a Rectangular Signal

A TTL (transistot–transistor logic) gate delivers the signal shown in Fig. 6.7. The duty cycle of this signal is 0.5. The amplitude spectrum is to be determined.

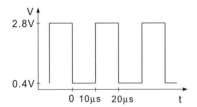

Fig. 6.7. Idealised rectangular pulses from a TTL circuit

Signals (1) or (2) from Table 6.3 are closest to the signal in Fig. 6.7. Defining the time origin $t = 0$ is a matter of choice in this example. The peak-to-peak amplitude of the digital signal is 2.4 V. This means that $A = 1.2$ V, and the DC component is 1.6 V.

The period is $T = 20$ µs, so $\omega = 2\pi \cdot 50$ kHz. The Fourier series of the signal above is, according to signal (2) from Table 6.3

$$g(t) = 1.6\,\text{V} + 1.2\,\text{V} \cdot \frac{4}{\pi} \cdot \left(\cos \omega t - \frac{1}{3} \cos 3\omega t + \ldots \right)$$

The amplitudes of the individual spectral components are

f (kHz)	0	50	100	150	200	250	300	350	400	450
A_n	1.6 V	1.53 V	0	0.51 V	0	0.31 V	0	0.22 V	0	0.17 V

Figure 6.8 gives a graphical representation of the amplitude spectrum. The small circles indicate that the respective spectral components vanish, even though they are harmonics. The term **line spectrum** is derived from this kind of representation.

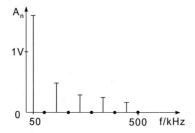

Fig. 6.8. Amplitude spectrum of the rectangular signal from Fig. 6.7 (DC-free)

Figure 6.9 shows the superposition of the harmonics with ω, 3ω and 5ω to a rectangular signal. As already shown in Table 6.4, the amplitude of the fundamental frequency is higher than the resulting rectangular signal.

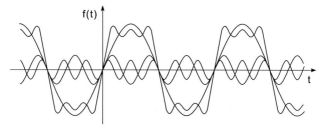

Fig. 6.9. Superposition of spectral components up to the fifth harmonic to compose a rectangular signal

6.2.7.2 Spectrum of a Sawtooth Signal

Figure 6.10 shows a sawtooth signal with falling slopes. The signal is composed of the inverted signal (8) from the table and a DC component of 1.5 V. The amplitude is $A = 1.5$ V, and the fundamental frequency of the signal is $f = 1/T = 4$ kHz.

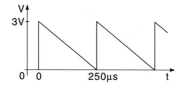

Fig. 6.10. Sawtooth signal with falling slopes and a DC component

The Fourier series of this sawtooth signal is

$$g(t) = 1.5 \text{ V} - 1.5 \text{ V} \cdot \frac{2}{\pi} \cdot \left(\sin \omega t + \frac{1}{2} \sin 2\omega t + \frac{1}{3} \sin 3\omega t + \ldots \right)$$

f (kHz)	0	4	8	12	16	20	24	28	32	36
A_n	1.5 V	0.95 V	0.48 V	0.32 V	0.24 V	0.19 V	0.16 V	0.14 V	0.12 V	0.11 V

The amplitude spectrum of this signal is shown in Fig. 6.11. Unlike the rectangular signal, this spectrum also contains even harmonics.

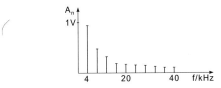

Fig. 6.11. Amplitude spectrum of the sawtooth signal shown in Fig. 6.10

6.2.7.3 Spectrum of a Composite Signal

The complicated signal shown in Fig. 6.12 is composed of the superposition of a rectangular signal with an amplitude of 2 V and a triangular signal with an amplitude of 1 V. Both signals correspond either to signals (1) and (6) or (2) and (7) in Table 6.3.

$$g(t) = 2\,\text{V} \cdot \frac{4}{\pi} \left[\cos\omega t + \frac{1}{3} \cos 3\omega t + \frac{1}{5} \cos 5\omega t + \ldots \right]$$

$$h(t) = 1\,\text{V} \cdot \frac{8}{\pi^2} \left[\cos\omega t + \frac{1}{3^2} \cos 3\omega t + \frac{1}{5^2} \cos 5\omega t + \ldots \right]$$

$$f(t) = \frac{8}{\pi}\,\text{V} \left[\left(1 + \frac{1}{\pi}\right) \cos\omega t + \left(\frac{1}{3} + \frac{1}{3^2\pi}\right) \cos 3\omega t + \left(\frac{1}{5} + \frac{1}{5^2\pi}\right) \cos 5\omega t + \ldots \right]$$

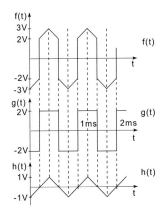

Fig. 6.12. Superposition of a rectangular signal and a triangular signal

The Fourier coefficients for each frequency are added (taking into account the proper signs). The fundamental frequency of the signal is 1 kHz. The amplitude spectrum is then:

f (kHz)	1	2	3	4	5	6	7	8	9
A_n	2.55 V	0	0.92 V	0	0.54 V	0	0.38 V	0	0.29 V

6.3 Systems

6.3.1 System Properties

Often in communications engineering the internal structure of a system is of no particular interest. It is more interesting to consider the behaviour of the input and output signals, which in most cases are voltages. Such systems are often called **black-box** systems (Fig. 6.13). The function of a system is described symbolically by a transformation of the input signal into the output signal.

$$v_{\text{out}} = T(v_{\text{in}})$$

Fig. 6.13. A black-box system with input and output signals

6.3.1.1 Linear Systems

Many systems can be modelled to a good approximation as **linear systems**. It then holds that

$$T(\alpha \cdot v_{\text{in}}) = \alpha \cdot T(v_{\text{in}}) \tag{6.17}$$

- The output signal is proportional to the input signal.

$$T(v_1 + v_2) = T(v_1) + T(v_2) \tag{6.18}$$

- Either of the two input signals can be considered individually while passing through the system as if the other were not present (Fig. 6.14).

Fig. 6.14. Principle of superposition in linear systems

The approach shown in Eq. (6.18) and Fig. 6.14 is called **principle of superposition**.

6.3.1.2 Causal Systems

Causal systems show no system response *before* the excitation.

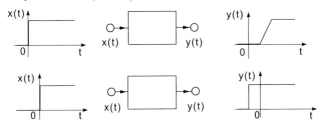

Fig. 6.15. A causal (*top*) and a noncausal system (*bottom*)

Mathematically this can be expressed by

$$x(t) = 0, \text{ for } t < t_0. \text{ It follows that } T[x(t)] = 0, \text{ for } t < t_0 \qquad (6.19)$$

NOTE: According to the definition of causal systems, causal *signals* are defined. Their values are always zero before the time $t = 0$ and can be nonzero after $t = 0$ (Fig. 6.15).

6.3.1.3 Time-Invariant Systems

Time-invariant systems do not change their inner properties. Their response to a specific input signal is always identical and does not depend on the time of its arrival.

Mathematically this can be expressed by

$$y(t) = T[x(t)]. \text{ It follows that } T[x(t - t_0)] = y(t - t_0) \qquad (6.20)$$

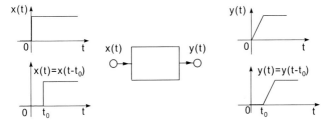

Fig. 6.16. System response of a time-invariant system

If the input signal is shifted in time, the corresponding output signal experiences the same shift (Fig. 6.16).

6.3.1.4 Stable Systems

Systems are considered **stable** if the system response to signals with a finite amplitude are signals with finite amplitudes. Mathematically this can be expressed by

$$|x(t)| < M < \infty \Rightarrow |T[x(t)]| < N < \infty, \quad \text{for all } t \tag{6.21}$$

6.3.1.5 LTI Systems

Linear time-invariant systems or **LTI systems** are of special interest.

It is normally assumed that the systems are causal, since noncausal systems cannot be realised in the time domain.

- Systems that are composed of resistors, inductors, capacitors, transformers and linearly controlled sources (transistors in small-signal operation) can be described to a good approximation as LTI systems. However, caution is required in the case of positive feedback.

6.3.2 Elementary Signals

In order to describe systems their response to typical test signals is evaluated. The most important test signals are described below. The use of the symbols varies in the literature.

6.3.2.1 The Step Function

The **step function** $s(t)$ is zero until $t = 0$ and is 1 for all $t > 0$ (Fig. 6.17).

$$s(t) = \begin{cases} 0 & \text{for } t < 0 \\ 1 & \text{else} \end{cases} \tag{6.22}$$

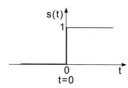

Fig. 6.17. The step function

- The **step function** is a **power signal**.

6.3.2.2 The Rectangular Pulse

The **rectangular pulse** $\text{rect}(t)$ is a rectangular signal that is symmetrical about the time $t = 0$ and that has unity area (Fig. 6.18).

$$\text{rect}(t) = \begin{cases} 1, & \text{for } |t| < 1/2 \\ 0, & \text{else} \end{cases} \tag{6.23}$$

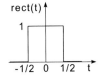

Fig. 6.18. The rectangular pulse

- The **rectangular pulse** is an **energy signal**.

6.3.2.3 The Triangular Pulse

The **triangular pulse** $\Lambda(t)$ is a triangular signal that is symmetrical about the time $t = 0$ and that has unity area (Fig. 6.19).

$$\Lambda(t) = \begin{cases} 1 - |t|, & \text{for } |t| < 1 \\ 0, & \text{else} \end{cases} \tag{6.24}$$

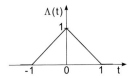

Fig. 6.19. The triangular pulse

- The **triangular pulse** is an **energy signal**.

6.3.2.4 The Gaussian Pulse

The **Gaussian pulse** $\Gamma(t)$ is a pulse that is symmetrical about the time $t = 0$ and that has unity area (Fig. 6.20).

$$\Gamma(t) = e^{-\pi t^2} \tag{6.25}$$

- The **Gaussian pulse** is an **energy signal**.

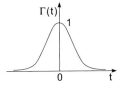

Fig. 6.20. The Gaussian pulse

6.3.2.5 The Impulse Function (Delta Function)

The **impulse function** is the limit of a family of realisable signals. The family being considered is the rectangular impulses

$$\text{rect}_n = n \cdot \text{rect}(n \cdot t)$$

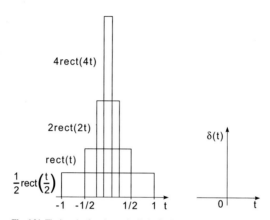

Fig. 6.21. The impulse function as the limit of a family of rectangular pulses

All impulses have unity area. The sequence of these impulses converges to the limit $\delta(t)$ as $n \to \infty$, with the following properties

$$\delta(t) = 0 \quad \text{for } t \neq 0, \qquad \int_{-\infty}^{\infty} \delta(t)\,dt = 1$$

Because the value for $t = 0$ is not defined, $\delta(t)$ is not a function in the usual sense. It is called the **delta function**, the **impulse function** or the **Dirac impulse** (Fig. 6.21). It is characterised by the following properties:

$$\int_{-\infty}^{\infty} \delta(t)\,dt = 1, \qquad \int_{-\infty}^{\infty} f(t)\delta(t-t_0)\,dt = f(t_0), \qquad \delta(t) = \delta(-t) \tag{6.26}$$

These properties mean:

- The delta function has unity area.
- The delta function filters out the function value under the integral of the function, where the argument of the delta function is zero.
- The delta function is an even function.

NOTE: The delta function can also be represented as the limit of a sequence of Gaussian functions. The functions become narrower and higher, while the area under the function is always unity.

The **derivative** (in a generalised sense) of the **step function** is the delta function:

$$[s(t)]' = \delta(t), \quad \int_{-\infty}^{t} \delta(\tau)\, d\tau = s(t) \tag{6.27}$$

EXAMPLE: Figure 6.22 (upper left) shows the function

$$f(t) = s(t-1) + s(t-2) - 2 \cdot s(t-3)$$

Its generalised derivative is

$$f'(t) = \delta(t-1) + \delta(t-2) - 2 \cdot \delta(t-3)$$

The signal is shown directly beneath.

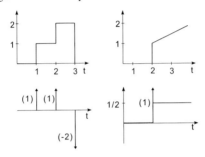

Fig. 6.22. Two signals (*top*) and their generalised derivatives (*bottom*)

The function $f(t) = s(t-2) \cdot \frac{t}{2}$ is shown on the right of Fig. 6.22. Its derivative is calculated by the product rule as

$$f'(t) = \left[s(t-2) \cdot \frac{t}{2} \right]' = \delta(t-2) \cdot \frac{t}{2} + s(t-2) \cdot \frac{1}{2} = \delta(t-2) + \frac{1}{2} \cdot s(t-2)$$

6.3.3 Shifting and Scaling of Time Signals

The function

$$f(t) = s(t - t_0)$$

represents a **time-shifted** step function in which the step is shifted or delayed by the time t_0 (Fig. 6.23).

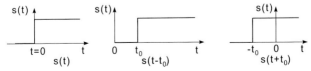

Fig. 6.23. Time-shifted step function

The signal
$$f(t) = \text{rect}\left(\frac{t}{a}\right) \qquad a > 0$$
represents a **time-scaled** rectangular pulse, where a is the time-scaling factor (Fig. 6.24). For $a > 1$ the pulse widens, and for $a < 1$ it narrows.

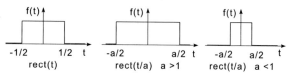

Fig. 6.24. Time-scaled rectangular impulse

EXAMPLE: The signal
$$f(t) = \frac{3}{2} \cdot \Lambda\left(\frac{t}{2} - 1\right) = \frac{3}{2} \cdot \Lambda\left(\frac{t-2}{2}\right)$$
represents a time-scaled and time-shifted triangular pulse (Fig. 6.25).

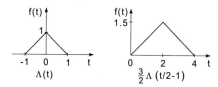

Fig. 6.25. Time-scaled and time-shifted triangular impulse

6.3.4 System Responses

- LTI systems respond to a harmonic input signal with a harmonic output signal of the **same frequency**, while its amplitude and phase are usually changed.

NOTE: Systems responding to harmonic input signals with nonharmonic output signals are called **nonlinear systems**. The output signal contains components with frequencies different from the input signal.

Systems can be characterised by their output signals for defined input signals.

6.3.4.1 Impulse Response

The **impulse response** is the output signal of a system excited by a delta function (Fig. 6.26).
$$g(t) = T\{\delta(t)\}$$
Function $g(t)$ is also known as the **weighting function** of the system.

Fig. 6.26. Impulse response of an LTI system

EXAMPLE: The impulse response of the RC circuit shown in Fig. 6.27 is a declining exponential function.

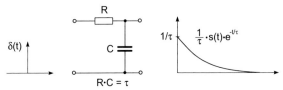

Fig. 6.27. Impulse response of an RC low-pass filter

The capacitor is charged instantaneously by the impulse and then discharges via the resistor with a time constant $\tau = RC$. Multiplication by the step function $s(t)$ enforces a nonzero response only after the time $t = 0$.

NOTE: In order to evaluate the impulse response of a real system the system is excited with narrow rectangular pulses. Delta pulses cannot be generated in reality. However, the narrower the pulses the lower is their energy content. The impulse amplitude cannot be increased arbitrarily for real systems since the systems' outputs are amplitude limited. Therefore the step response is often preferred.

6.3.4.2 Step Response

The **step response** is the output signal of a system that is fed with the step function (Fig. 6.28).

$$h(t) = T\{s(t)\}$$

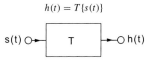

Fig. 6.28. Step response of an LTI system

- The step response is the integral of the impulse response.

$$h(t) = \int_{-\infty}^{t} g(\tau)\, d\tau$$

NOTE: In order to determine the impulse response of a system, the step response is often determined first and the impulse response is calculated as its derivative.

EXAMPLE: The RC circuit in Fig. 6.29 reacts to the step function with

$$h(t) = s(t) \cdot (1 - e^{-t/\tau})$$

Derivation of the step response yields the impulse response:

$$g(t) = h'(t) = s'(t) \cdot (1 - e^{-t/\tau}) + s(t) \cdot (1 - e^{-t/\tau})' \quad \text{(Product rule)}$$
$$= \underbrace{\delta(t) \cdot (1 - e^{-t/\tau})}_{(1-1)} + s(t) \cdot \tfrac{1}{\tau} e^{-t/\tau} = s(t) \cdot \tfrac{1}{\tau} e^{-t/\tau} = g(t)$$

The delta function effectively filters out all elements of the summation except the element at $t = 0$.

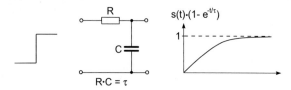

Fig. 6.29. Step response of an RC low-pass filter

6.3.4.3 System Response to Arbitrary Input Signals

The system shown in Fig. 6.30 with the impulse response $g(t)$ responds to an input signal $x(t)$ with an output signal given by

$$y(t) = \int_{-\infty}^{\infty} x(\tau) \cdot g(t - \tau) \, d\tau \tag{6.28}$$

The operation on $x(t)$ and $g(t)$ defined in Eq. (6.28) is known as **convolution**. This is often written symbolically

$$y(t) = x(t) * g(t)$$

Spoken as: x convolved with g.

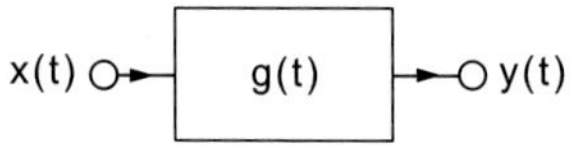

Fig. 6.30. Input and output signals of a system with the impulse response $g(t)$

NOTE: For the special input signal $x(t) = \delta(t)$ the system response is

$$y(t) = \int_{-\infty}^{\infty} \delta(\tau) \cdot g(t - \tau) \, d\tau = g(t)$$

which is the impulse response.

NOTE: In practice, the system response is rarely calculated using convolution. It is more efficient to calculate in the frequency domain. However, in time discrete-systems, such as digital filters, the convolution integral becomes a summation, which is calculated explicitly in signal processors.

6.3.4.4 Rules of Convolution

Let $f(t)$, $g(t)$ and $h(t)$ be arbitrary time functions.

It holds that

$$0 * f(t) = 0, \qquad \delta(t) * f(t) = f(t) \tag{6.29}$$

The delta function acts in convolution of functions as unity in the multiplication of numbers. This is also known as the **convolution product**, because of this property and the commutative, associative and distributive laws.

Commutative Law

$$f(t) * g(t) = g(t) * f(t) \tag{6.30}$$

- Input signal and impulse response can be exchanged (Fig. 6.31).

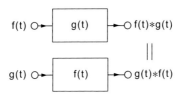

Fig. 6.31. Commutative law of convolution

Associative Law

$$f(t) * g(t) * h(t) = f(t) * \bigl[g(t) * h(t)\bigr] \tag{6.31}$$

- Two cascaded systems can be combined into a single system. The impulse response of the total system then is the convolution of the individual impulse responses (Fig. 6.32).

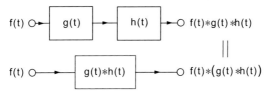

Fig. 6.32. Associative law of convolution

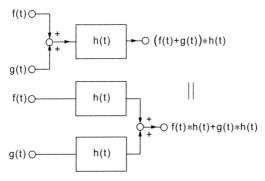

Fig. 6.33. Distributive law of convolution

Distributive Law

$$(f(t) + g(t)) * h(t) = f(t) * h(t) + g(t) * h(t) \quad (6.32)$$

- Each signal can be treated as if it passes through the system independently of other signals. The systems' outputs are finally added (Fig. 6.33) through the principle of superposition.

6.3.4.5 Transfer Function

An LTI system with an impulse response $g(t)$ reacts according to Eq. (6.28) to the special input signal $x(t) = e^{j\omega t}$ as follows:

$$y(t) = \int_{-\infty}^{\infty} e^{j\omega(t-\tau)} \cdot g(\tau) \, d\tau = e^{j\omega t} \cdot \underbrace{\int_{-\infty}^{\infty} g(\tau) \cdot e^{-j\omega\tau} \, d\tau}_{G(\omega)}$$

The input signal $e^{j\omega t}$ appears at the output weighted by the complex factor $G(\omega)$.

$$y(t) = G(\omega) \cdot x(t), \quad \text{for } x(t) = e^{j\omega t}$$

The factor $G(\omega)$ is called the **transfer function** of the system (Fig. 6.34).

$$\boxed{G(\omega) = \int_{-\infty}^{\infty} g(t) \cdot e^{-j\omega t} \, dt} \quad (6.33)$$

- The **transfer function** is the **Fourier transform** of the **impulse response**. Either is equivalent in the representation of a system.

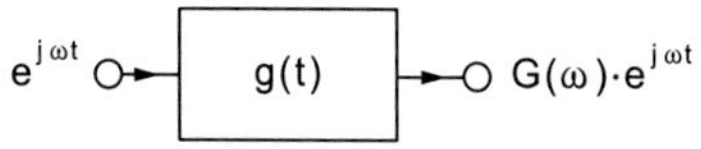

$$e^{j\omega t} \longrightarrow \boxed{g(t)} \longrightarrow G(\omega) \cdot e^{j\omega t}$$

Fig. 6.34. System response to a complex harmonic input signal

NOTE: The transfer function for LTI systems with a known internal structure can be determined using complex calculus.

6.3.4.6 System Response Calculation in the Frequency Domain

It is often more efficient to calculate the system response to an arbitrary excitation in the frequency domain rather than using convolution according to Eq. (6.28).

Convolution in the time domain is equivalent to multiplication in the frequency domain.

$$y(t) = x(t) * g(t), \quad Y(\omega) = X(\omega) \cdot G(\omega) \quad (6.34)$$

$X(\omega)$ and $Y(\omega)$ are the spectra of the input and output signals, respectively, and $G(\omega)$ is the transfer function of the system. The calculation of the system response is performed according to the following procedure (Fig. 6.35):

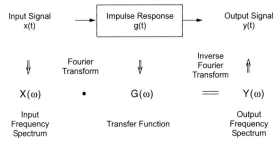

Fig. 6.35. Calculation of the system response in the time domain and in the frequency domain

- calculate the **input spectrum** $X(\omega)$ by applying the Fourier transform to the input signal $x(t)$;
- calculate the **transfer function** $G(\omega)$ using complex calculus;
- calculate the **output spectrum** $Y(\omega)$ by multiplying the input spectrum with the transfer function;
- calculate the **output signal** $y(t)$ by applying the inverse Fourier transform to the output spectrum $Y(\omega)$.

NOTE: Despite the fact that this calculation method appears much more complicated than convolution, it is often the most suitable for problems in Communications.

6.3.5 Impulse and Step Response Calculation

6.3.5.1 Normalisation of Circuits

For easier manipulation all signals in systems theory are considered without their units. Problems occur when impulse and step responses for a particular physical system are investigated. In this case **normalisation** helps.

Impedance normalisation: All resistances are referred to the reference resistance R_r. The normalised resistance values R_n are calculated as $R_n = R/R_r$. The reference resistance is chosen such that most resistances in the circuit are close to unity. The source resistance of the signal source or the load resistance are often suitable choices.

Frequency normalisation: All frequencies are referred to a reference frequency. Often a 'natural' frequency of the circuit is chosen, e.g. a corner or resonant frequency. The same applies to angular frequencies.

These two independent normalisations determine the normalisation of all other quantities. The complex impedances for inductances and capacitances are $j\omega L_n$ and $1/j\omega C_n$ respectively.

EXAMPLE: Normalise the circuit in Fig. 6.36.

> The reference resistance is chosen as $R_r = 200\ \Omega$. The reference angular frequency is $\omega_r = 1/\sqrt{LC} = 10^5\ \text{s}^{-1}$. It follows that $R_n = 1$, $C_n = 2$ and $L_n = 0.5$.

Table 6.5 summarises the normalisation relations for various circuit quantities.

Table 6.5. Summary of normalisation relations

Quantity	Normalised quantity	Denormalisation
R	$R_n = \dfrac{R}{R_r}$	$R = R_n \cdot R_r$
ω	$\omega_n = \dfrac{\omega}{\omega_r}$	$\omega = \omega_n \cdot \omega_r$
t	$t_n = t \cdot \omega_r$	$t = \dfrac{t_n}{\omega_r}$
C	$C_n = C \cdot \omega_r \cdot R_r$	$C = \dfrac{C_n}{\omega_r \cdot R_r}$
L	$L_n = L \cdot \dfrac{\omega_r}{R_r}$	$L = L_n \cdot \dfrac{R_r}{\omega_r}$
R_r : reference resistance ω_r : reference angular frequency		

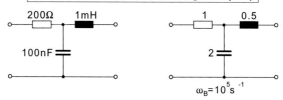

Fig. 6.36. A circuit and its corresponding normalised circuit

6.3.5.2 Impulse and Step Response of First-Order Systems

Linear first-order systems are RC or RL circuits with *one* independent energy-storing component (capacitor or inductor). The general form of the transfer function of such systems is

$$G(\omega) = \frac{a_0 + a_1 j\omega}{b_0 + j\omega}, \qquad a_i, b_0 \text{ real} \tag{6.35}$$

All coefficients a_i, b_0 are real. For stable systems $b_0 > 0$ must also hold. Function $G(\omega)$ is a rational, fractional function of ω. It can be expanded to

$$G(\omega) = a_1 + (a_0 - a_1 b_0) \cdot \frac{1}{b_0 + j\omega}$$

Each term can be transformed individually (see Table 6.7)

$$a_1 \circ\!\!-\!\!\bullet\, a_1 \cdot \delta(t), \qquad \text{and} \qquad \frac{1}{b_0 + j\omega} \bullet\!\!-\!\!\circ\, s(t) \cdot e^{-b_0 t}$$

The **impulse response** of a first-order system is then

$$\boxed{g(t) = a_1 \cdot \delta(t) + s(t) \cdot (a_0 - a_1 b_0) \cdot e^{-b_0 t}} \tag{6.36}$$

with coefficients a_0, a_1, b_0 according to Eq. (6.35).

The **step response** is the integral of the impulse response and is

$$\boxed{h(t) = s(t) \cdot \left(\frac{a_0}{b_0} - \frac{a_0 - a_1 b_0}{b_0} \cdot e^{-b_0 t} \right)} \tag{6.37}$$

EXAMPLE: The transfer function of the circuit in Fig. 6.37 is

$$G(\omega) = \frac{R}{R + j\omega L} = \frac{R/L}{R/L + j\omega}$$

Comparing the coefficients, it follows that $a_0 = R/L$, $a_1 = 0$ and $b_0 = R/L$.

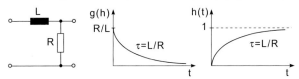

Fig. 6.37. A system and its impulse and step response

Substituting in Eq. (6.36) yields the impulse response

$$g(t) = s(t) \cdot \frac{R}{L} \cdot e^{-\frac{R}{L}t}$$

Using Eq. (6.37) yields the step response

$$h(t) = s(t) \cdot \left(1 - e^{-\frac{R}{L}t}\right)$$

EXAMPLE: Determine the impulse and step response of the circuit in Fig. 6.38.

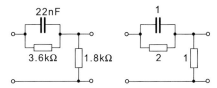

Fig. 6.38. A circuit and its normalised representation

The reference quantities are chosen as $R_r = 1.8$ kΩ and $\omega_r = 1/R_r C = (1.8$ k$\Omega \cdot 22$ nF$)^{-1} = 25\,252$ s^{-1}. The normalised quantities are shown on the right side of Fig. 6.38. The denormalised transfer function is

$$G_n(\omega) = \frac{1}{1 + 2||\frac{1}{j\omega}} = \frac{1/2 + j\omega}{3/2 + j\omega}$$

The coefficients $a_0 = 1/2$, $a_1 = 1$, $b_0 = 3/2$ can be derived from the transfer function. According to Eq. (6.36), it follows for the impulse response

$$g_n(t_n) = \delta(t_n) - s(t_n) \cdot e^{-\frac{3}{2}t_n}$$

According to Eq. (6.37) the step response is (Fig. 6.39)

$$h(t_n) = s(t_n) \cdot \left(\frac{1}{3} + \frac{2}{3} \cdot e^{-\frac{3}{2}t_n}\right)$$

The time axis of the impulse response is given in units of the normalised time t_n. Denormalisation yields $t = t_n/\omega_r = t_n R_r C = 39.6$ µs. The time constant of the exponential signal is then given by $\frac{2}{3} R_r C = 26.4$ µs.

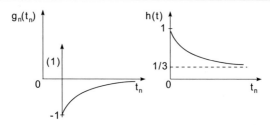

Fig. 6.39. Normalised impulse and step response of the circuit above

6.3.5.3 Impulse and Step Response of Second-Order Systems

Second-order systems are RLC circuits with two independent energy-storing components (capacitors and/or inductors). The transfer function has the form

$$G(\omega) = \frac{a_0 + a_1 j\omega - a_2 \omega^2}{b_0 + b_1 j\omega - \omega^2}, \qquad a_i, b_i \text{ real} \tag{6.38}$$

All coefficients a_i, b_i are real, and $b_0 > 0$ and $b_1 > 0$ are the conditions for stable systems. The transfer function can be expanded to

$$G(\omega) = a_2 + \frac{c_0 + c_1 j\omega}{b_0 + b_1 j\omega - \omega^2}, \qquad \text{with} \quad c_0 = a_0 - a_2 b_0, \quad \text{and} \quad c_1 = a_1 - a_2 b_1$$

It is useful to represent the denominator polynomial in a form where the roots are given explicitly:

$$b_0 + b_1 j\omega - \omega^2 = (j\omega - p_1) \cdot (j\omega - p_2)$$

$$p_{1/2} = -\frac{b_1}{2} \pm \sqrt{\frac{b_1^2}{4} - b_0}$$

NOTE: The roots p_1, p_2 are either both real valued or are complex conjugates.

At the zeros of the denominator polynomial the transfer function exhibits poles; therefore the roots (zeros) are designated p_1, p_2.

NOTE: The special case where both roots are equal, i.e. $p_1 = p_2$ (double-pole position), is excluded in further consideration.

The transfer function can be expressed as the sum of two partial fractions, which can be transformed individually. The **impulse resonse** is then

$$\boxed{\begin{array}{c} G(\omega) = a_2 + \dfrac{Z_1}{(j\omega - p_1)} + \dfrac{Z_2}{(j\omega - p_2)} \\ \downarrow \quad\quad \downarrow \quad\quad\quad \downarrow \quad\quad\quad\quad \downarrow \\ g(t) = a_2 \delta(t) + Z_1 \cdot s(t) \cdot e^{p_1 t} + Z_2 \cdot s(t) \cdot e^{p_2 t} \end{array}} \tag{6.39}$$

where

$$Z_1 = \frac{c_0 + c_1 p_1}{p_1 - p_2}, \quad Z_2 = \frac{c_0 + c_1 p_2}{p_2 - p_1}, \quad p_1 \neq p_2,$$

with $c_0 = a_0 - a_2 b_0$, and $c_1 = a_1 - a_2 b_1$

$$p_1 = -\frac{b_1}{2} + \sqrt{\frac{b_1^2}{4} - b_0}, \quad p_2 = -\frac{b_1}{2} - \sqrt{\frac{b_1^2}{4} - b_0}$$

The **step response** is

$$h(t_n) = a_2 \cdot s(t_n) + \frac{Z_1}{p_1} \cdot s(t_n) \cdot \left[\left(e^{p_1 t_n} - 1 \right) + \frac{Z_2}{p_2} \cdot \left(e^{p_2 t_n} - 1 \right) \right] \quad (6.40)$$

with the coefficients a_i, b_i given by Eq. (6.38).

EXAMPLE: Determine the impulse response of the circuit in Fig. 6.40. The reference quantities are chosen to be $R_r = 680 \, \Omega$, and $\omega_r = 1/\sqrt{LC} \approx 45\,000 \, \text{s}^{-1}$. The normalised quantities (with small rounding errors) can be derived from these and are given on the right side of the diagram.

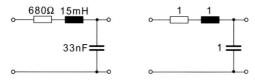

Fig. 6.40. Circuit and its normalised form

The normalised transfer function is given by

$$G_n(\omega) = \frac{\dfrac{1}{j\omega}}{\dfrac{1}{j\omega} + 1 + j\omega} = \frac{1}{1 + j\omega - \omega^2}$$

Therefore $a_0 = 1$, $a_1 = 0$, $a_2 = 0$, $b_0 = 1$, $b_1 = 1$. It follows that $c_0 = 1$, $c_1 = 0$ and $p_{1/2} = -\dfrac{1}{2} \pm j\dfrac{\sqrt{3}}{2}$, and $Z_1 = \dfrac{1}{j\sqrt{3}}$, $Z_2 = -Z_1$.

Substituting Eq. (6.39) gives the step response

$$g_n(t) = s(t) \cdot \left(\frac{1}{j\sqrt{3}} \cdot e^{p_1 t} - \frac{1}{j\sqrt{3}} \cdot e^{p_2 t} \right)$$

The real part of the exponents can be factored out

$$g_n(t) = s(t) \cdot e^{-1/2 \, t} \cdot \frac{1}{\sqrt{3}} \underbrace{\frac{1}{j} \left(e^{j\sqrt{3}/2 \, t} - e^{-j\sqrt{3}/2 \, t} \right)}_{2 \cdot \sin(\sqrt{3}/2 \, t)}$$

The variable t represents the normalised time. The impulse response of the circuit can be represented by oscillations exhibiting an exponential decay.

$$g_n(t_n) = s(t_n) \cdot \frac{2}{\sqrt{3}} \cdot e^{-1/2 \, t_n} \cdot \sin\left(\frac{\sqrt{3}}{2} t_n \right)$$

For large enough damping, there are no oscillations (Fig. 6.41). For the normalised quantities $R_n = 4$, $L_n = 1$ and $C_n = 1/3$, the normalised transfer function is given by

$$G_n(\omega) = \frac{3}{3 + 4j\omega - \omega^2}$$

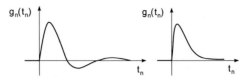

Fig. 6.41. Normalised impulse response of the circuit in Fig. 6.40e and the normalised impulse response of a circuit with greater damping

The roots of the denominator polynomial are purely real: $p_{1/2} = -2 \pm 1$. The calculation of the impulse response shows no imaginary part in the exponential terms that could be factored out, i.e. there is *no* periodic component in the signal. The impulse response is the difference between two decaying exponential functions (see the right side of Fig. 6.41).

$$g_n(t_n) = s(t_n) \cdot Z_1 \cdot \left(e^{p_1 t_n} - e^{p_2 t_n}\right) = s(t_n) \cdot \frac{3}{2} \cdot \left(e^{-t_n} - e^{-3t_n}\right)$$

6.3.6 Ideal Systems

Ideal systems are systems with idealised properties, which real systems can only approximate. Ideal systems are used as models to discuss the basic properties of real systems.

6.3.6.1 Distortion-Free Systems

A **distortion-free system** transmits a signal without changing its form. Changes in the amplitude and time shifts are permitted (Fig. 6.42). For an arbitrary input signal $x(t)$ it holds that

$$y(t) = k \cdot x(t - t_0), \qquad \text{for causal systems } t_0 \geq 0$$

where k is an arbitrary real amplitude factor, and t_0 is an arbitrary real **delay time**.

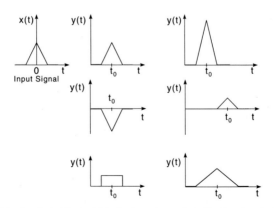

Fig. 6.42. Examples of output signals of distortion-free systems excited with a triangular pulse. The signals in the last row originate from nondistortion-free systems

The **transfer function** of the **distortion-free system** is

$$G(\omega) = k \cdot e^{-j\omega t_0}$$

It follows for the magnitude and the phase of the transfer function that

$$|G(\omega)| = k, \qquad \varphi(\omega) = -\omega \cdot t_0 \tag{6.41}$$

- A **distortion-free system** has a **constant attenuation** (or gain).
- A distortion-free system has a **linear phase response**.

NOTE: A system with constant gain over all frequencies is known as an **all-pass filter**.

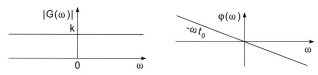

Fig. 6.43. Magnitude and phase responses of the transfer function of a distortion-free system

For the description of systems other quantities derived from the transfer function are used.

Attenuation constant

$$a(\omega) = -20 \log_{10} |G(\omega)| \text{ (dB)} \tag{6.42}$$

Phase constant

$$b(\omega) = -\varphi(\omega) \tag{6.43}$$

Phase delay

$$\tau_p = \frac{b(\omega)}{\omega} \tag{6.44}$$

Group delay

$$\tau_g = \frac{d b(\omega)}{d \omega} \tag{6.45}$$

A **distortion-free system** thus exhibits **constant attenuation** and **constant group delay** for all frequencies. This means that all frequency components of a signal are delayed by the same amount of time and therefore appear with the correct phase at the output of the system.

If this is not the case, it is known as linear **delay distortion** (phase distortion). If the attenuation is not constant this is known as linear **attenuation distortion**.

NOTE: **Nonlinear distortions** are different from **linear distortions**. The former can create frequency components that were not contained in the original input signal.

6.3.6.2 Ideal Low-Pass Filter

The **ideal low-pass filter** passes signals in the passband up to the critical frequency f_c (ω_c) without any distortion. Signal components above the critical frequency are suppressed completely. Its **transfer function** is

$$G(\omega) = \begin{cases} k \cdot e^{-j\omega t_0}, & \text{for } |\omega| \leq \omega_c \\ 0, & \text{else} \end{cases} \tag{6.46}$$

where k is a real amplitude factor, and t_0 is the group delay time (signal delay time) of the low-pass filter system (Fig. 6.44).

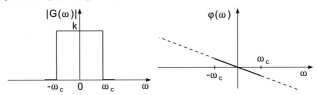

Fig. 6.44. Transfer function of the ideal low-pass filter

The **impulse response** of the ideal low-pass filter is

$$g(t) = k \cdot \frac{\omega_c}{\pi} \cdot \text{sinc}\left[\omega_c(t - t_0)\right] = 2kf_c \cdot \text{sinc}\left[2\pi f_c(t - t_0)\right] \tag{6.47}$$

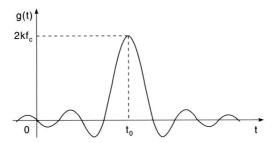

Fig. 6.45. Impulse response of the ideal low-pass filter

NOTE: The sinc function is defined as

$$\text{sinc}(x) = \begin{cases} 1, & \text{for } x = 0 \\ \frac{\sin x}{x}, & \text{else} \end{cases} \tag{6.48}$$

- The ideal low-pass filter is a **noncausal system**. The impulse response appears *before* the input signal arrives.

The impulse response assumes its maximum value at $t = t_0$, and $g(t_0) = k\omega_c/\pi = 2kf_c$. Therefore t_0 represents the propagation delay time (Fig. 6.45).

The **step response** of the ideal low-pass filter is (Fig. 6.46)

$$h(t) = \frac{k}{2} + \frac{k}{\pi} \cdot \text{Si}(\omega_c(t - t_0)) \tag{6.49}$$

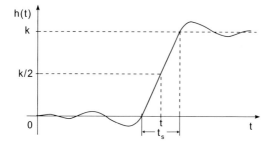

Fig. 6.46. Step response of the ideal low-pass filter

NOTE: The step response assumes nonzero values before the time $t = 0$.

NOTE: The Si function is defined as (Integral-Sine)

$$\text{Si}(x) = \int_0^x \frac{\sin \tau}{\tau} \, d\tau \qquad (6.50)$$

The Si function cannot be represented analytically. An approximation can be made by a power series:

$$\text{Si}(x) = x - \frac{x^3}{18} + \frac{x^5}{600} - \frac{x^7}{35280} \cdots$$

$$= x - \frac{x^3}{3 \cdot 3!} + \frac{x^5}{5 \cdot 5!} - \frac{x^7}{7 \cdot 7!} \cdots + \frac{(-1)^i x^{2i+1}}{(2i+1) \cdot (2i+1)!} \cdots$$

The **overshoot** is characteristic and amounts to 8.6% of the steady-state value, independent of the bandwidth of the system. Overshoot in band-limited systems is known as **Gibb's phenomenon**.

The impulse response approaches a steady state value of $h_\infty = k$. At the time $t = t_0$ it passes through $k/2$, half the final value, at which point the slope (rate of change) of the response is maximum.

The **settling time** is defined by the time taken to the intersection of the tangent of the turning point of the step response with the input signal start value, and the intersection of the tangent with the input signal steady-state value. The settling time t_s of the ideal low-pass filter is

$$\boxed{t_s = \frac{\pi}{\omega_c} = \frac{1}{2 f_c}} \qquad (6.51)$$

- The step response of the ideal low-pass filter reaches the steady-state value faster with increasing critical frequency.

Time–Bandwidth Product

The impulse response $g(t)$ of the ideal low-pass filter is infinitely long. However, it can be characterised by a pulse width. In order to do so, a rectangular impulse with an area equal to $g(t)$ is constructed, where the amplitude is equal to the maximum amplitude of the impulse response. Its width is defined as the **pulse width** Δt_p.

$$\Delta t_p = \frac{1}{g_{\max}} \cdot \int_{-\infty}^{\infty} g(t) \, dt$$

For the ideal low-pass filter this definition yields the pulse width of the impulse response

$$\Delta t_p = \frac{1}{2 f_c} \tag{6.52}$$

The settling time and the width of the impulse response are equal for the ideal low-pass filter.

- The width of the impulse response is inversely proportional to the bandwidth of the low-pass filter.

The **time–bandwidth product**

$$f_c \cdot \Delta t_p = \frac{1}{2}$$

is **constant**. This concept can be generalised. With the definition for the **bandwidth**

$$B = \sqrt{\int_{-\infty}^{\infty} \omega^2 |G(\omega)|^2 \, d\omega}$$

and for the **pulse width**

$$\Delta T = \sqrt{\int_{-\infty}^{\infty} t^2 |g(t)|^2 \, dt}$$

using the normalisation condition

$$\int_{-\infty}^{\infty} |g(t)|^2 \, dt = 1$$

the so-called **uncertainty principle** follows:

$$\boxed{B \cdot \Delta T \geq \sqrt{\frac{\pi}{2}}} \tag{6.53}$$

This relation holds for all kinds of low-pass filters. The smallest time–bandwidth product achieve filters with a Gaussian impulse response.

- The bandwidth and pulse width of the impulse response are inversely proportional for a given type of filter.

6.3.6.3 Ideal Bandpass Filter

The **ideal bandpass filter** passes signals within a frequency range Δf ($\Delta \omega$) without any distortion. In the stop-bands all signal components are completely suppressed. The **transfer function** is

$$G(\omega) = \begin{cases} k \cdot e^{-j\omega t_0}, & \text{for } |\omega - \omega_0| < \dfrac{\Delta\omega}{2} \\ 0, & \text{else} \end{cases} \qquad (6.54)$$

where ω_0 is the centre (angular) frequency of the bandpass filter, and $\Delta\omega$ is its bandwidth. This equation only makes sense if the centre frequency is at least twice the bandwidth ($\omega_0 > \Delta\omega/2$).

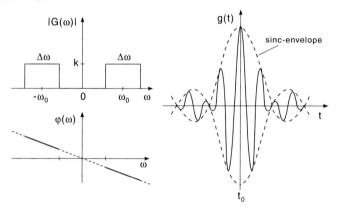

Fig. 6.47. Transfer function and impulse response of the ideal bandpass filter

The **impulse response** of the ideal bandpass filter is

$$\begin{aligned} g(t) &= k \cdot \Delta f \cdot \text{sinc}\big[\pi \Delta f (t - t_0)\big] \cdot 2\cos\big[2\pi f_0 (t - t_0)\big] \qquad (6.55) \\ &= k \cdot \frac{\Delta\omega}{2\pi} \cdot \text{sinc}\big[\Delta\omega (t - t_0)\big] \cdot 2\cos\big[2\omega_0 (t - t_0)\big] \end{aligned}$$

The impulse response resembles a signal with a centre frequency f_0 and with an envelope that corresponds to the impulse response of a low-pass filter with a cutoff frequency of $\Delta f/2$ (Fig. 6.47).

- The ideal bandpass filter is a **noncausal system**. The impulse response appears *before* the input signal arrives.

6.4 Fourier Transforms

6.4.1 Principle

The principle of the Fourier transform is to transform a signal $f(t)$ from the time domain into a signal $F(\omega)$ in the frequency domain such that this transform is reversible and unambiguous (Fig. 6.48).

The Fourier transform represents a time function as a superposition of an infinite number of harmonic exponential functions. Similar to the Fourier series, which describes a periodic function as a *summation* of infinitely discrete oscillations, the Fourier transform is the *integral* over an infinitely large number of oscillations. By expanding this concept to continuous spectra, nonperiodic functions can also be represented in the frequency domain.

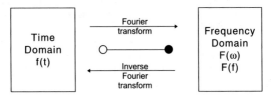

Fig. 6.48. Principle of the Fourier transform

It is often easier to calculate the effect of filters and transmission systems in the frequency domain. Problems requiring the solution of a linear differential equation in the time domain can be treated by solving an algebraic equation in the frequency domain.

Application of the inverse Fourier transform yields the corresponding signal in the time domain.

6.4.2 Definition

The **Fourier transform** of a function of time $f(t)$ is defined as

$$F(f) = \int_{-\infty}^{\infty} f(t) \cdot e^{-j2\pi ft} \, dt \qquad (6.56)$$

The **inverse Fourier transform** is

$$f(t) = \int_{-\infty}^{\infty} F(f) \cdot e^{j2\pi ft} \, df \qquad (6.57)$$

This is also written as

$$F(f) = \mathcal{F}\{f(t)\}, \qquad f(t) = \mathcal{F}^{-1}\{F(f)\}$$

or by using the correspondence symbol $\circ\!\!-\!\!\bullet$

$$f(t) \circ\!\!-\!\!\bullet F(f), \quad \text{or} \quad f(t) \circ\!\!-\!\!\bullet F(\omega)$$

This symbol can be read in both directions and thus illustrates the reversibility of the transformation. The filled circle corresponds to the **frequency domain**.

NOTE: Occasionally the representation of the transform with the *angular frequency* ω as the parameter in $F(\omega)$ is used. In Eq. (6.56) $2\pi f$ is replaced by ω. The **Fourier transform** is then

$$F(\omega) = \int_{-\infty}^{\infty} f(t) \cdot e^{-j\omega t} \, dt \qquad (6.58)$$

NOTE: When using this representation of the Fourier transform watch out for a factor 2π, since the **inverse Fourier transform** is then expressed by

$$f(t) = \frac{1}{2\pi} \cdot \int_{-\infty}^{\infty} F(\omega) \cdot e^{j\omega t} \, d\omega \qquad (6.59)$$

In this chapter both representations of the Fourier transform are used, if they differ from each other.

NOTE: In the literature a representation with $F(j\omega)$ is often found. It fully corresponds to the representation given by $F(\omega)$. It is found especially where the relationship with the Laplace transform is emphasised.

6.4.3 Representation of the Fourier Transform

The Fourier transform $S(f)$ of a real-valued time signal is a complex function and can therefore be represented as the sum of the **real** and **imaginary parts**.

$$S(f) = R(f) + j \cdot X(f)$$

For **real-valued functions of time** it holds that

$$R(f) = \mathrm{Re}\{S(f)\} = \int_{-\infty}^{\infty} f(t) \cdot \cos(2\pi f t) \, dt \qquad (6.60)$$
$$X(f) = \mathrm{Im}\{S(f)\} = -\int_{-\infty}^{\infty} f(t) \cdot \sin(2\pi f t) \, dt \qquad (6.61)$$

Furthermore

$$R(f) = R(-f), \qquad X(f) = -X(-f)$$

For the Fourier transform of **real-valued functions** of time it holds that

- The **real part** is an **even function**.
- The **imaginary part** is an **odd function**.

Like any complex function, the Fourier transform can also be represented in polar form:

$$S(f) = |S(f)| \cdot e^{j\varphi(f)}$$

with

$$|S(f)| = \sqrt{R^2(f) + X^2(f)}, \qquad \text{and} \qquad \varphi(f) = \arctan\left[\frac{X(f)}{R(f)}\right]$$

For **real-valued functions** of time it holds that

- The **magnitude** of the Fourier transform is an **even function**.
- The **phase** of the Fourier transform is an **odd function**.

NOTE: When dealing with the Fourier transform it may be useful to also work with *complex* functions of time, e.g. $f(t) = e^{j\omega t}$. The statements above about symmetries hold only for real-valued functions.

6.4.3.1 Symmetry Properties

For real-valued functions of time it holds that
- The Fourier transform of **even** functions of time is **purely real**.
- The Fourier transform of **odd** functions of time is **purely imaginary**.

EXAMPLE: The cosine function is an even function. Its Fourier transform is $\frac{1}{2}\delta(f+f_0) + \frac{1}{2}\delta(f-f_0)$. It is purely real (Fig. 6.49).

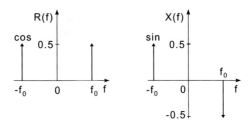

Fig. 6.49. Cosine and sine functions in the frequency domain

The sine-function is an odd function. Its Fourier transform is $-\frac{j}{2}\delta(f-f_0) + \frac{j}{2}\delta(f+f_0)$. It is purely imaginary (Fig. 6.49).

6.4.4 Overview: Properties of the Fourier Transform

Let $s(t)$ and $r(t)$ be abitrary functions of time, then $S(f)$ and $R(f)$ ($S(\omega)$ and $R(\omega)$, respectively) are their corresponding Fourier transforms. Whenever the two notations differ the spectrum with ω is also given. Table 6.6 summarises the properties of the Fourier transform.

Table 6.6. Properties of the Fourier transform

$s(t)$	○—•	$S(f)$
		$S(\omega)$
Fourier transform		
$s(t)$	○—•	$S(f) = \int\limits_{-\infty}^{\infty} s(t) \cdot e^{-j2\pi ft} \, dt$
Inverse Fourier transform		
$s(t) = \int\limits_{-\infty}^{\infty} S(f) \cdot e^{j2\pi ft} \, df$	○—•	$S(f)$
$s(t) = \dfrac{1}{2\pi} \int\limits_{-\infty}^{\infty} S(\omega) \cdot e^{j\omega t} \, d\omega$	○—•	$S(\omega)$
Complex conjugate		
$s^*(t)$	○—•	$S^*(-f)$

Table 6.6. (cont.)

Duality		
$S(t)$	○—•	$s(-f)$
	○—•	$2\pi \cdot s(-\omega)$
Multiplication		
$r(t) \cdot s(t)$	○—•	$R(f) * S(f)$
	○—•	$\dfrac{1}{2\pi} R(\omega) * S(\omega)$
Convolution		
$r(t) * s(t)$	○—•	$R(f) \cdot S(f)$
Superposition		
$a \cdot r(t) + b \cdot s(t)$	○—•	$a \cdot R(f) + b \cdot S(f)$
Time shift		
$s(t - t_0)$	○—•	$S(f) \cdot \mathrm{e}^{-\mathrm{j}2\pi f t_0}$
	○—•	$S(\omega) \cdot \mathrm{e}^{-\mathrm{j}\omega t_0}$
Frequency shift		
$s(t) \cdot \mathrm{e}^{\mathrm{j}2\pi f_0 t}$	○—•	$S(f - f_0)$
$s(t) \cdot \mathrm{e}^{\mathrm{j}\omega_0 t}$	○—•	$S(\omega - \omega_0)$
Time scaling		
$s\left(\dfrac{t}{a}\right)$	○—•	$\lvert a \rvert \cdot S(a \cdot f)$
Differentiation		
$\dfrac{\mathrm{d}}{\mathrm{d}t} s(t)$	○—•	$\mathrm{j}2\pi f \cdot S(f)$
	○—•	$\mathrm{j}\omega \cdot S(\omega)$
Integration		
$\displaystyle\int_{-\infty}^{t} s(\tau)\,\mathrm{d}\tau$	○—•	$\dfrac{1}{\mathrm{j}2\pi f} \cdot S(f) + \dfrac{1}{2} \cdot S(0) \cdot \delta(f)$
	○—•	$\dfrac{1}{\mathrm{j}\omega} \cdot S(\omega) + \pi \cdot S(0) \cdot \delta(\omega)$
For DC-free signals the terms with delta function are discarded		

6.4.5 Fourier Transforms of Elementary Signals

6.4.5.1 Spectrum of the Delta Function

The Fourier transform of the delta function is

$$S(f) = \int_{-\infty}^{\infty} f(t) \cdot \mathrm{e}^{-\mathrm{j}\omega t}\,\mathrm{d}t = \int_{-\infty}^{\infty} \delta(t) \cdot \mathrm{e}^{-\mathrm{j}\omega t}\,\mathrm{d}t = \mathrm{e}^0 = 1$$
$$\delta(t) \circ\!\!-\!\!\bullet\, 1 \tag{6.62}$$

- A delta impulse contains all frequencies with equal amplitudes (Fig. 6.50).

Because of the duality of time and frequency it also holds that a signal that is constant in time (DC signal) corresponds to a delta impulse in the spectrum.

$$1 \circ\!\!-\!\!\bullet\, \delta(f), \quad \text{or} \quad 1 \circ\!\!-\!\!\bullet\, 2\pi\delta(\omega) \tag{6.63}$$

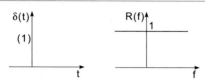

Fig. 6.50. Delta impulse and real part of its Fourier transform

6.4.5.2 Spectrum of the Signum and the Step Functions

The **signum function** is similar to the step function.

$$\text{sign}(t) = \begin{cases} 1 & t > 0 \\ 0 & t = 0 \\ -1 & t < 0 \end{cases}$$

The Fourier transform is

$$\text{sign}(t) \circ\!\!-\!\!\bullet -j\frac{1}{\pi f}, \quad \text{or} \quad \text{sign}(t) \circ\!\!-\!\!\bullet \frac{2}{j\omega} \qquad (6.64)$$

The signum function is an odd function; therefore its spectrum is purely imaginary (Fig. 6.51).

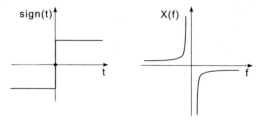

Fig. 6.51. Signum function and the imaginary part of its Fourier transform

Unlike the signum function the **step function** has a DC component. This can also be seen in its spectrum. The step function can be expressed using the signum function as

$$s(t) = \frac{1}{2} \cdot \text{sign}(t) + \frac{1}{2}$$

$$\circ\!\!-\!\!\bullet \qquad \circ\!\!-\!\!\bullet$$

$$\frac{-j}{2\pi f} \quad + \frac{1}{2}\delta(f),$$

$$\text{or} \quad \frac{1}{j\omega} \quad + \pi\delta(\omega)$$

NOTE: The representation of the step function through the signum function is not exact for $t = 0$ (Fig. 6.52). Generally, the inverse Fourier transform of the spectra of discontinuous functions at the discontinuity points is given by the average of the right- and left-side limits (in this case 0).

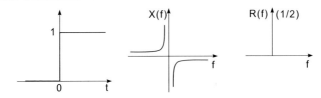

Fig. 6.52. Step function (*left*) and the imaginary (*centre*) and real parts (*right*) of its Fourier transform

6.4.5.3 Spectrum of the Rectangular Pulse

The spectrum of the rectangular pulse is (Fig. 6.53)

$$S(f) = \int_{-\infty}^{\infty} \text{rect}(t) \cdot e^{-j\omega t} \, dt = \int_{-1/2}^{1/2} e^{-j2\pi f t} \, dt = \frac{-1}{j2\pi f}\left(e^{-j\pi f} - e^{j\pi f}\right) \quad (6.65)$$

Applying the following representation of the sine function:

$$\sin x = \frac{1}{2j}\left(e^{jx} - e^{-jx}\right),$$

Equation (6.65) holds that

$$S(f) = \frac{\sin \pi f}{\pi f} = \text{sinc}(\pi f) \quad (6.66)$$

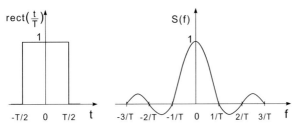

Fig. 6.53. Rectangular pulse and its amplitude spectrum

For pulses of arbitrary width, applying the similarity theorem, the following holds:

$$\text{rect}(t) \circ\!\!-\!\!\bullet \text{sinc}(\pi f)$$

$$\text{rect}\left(\frac{t}{T}\right) \circ\!\!-\!\!\bullet T \cdot \text{sinc}(\pi T f) \quad (6.67)$$

$$\text{or} \quad \text{rect}\left(\frac{t}{T}\right) \circ\!\!-\!\!\bullet T \cdot \text{sinc}\left(\frac{T\omega}{2}\right)$$

6.4.5.4 Spectrum of the Triangular Pulse

The triangular pulse $\Lambda(t)$ can be represented as the convolution of the rectangular pulse with itself.

$$\Lambda(t) = \text{rect}(t) * \text{rect}(t)$$

A convolution in the time domain corresponds to a multiplication in the frequency domain

$$\Lambda(t) = \text{rect}(t) * \text{rect}(t)$$
$$\downarrow \qquad \downarrow \qquad\qquad (6.68)$$
$$\text{sinc}(\pi f) \cdot \text{sinc}(\pi f)$$

It therefore holds that (Fig. 6.54)

$$\Lambda(t) \circ\!\!-\!\!\bullet \text{sinc}^2(\pi f) \qquad (6.69)$$

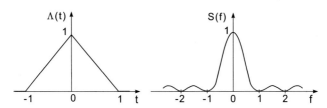

Fig. 6.54. Triangular pulse and its amplitude spectrum

6.4.5.5 Spectrum of the Gaussian Pulse

The Gaussian pulse

$$\Gamma(t) = e^{-\pi t^2}$$

again has a Gaussian amplitude spectrum.

$$S(f) = \int_{-\infty}^{\infty} e^{-\pi t^2} \cdot e^{-j\omega t} \, dt$$

Applying the Euler formula to $e^{j\omega t}$ yields

$$\int_{-\infty}^{\infty} e^{-\pi t^2} \cdot e^{-j\omega t} \, dt = \int_{-\infty}^{\infty} e^{-\pi t^2} \cdot \cos \omega t \, dt - \underbrace{\int_{-\infty}^{\infty} e^{-\pi t^2} \cdot \sin \omega t \, dt}_{=0}$$

The second integrand is the product of an even (Gaussian) with an odd (sinusoidal) function and is therefore itself an odd function. Its integral is zero. The first integrand is an even function. Therefore the integral over $[-\infty \ldots 0]$ is equal to the integral over $[0 \ldots \infty]$.

$$S(f) = 2 \cdot \int_{0}^{\infty} e^{-\pi t^2} \cdot \cos(2\pi f t) \, dt$$

The definite integral can be looked up in a table of integrals

$$\int_{0}^{\infty} e^{-a^2 t^2} \cdot \cos bt \, dt = \frac{\sqrt{\pi}}{2a} \cdot e^{-b^2/4a^2}$$

6.4 Fourier Transforms

Therefore $a^2 = \pi$ and $b = 2\pi f$ yields the spectrum of the Gaussian pulse

$$\Gamma(t) = e^{-\pi t^2} \circ\!\!-\!\!\bullet\, e^{-\pi f^2} = \Gamma(f) \tag{6.70}$$

Obviously, this function is converted to its spectrum by swapping the time and frequency variables. Functions with this property are called **self-reciprocal**.

6.4.5.6 Spectrum of Harmonic Functions

The Fourier transform of the complex harmonic function $e^{j2\pi f_0 t}$ is

$$e^{j2\pi f_0 t} \circ\!\!-\!\!\bullet \int_{-\infty}^{\infty} e^{j2\pi f_0 t} \cdot e^{-j2\pi f t}\, dt = \delta(f - f_0), \quad \text{or} \quad 2\pi\delta(\omega - \omega_0)$$

The real-valued harmonic cosine (Fig. 6.55) and sine (Fig. 6.56) functions of time can be composed of two periodic exponential functions.

$$\cos 2\pi f_0 t = \frac{1}{2} \cdot e^{j2\pi f_0 t} + \frac{1}{2} \cdot e^{-j2\pi f_0 t}$$

$$\updownarrow \qquad\qquad \updownarrow$$

$$\frac{1}{2} \cdot \delta(f - f_0) + \frac{1}{2} \cdot \delta(f + f_0)$$

$$\text{or } \pi \cdot \delta(\omega - \omega_0) + \pi \cdot \delta(\omega + \omega_0)$$

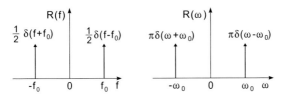

Fig. 6.55. Fourier transform of the cosine function

$$\sin 2\pi f_0 t = \frac{1}{2j} \cdot e^{j2\pi f_0 t} - \frac{1}{2j} \cdot e^{-j2\pi f_0 t}$$

$$\updownarrow \qquad\qquad \updownarrow$$

$$\frac{-j}{2} \cdot \delta(f - f_0) + \frac{j}{2} \cdot \delta(f + f_0)$$

$$\text{or } -j\pi\delta(\omega - \omega_0) + j\pi\delta(\omega + \omega_0)$$

The appearance of a pair of delta pulses in the spectrum indicates a periodic component in the signal. As is generally known, all periodic functions can be represented in a Fourier series as a sum of sine and cosine functions. Their spectrum is therefore always a discrete line spectrum, i.e. it consists of delta pulses in the frequency domain.

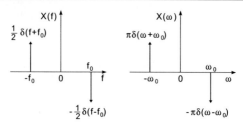

Fig. 6.56. Fourier transform of the sine function

6.4.6 Summary of Fourier Transforms

The graphs in Table 6.7 represent the functions of time $s(t)$ and the magnitude of their corresponding Fourier transform $|S(f)|$.

Table 6.7. Fourier transforms of elementary signals

	Signal $s(t)$	Spectrum $S(f), S(\omega)$
Delta impulse	$\delta(t)\ \circ\!\!-\!\!\bullet\ 1$	
DC signal	$1\ \circ\!\!-\!\!\bullet\ \delta(f),\quad 2\pi\delta(\omega)$	
Rectangular pulse	$\mathrm{rect}\left(\dfrac{t}{T}\right)\ \circ\!\!-\!\!\bullet\ T\cdot\mathrm{sinc}(\pi T f)$ $T\cdot\mathrm{sinc}(\omega T/2)$	
Si pulse	$\mathrm{sinc}(t)\ \circ\!\!-\!\!\bullet\ \mathrm{rect}(f)$ $\mathrm{rect}(\omega/2\pi)$	
Triangular pulse	$\Lambda(t)\ \circ\!\!-\!\!\bullet\ \mathrm{sinc}^2(\pi f) = \dfrac{\sin^2(\pi f)}{(\pi f)^2}$ $\mathrm{sinc}^2\left(\dfrac{\omega}{2}\right) = \dfrac{4\sin^2(\omega/2)}{\omega^2}$	

Table 6.7. (cont.)

	Signal $s(t)$	**Spectrum** $S(f), S(\omega)$
Gaussian pulse	$e^{-\pi t^2}$	$e^{-\pi f^2}$ $e^{-\omega^2/4\pi}$
Delta impulse sequence	$\sum_{n=-\infty}^{\infty} \delta(t-nT)$	$\sum_{n=-\infty}^{\infty} \delta\left(f - \frac{n}{T}\right)$ $2\pi \sum_{n=-\infty}^{\infty} \delta\left(\omega - \frac{2\pi n}{T}\right)$
Step function	$s(t)$	$\frac{1}{2}\delta(f) - j\frac{1}{2\pi f}$ $\pi\delta(\omega) + \frac{1}{j\omega}$
Signum function	$\text{sign}(t)$	$-j\frac{1}{\pi f}$ $\frac{2}{j\omega}$
Cosine waveform	$\cos(2\pi f_0 t)$ $\cos \omega_0 t$	$\frac{1}{2}\delta(f+f_0) + \frac{1}{2}\delta(f-f_0)$ $\pi\delta(\omega+\omega_0) + \pi\delta(\omega-\omega_0)$
Sine waveform	$\sin(2\pi f_0 t)$ $\sin \omega_0 t$	$\frac{j}{2}\delta(f+f_0) - \frac{j}{2}\delta(f-f_0)$ $j\pi\delta(\omega+\omega_0) - j\pi\delta(\omega-\omega_0)$

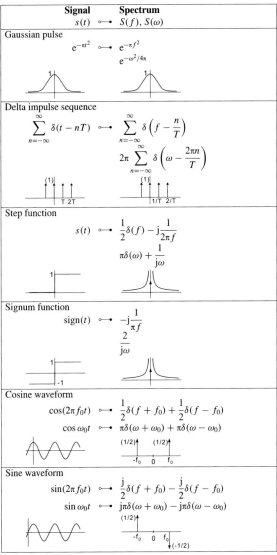

Table 6.7. (cont.)

Signal	Spectrum
$s(t)$ $\circ\!\!-\!\!\bullet$	$S(f), S(\omega)$

Single-sided cosine

$s(t) \cdot \cos(2\pi f_0 t)$ $\circ\!\!-\!\!\bullet$ $\dfrac{1}{4}\delta(f - f_0) + \dfrac{1}{4}\delta(f + f_0) + \dfrac{j}{2\pi}\dfrac{f}{f_0^2 - f^2}$

$s(t) \cdot \cos \omega_0 t$ $\circ\!\!-\!\!\bullet$ $\dfrac{\pi}{2}\delta(\omega - \omega_0) + \dfrac{\pi}{2}\delta(\omega + \omega_0) + j\dfrac{\omega}{\omega_0^2 - \omega^2}$

Single-sided sine

$s(t) \cdot \sin(2\pi f_0 t)$ $\circ\!\!-\!\!\bullet$ $\dfrac{j}{4}\delta(f + f_0) - \dfrac{j}{4}\delta(f - f_0) + \dfrac{1}{2\pi}\dfrac{f_0}{f_0^2 - f^2}$

$s(t) \cdot \sin(\omega_0 t)$ $\circ\!\!-\!\!\bullet$ $\dfrac{\pi}{2j}\delta(\omega + \omega_0) - \dfrac{\pi}{2j}\delta(\omega - \omega_0) + \dfrac{\omega_0}{\omega_0^2 - \omega^2}$

Single-sided exponential pulse

$s(t) \cdot e^{-at}$ $\circ\!\!-\!\!\bullet$ $\dfrac{1}{a + j\omega}$ for $\text{Re}\{a\} > 0$

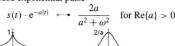

Double-sided exponential pulse

$s(t) \cdot e^{-a|t|}$ $\circ\!\!-\!\!\bullet$ $\dfrac{2a}{a^2 + \omega^2}$ for $\text{Re}\{a\} > 0$

Decaying cosine waveform

$s(t) \cdot e^{-at} \cdot \cos \omega_0 t$ $\circ\!\!-\!\!\bullet$ $\dfrac{j\omega + a}{(j\omega + a)^2 + \omega_0^2}$ for $\text{Re}\{a\} > 0$

Decaying sine waveform

$s(t) \cdot e^{-at} \cdot \sin \omega_0 t$ $\circ\!\!-\!\!\bullet$ $\dfrac{\omega_0}{(j\omega + a)^2 + \omega_0^2}$ for $\text{Re}\{a\} > 0$

6.5 Nonlinear Systems

6.5.1 Definition

Systems with a nonlinear relationship between the input and output signal are called **nonlinear systems**.

NOTE: In practice, there is no such thing as a linear system since any real system has output swing limits. Linear systems are in many cases a good approximation of real systems.

The following definition is more suitable for a practical characterisation of nonlinear systems:

- A system that responds to a harmonic input signal with a nonharmonic output signal is called **nonlinear system**.

6.5.2 Characterisation of Nonlinear Systems

Examples of components with a distinctive nonlinear $I(V)$-characteristic are rectifier diodes, Zener diodes, tunnel diodes and varistors (voltage-dependent resistors). Considering the thermal behaviour over time, this also holds for conductors with negative or positive temperature coefficients (NTC/PTC), and also for filament light bulbs.

Often (not always) the interest focuses on the realisation of systems with a broad-ranged linear response. Certain nonlinearities are then accepted within limits. The deviation from the desired linearity is characterised by quantities.

6.5.2.1 Characteristic Equation

One way to describe nonlinear characteristics is a polynomial equation

$$v_2 = a \cdot v_1 + b \cdot v_1^2 + c \cdot v_1^3 + \ldots \tag{6.71}$$

The order of the polynomial of the characteristic equation is called the **order of the nonlinear system**.

EXAMPLE: For a nonlinear second-order system the output voltage for the harmonic input voltage $v_1 = \hat{v}_1 \cdot \cos \omega t$ is

$$v_2 = a \cdot \hat{v}_1 \cdot \cos \omega t + b \cdot \hat{v}_1^2 \cdot \cos^2 \omega t$$

The square of the cosine function can be resolved using the following relationship

$$\cos^2 \omega t = \frac{1}{2} \cdot (1 + \cos 2\omega t) \tag{6.72}$$

Therefore the output voltage is

$$v_2 = \frac{b}{2} \cdot \hat{v}_1^2 + a \cdot \hat{v}_1 \cdot \cos \omega t + \frac{b}{2} \cdot \hat{v}_1^2 \cdot \cos 2\omega t$$

The output signal contains components having twice the frequency of the input signal. These components are called **harmonics**.

In general it holds that

- An nth-order nonlinear system produces harmonics up to n times the frequency of the input signal. The amplitude of each individual harmonic depends on the coefficients of the characteristic equation.

NOTE: The first harmonic is the angular frequency ω, which is the fundamental frequency. The second harmonic has a frequency of 2ω.

The description of nonlinear systems with the coefficients of their characteristic curves is not very suitable. Of greater interest is the effect on the distortion products. The total harmonic distortion is used for this analysis.

6.5.2.2 Total Harmonic Distortion

The **total harmonic distortion** of a signal is defined as

$$THD = \frac{\text{RMS value of the harmonics}}{\text{RMS value of the complete signal}} = \frac{\sqrt{\sum_{n=2}^{\infty} A_n^2}}{\sqrt{\sum_{n=1}^{\infty} A_n^2}} \quad (6.73)$$

The A_n are the Fourier coefficients of the amplitude spectrum of the related signal. The factor $\sqrt{2}$ relating the amplitude and RMS value of each component cancels out.

EXAMPLE: The signal $v(t) = 2 \text{ V} \cdot \cos \omega t + 0.2 \text{ V} \cdot \sin 3\omega t - 0.4 \text{ V} \cdot \sin 4\omega t$ has the total harmonic distortion

$$THD^2 = \frac{0.2^2 + 0.4^2}{2^2 + 0.2^2 + 0.4^2} = 0.0476 \Rightarrow THD = 0.218 \approx 22\%$$

NOTE: It is usually easier to calculate k^2 and then take the square root instead of applying the definition directly.

- The total harmonic distortion of a purely harmonic (sinusoidal) signal is zero.

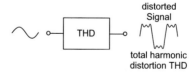

Fig. 6.57. Representation of the total harmonic distortion factor of a transmission system

If a system produces a total harmonic distortion k in the output signal for a purely harmonic input signal, the system distortion is quantified and denoted by THD (Fig. 6.57).

NOTE: It is not possible to determine the total harmonic distortion if the input signal already contains harmonics.

- The total harmonic distortion of the output signal depends on the output voltage swing. Providing the total harmonic distortion of a transmission system only makes sense if the measurement conditions are given.

Table 6.8. Typical total harmonic distortions

THD	Example
33%	Total harmonic distortion of a square-wave oscillation
10%	Voice signal still intelligible
1%	Maximum total harmonic distortion of HiFi amplifier, distortions just perceptible
0.1%	Total harmonic distortion of a good HiFi amplifier distortions imperceptible

Occasionally, only the amplitude of an individual harmonic is of interest. The **total harmonic distortion of nth order** is used for this.

$$THD_n = \frac{\text{RMS value of the } n\text{-th harmonic}}{\text{RMS-value of the complete signal}} \tag{6.74}$$

The **total harmonic distortion attenuation** is defined as

$$a_k = -20 \log THD \tag{6.75}$$

or the **total harmonic distortion attenuation of nth order** as (Table 6.9)

$$a_{kn} = -20 \log THD_n \tag{6.76}$$

Table 6.9. Total harmonic distortion attenuation values

Total harmonic distortion	Total harmonic distortion attenuation
10%	20 dB
1%	40 dB
0.1%	60 dB

Sine wave generators, spectrum analysers and selective level meters must have values of total harmonic distortion attenuation that are as high as possible (i.e. distort very little).

6.5.2.3 Signal-to-Intermodulation Ratio

Other effects resulting from nonlinearities are **intermodulation distortions**.

EXAMPLE: A nonlinear second-order system with a characteristic $v_2 = a \cdot v_1 + b \cdot v_1^2$ is excited with the **two-tone signal** $v_1(t) = \cos \omega_1 t + \cos \omega_2 t$. The output signal is

$$\begin{aligned}
v_2 =\ & \frac{b}{2} & \text{(DC component)} \\
& + a(\cos \omega_1 t + \cos \omega_2 t) & \text{(intended signal)} \\
& + \frac{b}{2}(\cos 2\omega_1 t + \cos 2\omega_2 t) & \text{(components at double the frequency)} \\
& + b \cdot \cos(\omega_1 + \omega_2)t & \text{(sum)} \\
& + b \cdot \cos(\omega_1 - \omega_2)t & \text{(and difference frequencies)}
\end{aligned}$$

Generally, for nonlinear systems of nth order, there will be signal components at frequencies

$$|p \cdot f_1 \pm q \cdot f_2|, \quad \text{with} \quad p, q = 0, 1 \ldots n \quad \text{and} \quad p + q \leq n \tag{6.77}$$

EXAMPLE: A nonlinear third-order system is excited by a two-tone signal with the frequencies 5 kHz and 7 kHz. The output signal contains the following frequencies:

p	0	0	0	0	1	1	1	2	2	3	5 kHz
q	0	1	2	3	0	1	2	0	1	0	7 kHz
$p \cdot f_1 + q \cdot f_2$	0	7	14	21	5	12	19	10	17	21	(kHz)
$|p \cdot f_1 - q \cdot f_2|$						2	9		3		(kHz)

For this system the distortion products are shown using logarithmic scales in Fig. 6.58.

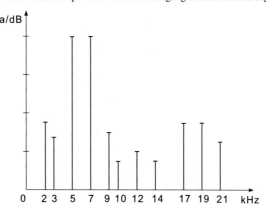

Fig. 6.58. Distortion products of a nonlinear system under excitation by a two-tone signal of 5 kHz and 7 kHz using logarithmic scales

Intermodulation also occurs in narrowband systems, which do not transmit any harmonics because of their inherent bandwidth limitations. Particularly disturbing during the transmission of the useful signals are distortions of third-order with frequencies of $2f_1 - f_2$ and $2f_2 - f_1$ (3 kHz and 9 kHz, respectively, in the example), since these are closest to the useful signal and therefore most difficult to suppress. For systems with small nonlinear distortions, it approximately holds for the distortion products of second and third-order that

$$v_2^{(2)} = const \cdot v_1^2, \quad \text{and} \quad v_2^{(3)} = const \cdot v_1^3 \tag{6.78}$$

The different constants are given by the coefficients of the characteristic equation.

- The amplitude of the distortion products of second order increases approximately quadratically with the input signal. The dependence is cubic for intermodulation products of third-order.

Taking the logarithm of both sides of Eq. (6.78) yields

$$\underbrace{20 \log_{10} v_2^{(2)}}_{L_2^{(2)}} = const + 2 \cdot \underbrace{20 \log_{10} v_1}_{L_1}$$

where L_1 is the input voltage level, and $L_2^{(n)}$ is the output voltage level of the intermodulation product of nth order.

$$L_2^{(1)} = const + L_1, \qquad L_2^{(2)} = const + 2 \cdot L_1, \qquad L_2^{(3)} = const + 3 \cdot L_1$$

In logarithmic representation the output voltage level of all signal components depends linearly on the input voltage level, and only the slope changes.

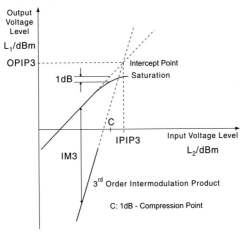

Fig. 6.59. Definition of the intermodulation margin, intercept point and 1 dB compression point (denoted by C)

The **intermodulation margin** is the logarithm of the ratio between the useful signal and the intermodulation product. This is denoted by $IM2$ or $IM3$, respectively. The intermodulation margin decreases with increasing output voltage swing. The input power, where the intermodulation margin vanishes, is called the **intercept point** (Fig. 6.59). Knowledge of the intercept point (IP) leads to the value for the intermodulation margin (IM) for a given input power L_1. The following expression may be used

$$IM3(L_1) = 2 \cdot (IP3 - L_1) \tag{6.79}$$

EXAMPLE: A microwave amplifier has an intermodulation margin IM of 34 dB for a given input voltage level of -15 dBm. To what level must the input voltage be reduced to produce an intermodulation margin of 40 dB?

The intercept point of the system is at 2 dBm according to Eq. (6.79). A drop of the input voltage level to -18 dBm, i.e. a drop of 3 dB, yields the desired intermodulation margin.

NOTE: There are two ways to identify the intercept point, referring to input or output power, respectively. Either the *input intercept point* ($IPIP$) or the *output intercept point* ($OPIP$) is given.

For practical systems, the intercept point cannot be achieved, because the output signal saturates. The **1 dB compression point** is used to characterise the output voltage swing limits. This is the input power for which the actual output power lies 1 dB below the theoretically expected value.

6.6 Notation Index

a	time scaling factor
a_0, a_1, a_2	coefficients of nominator polynomial
$a(\omega)$	damping ratio (dB)
$a_0/2$	DC component of a signal
a_k	total harmonic distortion attenuation (dB)
a_{kn}	total harmonic distortion attenuation nth-order (dB)
a_n	Fourier coefficients
A_n	Fourier coefficients of the amplitude spectrum
b_0, b_1	coefficients of denominator polynomial
b_n	Fourier coefficients
$b(\omega)$	phase response
B	bandwidth
c_n	complex Fourier coefficients
C_n	normalised capacitance value
E	energy of a signal
Δf	bandwidth of the ideal LPF
f_c	critical frequency
$f(t)$	function of time
$F(f), F(\omega)$	Fourier transform
$\mathcal{F}\{\}$	Fourier transform
$\mathcal{F}^{-1}\{\}$	inverse Fourier transform
$g(t)$	impulse response, weighting function of a system
$g_n(t)$	normalised impulse response
$G(\omega)$	transfer function
G_n	normalised transfer function
$h(t)$	step response of a signal
IM	intermodulation margin (dB)
k_n	total harmonic distortion of nth order
L_1	input voltage level (dBm)
$L_2^{(n)}$	output voltage level of the intermodulation product nth order
L_n	normalised inductance value
M, N	upper limits for the magnitude of a signal
p_1, p_2	zeros of the denominator polynomial, poles
P	power of a signal
$\text{rect}(t)$	rectangular waveform, rectangular pulse
$R(f)$	real part of the Fourier transform
R_n	normalised resistance value
R_r	reference resistance for the impedance normalisation (Ω)
$s(t)$	step function
$\text{sign}(t)$	signum function
sinc	sinc function
$|S(f)|$	magnitude of the Fourier transform

Si	integral sine function
t_0	delay time
t_n	normalised time
Δt_p	pulse width (ideal LPF)
t_s	settling time
T	period of a periodic signal
T	transformation through a system
ΔT	pulse width
THD	total harmonic distortion
$\hat{v}$	amplitude of the voltage
v_1	input voltage
v_2	output voltage
v_{in}	input voltage
v_{out}	output voltage
$X(f)$	imaginary part of the Fourier transform
$X(\omega)$	Fourier transform of the input signal
$Y(\omega)$	Fourier transform of the output signal
$\delta(t)$	impulse function, delta impulse (s^{-1})
$\varphi(f)$	phase component of the Fourier transform
φ_n	Fourier coefficients of the phase spectrum
$\varphi(\omega)$	phase response
$\Gamma(t)$	Gaussian pulse
$\Lambda(t)$	triangular pulse
$\Delta\omega$	bandwidth of the ideal LPF
ω_0	angular centre frequency (s^{-1})
ω_c	critical angular frequency
ω_n	normalised frequency
ω_r	reference frequency for frequency normalisation (s^{-1})
τ	integration variable
τ_g	group delay time
τ_p	phase delay time

6.7 Further Reading

CARLSON, G. E.: *Signal and Linear Systems Analysis, 2nd Edition*
John Wiley & Sons (1997)

CHAPRA, S. C.; CANALE, R. P.: *Numerical Methods for Engineers, 3rd Edition*
McGraw-Hill (1998)

CHEN, C. T.: *Linear System Theory and Design, 3rd Edition*
Oxford University Press (1998)

DORF, R. C.: *The Electrical Engineering Handbook, Section II*
CRC press (1993)

DORF, R. C.; BISHOP, R. H.: *Modern Control Systems, 8th Edition*
Addison-Wesley (1997)

KENNEDY, G.; DAVIS, B.: *Electric Communication Systems*
McGraw-Hill (1992)

LINDNER, D. K.: *Introduction to Signals and Systems*
McGraw-Hill (1999)

O'NEIL, P. V.: *Advanced Engineering Mathematics, 4th Edition*
Brooks/Cole Publishing Company (1997)

OPPENHEIM, A. V.; SCHAFER, R. W.: *Digital Signal Processing, 1st Edition*
Prentice Hall (1975)

OPPENHEIM, A. V.; SCHAFER, R. W.; BUCK, J. R.: *Discrete-Time Signal Processing, 2nd Edition*
Prentice Hall (1999)

OPPENHEIM, A. V.; WILLSKY, A. S.: *Signals & Systems, 2nd Edition*
Prentice Hall (2000)

7 Analogue Circuit Design

This chapter on **analogue circuit design** describes electric circuits that are used for the processing of analogue signals. **Analogue signals** have a continuous progression and can have any arbitrary value within certain limits.

7.1 Methods of Analysis

Calculations in analogue circuit design are made to identify the circuit configuration and to derive the component values. Often calculations can only be reasonably carried out by making simplifying assumptions. Therefore the equivalent circuits are strongly simplified and only represent the characteristics of the required function. Circuit analysis methods can describe the actual circuit conditions with an accuracy of approximately 10–20 %. Since values of semiconductors can vary by a factor of 2, and resistors and capacitors by 5–10 %, it is necessary to design circuits independent of the large tolerances of the components. To achieve this, methods from control engineering, especially negative feedback, are employed.

7.1.1 Linearisation at the Operating Point

The relationships between current and voltage in semiconductors are usually nonlinear.

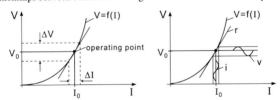

Fig. 7.1. Linearisation at the operating point

Provided that the voltages and currents vary only marginally about the operating point V_0, I_0 the function $V = f(I)$ can be linearised at V_0, I_0. The small amplitude variations of the signal around the operating point is shown in Fig. 7.1 by ΔV, ΔI. This signal is called a **small signal** because its amplitude is small compared to the operating point values. In order to replace the real nonlinear function with a linear function, all calculations concerning the small signal around the operating point will be simplified. The smaller the signal is compared to the operating point values, the more valid the linearisation assumption is. Linearisation is especially useful in small signal amplifier analysis, where the signal to be amplified, e.g. an audio signal, is small compared to the operating point values of the semiconductor circuit.

CALCULATION:

The function $V = f(I)$ is substituted with its slope in the operating point. For a small change ΔI of the current I around the operating point it then holds that:

$$\Delta V = \left.\frac{dV}{dI}\right|_{I_0} \cdot \Delta I \qquad (7.1)$$

For the small signal v, i it holds respectively that:

$$v = \left.\frac{dV}{dI}\right|_{I_0} \cdot i, \quad \text{or} \quad v = r \cdot i \qquad (7.2)$$

The resistance r is called the **dynamic resistance**, the **incremental resistance** or the **small signal resistance**. It is dependent on the operating point. The representation $v = r \cdot i$ means that the origin of the small signal $v = 0$, $i = 0$ has been moved to the operating point V_0, I_0.

NOTE: The principle of the linearisation in the operating point can be also applied to other nonlinear physical relationships.

7.1.2 AC Equivalent Circuit

Circuits for small-signal amplification usually have a DC supply voltage, while the signal itself is an AC voltage. In order to simplify the calculation only the quantities relevant to the signal are considered.

According to the **principle of superposition**, the effect of a voltage in a linear circuit can be calculated by eliminating all other voltage and current sources (voltage sources were replaced by a short circuit, current sources by an open circuit). If a real circuit with semiconductors is linearised at the operating point, the assumption for the superposition is met, i.e. the relationship between cause and effect is linear. A circuit where all supply voltages are replaced with a short circuit, so that only the small-signal source remains, is called the **small-signal equivalent circuit**.

EXAMPLE: The voltage V_2 consists of an AC and a DC part (Fig. 7.2). The following calculation determines the AC part v_2 of V_2: V_0 is replaced with a short circuit. It follows for $v_2 = f(v_1)$:

$$v_2 = \frac{R_2 \| R_3}{R_1 + (R_2 \| R_3)} \cdot v_1$$

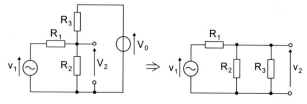

Fig. 7.2. Generation of an AC equivalent circuit

7.1.3 Input and Output Impedance

7.1.3.1 Determination of the Input Impedance

The **input impedance** $\underline{Z}_{in}$ of a small-signal circuit is the impedance between the input terminals for a small AC signal.

$$\underline{Z}_{in} = \frac{\underline{v}_{in}}{\underline{i}_{in}}$$

Fig. 7.3. Definition of the input impedance $\underline{Z}_{in}$

- For **passive circuits** the resulting impedance $\underline{Z}_{in}$ is obtained by combining all impedances occurring in the circuit.

- For **active circuits with unregulated/uncontrolled sources** $\underline{Z}_{in}$ is the resulting impedance at the input terminals, if all internal voltage sources are shorted and all current sources are opened/interrupted.

- For **active circuits with controlled sources** $\underline{Z}_{in}$ is determined by applying $\underline{v}_{in}$ across the input terminals and measuring $\underline{i}_{in}$ or calculated by using nodal and mesh analysis. Controlled sources are sources where the output values are determined by an other electrical quantity.

In practice the last case is the most relevant for the majority of applications with semiconductor amplifiers.

EXAMPLE: Calculate the input impedance of a circuit with a controlled current source (Fig. 7.4):

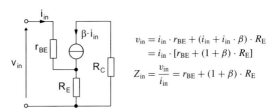

$$v_{in} = i_{in} \cdot r_{BE} + (i_{in} + i_{in} \cdot \beta) \cdot R_E$$
$$= i_{in} \cdot [r_{BE} + (1 + \beta) \cdot R_E]$$
$$Z_{in} = \frac{v_{in}}{i_{in}} = r_{BE} + (1 + \beta) \cdot R_E$$

Fig. 7.4. Calculation of the input impedance $\underline{Z}_{in}$

7.1.3.2 Determination of the Output Impedance

The idea behind determining **output impedance, equivalent source resistance** $\underline{Z}_{out}$ is to regard the active circuit as a voltage or current source with a source impedance (Fig. 7.5).

The output impedance $\underline{Z}_{out}$ is calculated as:

$$\underline{Z}_{out} = \frac{\text{open circuit voltage}}{\text{short circuit current}} = \frac{\underline{v}_{o/c}}{\underline{i}_{s/c}} \qquad (7.3)$$

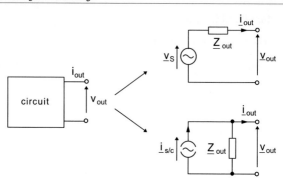

Fig. 7.5. Calculation of the output impedance $\underline{Z}_\text{out}$

Technically, the output impedance can be determined by measuring two different load states:

$$\underline{Z}_\text{out} = \frac{\underline{v}_1 - \underline{v}_2}{\underline{i}_2 - \underline{i}_1} \tag{7.4}$$

7.1.3.3 Combination of Two-Terminal Networks

When two two-terminal networks (circuits) are combined into one circuit, the current depends on the output impedance of one two-terminal network and the input impedance of the other two-terminal network (Fig. 7.6).

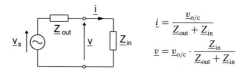

Fig. 7.6. Combination of two circuits

For such combinations three different cases are distinguished:

1. $\underline{Z}_\text{out} = \underline{Z}_\text{in}^*$ ($\underline{Z}_\text{in}$ is the complex conjugate of $\underline{Z}_\text{out}$)
   This is called **power matching**: $v = \dfrac{v_s}{2}$

2. $Z_\text{out} \ll Z_\text{in}$
   The impedance of the voltage source is much lower than the load impedance. The input voltage of the load impedance is approximately equal to the open-circuit voltage of the voltage source. In this case, the voltage v is approximately independent of the load impedance.

3. $Z_\text{out} \gg Z_\text{in}$
   The impedance of the voltage source is much larger than the load impedance. The current is mainly determined by the output impedance and is approximately independent of the load impedance.

7.1.4 Two-Port Networks

Two-port networks are circuits with four accessible terminals, where two terminals are the input (v_1, i_1) and two terminals are the output (v_2, i_2), see Fig. 7.7.

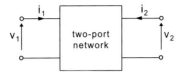

Fig. 7.7. Two-port network

Classification:

- Two-port networks are **active** if they contain sources (also controlled sources, e.g. by the input current); otherwise they are **passive**.

- Two-port networks are **symmetrical** if the input and output terminals can be swapped; otherwise they are **asymmetrical**.

- Two-port networks are **linear** if currents and voltages have linear relationships; otherwise they are **nonlinear**.

- Two-port networks are **reciprocal, reversible**, if the ratio of the input voltage to the output voltage is not affected by exchanging the input and output terminals; otherwise they are **irreversible, nonreciprocal**. All linear passive two-port networks are reversible.

- Two-port networks are **nonreactive** if they do *not* change the relevant output quantity of the previous and the relevant input quantity of the following two-port network. This is, for instance, the case if two-port networks are combined in a chain, where the inputs have a high impedance and the outputs have a low impedance.

7.1.4.1 Two-Port Network Equations

The electrical characteristics of a linear two-port network can be described unambiguously by means of their **two-port network equations**. The coefficients of the electrical quantities are called **two-port network parameters**. The two-port network equations are used to describe the small-signal behaviour of analogue circuits. They are especially useful in small-signal analysis of basic transistor circuits. Particularly significant are the hybrid and admittance forms of the two-port network equations.

7.1.4.2 Hybrid Parameters (h-Parameters)

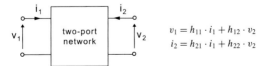

Fig. 7.8. The two-port network equations in hybrid form

The two-port network equations can be expressed in hybrid form using hybrid parameters (h-Parameters), Fig. 7.8. The parameters have the following meaning:

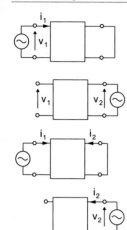

input resistance with shorted input
$$h_{11} = \frac{v_1}{i_1}, \quad \text{for} \quad v_2 = 0$$

reverse voltage transfer ratio with open input
$$h_{12} = \frac{v_1}{v_2}, \quad \text{for} \quad i_1 = 0$$

forward current gain with shorted output
$$h_{21} = \frac{i_2}{i_1}, \quad \text{for} \quad v_2 = 0$$

output admittance with open input
$$h_{22} = \frac{i_2}{v_2}, \quad \text{for} \quad i_1 = 0$$

The parameters are measured or calculated while the output is shorted or the input is open. The formal contexts of the two-port equations can be represented in an equivalent circuit as shown in Fig. 7.9.

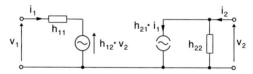

Fig. 7.9. Two-port network equivalent circuit diagram for the h-parameters

7.1.4.3 Admittance Parameters (y-Parameters)

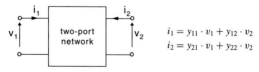

$$i_1 = y_{11} \cdot v_1 + y_{12} \cdot v_2$$
$$i_2 = y_{21} \cdot v_1 + y_{22} \cdot v_2$$

Fig. 7.10. The two-port equations in admittance form

The two-port network equations can be expressed in hybrid form using hybrid parameters (y-Parameters), Fig. 7.10. The parameters have the following meaning:

The parameters are measured or calculated while the input and output are shorted. The formal contexts of the two-port network equations can be represented in an equivalent circuit as shown in Fig. 7.11.

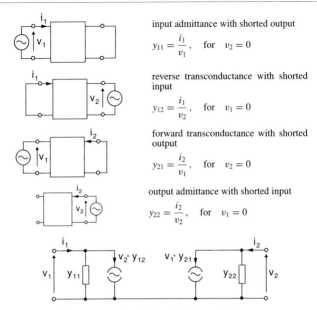

input admittance with shorted output

$$y_{11} = \frac{i_1}{v_1}, \quad \text{for} \quad v_2 = 0$$

reverse transconductance with shorted input

$$y_{12} = \frac{i_1}{v_2}, \quad \text{for} \quad v_1 = 0$$

forward transconductance with shorted output

$$y_{21} = \frac{i_2}{v_1}, \quad \text{for} \quad v_2 = 0$$

output admittance with shorted input

$$y_{22} = \frac{i_2}{v_2}, \quad \text{for} \quad v_1 = 0$$

Fig. 7.11. Two-port network equivalent circuit diagram for the y-parameters

7.1.5 Block Diagrams

Block diagrams are used for the representation and calculation of complex analogue circuits. Individual parts of the circuit are represented by a **block**, where the transfer characteristics between output $X_{\text{out}}(s)$ and input $X_{\text{in}}(s)$ can be described unambiguously by a transfer function $F(s)$, see Fig. 7.12.

$$X_{\text{out}}(s) = F(s) \cdot X_{\text{in}}(s)$$

Fig. 7.12. Representation of a block

Input and output quantities as well as the **transfer function** are represented in the Laplace frequency domain, i.e. as a function of the complex frequency s. The **transfer function** is written into the block. The values of $X_{\text{out}}(s)$ and $X_{\text{in}}(s)$ may have different physical units. The combination of circuit parts is represented with the connection of the corresponding blocks. The signal direction is marked by arrows on the connecting lines. Addition and subtraction of signals are represented by **summation points**.

NOTE: This representation becomes very clear if the individual blocks are nonreactive, i.e. the following block does not influence the previous block (no loading effect). This is achieved if the individual circuit parts have low-impedance outputs and high-impedance inputs, or are separated by an impedance converter.

7.1.5.1 Calculation Rules for Block Diagrams

The transfer function of a complex circuit can be calculated using the following (Fig. 7.13) calculation rules:

Combination of two blocks connected in series

Combination of two blocks connected in parallel

Elimination of a feed back loop

Translation of a summation point

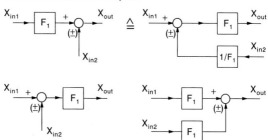

Fig. 7.13. Block diagram algebra

EXAMPLE: Calculation of a transfer function using block diagram algebra (Fig. 7.14):

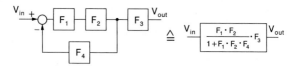

Fig. 7.14. Example of block diagram algebra

7.1.6 Bode Plot

The **Bode plot** represents the transfer characteristics of two-port networks with identical physical units at the input and the output (e.g. amplifiers, attenuators, Fig. 7.15). A distinction is made between the **frequency response** and the **phase response**. The frequency response represents the gain as a function of the angular frequency in a diagram, where both axes have logarithmic scales. The phase response shows the phase difference between the output and the input as a function of the angular frequency, where the frequency axis is logarithmic.

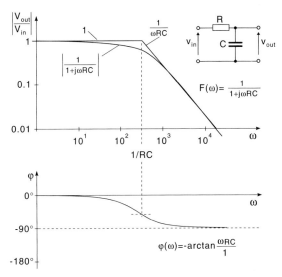

Fig. 7.15. Bode-plot of a low-pass filter

NOTE: The Bode plot is very useful for nonreactive circuits, which are combined in series (see Sect. 7.1.5). In this case the transfer functions must be multiplied, i.e. the magnitudes of the gain must be multiplied while the phase shift must be added. In the Bode plot this multiplication can be done graphically through linear geometric addition.

7.2 Silicon and Germanium Diodes

Diodes are semiconductors with a single p–n junction, which in general allow currents to pass in one direction only (rectification). Diodes may also be used for other purposes such as signal mixing, variable capacitors and voltage biasing.

7.2.1 Current–Voltage Characteristic of Si and Ge Diodes

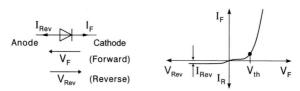

Fig. 7.16. Circuit symbol and characteristic diagram of Si and Ge diodes

The **reverse current** I_{Rev} of silicon diodes is approximately 10 pA and that of germanium diodes is approximately 100 nA. The **threshold voltage** is defined as the forward voltage across the diode when the forward current reaches 10% of the maximum permanent DC current. For silicon diodes the threshold voltage is approximately 0.7 V, and for germanium diodes this value is approximately 0.3 V. Because of the sharp rise in the forward bias characteristic curve, in approximate calculations it is presumed that the voltage drop V_F is 0.7 V for silicon diodes and 0.3 V for germanium diodes (Fig. 7.16).

The analytical function of the characteristic curve is given by:

$$I_F = I_{Rev} \cdot \left(e^{\frac{V_F}{V_T}} - 1 \right) \tag{7.5}$$

with I_{Rev} : reverse current
 $V_T = \dfrac{kT}{e}$: thermal voltage
 k : Boltzmann's constant $= 1.38 \cdot 10^{-23} \dfrac{J}{K}$
 T : absolute temperature
 e : elementary charge

The thermal voltage V_T is approximately 25 mV at $T = 300$ K (approx. 25°C, room temperature).

Approximations:

For **forward operation**: $e^{\frac{V_F}{V_T}} \gg 1$. The diode characteristic then simplifies to

$$I_F \approx I_{Rev} \cdot e^{\frac{V_F}{V_T}}. \tag{7.6}$$

For **reverse operation**: $e^{\frac{V_F}{V_T}} \ll 1$. Therefore the reverse current is approximately $I_{Rev} = $ const. for the whole reverse operation range.

7.2.2 Temperature Dependency of the Threshold Voltage

The **threshold voltage** of a p–n junction decreases with increasing temperature by 2 to 2.5 mV/K.

NOTE: Because diodes have a negative temperature coefficient they must *not* be combined in parallel in order to increase the maximum rectification current. Minimum differences in the temperature would cause a lower forward voltage at the warmer diode. Therefore the warmer diode would take over a higher current than the cooler diode. This would lead to a further temperature increase in the warmer diode, which then would again take over a bigger part of the total current. This results in the hotter diode taking over the entire current.

7.2.3 Dynamic Resistance (Differential Resistance)

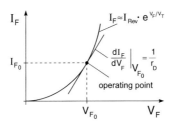

Fig. 7.17. Dynamic resistance r_D of Si and Ge diodes

The **dynamic resistance** r_D of the diode is the slope of the characteristic in the operating point.

$$r_D = \left.\frac{dV_F}{dI_F}\right|_{V_{F0}} \tag{7.7}$$

$$\frac{1}{r_D} = \left.\frac{dI_F}{dV_F}\right|_{I_{F0}} = \frac{1}{V_T} \cdot \underbrace{I_{Rev} \cdot e^{\frac{V_F}{V_T}}}_{I_{F0}} = \frac{I_{F0}}{V_T} \tag{7.8}$$

- The dynamic resistance of the diode is equal to the thermal voltage V_T divided by the forward current I_{F0} in the operating point (Fig. 7.17). Therefore the dynamic resistance r_D is reciprocally proportional to the forward current I_F.

$$\boxed{r_D = \frac{V_T}{I_{F0}}} \tag{7.9}$$

7.3 Small-Signal Amplifier with Bipolar Transistors

Small-signal amplifiers are circuits used for the amplification of small alternating signals, where the signal amplitude is much smaller than the operating point values (i.e. the DC voltages across the components). The operating frequencies are supposed to be low, so that group and phase propagation delays from parasitic elements do not have to be considered (however, it is noted in the case where special operating conditions apply).

7.3.1 Transistor Characteristics

7.3.1.1 Symbols, Voltages and Currents for Bipolar Transistors

A distinction is made between n–p–n and p–n–p transistors

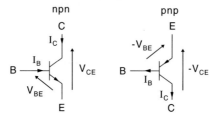

Fig. 7.18. Symbol, voltages and currents for bipolar transistors

The terminals are called the **base (B)**, the **collector (C)** and the **emitter (E)**. The base–emitter junction and the base–collector junction are p-n junctions. In normal operation the base–emitter diode is used in forward bias operation while the base–collector diode is used in reverse bias operation. The direction of the arrow in the circuit symbol gives the forward direction of the diode. The positive base current flows into the base for a n–p–n transistor and comes out of the base for the p–n–p transistor. The base current causes a voltage drop of appr. 0.7 V across the base–emitter diode. The base current controls the collector current, provided that the applied collector–emitter voltage drives the collector-base diode in reverse bias operation (Fig. 7.18). Then the collector current is approximately proportional to the base current.

NOTE: The type of an unknown transistor can be determined by checking the direction of the base–emitter and the base–collector diode with an ohmmeter.

NOTE: A transistor can be checked for defects by

a) checking the base–emitter and the base–collector diodes, and

b) measuring if the collector–emitter path has a high resistance (is not conducting) when the base is open.

7.3.1.2 Output Characteristics

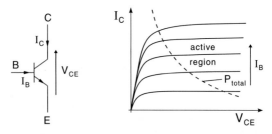

Fig. 7.19. Output characteristics $I_C = f(V_{CE})$, I_B=Parameter

The **output characteristics** show the collector current as a function of the collector–emitter voltage (Fig. 7.19). The base current acts as a parameter. The output characteristics provide all essential information necessary for the design of a circuit. In the so-called active region the output characteristics are almost horizontal. Within this region the collector current is approximately proportional to the base current. In this active region the transistor may be used as a small-signal amplifier. Often in these characteristics P_{total}, which is a hyperbolic function, is given as well. This shows which current–voltage values are permitted given the maximum allowed temperature of the transistor.

7.3.1.3 Transfer Characteristic

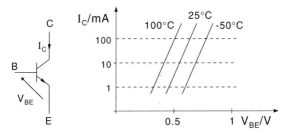

Fig. 7.20. Transfer characteristic $I_C = f(V_{BE})$

The **transfer characteristic** shows the collector current as a function of the base–emitter voltage (Fig. 7.20). Because of the diode characteristic of the base–emitter junction $I_C = f(V_{BE})$ is also an exponential function since $I_C \propto I_B$. It appears linearly in a diagram with a logarithmic ordinate. Often several characteristics are given with the temperature as a parameter.

7.3.1.4 Input Characteristic

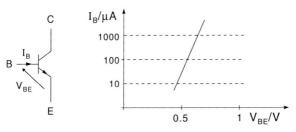

Fig. 7.21. Input characteristic $I_B = f(V_{BE})$

The **input characteristic** is the diode characteristic of the base–emitter junction (Fig. 7.21).

7.3.1.5 Static Current Gain β_{DC}

The static **current gain** β_{DC} is the relationship between the collector current and the base current in the active region:

$$\beta_{DC} = \frac{I_C}{I_B} \tag{7.10}$$

Common values are between 100 and 1000 for small-signal transistors, and between 10 and 200 for power transistors.

7.3.1.6 Differential Current Gain β

The **differential current gain** β is the current gain for small signals around the operation point. It is the derivative of the collector current with respect to the base current. A small change of the base current ΔI_B causes a small change $\beta \cdot \Delta I_B$ in the collector current. Therefore a small signal is amplified with this differential current gain. The differential current gain is also called the **AC current gain** or the **small-signal current gain**.

A distinction is made between β and β_0. While β is a common expression for the differential current gain, β_0 is a certain differential current gain, which is called the **forward current gain with shorted output** (Fig. 7.22). This is the differential current gain for low frequencies (propagation delay times and phase shifts caused by parasitic elements need not be considered), and where the collector–emitter voltage is kept constant ($V_{CE} = $ const.) Keeping the collector–emitter voltage constant means that the AC signal is shorted, therefore the term *shorted output*.

$$\beta_0 = \left.\frac{dI_C}{dI_B}\right|_{V_{CE}= \text{const}} \approx \left.\frac{\Delta I_C}{\Delta I_B}\right|_{V_{CE}= \text{const}} \tag{7.11}$$

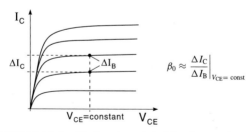

Fig. 7.22. Determination of the forward current gain β_0 from the output characteristics

Common values for β_0 are between 100 and 1000 for small-signal transistors, and between 10 and 200 for power transistors.

NOTE: In case a data sheet is not available, it is recommended to assume $\beta_0 = 100$ for further calculations.

7.3 Small-Signal Amplifier with Bipolar Transistors

7.3.1.7 Transconductance g_m

The **transconductance** g_m is the change in the collector current I_C depending on the change in the base–emitter voltage V_{BE}. It is the slope of the transfer characteristic.

$$g_m = \left. \frac{dI_C}{dV_{BE}} \right|_{V_{CE} = \text{const}} \approx \left. \frac{\Delta I_C}{\Delta V_{BE}} \right|_{V_{CE} = \text{const}} \tag{7.12}$$

NOTE: The use of the transconductance g_m for bipolar transistor circuit design is not recommended, since bipolar transistors are *current-controlled* devices. The transconductance is mainly used for field-effect transistor circuit design, since field-effect transistors are *voltage-controlled* components. However, occasionally the *very high transconductance of bipolar transistors* is pointed out, because the collector current I_C changes drastically for small changes in the base–emitter voltage V_{BE}.

7.3.1.8 Thermal Voltage Drift

Thermal voltage drift is the base–emitter voltage change ΔV_{BE} that depends on the junction temperature. The base–emitter voltage decreases with increasing temperature. The change is $|\Delta V_{BE}| = 2\text{–}2.5$ mV/K.

7.3.1.9 Differential Input Resistance r_{BE}

The **small-signal differential input resistance** is the slope of the input characteristic curve at the operating point. This is the differential resistance of the base–emitter diode (see Sect. 7.2.3).

$$r_{BE} = \frac{dV_{BE}}{dI_B} \approx \frac{\Delta V_{BE}}{\Delta I_B} \approx \frac{V_T}{I_B} \tag{7.13}$$

where V_T is the thermal voltage (about 25 mV at $T = 300$ K)

7.3.1.10 Differential Output Resistance r_{CE}

The **small-signal differential output resistance** defines the change in the collector current as a function of the collector–emitter voltage for a constant base current (Fig. 7.23). This can be calculated from the output characteristic curve.

$$r_{CE} = \left. \frac{dV_{CE}}{dI_C} \right|_{I_B = \text{const}} \approx \left. \frac{\Delta V_{CE}}{\Delta I_C} \right|_{I_B = \text{const}} \tag{7.14}$$

- If the output characteristic is horizontal then $r_{CE} \to \infty$.

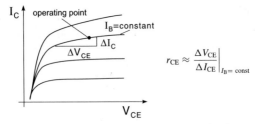

Fig. 7.23. Calculation of the small-signal differential output resistance r_{CE} from the output characteristic curve

7.3.1.11 Reverse Voltage Transfer Ratio A_r

The **reverse voltage-transfer ratio** defines the change in the input voltage as a function of the output voltage for a constant base current.

$$A_r = \left.\frac{d V_{BE}}{d V_{CE}}\right|_{I_B = \text{const}} \approx \left.\frac{\Delta V_{BE}}{\Delta V_{CE}}\right|_{I_B = \text{const}} \tag{7.15}$$

The reverse voltage-transfer ratio is negligible at lower frequencies. For higher frequencies this can be taken into account either through the relevant value from a data sheet, or by the addition of a capacitor between the collector and the emitter (Miller capacitance). In this manner it need not be considered in the transistor AC equivalent circuit.

7.3.1.12 Unity Gain and Critical Frequencies

The **unity gain frequency** is the frequency at which the current gain β has a value of 1.

The **critical frequency** f_β is the frequency at which β has fallen 3 dB below β_0. This is also known as the cutoff or corner frequency. For transistors whose short-circuit current gain is considerably greater than 1 ($\beta_0 \gg 1$), it can be approximated that:

$$f_\beta = \frac{f_T}{\beta_0} \tag{7.16}$$

NOTE: In circuits without negative feedback the useful frequency range lies between $0 < f < f_\beta$. Negative feedback increases the frequency range roughly by the feedback factor.

7.3.2 Equivalent Circuits

7.3.2.1 Static Equivalent Circuit

In order to design or understand electronic circuits, the following **static equivalent circuit** is useful.

The bipolar transistor consists of two back-to-back p–n junctions. In normal operation the base–collector diode is reverse biased, and the base–emitter diode is forward biased. The base–collector diode can be assumed to be a current source whose current is proportional to the base current. The voltage drop at the base–emitter junction is approximately 0.7 V from the diode characteristic.

The difference between the n–p–n and p–n–p transistor is that currents and voltages are in opposite directions (Fig. 7.24).

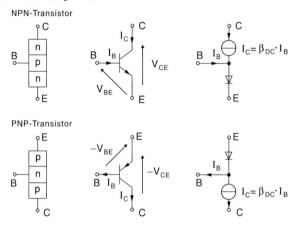

Fig. 7.24. Topology, circuit symbol and static equivalent circuit of the bipolar transistor

7.3.2.2 AC Equivalent Circuit

In the **AC equivalent circuit** (Fig. 7.25) only alternating quantities that are small about the operating point are considered. The operating point must lie within the active region of the output characteristic.

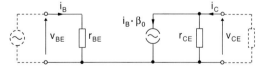

Fig. 7.25. AC equivalent circuit of the bipolar transistor

The base current controls the collector current. The base resistance r_{BE} is equal to the dynamic resistance of the base–emitter diode. The base current i_B controls the internal collector current $i_B \cdot \beta_0$. A small portion of the current $i_B \cdot \beta_0$ flows away through r_{CE} and therefore does not appear at the collector terminal. The resistance r_{CE} is high impedance (see the output characteristic curve where the parameter i_B is represented by constant lines, which means $r_{CE} \to \infty$). It can be neglected in approximate calculations.

7.3.2.3 The Giacoletto Equivalent Circuit

The **Giacoletto equivalent circuit** is an AC equivalent circuit (Fig. 7.26). It describes the AC characteristics of the transistor up to approximately half of the unity gain frequency.

Physical Explanation:

The internal collector current $v_{B'E} \cdot g_{mB'E}$ is proportional to the internal base voltage $v_{B'E}$. The output voltage v_{CE} is fed back in anti-phase to the internal base voltage $v_{B'E}$ through the feedback capacitor $C_{B'C}$. The feedback effect through $C_{B'C}$ increases with increasing frequency as the feedback impedance $1/\omega C_{B'C}$ decreases. Accordingly, the transistor gain i_C/i_B decreases with increasing frequency.

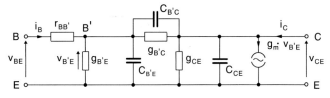

Fig. 7.26. Giacoletto AC equivalent circuit

7.3.3 Darlington Pair

At low frequencies the **Darlington pair** shows the characteristics of a bipolar transistor whose current gain is approximately the product of the two individual current gains (Fig. 7.27).

The **static current gain** or **DC current gain** β_{DC} of the Darlington pair is given by:

$$I_C = I_{C1} + I_{C2} = I_{B1}\beta_{DC1} + I_{B2}\beta_{DC2} = I_{B1}\beta_{DC1} + I_{B1}(1 + \beta_{DC1})\beta_{DC2}$$
$$= I_{B1}(\beta_{DC1} + \beta_{DC2} + \beta_{DC1}\beta_{DC2})$$

For $\beta_{DC1} \gg 1$ and $\beta_{DC2} \gg 1$ it holds that:

$$\boxed{\beta_{DC} \approx \beta_{DC1} \cdot \beta_{DC2}} \tag{7.17}$$

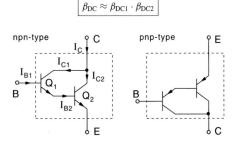

Fig. 7.27. Darlington pair

The **dynamic current gain** or **small-signal current gain** β_0 of the Darlington pair is given by:

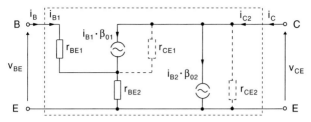

Fig. 7.28. AC equivalent circuit of the Darlington pair

Neglecting the collector–emitter resistances, the circuit diagram in Fig. 7.28 yields:

$$i_C = i_{B1}\beta_{01} + i_{B2}\beta_{02} = i_{B1}\beta_{01} + i_{B1}(1+\beta_{01})\beta_{02}$$
$$= i_{B1}(\beta_{01} + \beta_{02} + \beta_{01}\beta_{02})$$

For $\beta_{01} \gg 1$ and $\beta_{02} \gg 1$ it holds that:

$$\beta_0 \approx \beta_{01} \cdot \beta_{02} \qquad (7.18)$$

For the **differential input resistance** r_{BE} it holds that:

$$r_{BE} = \frac{v_{BE}}{i_B} = r_{BE1} + \beta_{01} \cdot r_{BE2} \approx r_{BE1} + \frac{\beta_{01}}{\beta_{DC1}} \cdot r_{BE1}, \qquad \text{with} \qquad r_{BE2} = \frac{r_{BE1}}{\beta_{DC1}}$$

Using the approximation that $\beta_{01} \approx \beta_{DC1}$ it follows that:

$$r_{BE} \approx 2 \cdot r_{BE1} \approx 2\frac{V_T}{I_{B1}} \qquad (7.19)$$

- The input impedance of the Darlington pair is approximately twice the thermal voltage V_T divided by the quiescent input current I_B.

The Darlington pair is employed where a high output power has to be controlled by a small control power. The high current gain of the Darlington pair results in high input impedances in amplifiers. In power electronics three- or four-fold Darlington connections are even used for switching high currents.

7.3.3.1 Pseudo-Darlington Pair

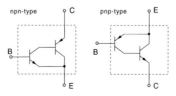

Fig. 7.29. Pseudo-Darlington pair

The Current gain and the input impedance of the pseudo-Darlington pair (Fig. 7.29) is given by:

$$\beta_{DC} \approx \beta_{DC1} \cdot \beta_{DC2}, \qquad \beta_0 \approx \beta_{01} \cdot \beta_{02}, \qquad r_{BE} \approx r_{BE1} \approx \frac{V_T}{I_B} \qquad (7.20)$$

7.3.4 Basic Circuits with Bipolar Transistors

Small-signal operation of bipolar transistors is divided into three operating groups, namely **common-emitter, common-collector and common-base** (Fig. 7.30). Their circuits each have different input and output voltage terminations. The input and output voltages are measured with respect to the common line. This common line gives the circuit its name, e.g. common-emitter circuit. Each circuit has different gain and impedance properties.

	Common Emitter circuit	Common Collector circuit	Common Base circuit
Circuit	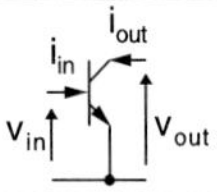 (diagram)	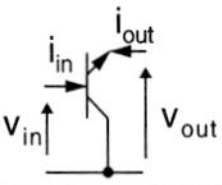 (diagram)	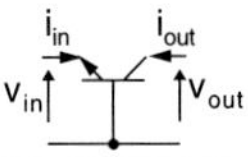 (diagram)
voltage gain A_v	> 1	≈ 1	> 1
current gain A_i	> 1	> 1	≈ 1
input impedance r_{in}	medium	very high	very small
output impedance r_{out}	high	very small	high

Fig. 7.30. Basic bipolar transistor circuits

7.3.5 Common-Emitter Circuit

The **common-emitter circuit** has a high power, current and voltage gain. The output voltage has the opposite phase to the input voltage.

The transistor operating point in the circuit in Fig. 7.31 is adjusted by the resistors R_1, R_2, R_C and R_E so that it lies in the active region of the output characteristic. The AC signal is coupled into the circuit through C_1 and out of the circuit through C_2. The values of C_1 and C_2 are chosen so that they appear as short circuits in the relevant frequency range (Fig. 7.31).

The capacitor C_E is also approximately a short circuit in the relevant frequency range, so that the AC emitter is approximately at ground. Occasionally, the capacitor C_E is not used so that the emitter is *not* grounded. In this case, the circuit does *not* correspond to the definition of a common-emitter circuit, but is nonetheless described as such.

The use of common-emitter circuits is limited to low- and mid-range frequencies, because at high frequencies negative feedback occurs between the antiphase input and output voltages, through the collector–base parasitic capacitor (**Miller capacitance**).

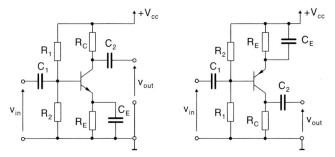

Fig. 7.31. Common-emitter circuit with n–p–n and p–n–p transistor

7.3.5.1 Common-Emitter Circuit Two-Port Network Equations

The common-emitter two-port network parameters are usually given as h-parameters (Fig. 7.32).

$$v_{BE} = h_{11E} \cdot i_B + h_{12E} \cdot v_{CE} \tag{7.21}$$
$$i_C = h_{21E} \cdot i_B + h_{22E} \cdot v_{CE} \tag{7.22}$$

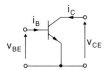

Fig. 7.32. h-parameters of the common-emitter circuit

Short circuit-input resistance:

$$h_{11E} = r_{BE} = \left.\frac{dV_{BE}}{dI_B}\right|_{V_{CE}=\text{const}} = \left.\frac{v_{BE}}{i_B}\right|_{v_{CE}=0} \approx \frac{V_T}{I_B} \tag{7.23}$$

Parameter h_{11E} is called the short-circuit input resistance. It is equal to the AC input voltage v_{BE} divided by the AC input current i_B (see also Sect. 7.2.3). The $V_{CE} = $ const. condition is meaningless in the calculation or measurement of the input impedance at lower frequencies, as the output voltage has hardly any influence on the input voltage (see h_{12E}).

Reverse voltage transfer ratio:

$$h_{12E} = \left.\frac{dV_{CE}}{dV_{BE}}\right|_{I_B=\text{const}} \approx 0 \tag{7.24}$$

Parameter h_{12E} is called the reverse voltage-transfer ratio with open input (actually only the AC input is open, and a DC quiescent current has to be present). For low frequencies it is approximately zero. For higher frequencies the feedback voltage can be modelled by an equivalent capacitance C_{CB}, so that does not appear in the transistor AC equivalent circuit (see also Sect. 7.3.1).

Forward current gain:

$$h_{21E} = \beta_0 = \left.\frac{dI_C}{dI_B}\right|_{V_{CE}=\text{const}} = \left.\frac{i_C}{i_B}\right|_{v_{CE}=0} \approx \left.\frac{\Delta I_C}{\Delta I_B}\right|_{V_{CE}=\text{const}} \quad (7.25)$$

Parameter h_{21E} is called the forward current gain with shorted output (AC-shorted only). It is the AC current gain β_0. This gives the relationship between the AC collector and base currents for an AC-shorted collector–emitter junction. The collector–emitter path is shorted with a capacitor to measure h_{21E}. Graphically, h_{21E} can be determined from the output characteristic curve (Fig. 7.22).

Output admittance:

$$h_{22E} = \frac{1}{r_{CE}} = \left.\frac{dI_C}{dV_{CE}}\right|_{I_B=\text{const}} = \left.\frac{i_C}{v_{CE}}\right|_{i_B=0} \approx \left.\frac{\Delta I_C}{\Delta V_{CE}}\right|_{I_B=\text{const}} \quad (7.26)$$

Parameter h_{22E} is called the output admittance with open input (actually, only the AC input is open, and a DC quiescent current has to be present). This corresponds to the output impedance r_{CE} and can be determined from the output characteristic curve (Fig. 7.23).

7.3.5.2 Common-Emitter AC Equivalent Circuit

Figures 7.33 and 7.34 show the common-emitter circuit *with* (i.e. emitter grounded) and without the emitter capacitor C_E.

The capacitors are chosen to be short circuits in the relevant frequency range. Here V_{CC} is the DC supply voltage, R_{int} is the source resistance of the AC input voltage source, and R_L is the load resistance (e.g. the input resistance of a following circuit). Note that r_{CE} is not shown in the equivalent circuit in Fig. 7.34, because of its high impedance. For clarity it is not considered in the following calculations.

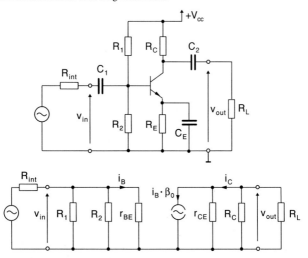

Fig. 7.33. Common-emitter circuit with bypass capacitor and its equivalent circuit

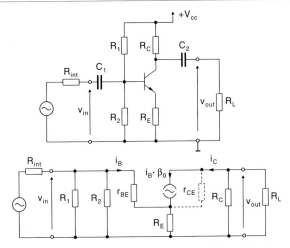

Fig. 7.34. Common-emitter circuit without bypass capacitor and its equivalent circuit

7.3.5.3 Common-Emitter Circuit Input Impedance

The input impedance is different for Fig. 7.33 and Fig. 7.34 (with and without emitter resistor bypassing):

In Fig. 7.33:

$$r_{in} = \frac{v_{in}}{i_{in}} = R_1 \| R_2 \| r_{BE} \tag{7.27}$$

If R_1 and R_2 have a high impedance compared to r_{BE}, the input impedance simplifies as follows:

$$\boxed{r_{in} \approx r_{BE}} \tag{7.28}$$

In Fig. 7.34:

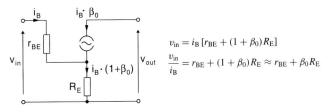

$$v_{in} = i_B \left[r_{BE} + (1 + \beta_0) R_E \right]$$
$$\frac{v_{in}}{i_B} = r_{BE} + (1 + \beta_0) R_E \approx r_{BE} + \beta_0 R_E$$

The input impedance is:

$$\boxed{r_{in} \approx R_1 \| R_2 \| (r_{BE} + \beta_0 R_E)} \tag{7.29}$$

- If the circuit is realised *without* the bypass capacitor C_E, the input impedance increases drastically. The emitter resistor R_E multiplied by a factor β_0 influences this value! However, the voltage gain decreases by the same factor (see Sect. 7.3.5.5).

7.3.5.4 Common-Emitter Circuit Output Impedance

The output impedance r_{out} is calculated by considering the circuit as a voltage or a current source with an internal source resistance (of course, both yield the same result). See also Sect. 7.1.3.2.

The output impedance is then

$$r_{out} = \frac{\text{open-circuit voltage}}{\text{AC short-circuit current}} = \frac{v_{o/c}}{i_{s/c}} \tag{7.30}$$

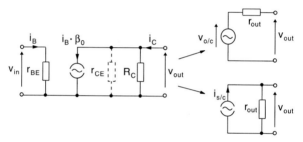

Fig. 7.35. Common-emitter circuit: calculation of the output impedance

Assume the input voltage v_{in} in Fig. 7.35 is known. Then

$$r_{out} = \frac{v_{o/c}}{i_{s/c}} = \frac{-\dfrac{v_{in}}{r_{BE}}\beta_0\,(r_{CE}\|R_C)}{-\dfrac{v_{in}}{r_{BE}}\cdot \beta_0} = r_{CE}\|R_C \tag{7.31}$$

Usually r_{CE} can be neglected because of its high impedance. Then

$$\boxed{r_{out} \approx R_C} \tag{7.32}$$

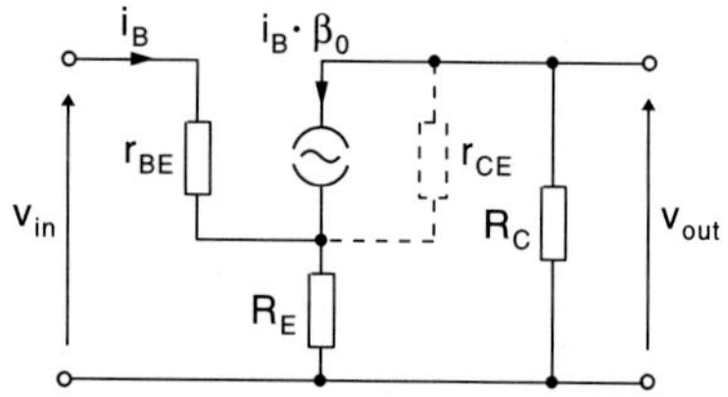

Fig. 7.36. Common-emitter circuit without emitter capacitor C_E

For the common-emitter circuit *without* the emitter capacitor C_E (Fig. 7.36) and neglecting r_{CE}:

$$r_{out} = \frac{v_{o/c}}{i_{s/c}} \approx \frac{-\dfrac{v_{in}}{r_{BE} + (1+\beta_0)R_E}\beta_0 R_C}{-\dfrac{v_{in}}{r_{BE} + (1+\beta_0)R_E}\beta_0} = R_C \quad (7.33)$$

The output of the common-emitter circuit is considered – physically correctly – a current source. The higher the value R_C, the higher the circuit efficiency. Unfortunately, as the operating point is also defined by R_C, the choice of R_C is not entirely free. It is possible to couple the alternating part of the collector current with high impedance using an (ideal) transformer in the collector arm (Fig. 7.37). In this case, the whole alternating part of the collector current flows through R_L. The output impedance is then very high and is given by $r_{out} = r_{CE}$.

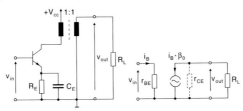

Fig. 7.37. Common-emitter circuit with transformer coupling of the output current and its corresponding AC equivalent circuit

7.3.5.5 Common-Emitter Circuit AC Voltage Gain

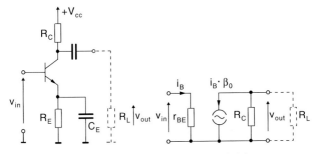

Fig. 7.38. AC equivalent circuit for voltage gain calculation

Calculation of the AC (small-signal) voltage gain G_v (Fig. 7.38):

$$A_v = \frac{v_{out}}{v_{in}} = \frac{-\dfrac{v_{in}}{r_{BE}}\beta_0 R_C}{v_{in}} = -\frac{\beta_0 R_C}{r_{BE}} \quad (7.34)$$

The voltage gain is negative. This means that the input and output voltage have opposite phases. If the output is loaded by a resistance R_L then the voltage gain decreases, as the current $i_B \cdot \beta_0$ divides between R_C and R_L. Then

$$A_v = -\frac{\beta_0}{r_{BE}}(R_C \| R_L) \qquad (7.35)$$

For the common-emitter circuit without C_E (Fig. 7.39):

$$A_v = \frac{v_{out}}{v_{in}} = \frac{-\dfrac{v_{in}}{r_{BE} + (1+\beta_0)R_E}}{v_{in}} \beta_0 R_C = -\frac{\beta_0 R_C}{r_{BE} + (1+\beta_0)R_E} \approx -\frac{R_C}{\dfrac{r_{BE}}{\beta_0} + R_E} \qquad (7.36)$$

When $\dfrac{r_{BE}}{\beta_0} \ll R_E$, then the gain A_v is given by $\dfrac{R_C}{R_E}$. If the circuit is loaded by R_L, then

$$A_v \approx \frac{-R_C \| R_L}{\dfrac{r_{BE}}{\beta_0} + R_E} \approx -\frac{R_C \| R_L}{R_E} \qquad (7.37)$$

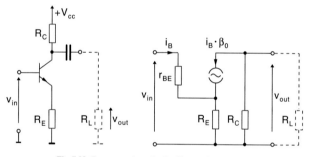

Fig. 7.39. Common-emitter circuit without emitter capacitor

The resistor R_E is called a **negative-feedback resistor**. Its voltage, which is proportional to i_C, is subtracted from the input voltage, i.e. negatively fed back.

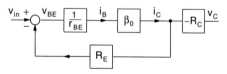

Fig. 7.40. Block diagram of the common-emitter circuit without emitter capacitor

The block diagram of the common-emitter circuit (Fig. 7.40) yields the same result (Eq. (7.36)).

7.3.5.6 Operating Point Biasing

The operating point, i.e. the point defined by the transistor DC values V_{CE} and I_C, should be in the active region of the output characteristic curve and also beneath the power dissipation

7.3 Small-Signal Amplifier with Bipolar Transistors

hyperbolic curve. The operating point should be stable with respect to thermal runaway and with respect to the production variations in the current gain β_{DC}.

The operating point is calculated as follows for the circuit shown in Fig. 7.41:

1. V_{CE} and I_C are chosen first. The voltage V_{CE} is chosen as a little less than half of the supply voltage V_{CC}. The choice of the collector current I_C has an upper limit given by permitted transistor power dissipation P_{total}.

$$V_{CE} \approx 0.3 \cdots 0.5\, V_{CC}, \qquad I_C : P_{total} < V_{CE} \cdot I_C$$

2. The resistor R_E stabilises the operating point. It is chosen so that approximately $1--2$ V is dropped across it.

3. The voltage $V_{RC} = V_{CC} - V_{CE} - V_{RE}$ drops across the resistor R_C. Then

$$R_C \approx \frac{V_{CC} - V_{CE} - V_{RE}}{I_C} \tag{7.38}$$

4. The base voltage is fixed by the resistors R_1 and R_2. They are also known as a base voltage divider. The choice of R_E and I_C means that the base voltage cannot be freely chosen.

$$V_{B0} = V_{RE} + V_{BE} = V_{RE} + 0.7\text{ V} \tag{7.39}$$

The voltage divider current I_s is chosen to be approximately 10 times the base current. This means that the base current only slightly loads the voltage divider, and thus the production variation of the current gain β_{DC} does not change the operating point.

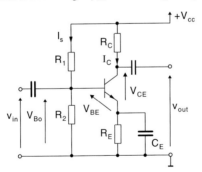

Fig. 7.41. Operating point biasing

EXAMPLE: Fixing the operating point of a common-emitter circuit:

1. Choice of V_{CE}: $V_{CE} = 4.5$–7.5 V , choose $V_{CE} = 6$ V
   Choice of I_C: $I_{C\,max} = 500$ mW/6 V $= 83$ mA , choose $I_C = 50$ mA
2. $R_E = 1$ V/50 mA $= 20\,\Omega$,
   choose $R_E = 22\,\Omega$, it follows that $V_{RE} = 1.1$ V
3. $R_C = (V_{CC} - V_{CE} - V_{R_E})/I_C = 7.9$ V/50 mA $= 158\,\Omega$, choose $R_C = 150\,\Omega$
4. $I_s \approx 10 \cdot I_B \approx 10 \cdot I_C/B = 10 \cdot 50$ mA/200 $= 2.5$ mA
   $\Longrightarrow R_2 \approx (V_{R_E} + V_{BE})/I_s = (1.1\text{ V} + 0.7\text{ V})/2.5\text{ mA} = 720\,\Omega$

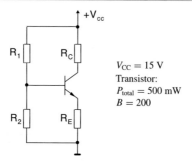

$V_{CC} = 15$ V
Transistor:
$P_{total} = 500$ mW
$B = 200$

$\Rightarrow R_1 \approx [V_{CC} - (V_{RE} + V_{BE})]/I_s = (15 \text{ V} - 1.9 \text{ V})/2.5 \text{ mA} = 5.2 \text{ k}\Omega$
choose $R_1 = 5.6$ kΩ and $R_2 = 820$ Ω

The calculation just described yields realistic component values. It is, however, not the only solution. For example, the voltage divider could have a higher impedance to increase the input impedance. Alternatively, R_C could be larger in order to have a greater open circuit voltage gain (see Fig. above). It is also not the only means to do the calculation. If, for example, the output impedance R_C should be equal to the load impedance R_L, then it is better to begin as follows: $R_C = R_L$
$\rightarrow I_C \approx (V_{CC}/2)/R_C \rightarrow R_E \approx (1\cdots 2 \text{ V})/I_C \rightarrow V_{B0} \approx V_{BE} + V_{RE}, I_s \approx 10 \cdot I_C/B \rightarrow R_2 = V_{B0}/I_s$ and $R_1 = (V_{CC} - V_{B0})/I_s$.

In general:

- The base voltage divider and R_E define the collector current. The collector resistor R_C defines the collector–emitter voltage.

7.3.5.7 Operating Point Stabilisation

Changes in the transistor data lead to a shift in the operating point. Thermal runaway ΔV_{BE} and production sample variation of the current gain β_{DC} are important in this context.

- All steps to stabilise the operating point must focus on keeping the collector current constant.

Stabilising the operating point using current feedback:

The resistor R_E is called a feedback resistor.

The **feedback mechanism:** if the base–emitter voltage V_{BE} decreases by an amount ΔV_{BE} because of a temperature increase, then the voltage V_{RE} increases (for $V_{B0} =$ constant). The difference of these two changes appears across the differential input impedance r_{BE} and produces a change ΔI_B in the base current. This is multiplied by the current gain β_0, yielding the change in collector current ΔI_C. This in turn produces a change in the voltage drop across R_E. The change ΔI_C with feedback present can be used to calculate the voltage change V_{RC} and thus V_{CE}: $\Delta V_{CE} = -\Delta V_{RC} = -\Delta I_C \cdot R_C$.

The relationship $\Delta I_C = f(\Delta V_{BE})$ can be derived using the block diagram in Fig. 7.42.

$$\frac{\Delta I_C}{\Delta V_{BE}} = \frac{-1}{\dfrac{r_{BE}}{\beta_0} + R_E} \qquad (7.40)$$

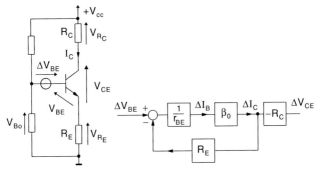

Fig. 7.42. Stabilising the operating point using current feedback

The voltage change ΔV_{BE} is considered as an extra voltage source at the base.

The relationship

$$\frac{\Delta V_{CE}}{\Delta V_{BE}} = A_{DR} = +\frac{1}{\frac{r_{BE}}{\beta_0} + R_E} \cdot R_C \approx +\frac{R_C}{R_E} \qquad (7.41)$$

is the **thermal voltage drift gain**. It shows how much the collector voltage changes as a result of thermal drift. It decreases as R_E increases. Usual values for A_{DR} lie in the range 5–10.

- The stabilising effect improves the larger R_E becomes.

NOTE: The recommendations made in 7.3.5.6 for the measurement of R_E are directly related to the thermal voltage drift gain. For normal supply voltages, the voltage drop across R_E lies between 1–2 V.

Stability for production sample variations of current gain is achieved using low impedance base voltage dividers. This means that sample variations in the base current do not influence the base quiescent current.

Operating-point stabilisation using voltage feedback:

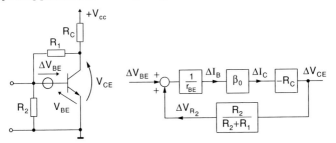

Fig. 7.43. Operating point stabilisation using voltage feedback

Feedback mechanism: if the voltage V_{BE} decreases by an amount ΔV_{BE} because of a temperature increase, then the base current I_B increases. As I_B increases, the collector current I_C also increases, causing the collector voltage to decrease. The base voltage defined by the voltage divider R_1, R_2 also decreases, and the base current (which was increased by the temperature increase) decreases again. This is represented in the block diagram in Fig. 7.43.

$$\frac{\Delta V_{CE}}{\Delta V_{BE}} = \frac{\dfrac{\beta_0}{r_{BE}} \cdot R_C}{1 + \dfrac{\beta_0 R_C}{r_{BE}} \dfrac{R_2}{R_1 + R_2}} = \frac{1}{\dfrac{r_{BE}}{\beta_0 R_C} + \dfrac{R_2}{R_1 + R_2}} \approx \frac{R_1 + R_2}{R_2} \qquad (7.42)$$

NOTE: The voltage feedback has the disadvantage that an AC current will also experience negative feedback. Therefore the AC current gain is the same as the thermal voltage drift gain. Alternatively, the AC voltage gain and thermal voltage drift gain with current feedback can be different, as the feedback resistor R_E can be AC-shorted by a capacitor C_E placed in parallel. Capacitor C_E is chosen so that it is a short circuit for the AC signal to be amplified, but exhibits a high impedance for the much slower changing thermal voltage drift.

Nonlinear stabilisation of the operating point:

The stabilisation of the operating point using current feedback can be further improved if a p–n junction is placed in the base voltage divider, which is thermally coupled with the transistor Q_1 (Fig. 7.44). Any thermal drift of the transistor Q_1 is therefore directly compensated for in the base voltage divider.

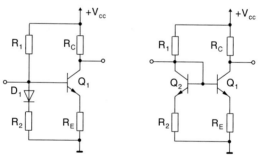

Fig. 7.44. Nonlinear operating point stabilisation

7.3.5.8 Load Line

The mesh equation $V_{CC} = I_C \cdot (R_C + R_E) + V_{CE}$ is a linear equation.

$$I_C = \frac{V_{CC} - V_{CE}}{R_C + R_E} = \underbrace{-\frac{1}{R_C + R_E}}_{\text{slope}} \cdot V_{CE} + \underbrace{\frac{V_{CC}}{R_C + R_E}}_{\text{constant}} \qquad (7.43)$$

Equation (7.43) is called the **static load line**. V_{CE} and I_C can only take on values that lie on the static load line. The operating point can be so chosen, by using the load line, that a maximum output range is achieved that uses the entire active range of the transistor.

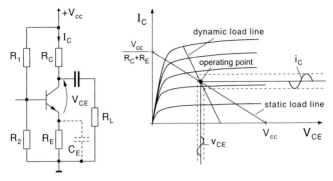

Fig. 7.45. Static and dynamic load line in the output characteristic

Bypassing R_E with a capacitor C_E leads to the **dynamic load line** (Fig. 7.45). The slope is given by $\dfrac{dI_C}{dV_{CE}} = -\dfrac{1}{R_C}$, or if the load resistance is included $\dfrac{dI_C}{dV_{CE}} = -\dfrac{1}{R_C \| R_L}$. This represents the relationship between the AC quantities v_{CE} and i_C.

7.3.5.9 Common-Emitter Circuit at High Frequencies

The collector AC voltage is in antiphase to the base voltage. A frequency dependent feedback exists through the parasitic collector–base capacitance (Miller capacitance). This increases with increasing frequency. The amount of feedback also depends on the internal resistance of the input voltage source. The smaller the resistance, the smaller the amount of feedback.

Current feedback increases the critical frequency of the circuit. The voltage gain is decreased, and so the voltage feedback is less. Also, the current gain β in the expression for the AC voltage gain is equally frequency dependent and thus decreases in value.

A value for the critical frequency can be measured or predicted with a suitable simulation system.

- A high critical frequency can be achieved by using current feedback and a small internal resistance in the input voltage source.

7.3.6 Common-Collector Circuit (Emitter Follower)

The common-collector circuit has a voltage gain of about 1. The output voltage range is from around $0.7 \text{ V} \leq V_B \leq V_{CC}$, i.e. the output voltage range practically extends to the supply voltage (Fig. 7.46).

The emitter voltage is always about 0.7 V below the base voltage. Hence the name **emitter follower**, as the emitter voltage follows the base voltage and differs by the fixed amount of 0.7 V.

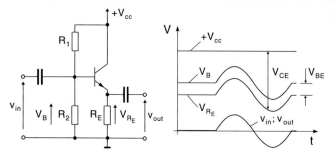

Fig. 7.46. The common-collector circuit and its voltages

The common-collector circuit has a very high input impedance and a small output impedance. Therefore it is used as an **impedance converter**, e.g. in combination with a common-emitter circuit (Fig. 7.47).

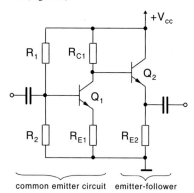

Fig. 7.47. Common-collector circuit as an impedance converter for a common-emitter circuit

7.3.6.1 Common-Collector AC Equivalent Circuit

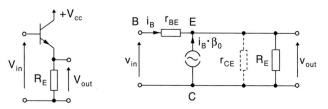

Fig. 7.48. Common-collector circuit and its AC equivalent circuit

7.3.6.2 Common-Collector Circuit Input Impedance

The common-collector input-impedance r_in can be expressed as

$$r_\text{in} = \frac{v_\text{in}}{i_\text{B}} = \frac{i_\text{B} \cdot r_\text{BE} + i_\text{B} \cdot (1 + \beta_0) \cdot R_\text{E}}{i_\text{B}} = r_\text{BE} + (1 + \beta_0) \cdot R_\text{E} \approx \beta_0 \cdot R_\text{E} \tag{7.44}$$

and with a load resistance R_L:

$$\boxed{r_\text{in} \approx \beta_0 \cdot (R_\text{E} \| R_\text{L})} \tag{7.45}$$

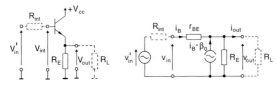

Fig. 7.49. AC equivalent circuit for the calculation of the input and output resistance

7.3.6.3 Common-Collector Circuit Output Impedance

The output impedance is given by:

$$r_\text{out} = \frac{\text{open-circuit AC voltage}}{\text{short-circuit AC current}} = \frac{v_{\text{o/c}}}{i_{\text{s/c}}} \tag{7.46}$$

The input AC voltage v_in is supplied.

This yields (Fig. 7.49):

$$v_{\text{o/c}} = i_\text{B} \cdot (1 + \beta_0) \cdot R_\text{E} = \frac{v_\text{in}}{r_\text{BE} + (1 + \beta_0) R_\text{E}} (1 + \beta_0) R_\text{E} \approx v_\text{in} \tag{7.47}$$

$$i_{\text{s/c}} = i_\text{B} \cdot \beta_0 = \frac{v_\text{in}}{r_\text{BE}} \beta_0 \tag{7.48}$$

This further yields:

$$r_\text{out} \approx \frac{r_\text{BE}}{\beta_0} \tag{7.49}$$

If the common-collector circuit is fed by a voltage source with an internal resistance R_int (e.g. by a common-emitter circuit with $R_\text{int} = R_\text{C}$), then this internal resistance appears at the output impedance reduced by a factor β_0.

$$\boxed{r_\text{out} \approx \frac{r_\text{BE} + R_\text{int}}{\beta_0}} \tag{7.50}$$

NOTE: The output impedance of a common-emitter stage can be reduced by a factor β_0 by the addition of an emitter follower (only two components!) as shown in Fig. 7.47.

7.3.6.4 Common-Collector Circuit AC Current Gain

The AC current gain is given by (Fig. 7.49):

$$A_i = \frac{i_{\text{out}}}{i_{\text{in}}} = \beta_0 \cdot \frac{R_E}{R_E + R_L} \tag{7.51}$$

The AC current gain of the emitter follower is unimportant in analogue design. Because of its large input impedance, a voltage is applied to the input of the emitter follower and is subsequently coupled to the output with a low output impedance.

7.3.6.5 Common-Collector Circuit at High Frequencies

The common-collector circuit has its critical frequency f_c approximately at the critical frequency of the current gain f_β (see also Sect. 7.3.1.12).

$$f_c \approx f_\beta \approx \frac{f_T}{\beta_0} \tag{7.52}$$

7.3.7 Common-Base Circuit

The **common-base circuit** has a current gain of 1 and a voltage gain similar to the common-emitter. The output voltage has the same phase as the input voltage. The input impedance is very small, so a transformer coupling is often used, which, depending on the winding ratio, can be very low impedance and deliver a large current for a small voltage (Fig. 7.50).

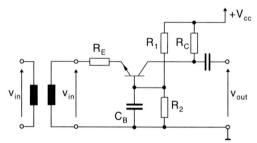

Fig. 7.50. Common-base circuit with transformer coupling

The common-base is suitable for very high frequencies. As its current gain $A_i = 1$ and the output voltage is in phase with the input voltage, it can be used up to approximately the unity gain frequency f_T.

NOTE: The importance of the common-base diminished greatly with the introduction of field-effect transistors, because common-source circuits (comparable to common-emitter circuits) are suitable up to frequencies that are achievable with bipolar transistors only with the common-base circuit.

Operation of the common-base: The base–emitter voltage is the controlling voltage. Given that the base is at AC ground, the input voltage must control the emitter voltage.

This has the disadvantage that the input voltage source must supply the emitter current and not the base current as in the case of the common emitter. For a positive change in the input voltage, the base–emitter voltage decreases. The collector current decreases, and the collector voltage increases (Fig. 7.50).

7.3.7.1 Common-Base AC Equivalent Circuit

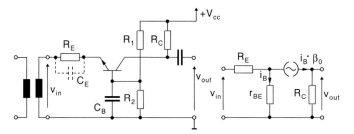

Fig. 7.51. Common-base circuit and its AC equivalent circuit

7.3.7.2 Common-Base Circuit Input Impedance

The input impedance is (Fig. 7.51):

$$r_{in} = \frac{v_{in}}{i_{in}}, \quad v_{in} = i_{in} \cdot R_E + i_B \cdot r_{BE} = i_{in} \left(R_E + \frac{r_{BE}}{1 + \beta_0} \right)$$

$$\implies \boxed{r_{in} = R_E + \frac{r_{BE}}{1 + \beta_0} \approx R_E + \frac{r_{BE}}{\beta_0}} \quad (7.53)$$

If R_E is bypassed by a capacitor (Fig. 7.51), then the input impedance reduces to

$$r_{in} \approx \frac{r_{BE}}{\beta_0} \quad (7.54)$$

7.3.7.3 Common-Base Circuit Output Impedance

The output impedance is given by:

$$r_{out} = \frac{\text{open-circuit AC voltage}}{\text{short-circuit AC current}} = \frac{v_{o/c}}{i_{s/c}} \quad (7.55)$$

$$v_{o/c} \approx i_B \cdot \beta_0 \cdot R_C = v_{in} \frac{\beta_0 R_C}{r_{BE}}, \quad i_{s/c} \approx i_B \cdot \beta_0 = \frac{v_{in}}{r_{BE}} \beta_0 \quad (7.56)$$

$$\implies \boxed{r_{out} = \frac{v_{o/c}}{i_{s/c}} \approx R_C} \quad (7.57)$$

- The common-base output impedance is the same as for the common emitter.

7.3.7.4 Common-Base Circuit AC Voltage Gain

The AC voltage gain of the common-base circuit is:

$$A_v = \frac{v_{out}}{v_{in}}, \qquad v_{out} = i_B \beta_0 R_C, \qquad v_{in} = i_B(1+\beta_0)R_E + i_B \cdot r_{BE}$$

$$\Longrightarrow \boxed{V_v = \frac{v_{out}}{v_{in}} \approx \frac{R_C}{\dfrac{r_{BE}}{\beta_0} + R_E}} \qquad (7.58)$$

If R_E is bypassed by a capacitor, the gain increases to:

$$A_v = \frac{v_{out}}{v_{in}} = \frac{\beta_0 \cdot R_C}{r_{BE}} \qquad (7.59)$$

- The AC voltage gain of the common-base is as for the common-emitter circuit.

7.3.7.5 Common-Base Circuit at High Frequencies

The common-base current gain is 1. Thus the current gain does not create any unwanted negative feedback. The output voltage is in phase with the input voltage, so that in this case feedback over parasitic capacitances is also not a problem. For these reasons the common-base can be operated up to approximately the unity gain frequency f_T.

7.3.8 Overview: Basic Bipolar Transistor Circuits

Fig. 7.52 gives an overview of basic bipolar transistor circuits.

7.3.9 Bipolar Transistor Current Sources

Real current sources can be represented by a circuit diagram consisting of an ideal current source I_s and a source resistor R_{int}, (Fig. 7.53).

Current sources in circuit theory should deliver a defined current:

- independent of the terminal voltage V_{out}, and
- independent of the supply voltage V_{CC} (in particular repressing any mains hum present)

Bipolar transistor current source:

The mesh equation $-V_z + V_{BE} + I_s \cdot R_E = 0$ yields an expression for the current I_s (Fig. 7.54):

$$\boxed{I_s \approx \frac{V_z - 0.7\,\text{V}}{R_E} \qquad \text{for} \qquad 0 < V_{out} < (V_{CC} - V_z)} \qquad (7.60)$$

The current I_s is independent of the output voltage V_{out} with the choice of the Zener voltage V_z and the emitter resistor R_E. The Zener diode works as a constant voltage source. Other

	Common-emitter circuit	Common-collector circuit (Emitter follower)	Common-base circuit
	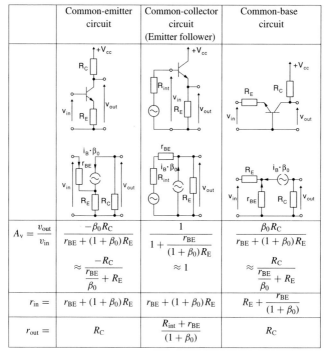		
$A_v = \dfrac{v_{out}}{v_{in}}$	$\dfrac{-\beta_0 R_C}{r_{BE} + (1+\beta_0)R_E}$ $\approx \dfrac{-R_C}{\dfrac{r_{BE}}{\beta_0} + R_E}$	$\dfrac{1}{1 + \dfrac{r_{BE}}{(1+\beta_0)R_E}}$ ≈ 1	$\dfrac{\beta_0 R_C}{r_{BE} + (1+\beta_0)R_E}$ $\approx \dfrac{R_C}{\dfrac{r_{BE}}{\beta_0} + R_E}$
$r_{in} =$	$r_{BE} + (1+\beta_0)R_E$	$r_{BE} + (1+\beta_0)R_E$	$R_E + \dfrac{r_{BE}}{(1+\beta_0)}$
$r_{out} =$	R_C	$\dfrac{R_{int} + r_{BE}}{(1+\beta_0)}$	R_C

Fig. 7.52. Comparison of basic bipolar transistor circuits

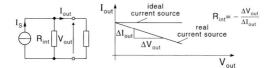

Fig. 7.53. Current source representation

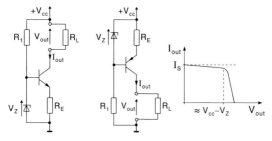

Fig. 7.54. Bipolar transistor current source

voltage sources could be used in place of the diode, such as reference elements, LEDs or series connected silicon diodes.

The **source resistance of the current source** can be determined from the AC equivalent circuit (Fig. 7.55).

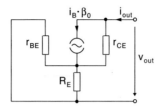

Fig. 7.55. AC equivalent circuit for bipolar transistor current sources

$$r_i = -\frac{v_{out}}{i_{out}} = \frac{R_E(r_{BE} + \beta_0 r_{CE}) + r_{CE}(R_E + r_{BE})}{R_E + r_{BE}} \tag{7.61}$$

- The source resistance lies between r_{CE} and $\beta_0 \cdot r_{CE}$, depending on the circuit layout.

$$\boxed{r_{CE} < r_i < \beta_0 \cdot r_{CE}} \tag{7.62}$$

For a normal choice of voltage V_z (V_z = a few volts) the source resistance is approximately 10–20 times r_{CE}.

- In general, r_i increases with increasing V_z and R_E.

Current source stabilisation against voltage supply variations only partially depends on the source resistance. The source resistance of the voltage source V_z has a similar influence. It causes V_z to change with the supply voltage V_{CC} and thus also the current I_s.

To reduce 100 Hz **mains ripple**, V_z can be stabilised by using a low-pass filter (Fig. 7.56).

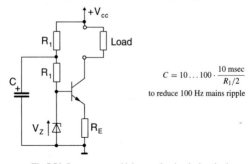

$$C = 10 \ldots 100 \cdot \frac{10 \text{ msec}}{R_1/2}$$

to reduce 100 Hz mains ripple

Fig. 7.56. Current source with improved mains ripple reduction

7.3.10 Bipolar Transistor Differential Amplifier

The **differential amplifier** amplifies the difference of the input voltages (Fig. 7.57):

$$-v_{out1} = v_{out2} = (v_{in1} - v_{in2}) \cdot A_d = v_d \cdot A_d \tag{7.63}$$

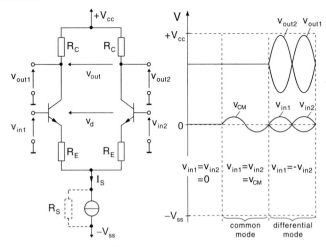

Fig. 7.57. Differential amplifier with bipolar transistors

Differential amplifiers are used mostly as addition points in a feedback loop.

Differential amplifiers are usually used with a symmetric plus/minus-supply voltage. The quiescent input voltage is at ground. The quiescent collector voltage (operating point) is chosen to be at $V_{CC}/2$ for n–p–n transistors or at $-V_{SS}/2$ for p–n–p transistors. The quiescent collector current is equal to *half* the source current I_s ($I_C = I_s/2$). The emitter resistor R_E (**current feedback**) can be chosen to be very small, as the thermal drift effects both transistors. For selected transistors, which differ only slightly in their parameters (matched transistors), R_E can be discarded. If the output voltage is taken between the collector terminals, then $V_{out} \propto V_d$ also in DC.

A distinction is made between **common mode** and **differential mode**. If the input voltages have the *same amplitude and phase*, then they are in *common-mode*. If the input voltages have the *same amplitude and are in antiphase*, then they are in *differential* mode. If v_{in1} and v_{in2} are not *equal*, then they can be broken into common-mode and differential mode constituent parts.

In theory, a common-mode signal $v_{in1} = v_{in2} = v_{CM}$ does not cause an output signal, as the current I_s is defined and should divide equally between the two transistor arms because of the input voltage symmetry. A common-mode signal produces an output signal only as a result of the finite source resistance of the current source I_s.

The ratio

$$\frac{\text{common-mode output voltage}}{\text{common-mode input voltage}} = \frac{v_{out1}}{v_{CM}} = \frac{d V_{out1}}{d A_{CM}} = V_{CM} \tag{7.64}$$

is known as **common-mode gain**. Ideally it is zero.

A differential signal $v_{in1} = -v_{in2} = v_d/2$ produces an output signal $v_{out1} = -v_{out2}$, as it causes the current I_s to divide unequally between the two transistor arms.

The ratio

$$\frac{\text{differential output voltage}}{\text{differential input voltage}} = \frac{v_{out1}}{v_{in1} - v_{in2}} = \frac{v_{out1}}{v_d} = \frac{d\,V_{out1}}{d\,(V_{in1} - V_{in2})} = A_d \quad (7.65)$$

is known as **differential-mode gain**.

7.3.10.1 Differential Mode Gain

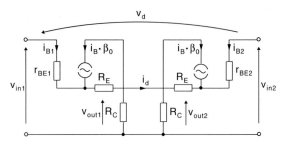

Fig. 7.58. Calculation of the differential mode gain using an AC equivalent circuit of the differential amplifier

The AC equivalent circuit in Fig. 7.58 yields:

$$v_d = i_{B1} \cdot r_{BE1} + i_d \cdot 2R_E - i_{B2} \cdot r_{BE2} \quad (7.66)$$

$$v_{out1} = -i_{B1} \cdot \beta_0 \cdot R_C \quad (7.67)$$

$$i_d = i_{B1} \cdot (1 + \beta_0) = -i_{B2} \cdot (1 + \beta_0) \quad (7.68)$$

It follows that:

$$A_d = \frac{v_{out1}}{v_d} = -\frac{v_{out2}}{v_d} = -\frac{1}{2} \frac{\beta_0 \cdot R_C}{r_{BE} + (1 + \beta_0) R_E} \quad (7.69)$$

or, alternatively:

$$\boxed{A_d = \frac{v_{out1}}{v_d} = -\frac{v_{out2}}{v_d} \approx -\frac{1}{2} \frac{R_C}{\frac{r_{BE}}{\beta_0} + R_E}} \quad (7.70)$$

- The lower the resistance of R_E, the greater the differential gain A_d.

NOTE: In order to be able to choose a small feedback resistor R_E, the transistors must be as similar as possible and be exposed to the same temperature. For this reason monolithic transistors (dual transistors in the same housing) are produced. These are manufactured in the same process (on the same chip) and are thus very similar, and they are at the same temperature because of the common casing. In this case R_E can be discarded.

7.3.10.2 Common-Mode Gain

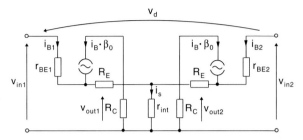

Fig. 7.59. Differential amplifier AC equivalent circuit for the common-mode gain calculation

Calculation of the common-mode gain, taking into account the source resistance r_{is} of the current source I_s the AC equivalent circuit yields (Fig. 7.59):

$$v_{Gl} = i_{B1} \cdot r_{BE1} + i_{B1} \cdot (1 + \beta_0) R_E + i_s \cdot r_{is} \tag{7.71}$$

$$v_{out1} = -i_{B1} \cdot \beta_0 \cdot R_C \tag{7.72}$$

$$i_s = (i_{B1} + i_{B2}) \cdot (1 + \beta_0) = 2 \cdot (1 + \beta_0) \cdot i_{B1} \tag{7.73}$$

It follows that:

$$A_{CM} = \frac{v_{out1}}{v_{CM}} = -\frac{\beta_0 R_C}{r_{BE} + (1 + \beta_0) R_E + (1 + \beta_0) \cdot 2 r_{is}} \tag{7.74}$$

With $2 r_{is} \gg R_E$ it follows that:

$$\boxed{A_{CM} = \frac{v_{out1}}{v_{CM}} = -\frac{v_{out2}}{v_{CM}} \approx -\frac{R_C}{2 r_{is}}} \tag{7.75}$$

- The higher the source impedance of the current source, the smaller the common mode gain is.

7.3.10.3 Common-Mode Rejection Ratio

The common-mode rejection ratio CMRR is the quotient of differential-mode gain and common-mode gain.

$$\boxed{CMRR = \frac{A_d}{A_{CM}}} \tag{7.76}$$

Usually it is expressed in dB. The common-mode rejection ratio is:

$$\boxed{CMRR = \frac{A_d}{A_{CM}} = \frac{r_{is}}{\frac{r_{BE}}{\beta_0} + R_E}} \tag{7.77}$$

7.3.10.4 Differential Amplifier Input Impedance

Differential-mode input resistance r_d (Fig. 7.58):

$$r_\text{d} = \frac{v_\text{d}}{i_{\text{B}1}} = 2r_\text{BE} + (1+\beta_0)2R_\text{E} \tag{7.78}$$

$$\boxed{r_\text{d} \approx 2(r_\text{BE} + \beta_0 R_\text{E})} \tag{7.79}$$

Common-mode input resistance r_CM (Fig. 7.59):

$$r_\text{CM} = \frac{v_\text{CM}}{i_{\text{B}1}} = r_\text{BE} + (1+\beta_0)\cdot R_\text{E} + 2\cdot(1+\beta_0)\cdot r_\text{is} \tag{7.80}$$

$$\boxed{r_\text{CM} \approx 2\beta_0 \cdot r_\text{is}} \tag{7.81}$$

7.3.10.5 Differential Amplifier Output Impedance

The output impedance r_out is (as for the common-emitter circuit):

$$\boxed{r_\text{out} = R_\text{C}} \tag{7.82}$$

7.3.10.6 Offset Voltage of the Differential Amplifier

The offset voltage V_0 (**input offset voltage**) is the differential input voltage that must be applied so that the output voltages V_out1 and V_out2 are equal.

$$\boxed{V_0 = (V_\text{in1} - V_\text{in2})|_{V_\text{out1}=V_\text{out2}}} \tag{7.83}$$

The offset voltage is a tolerance value. The value given in the data sheet is the worst case.

7.3.10.7 Differential Amplifier Offset Current

The offset current I_0 (**input offset current**) is the differential input current that must be supplied so that V_out1 and V_out2 are equal.

$$\boxed{I_0 = (I_\text{in1} - I_\text{in2})|_{V_\text{out1}=V_\text{out2}}} \tag{7.84}$$

7.3.10.8 Input Offset Voltage Drift

The **thermal voltage drift** of both differential amplifier transistors effectively cancels out because of their matched fabrication. Only the tolerance-defined differences in the thermal drift have an effect. The **offset voltage drift** (also described as the **temperature coefficient of the input offset voltage**) is the change in the offset voltage caused by the different temperature responses of the transistors. It lies several decades below the thermal voltage drift ΔV_BE. The input offset voltage drift is given in units of $\frac{\mu\text{V}}{\text{K}}$.

7.3.10.9 Differential Amplifier Examples

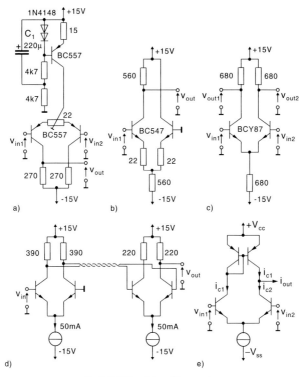

Fig. 7.60. Differential amplifier examples

A number of examples of differential amplifiers are given in Fig. 7.60:

a) Differential amplifier with current source, good mains-ripple repression and current feedback with a potentiometer for symmetry;
b) Differential amplifier with a single output voltage. A collector resistor can therefore be discarded. Disadvantage: The power dissipated in the transistors is different, causing the transistors' thermal symmetry to be lost.
c) Differential amplifier without current feedback. The transistor BCY87 is a dual transistor in a single casing especially suitable for differential amplifiers.
d) Symmetrical analogue signal transmission. Electromagnetically coupled interferences in the transmission channel cancel each other out in the receiver circuit.
e) Differential circuit with current mirror to couple current out, $i_{out} = (v_{in1} - v_{in2})\dfrac{\beta_0}{r_{BE}}$. This circuit is particularly important in IC design because of the required thermal coupling and the required small deviation in the transistor parameters.

7.3.11 Overview: Bipolar Transistor Differential Amplifiers

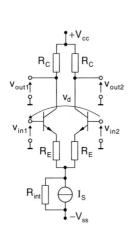

Differential-mode gain:
$$A_d = \frac{v_{out1}}{v_d} = -\frac{v_{out2}}{v_d}$$
$$\approx -\frac{1}{2}\frac{R_C}{\frac{r_{BE}}{\beta_0} + R_E}$$
with $v_{in1} - v_{in2} = v_d$

common-mode gain:
$$A_{CM} = \frac{v_{out1}}{v_{CM}} = \frac{v_{out2}}{v_{CM}} \approx -\frac{R_C}{2r_{is}}$$
with $v_{in1} = v_{in2} = v_{CM}$

common-mode rejection ratio:
$$CMRR = \frac{A_d}{A_{CM}} = \frac{r_{is}}{\frac{r_{BE}}{\beta_0} + R_E}$$

Differential-mode input impedance:
$$r_d \approx 2(r_{BE} + \beta_0 R_E)$$

Output impedance:
$$r_{out} = R_C$$

7.3.12 Current Mirror

The current mirror produces an output current I_{out}, which is equal to the input current I_1. The circuit output has the qualities of a current source, i.e. it has a very high impedance source resistance.

In Fig. 7.61 current I_1 is the input quantity. Transistors T_1 and T_2 are equal and are at the same temperature. It follows that:

$$I_1 = I_{C1} + I_B, \qquad I_{B1} = I_{B2} = \frac{I_B}{2}, \qquad I_{C1} = \beta_{DC} \cdot I_{B1} \tag{7.85}$$

$$\left.\begin{array}{l} I_1 = \beta_{DC} \cdot I_{B1} + 2I_{B1} = (2 + \beta_{DC}) \cdot I_{B1} \\ I_{out} = \beta_{DC} \cdot I_{B2} = \beta_{DC} \cdot I_{B1} \end{array}\right\} \quad \boxed{I_{out} \approx I_1} \tag{7.86}$$

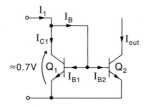

Fig. 7.61. Current mirror circuit

7.3.12.1 Current Mirror Variations

Fig. 7.62 shows how to multiply or to divide an input current.

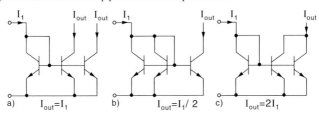

Fig. 7.62. Current mirror variation

7.4 Field-Effect Transistor Small-Signal Amplifiers

Small-signal amplifiers are circuits that amplify small AC signals, where the signal amplitude is much smaller than the operating point values (i.e. the DC values applied to the components). The operating frequencies should be low, so that propagation delays and phase changes caused by parasitic elements can be neglected (otherwise it is pointed out in the operating conditions).

7.4.1 Transistor Characteristics and Ratings

7.4.1.1 Symbols, Voltages and Currents for Field-Effect Transistors

The transistor terminals are called the **drain**, the **source** and the **gate**. Field-effect transistors are voltage-controlled components. The drain–source current is controlled by the gate–source voltage. At low frequencies the control requires no power. This means that the gate current is insignificantly small.

A distinction is made between **junction field-effect transistors** (**JFET** or junction-FET) and **insulated gate field-effect transistors** (**IGFET** or insulated-gate FET while MOSFET). JFETs are always depletion types, while IGFETs can be depletion type or enhancement type. **Depletion** means that the drain–source path conducts for $V_{GS} = 0$. **Enhancement** means that the drain–source path does not conduct for $V_{GS} = 0$.

A further distinction is made between **n-channel** and **p-channel** types. In n-channel types the drain current flows into the drain. The drain current increases if the gate–source voltage is changed in a positive sense. In p-channel types the drain current flows from the drain. It increases if the gate–source voltage is changed in a negative sense.

NOTE: In terms of the current and voltage directions, an n-channel FET corresponds to an n–p–n transistor and a p-channel FET to a p–n–p transistor.

Figure 7.63 summerizes informtion for FETs

For **JFETs** the gate–source path is a silicon diode which is reverse-biased in normal operation. Forward biasing can easily lead to the destruction of the FET, as the current follows the forward-bias diode characteristic.

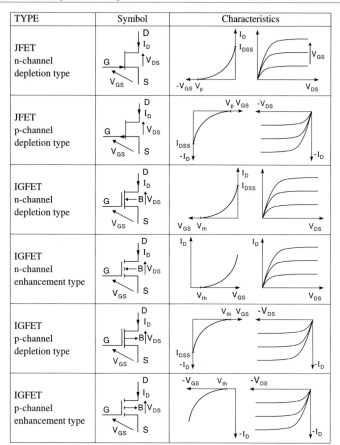

Fig. 7.63. Classification, voltages, currents and characteristics of FETs

For **IGFETs** or MOSFETs the gate is isolated with respect to the drain and source. The maximum rating for the gate–source voltage is in the region of ±20 V.

MOSFETs are often used as 'electronic switches'. The smallest **drain–source resistance** $R_{DS(ON)}$ (ON resistance) is given in the switched on state ($V_{GS} > 10$ V).

MOSFETs often have a silicon diode in parallel (reverse-current diode). In the non-conducting state the source–drain path behaves like a forward-biased silicon diode. This diode must be a fast rectifier for applications in frequency inverters and in push–pull amplifiers.

In **MOSFETs** the termination is occasionally accessible. It is described by the term **BULK** (B). It has a similar controlling influence as the gate.

NOTE: To test whether an IGFET (MOSFET) is defective or not, a continuity tester can be used on the drain–source path in conjunction with a voltage source (about 10 V), which controls the gate–source voltage. The state of the drain–source path must maintain its state (conducting or not), even if the control voltage V_{GS} is removed.

7.4.1.2 JFET Characteristic Curves

The **transfer characteristic** $I_D = f(V_{GS})$ and the **output characteristic** $I_D = f(V_{DS})$, where V_{GS} is a parameter, represent the relationship between all voltages and currents of the field-effect transistor (Fig. 7.64).

V_P is the **pinch-off voltage**. At $V_{GS} = V_P$ the drain current I_D becomes practically zero. The value of V_P is governed by sample variations and temperature dependence.

The **input voltage** of the gate–source voltage lies between $V_P < V_{GS} < 0$ V for the JFET. For $V_{GS} > 0$ V the high impedance of the gate is lost.

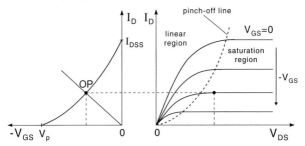

Fig. 7.64. Transfer and output characteristics of a JFET (here: n-channel JFET)

The analytical form of the transfer characteristic is

$$I_D = I_{DSS}\left(1 - \frac{V_{GS}}{V_P}\right)^2 \tag{7.87}$$

The output characteristic curve is divided into two regions, the **saturation region** and the **linear region** (or triode region). In the saturation region the characteristic curve is almost horizontal, and the drain current only depends on the gate–source voltage and is almost independent of the applied drain–source voltage. In the linear region the drain current increases approximately proportionally to the drain–source voltage. The increase depends on V_{GS}. Both regions are separated by a pinch-off curve given by

$$V_k = (V_{GS} - V_P) \tag{7.88}$$

7.4.1.3 IGFET Characteristic Curves

The **threshold voltage** V_{th} of the gate–source voltage for enhancement-type FETs lies in a positive voltage range and in a negative range for depletion types. The threshold voltage, like V_P for the JFET, varies greatly due to manufacturing tolerances. The gate isolation

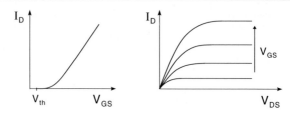

Fig. 7.65. IGFET (MOSFET) transfer and output characteristics; here: enhancement-type n-channel IGFET

from the conducting channel means that relatively high gate–source voltages may be used. Usual values are ± 20 V.

The analytical form of the transfer characteristic of IGBTs is the same as for JFETs:

$$I_D = I_{DSS}\left(1 - \frac{V_{GS}}{V_P}\right)^2 \tag{7.89}$$

For enhancement-type FETs the current $I_D = I_D(V_{GS} = 2V_{th})$ is substituted for I_{DSS} (Fig. 7.65).

7.4.1.4 Transconductance

Transconductance g_m is given by the slope of the transfer characteristic curve $I_D = f(V_{GS})$, see Fig. 7.66.

$$g_m = \left.\frac{dI_D}{dV_{GS}}\right|_{V_{DS}=\text{const}} \approx \left.\frac{\Delta I_D}{\Delta V_{GS}}\right|_{V_{DS}=\text{const}} \tag{7.90}$$

The drain–source voltage feedback to the gate is small at low frequencies, so that the measurement condition $V_{DS} = \text{const}$ is practically meaningless.

Transconductance g_m is given in siemens or millisiemens.

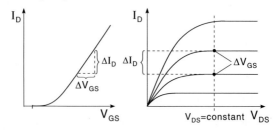

Fig. 7.66. Definition of the forward transconductance in the transfer characteristic and in the output characteristic

7.4.1.5 Dynamic Output Resistance

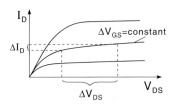

Fig. 7.67. Definition of the differential output resistance

The **dynamic output resistance** defines the change in the drain current as a function of the drain–source voltage for a constant gate–source voltage (Fig. 7.67).

$$r_{DS} = \left. \frac{dV_{DS}}{dI_D} \right|_{V_{GS}=\text{const}} \approx \left. \frac{\Delta V_{DS}}{\Delta I_D} \right|_{V_{GS}=\text{const}} \tag{7.91}$$

- r_{DS} is extremely high, especially for MOSFETs (the output characteristic curve is almost horizontal).

7.4.1.6 Input Impedance

The **input impedance** of a field-effect transistor is the impedance of the gate–source junction, which is capacitive. It is given in data sheets either by C_{iss} or by their two-port parameters C_{11S}. The value is in the range of a few picofarads to several nanofarads.

7.4.2 Equivalent Circuit

7.4.2.1 Equivalent Circuit for Low Frequencies

The gate–source voltage controls the drain current. The value of r_{DS} is usually so high that it can be neglected.

It then holds:

$$i_D \approx g_m \cdot v_{GS}, \quad \text{and} \quad \Delta I_D \approx \Delta g_m \cdot V_{GS}, \quad \text{respectively} \tag{7.92}$$

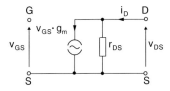

Fig. 7.68. FET AC equivalent circuit for low frequencies

7.4.2.2 Equivalent Circuit for High Frequencies

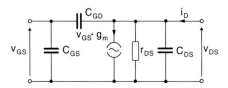

Fig. 7.69. FET AC equivalent circuit for high frequencies

At higher frequencies the parasitic capacitances between each of the terminals begin to take effect (Fig. 7.69). The gate–source capacitance loads the input voltage source. The gate–drain capacitance causes feedback in the common-source circuit, the amount of which depends on the source resistance of the input voltage source. The frequency-dependent feedback decreases with decreasing source resistance.

The following connections exist between the data sheet parameters C_{iss}, C_{rss} and C_{oss} (also denoted by C_{11S}, C_{12S} and C_{22S}) and the equivalent circuit values:

Input capacitance: $\quad C_{iss} = C_{11S} \approx C_{GS} + C_{GD}$
Reverse transfer capacitance: $\quad C_{rss} = C_{12S} \approx C_{GD}$
Output capacitance: $\quad C_{oss} = C_{22S} \approx C_{DS} + C_{GD}$

7.4.2.3 Critical Frequency of Transconductance

The **critical frequency of transconductance** is very high for field-effect transistors (for the BF245, a particularly popular JFET, it is at 700 MHz). The FET is thus particularly suitable as a high-frequency amplifier.

The critical frequency of transconductance is only given for those FETs that are intended for analogue usage. It is not given, therefore, for most MOSFETs, which are intended for use in fast switching.

7.4.3 Basic Circuits using Field-Effect Transistors

Similar to bipolar transistors, there are three different small signal modes of operation. These are the common-source, the common-gate and the common-drain (Fig. 7.70, see also Sect. 7.3.4).

7.4.4 Common-Source Circuit

The common-source circuit is an amplifier circuit for voltage and current amplification (Fig. 7.71).

Common-Source Circuit with JFET

Figure 7.71a shows the common-source circuit with a JFET. The transistor operating point is selected so that it lies in the saturation region of the output characteristic curve. V_{GS}

7.4 Field-Effect Transistor Small-Signal Amplifiers

	Common-source circuit	Common-drain circuit	Common-gate circuit
Circuit			
Voltage gain A_v	> 1	< 1	> 1
Current gain A_i	$\to \infty$	$\to \infty$	1
Input impedance r_{in}	Very high	Very small	Small
Output impedance r_{out}	Intermediate	Small	Intermediate

Fig. 7.70. Basic FET circuits

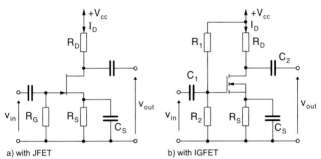

Fig. 7.71. Common-source circuit

must be therefore negative for depletion-type n-channel FETs. R_G connects the gate via a high impedance to Ground, while the drain current passing through R_S causes the source voltage to have a positive value. The source is connected to AC ground via the capacitor C_S. R_D defines the DC drain–source voltage. The output voltage is taken from this point. The output voltage v_{out} has the opposite phase to the input voltage v_{in} (see Sect. 7.4.4.8).

Common-Source Circuit with IGFET

Figure 7.71b shows the common-source circuit with an enhancement-type IGFET. The configuration is similar to the common-emitter configuration. The gate voltage must be positive with respect to the source voltage. The stabilisation of the operating point is achieved using R_S (see also operating-point stabilisation for the IGFET).

7.4.4.1 Common-Source Two-Port Parameters

The common-source two-port parameters are usually given as y-parameters (Fig. 7.72 and Tab. 7.1).

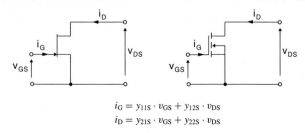

$$i_G = y_{11S} \cdot v_{GS} + y_{12S} \cdot v_{DS}$$
$$i_D = y_{21S} \cdot v_{GS} + y_{22S} \cdot v_{DS}$$

Fig. 7.72. Definition of the two-port parameters for the common-source circuit

Table 7.1. FET y-parameters of the common-source cicuit

	low frequencies	high frequencies
Input admittance with shorted output: $Y_{11S} = \left.\dfrac{dI_G}{dV_{GS}}\right|_{V_{DS}=\text{const}} = \left.\dfrac{i_G}{v_{GS}}\right|_{v_{DS}=0} \approx$	0	ωC_{11S}
Reverse transconductance with shorted input: $Y_{12S} = \left.\dfrac{dI_G}{dV_{DS}}\right|_{V_{GS}=\text{const}} = \left.\dfrac{i_G}{v_{DS}}\right|_{v_{GS}=0} \approx$	0	ωC_{12S}
Forward transconductance with shorted output: $Y_{21S} = \left.\dfrac{dI_D}{dV_{GS}}\right|_{V_{DS}=\text{const}} = \left.\dfrac{i_D}{v_{GS}}\right|_{v_{DS}=0} \approx$	g_m	g_m
Output admittance with shorted input: $Y_{22S} = \left.\dfrac{dI_D}{dV_{DS}}\right|_{V_{GS}=\text{const}} = \left.\dfrac{i_D}{v_{DS}}\right|_{v_{GS}=0} \approx$	0	ωC_{22S}

7.4.4.2 AC Equivalent Circuit of the Common-Source Circuit

The resistance of R_G is usually selected to be very large, so it is not considered in the equivalent circuit.

The equivalent circuit for low frequencies can, in most cases, be the one used with field-effect transistors (Fig. 7.73a).

The equivalent circuit for high frequencies is valid when the parasitic reactances $1/\omega C_{GD}$, $1/\omega C_{GS}$ and $1/\omega C_{DS}$ are not negligible compared with R_D, the source resistance of the voltage source, R_{int} or the load (Fig. 7.73b). This could be the case even at lower frequencies, especially if the source resistance R_{int} is large enough to effect feedback over the feedback capacitor C_{GD} (Miller effect).

The equivalent circuit in Fig. 7.73c is valid if R_S is *not* AC-bypassed by the capacitor C_S.

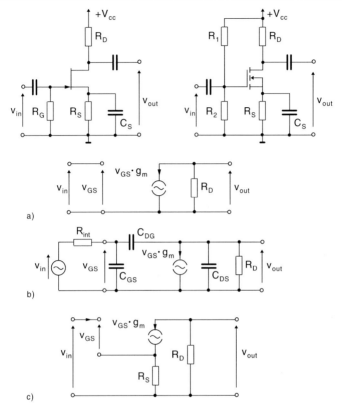

Fig. 7.73. Equivalent circuits of the common-source: **a** for low frequencies; **b** for high frequencies; **c** for low frequencies and without C_S

7.4.4.3 Input Impedance of the Common-Source Circuit

The input impedance is

$$z_{in} = \frac{1}{Y_{11S}} \approx \begin{cases} \infty, & \text{for low frequencies} \\ \omega C_{11S} = \omega C_{iSS}, & \text{for high frequencies} \end{cases} \tag{7.93}$$

7.4.4.4 Output Impedance of the Common-Source Circuit

The output impedance is

$$z_{out} = \frac{\text{open-circuit AC voltage}}{\text{AC short-circuit current}} = \frac{v_{o/c}}{i_{s/c}} = \frac{v_{o/c} S R_D}{v_{o/c} S}$$

this yields

$$z_{\text{out}} = \begin{cases} \dfrac{v_{\text{o/c}} g_m R_D}{v_{\text{o/c}} g_m} = R_D, & \text{for low frequencies} \\[2ex] \dfrac{v_{\text{o/c}} g_m (R_D || \dfrac{1}{j\omega C_{22S}})}{v_{\text{o/c}} g_m} = R_D || \dfrac{1}{j\omega C_{22S}}, & \text{for high frequencies} \end{cases} \quad (7.94)$$

7.4.4.5 AC Voltage Gain

The small-signal AC gain A_v according to Fig. 7.73a is:

$$A_v = \frac{v_{\text{out}}}{v_{\text{in}}} = -\frac{v_{GS} \cdot g_m \cdot R_D}{v_{GS}} = -g_m \cdot R_D \quad (7.95)$$

If the load resistance R_L at the circuit output is considered, then this yields:

$$\boxed{A_v = -g_m \cdot (R_D || R_L)} \quad (7.96)$$

- The output voltage has the opposite phase from the input voltage.

- The voltage gain of the common-source circuit is significantly smaller than of the common-emitter circuit. This can be seen in the significantly smaller transconductance in the FET compared to the bipolar transistor.

The small-signal AC gain A_v according to Fig. 7.73c is:

$$\boxed{A_v = -\frac{g_m \cdot R_D}{1 + g_m \cdot R_S}} \quad (7.97)$$

At high frequencies the antiphase output voltage is returned to the gate–source voltage via the **feedback capacitance** C_{GD}. The higher the value of R_{int}, the more this reduces the amplification. Furthermore, at high frequencies the impedance of the output capacitance reaches the region of the drain impedance, which causes a further decrease in the gain. An exact analysis of the gain at high frequencies in practice should be carried out by measurements or with a suitable simulation system.

7.4.4.6 Operating-Point Biasing

Operating-point biasing for depletion-type FETs:

The transconductance curve and the output characteristic curve are used in the selection of the resistances R_D, R_S and R_G (Fig. 7.74). The operating point V_{DS0}, I_{D0} is chosen as follows

- $I_{D0} = 0.3 \ldots 0.5 \cdot I_{DSS}$, and $V_{DS0} \approx 0.3 \ldots 0.5 V_{CC}$
- in the saturation region taking into account the gate voltage
- under the curve of the power loss.

R_S and R_D are then:

$$\boxed{R_S = \frac{-V_{GS0}}{I_{D0}}, \quad \text{and} \quad R_D = \frac{V_{CC} - V_{DS0}}{I_{D0}}} \quad (7.98)$$

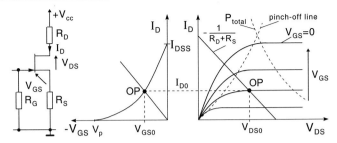

Fig. 7.74. Operating-point biasing for depletion-type FETs

R_D determines the voltage gain. To get a large voltage gain, small values of I_D0 should be chosen and large values of V_CC should be used.

R_G is present to connect the gate to ground. R_G can be chosen in the megohm range because of the large impedance of the gate.

The temperature dependence of V_P and its sample variations causes the operating point to shift on the bias line, whose slope is $-1/R_\mathrm{S}$. This means that there is an acceptable operating point over a large range of tolerances.

Operating-point definition for enhancement-type FETs:

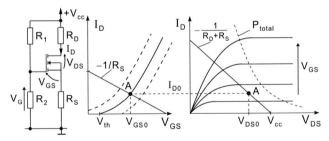

Fig. 7.75. Operating-point definition for enhancement-type FETs

The operating point V_DS0 and I_D0 is chosen from the output characteristic curve:

- $V_\mathrm{DS0} \approx 0.3 \ldots 0.5 V_\mathrm{CC}$;
- in the saturation region;
- underneath the power loss curve (Fig. 7.75).

The drain current is defined by the gate voltage divider R_1, R_2 and the source resistor R_S. To determine the three resistors, the possible variations in the transconductance curve $I_\mathrm{D} = f(V_\mathrm{GS})$ are considered. The bias line $-1/R_\mathrm{S}$ is chosen so that, despite the variations in V_th, the operating point remains in a valid position in the output characteristic curve. The choice of R_S and V_G may be carried out graphically from the transconductance curve.

The resistor values can be calculated thus:

$$R_D = \frac{V_{CC} - V_{DS0}}{I_{D0}} - R_S, \qquad R_S = \frac{V_G - V_{GS0}}{I_{D0}}, \qquad \frac{R_1}{R_2} = \frac{V_{CC} - V_G}{V_G} \qquad (7.99)$$

The voltage divider R_1/R_2 can be chosen in the megohm range.

7.4.4.7 Common-Drain Circuit, Source Follower

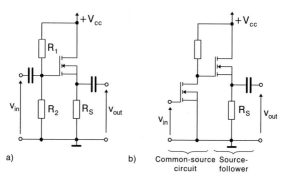

Fig. 7.76. Common-drain circuit (source follower)

The common-drain is similar to the emitter follower, but has a smaller gain $A_v < 1$ because of the smaller transconductance of the FET compared to the bipolar transistor. The gate voltage divider R_1, R_2 can be discarded if the common-drain is used as a stage after a common-source stage (Fig. 7.76).

The common-drain input impedance is extremely high. The output impedance is small, so the common-drain circuit is particularly useful as an impedance converter.

7.4.4.8 AC Equivalent Circuit of the common-drain Circuit

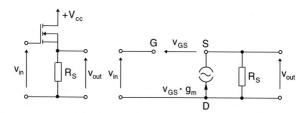

Fig. 7.77. Common-drain circuit (source follower) and its AC equivalent circuit

7.4.4.9 Input Impedance of the Common-Drain Circuit

The input impedance of the common-drain circuit is extremely high:

$$r_{\text{in}} \to \infty \tag{7.100}$$

7.4.4.10 Output Impedance of the Common-Drain Circuit

$$r_{\text{out}} = \frac{\text{open-circuit AC voltage}}{\text{AC short-circuit current}} = \frac{v_{\text{o/c}}}{i_{\text{s/c}}}$$

According to Fig. 7.77 it holds that:

Open circuit: $v_{\text{in}} = v_{\text{GS}} + v_{\text{GS}} \cdot g_{\text{m}} \cdot R_{\text{S}}$, $\quad v_{\text{o/c}} = v_{\text{GS}} \cdot g_{\text{m}} \cdot R_{\text{S}}$, $\quad v_{\text{o/c}} = \dfrac{g_{\text{m}} \cdot R_{\text{S}}}{1 + g_{\text{m}} \cdot R_{\text{S}}} v_{\text{in}}$

Short circuit: $i_{\text{s/c}} = v_{\text{in}} \cdot g_{\text{m}}$

This yields:

$$r_{\text{out}} = \frac{R_{\text{S}}}{1 + g_{\text{m}} \cdot R_{\text{S}}} \tag{7.101}$$

7.4.4.11 Voltage Gain of the Common-Drain Circuit

According to Fig. 7.77 it holds that:

$$v_{\text{in}} = v_{\text{GS}} + v_{\text{GS}} \cdot g_{\text{m}} \cdot R_{\text{S}}, \qquad v_{\text{out}} = v_{\text{GS}} \cdot g_{\text{m}} \cdot R_{\text{S}}$$

This yields:

$$A_{\text{v}} = \frac{v_{\text{out}}}{v_{\text{in}}} = \frac{g_{\text{m}} \cdot R_{\text{S}}}{1 + g_{\text{m}} \cdot R_{\text{S}}} \tag{7.102}$$

7.4.4.12 Common-Drain Circuit at High Frequencies

The common-drain circuit is suitable for operation up to the transconductance critical frequency f_{y21S}.

7.4.5 Common-Gate Circuit

The common-gate is similar to the common-base in bipolar transistors (Fig. 7.78). The current gain is 1, and the voltage gain corresponds to that of the common-source circuit. The input impedance is small, and the output impedance is R_{D}. The circuit is suitable as a voltage amplifier for high frequencies, as the output voltage has the same phase as the input voltage, and thus there can be no undesired frequency-dependent feedback. The operating point is determined and stabilised by R_{S}.

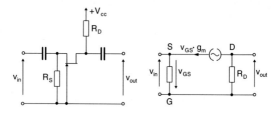

Fig. 7.78. Common-gate and its AC equivalent circuit

7.4.5.1 Input Impedance of the Common-Gate Circuit

The impedance is

$$r_{in} = \frac{1}{g_m} \| R_S \approx \frac{1}{g_m} \qquad (7.103)$$

7.4.5.2 Output Impedance of the Common-Gate Circuit

The output impedance is

$$r_{out} = R_D \qquad (7.104)$$

7.4.5.3 Voltage Gain of the Common-Gate Circuit

The voltage gain is

$$A_v = g_m \cdot R_D \qquad (7.105)$$

7.4.6 Overview: Basic Circuits using Field-Effect Transistors

	Common-source circuit	Common-drain circuit (source follower)	Common-gate circuit
Circuit			
AC equivalent circuit			
A_v	$-g_m R_D$	$\dfrac{g_m R_S}{1 + g_m R_S}$	$g_m R_D$
A_i	$\to \infty$	$\to -\infty$	-1
r_{in}	$\to \infty$	$\to \infty$	$1/g_m$
r_{out}	R_D	$\dfrac{R_S}{1 + g_m R_S}$	R_D

Fig. 7.79. Comparison of basic FET circuits

7.4.7 FET Current Source

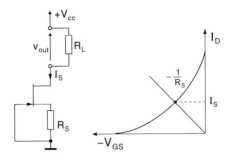

Fig. 7.80. JFET Current source

FET current sources are mainly realised using depletion-type FETs (Fig. 7.81). They have the advantages over bipolar transistors that they

- consist of only two components, and
- have a very high mains-ripple repression, as they do not require a reference voltage, which is supplied by the 'humming' supply voltage

A disadvantage is that the source current I_s can vary considerably because of production tolerances.

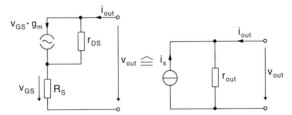

Fig. 7.81. AC equivalent circuit of the current source (equivalent circuit for small/incremental changes i_{out} of the source current I_s)

The differential internal impedance r_{out} is (Fig. 7.81):

$$r_{out} = -\frac{dV_{out}}{dI_{out}} = -\frac{v_{out}}{i_{out}} = r_{DS}(1 + g_m \cdot R_S) + R_S \approx r_{DS}(1 + g_m \cdot R_S) \tag{7.106}$$

- A horizontal progression of the output characteristic curve means that r_{DS} has a very large resistance, and thus that the current source has a very high impedance.

7.4.8 Differential Amplifier with Field-Effect Transistors

The differential amplifier with field-effect transistors operates similarly to the differential amplifier with bipolar transistors, as described in Sect. 7.3.10. The current I_s is divided up evenly between the two transistor arms, because of the symmetry of the input voltages.

7 Analogue Circuit Design

The source voltage and thus the gate–source voltage is defined by the corresponding transconductance curve, while the drain currents are $I_D = I_S/2$ for symmetry reasons (Fig. 7.82). In order to guarantee the required symmetry of the transistor parameters, monolithic dual FETs should be used. Differential amplifiers with field-effect transistors are employed where an extremely high input impedance is required.

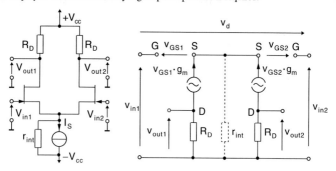

Fig. 7.82. Differential amplifier with FETs and the AC equivalent circuit, where r_{int} is the differential resistance of the current source I_S

7.4.8.1 Differential Mode Gain

Differential amplification is the amplification that exists for antiphase, equal amplitude input voltages.

From the AC equivalent circuit in Fig. 7.82 it follows that:

$$V_d = v_{GS1} - v_{GS2}, \quad v_{out1} = -v_{GS1} \cdot g_m \cdot R_D, \quad v_{GS1} = -v_{GS2}, \quad A_d = \frac{v_{out1}}{v_d} = -\frac{v_{out2}}{v_d}$$

$$\boxed{A_d = \frac{v_{out1}}{v_d} = -\frac{1}{2} g_m \cdot R_D} \tag{7.107}$$

7.4.8.2 Common-Mode Gain

Common-mode gain is the gain that exists for input voltages that are in phase and have equal amplitudes.

The source resistance r_{int} of the current source is now inserted (Fig. 7.82):

$$v_{in1} = v_{in2}, \quad v_{in1} = v_{GS1} + i_s \cdot r_{int}, \quad v_{out1} = -v_{GS1} \cdot g_m \cdot R_D,$$
$$i_s = v_{GS1} \cdot g_m + v_{GS2} \cdot g_m$$

$$\boxed{A_{CM} = \frac{v_{out1}}{v_{in1}} = \frac{v_{out2}}{v_{in1}} = -\frac{R_D}{2r_{int}}} \tag{7.108}$$

7.4.8.3 Common-Mode Rejection Ratio

The common-mode rejection ratio is:

$$CMRR = \frac{A_d}{A_{CM}} \approx g_m \cdot r_{int} \qquad (7.109)$$

7.4.8.4 Input Impedance

Differential mode input impedance:

$$z_d = \begin{cases} \to \infty, & \text{for low frequencies} \\ 2\dfrac{1}{j\omega C_{11S}}, & \text{for high frequencies} \end{cases} \qquad (7.110)$$

Common-mode input impedance:

$$r_{CM} \to \infty \qquad (7.111)$$

7.4.8.5 Output Impedance

$$r_{out} = R_D \qquad (7.112)$$

7.4.9 Overview: Differential Amplifier with Field-Effect Transistors

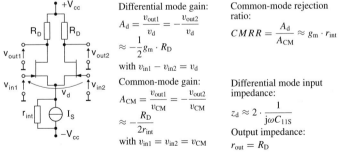

Differential mode gain:
$$A_d = \frac{v_{out1}}{v_d} = -\frac{v_{out2}}{v_d}$$
$$\approx -\frac{1}{2} g_m \cdot R_D$$
with $v_{in1} - v_{in2} = v_d$

Common-mode gain:
$$A_{CM} = \frac{v_{out1}}{v_{CM}} = -\frac{v_{out2}}{v_{CM}}$$
$$\approx -\frac{R_D}{2 r_{int}}$$
with $v_{in1} = v_{in2} = v_{CM}$

Common-mode rejection ratio:
$$CMRR = \frac{A_d}{A_{CM}} \approx g_m \cdot r_{int}$$

Differential mode input impedance:
$$z_d \approx 2 \cdot \frac{1}{j\omega C_{11S}}$$

Output impedance:
$$r_{out} = R_D$$

7.4.10 Controllable Resistor FETs

The FET as a controllable resistor is operated in the linear region of the output characteristic curve. This means that in this case the FET is operated with a very small drain–source voltage ($V_{DS} < V_k$) (Fig. 7.83).

The resistance of small-signal FETs is in the range of tens to several hundred Ohms.

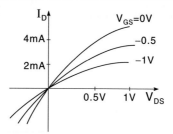

Fig. 7.83. The resistive range of the output characteristics

A linearisation of the curved characteristics is achieved with the circuit for the adjustable voltage divider in Fig. 7.84. The linearisation works as follows: for increasing output voltages the gate–source voltage is increased and thus the nonlinear characteristic curve is compensated. The resistors are chosen so that $R_2 = R_3 \gg R_{DS}$.

It then holds that:

$$\frac{V_{out}}{V_{in}} = \frac{R_{DS}}{R_1 + R_{DS}} \qquad (7.113)$$

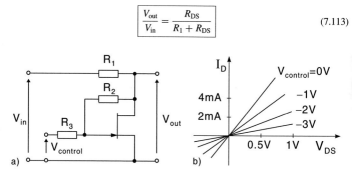

Fig. 7.84. Linearised voltage-controlled voltage divider

The FET as a controllable resistance is used, for example, in

- automatic voltage-level control,
- adjustable voltage dividers,
- amplitude stabilisation of oscillators, and
- circuits with variable gain.

7.5 Negative Feedback

Feedback is the term used when a circuit's output signal is fed back to the input. The term **negative feedback** is used when a part of the output signal is subtracted from the input signal, while for **positive feedback** the portion of the output signal is added to the input signal. For AC voltages negative feedback means that a part of the output signal is added in *antiphase* to the input signal, and positive feedback means that the portion of the output signal is added *in phase* to the input signal.

7.5 Negative Feedback

Positive-feedback systems are usually unstable, i.e. an independent oscillation exists or the output voltage saturates to the positive or negative rail. Positive feedback is important in the area of oscillators.

Systems with negative feedback are stable. Instabilities exist only if unwanted positive feedback occurs as a result of output signals that are phase-shifted at certain frequencies with respect to the input signal.

The purpose of negative feedback is

- to improve the linearity of an amplifier,
- to make the gain independent of the semiconductor parameters,
- to stabilise the output signal against load variations,
- to reduce the load on the source, and
- to improve the frequency response of an amplifier.

In general, negative feedback can be represented in a block diagram (Fig. 7.85). The output signal is multiplied by the **feedback factor** β and then subtracted from the input signal. The difference is amplified by A_{OL}. Such a system with negative feedback is also known as a **closed-loop system**.

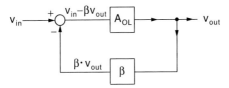

Fig. 7.85. System with negative feedback

The **closed-loop gain** $A_{\text{CL}} = v_{\text{out}}/v_{\text{in}}$ of the negative feedback system with

$$v_{\text{out}} = (v_{\text{in}} - \beta \cdot v_{\text{out}}) \cdot A_{\text{OL}} \tag{7.114}$$

is:

$$\boxed{A_{\text{CL}} = \frac{A_{\text{OL}}}{1 + \beta \cdot A_{\text{OL}}}} \tag{7.115}$$

The expression βA_{OL} is known as the **loop gain**.

The **amount of feedback** is given by $1 + \beta A_{\text{OL}}$. The closed loop gain A_{CL} decreases with increasing amounts of feedback. In this context the gain A_{OL} is called **open-loop gain**. This is the effective gain if the feedback loop is removed.

Transforming Eq. (7.115) yields:

$$A_{\text{CL}} = \frac{1}{\dfrac{1}{A_{\text{OL}}} + \beta} \tag{7.116}$$

It can be seen that the closed-loop gain becomes approximately independent of A_{OL} if A_{OL} is very high.

$$\boxed{\text{For} \quad A_{\text{OL}} \gg \frac{1}{\beta}, \quad follows \quad A_{\text{CL}} \approx \frac{1}{\beta}}$$

- If the open-loop gain is very high, the closed-loop gain becomes approximately $1/\beta$.
- The feedback circuit is usually a linear resistor network. If βA_{OL} is very large, then the amplifier with negative feedback is independent of the nonlinearities and tolerances of the semiconductor parameters of the amplifier A_{OL} and depends only on the feedback circuit.

7.5.1 Feedback Topologies

A distinction is made between four different kinds of negative feedback, depending on whether the input and output quantities are 'current' or 'voltage' (Fig. 7.86).

The description of the different types of feedback depends on the manner in which the output is sampled and fed back to the input. The first term of the description refers to the connection at the input and the second term refers to the connection at the output. So, for example, in series–parallel feedback the input of the corresponding circuit receives feedback in series and the output is sampled in parallel. Then the output appears to be a voltage source, and the input should be fed by a voltage source. The term shunt is often also used instead of parallel. The kinds of negative feedback are summarised as follows (Fig. 7.86):

a) **series–parallel feedback**: the output is sampled in parallel to give a series voltage feedback at the input.
   input: voltage
   stabilised output: voltage
   type of amplifier: voltage amplifier

b) **parallel–parallel feedback**: the output voltage is sampled in parallel to give a parallel current feedback at the input.
   input: current
   stabilised output: voltage
   type of amplifier: transimpedance amplifier, current–voltage converter

c) **series–series feedback**: the output is sampled in series to give a series voltage feedback at the input.
   input: voltage
   stabilised output: current
   type of amplifier: transconductance amplifier, voltage-current converter

d) **parallel–series feedback**: the output current is sampled in series to give a parallel current feedback at the input.
   input: current
   stabilised output: current
   type of amplifier: current amplifier

EXAMPLE: The current of a photodiode is to be converted into a voltage. Photodiodes behave approximately like current source. In order to convert this current into a voltage, a transimpedance amplifier is required (parallel–parallel feedback, Fig. 7.87 a).

EXAMPLE: The sensitive measurement voltage in a strain gauge is to be converted into a current in order to transmit the analogue measurement result over a greater distance. In this case series–series feedback is applied (Fig. 7.87 b).

7.5 Negative Feedback

Parallel-series feedback

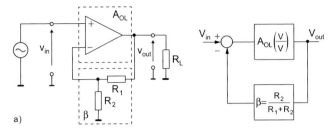

Parallel-parallel feedback

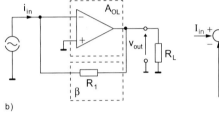

Series-parallel feedback

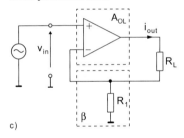
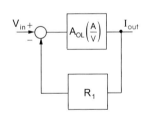

Series-parallel feedback

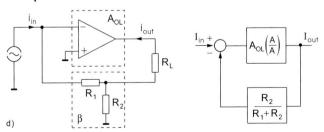

Fig. 7.86. Different types of feedback

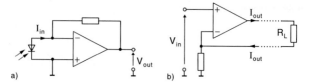

Fig. 7.87. Examples of **a** parallel–parallel feedback and **b** series–series feedback

7.5.2 Influence of Negative Feedback on Input and Output Impedance

The influence of negative feedback on the input and output impedance is calculated in the example on series–parallel feedback (for a noninverting amplifier), see Fig. 7.88.

Input Impedance

The input impedance of the open loop amplifier is assumed to be r'_{in}.

$$r_{in} = \frac{v_{in}}{i_{in}}, \qquad v_{out} = A_{OL} \cdot v'_{in}, \qquad r_{in} = \frac{v'_{in} + \beta \cdot v_{out}}{i_{in}} = \frac{v'_{in} + v'_{in} \cdot \beta \cdot A_{OL}}{i_{in}}$$

$$\boxed{r_{in} = r'_{in}(1 + \beta \cdot A_{OL})} \tag{7.117}$$

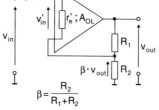

Fig. 7.88. Input configuration of the system with series–parallel feedback

- The input impedance increases in the case of series–parallel feedback by the amount of feedback.

Output Impedance

The output impedance of the open-loop amplifier is assumed to be r'_{out} (Fig. 7.89).

$$r_{out} = \frac{\text{open-circuit voltage}}{\text{short-circuit current}} = \frac{v_{o/c}}{i_{s/c}} = \frac{v_{out}}{i_{s/c}}, \qquad v_{out} = v_{in}\frac{A_{OL}}{1 + \beta \cdot A_{OL}},$$

$$i_{s/c} = \frac{v_{in} \cdot A_{OL}}{r'_{out}}$$

(for the short circuit $v_{out} = 0$ and $v_{in} = v'_{in}$)

$$r_{out} = \frac{r'_{out}}{1 + \beta \cdot A_{OL}} \qquad (7.118)$$

- The output impedance decreases in the case of series–parallel feedback by the amount of the feedback.

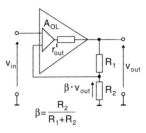

Fig. 7.89. Output configuration of the system with series-shunt feedback

7.5.2.1 Input and Output Impedance of the Four Kinds of Feedback

	$\dfrac{r_{in}}{r'_{in}}$	$\dfrac{r_{out}}{r'_{out}}$
a) Series–parallel feedback	$1 + \beta \cdot A_{OL}$	$\dfrac{1}{1 + \beta \cdot A_{OL}}$
b) Parallel–parallel feedback	$\dfrac{1}{1 + \beta \cdot A_{OL}}$	$\dfrac{1}{1 + \beta \cdot A_{OL}}$
c) Series–series feedback	$1 + \beta \cdot A_{OL}$	$1 + \beta \cdot A_{OL}$
d) Parallel–series feedback	$\dfrac{1}{1 + \beta \cdot A_{OL}}$	$1 + \beta \cdot A_{OL}$

- Negative feedback has always a positive effect on the input and output impedance: voltage outputs get a lower impedance, and current outputs get a higher impedance; voltage-driven inputs get a higher impedance, and current-driven inputs get a lower impedance.

7.5.3 Influence of Negative Feedback on Frequency Response

The amplifier A_{OL} is assumed to have low-pass characteristics:

$$A_{OL}(f) = \frac{A_{OL0}}{1 + jf/f'_c}$$

A_{OL0}: DC gain or low-frequency gain
f'_c: critical frequency

The transfer function of the system with negative feedback is then:

$$A_{CL}(f) = \frac{A_{OL}(f)}{1 + \beta \cdot A_{OL}(f)} = \underbrace{\frac{A_{OL0}}{1 + \beta \cdot A_{OL0}}}_{\text{gain}} \cdot \underbrace{\frac{1}{1 + j\dfrac{f}{f'_c}\dfrac{1}{1 + \beta \cdot A_{OL0}}}}_{\text{frequency response}} \quad (7.119)$$

- The **critical frequency** of the closed-loop system increases with respect to the critical frequency of the open-loop system by the amount of feedback $(1 + \beta A_{OL0})$.
- The **gain** decreases by the amount of feedback (Fig. 7.90).

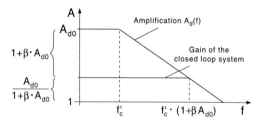

Fig. 7.90. Frequency response of the amplifier A_{OL} and of the closed-loop system

7.5.4 Stability of Systems with Negative Feedback

In theory systems with negative feedback are always stable. However, real amplifier gains A_{OL} have low-pass filter properties. This means that the gain decreases with increasing frequency and the phase is shifted between the input and output signal. Each pole rotates the input signal by 90°. Positive feedback will occur at the frequencies at which the phase is shifted by 180°, so that the output signal is added in phase to the input signal. If the loop gain $\beta \cdot A_{OL}$ is greater than 1 at any of these frequencies the signal is further amplified, the system becomes unstable and oscillations can occur (see also Sect. 7.6.2).

The **oscillation criterion** (**barkhausencriterion**) for feedback systems is given in general by:

Amplitude criterion: $\beta \cdot A_{OL} \geq 1$, and
Phase criterion: $\varphi = n \cdot 360°$, $n = 0, 1, 2, \ldots$

- An oscillation is generated in a closed-loop system if the phase shift is 0° or multiples of 360° and the loop gain is greater than 1.

The stability of a negative feedback system can be verified in a Bode plot: at a certain frequency (in Fig. 7.91 called f_1) the amplifier shifts the phase by 180°, and the negative feedback turns into positive feedback (Fig. 7.91, phase response). Thus, at this frequency the phase criterion has been fulfilled. The magnitude response $A_d(f)$ is split into the loop gain $\beta \cdot A_{OL}(f)$ and $1/\beta$. If the loop gain at the frequency f_{critical} is larger than 1 the closed-loop system is unstable. If the loop gain at f_{critical} is smaller than unity the closed-loop system is stable.

NOTE: In a negative-feedback system the critical phase shift (360°) occurs when the amplifier A_{OL} shifts the phase by 180°. A further shift of 180° occurs at the summation point where the feedback signal is subtracted from the input signal.

NOTE: The smaller the value of β is, the smaller is the portion of the output signal that is fed back, and the smaller is the risk of oscillation. A system with a large A_{OL} and large feedback, i.e. with a small overall gain A_{CL} is more likely to have problems with oscillations.

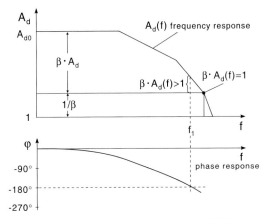

Fig. 7.91. Oscillation criterion in the Bode plot: at a frequency $f(\varphi = -180°)$ the closed loop gain is $\beta \cdot A_{OL} > 1$, the system starts oscillating (i.e. it is unstable)

7.6 Operational Amplifiers

An operational amplifier (op-amp) is an amplifier with a very high open-loop gain (Fig. 7.93). They are usually employed with negative feedback. Because of the high gain of the operational amplifier the amplification of the negative-feedback/closed-loop circuit depends only on the feedback circuit (see Sect. 7.5).

The operational amplifier input is a differential amplifier. One input is called the **inverting input** (V_n), and the other is the **noninverting input** (V_p). The differential voltage V_d is amplified with a gain of A_d. The output voltage is $V_{out} = A_d V_d$. The gain A_d usually falls in the range of 10^4–10^5. The output voltage can vary between the positive and the negative supply voltages. In order to obtain positive and negative output voltages, the operational amplifier requires a positive and a negative supply voltage (usually: ± 15 V).

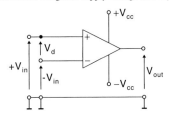

Fig. 7.92. Circuit symbol of the operational amplifier

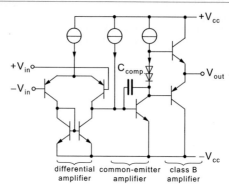

Fig. 7.93. Simplified circuit of an operational amplifier

7.6.1 Characteristics of the Operational Amplifier

7.6.1.1 Output Voltage Swing

The range of values that the output voltage can have is called the **output voltage swing**. The **maximum peak output voltage swing** lies about 1–3 V below the supply voltages (Fig. 7.94).

It is also possible to find so-called single supply op-amps which are supplied by a single positive supply voltage and whose output voltage swing is from 0 V up to approximately 1 V below the positive supply voltage value.

There are also so-called rail-to-rail op-amps whose output voltage swing is from exactly the negative to the positive supply voltage values.

7.6.1.2 Offset Voltage

The **offset voltage** V_0 (input offset voltage) is the input differential voltage V_d that has to be applied at the operational amplifier in order to obtain an output voltage of 0 V. V_0 is a worst case tolerance.

The transfer characteristic $V_{out} = f(V_d)$ of the ideal operational amplifier goes through the origin. For real op-amps the zero crossing is at V_0 (Fig. 7.94).

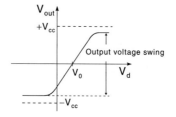

Fig. 7.94. Transfer characteristic of an operational amplifier

7.6.1.3 Offset Voltage Drift

The offset voltage V_0 is temperature dependent. The change in the offset voltage with temperature $\Delta V_{GI}/\Delta \vartheta$ is called the input offset voltage drift. It is in the range of 3–10 μV/K.

7.6.1.4 Common-Mode Input Swing

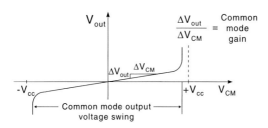

Fig. 7.95. Common-mode input swing

Common-mode amplification occurs when $V_{\text{in}-} = V_{\text{in}+} = V_{\text{CM}}$. Then $V_d = 0$ V. The ideal operational amplifier output voltage is also 0 V, independent of the value of V_{CM}. For real op-amps the common-mode input swing V_{CM} is given to define the range in which $V_{\text{out}} = 0$ V (Fig. 7.95).

7.6.1.5 Differential Mode Gain

The differential mode gain is The differential mode gain is usually in the region of 100 000,

$$A_d = \frac{V_{\text{out}}}{V_d} \qquad (7.120)$$

i.e. 100 dB.

7.6.1.6 Common-Mode Gain

The common-mode gain is

$$G_{\text{CM}} = \frac{V_{\text{out}}}{V_{\text{CM}}} \qquad (7.121)$$

7.6.1.7 Common-Mode Rejection Ratio

The common-mode rejection ratio (CMRR) is

$$CMRR = \frac{A_d}{A_{CM}} \tag{7.122}$$

Often this value is expressed in decibels. The range is 10^4–10^5, or between 80 and 100 dB, respectively.

7.6.1.8 Power Supply Rejection Ratio

The power supply rejection ratio (PSRR) is a measure of the influence of the supply voltage on the output voltage. It is defined via the offset voltage V_0. Its value expresses by how much the offset voltage has to be corrected in order to keep the output voltage at 0 V, when one of the supply voltages changes. The power supply rejection ratio is in the range of 10–100 μV/V. It is also expressed in dB.

7.6.1.9 Input Impedance

A distinction is made between the differential input impedance r_d and the common-mode input impedance r_{CM}. With bipolar operational amplifiers the differential input impedance r_d lies in the megohm range. Operational amplifiers with FET input stages have a differential input impedance of 10^{12} Ω. The common-mode input impedance is in the range 10^9 Ω to 10^{12} Ω.

NOTE: The input impedance is changed by the amount of feedback in the case of negative feedback (see Sect. 7.5.2):

$$r_{in} = r_d(1 + \beta A_{OL}), \quad \text{or} \quad \frac{r_d}{(1 + \beta A_{OL})}$$

7.6.1.10 Output Impedance

The output impedance of operational amplifiers is in the range of several hundred ohms to a few kilohms.

- This value is changed by negative feedback, so that, depending on the form of feedback used, the output can be regarded approximately as an ideal voltage source or as an ideal current source (see Sect. 7.5.2).

7.6.1.11 Input Bias Current

The input bias currents are the base currents absorbed by the differential amplifier. They are in the range of some tens to hundreds nanoamperes. In FET input stages the input bias currents are practically zero.

NOTE: Negative feedback does not influence the input bias currents.

Input bias current compensation: see Sect. 7.6.4.3

7.6.1.12 Gain–Bandwidth Product (Unity Gain Frequency)

The differential gain A_d has low-pass filter characteristics (Fig. 7.96):

$$A_d = \frac{A_{d0}}{1 + jf/f_c}$$

Above the critical frequency it approximately holds that:

$$A_d \approx \frac{A_{d0}}{jf/f_c}$$

Therefore:

$$\boxed{A_d \cdot f = A_{d0} \cdot f_c = f_T} \tag{7.123}$$

- At the unity gain frequency f_T the differential gain of the amplifier is 1. The unity gain frequency for op-amps is often given as the **gain–bandwidth product** because of the following relationship: $f_T = A_{d0} \cdot f_c$.

7.6.1.13 Critical Frequency

The critical frequency for frequency-compensated op-amps (see Sect. 7.6.2) lies between a few hertz and a few hundred hertz. With negative feedback this increases by the amount of feedback (see Sect. 7.5.3).

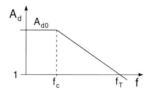

Fig. 7.96. Frequency response of the operational amplifier

7.6.1.14 Slew Rate of the Output Voltage

The slew rate defines the maximum rate of change of the output voltage. It is given in V/µs.

7.6.1.15 Equivalent Circuit of the Operational Amplifier

Figure 7.97 shows an equivalent circuit for a real operational amplifier. For standard frequency-compensated op-amps the following values are used:

- Input bias current I_B: for bipolar op-amps in the nanoampere range, negligible in FET op-amps.
- Differential input resistance r_d: for bipolar op-amps in the megohm range, for FET op-amps extremely high.
- Common-mode input resistance r_{CM}: almost always extremely high.

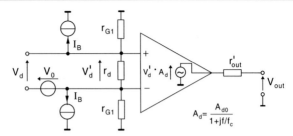

Fig. 7.97. Equivalent circuit of an operational amplifier

- Differential gain A_d: Characteristic curve like a low-pass filter. DC voltages gain A_{d0} lies around 10^5 (100 dB), and the critical frequency f_c lies between 10 and 100 Hz.
- Output resistance r'_{out}: lies between 100–1000 Ω.
- Offset voltage V_0: lies between 1 and a few millivolts.
- Common-mode rejection ratio $CMRR$ (not considered in the equivalent circuit): for DC it is about 80 dB and decreases dramatically with increasing frequency.

NOTE: The critical frequency and the output impedance of the op-amp with *negative feedback* depend on the amount of feedback $(1+\beta A_d)$ where β is the amount of feedback. The critical frequency increases by the amount of feedback, while the output impedance decreases by the amount of feedback. For an amplification in an op-amp with feedback of, for example, $A = 100$ the critical frequency would be about 10–100 kHz and the output impedance about 0.1–1 Ω! (Sect. 7.5.2).

NOTE: As well as the op-amp characteristics shown here, there are numerous designs with special characteristics, such as, for example, offset voltage in the microvolt range, input bias current in the picoampere range or a critical frequency in the megahertz range.

7.6.2 Frequency Compensation

Op-amps are **frequency compensated** for stability reasons. The low-pass filter characteristic is altered so that the critical frequency is shifted to lower frequencies. This is achieved by inserting a capacitor C_{comp} as a means of feed-back from the collector to the base of the voltage-amplifying emitter stage (see Fig. 7.93). This causes the gain A_d to dramatically decrease at high frequencies, so that in the system with feedback the loop gain βA_d is lower than unity when the phase shift reaches $\varphi = 180°$ (see Sect. 7.5.4).

For op-amps a distinction is made between frequency-compensated op-amps (**internally compensated**) and **uncompensated** op-amps. Uncompensated op-amp have external contacts, which can be connected to a capacitor C_{comp}. The choice of capacitance depends greatly on the amount of feedback chosen. The smaller the desired feedback, then the smaller is the required capacitance, i.e. the greater is the gain of the feedback system. The determination of C_{comp} can be carried out in an iterative manner. A square-wave input voltage can be applied to the system and the step response may be measured on an oscilloscope. Values of C_{comp} that are too large lead to a damping of the square-wave, while values of C_{comp} that are too small lead to oscillations and to instability in the circuit.

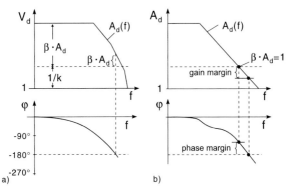

Fig. 7.98. Bode plot for frequency compensation: **a** uncorrected: $\beta A_d > 1$ at $f(\varphi = -180°)$, the circuit is unstable; **b** corrected: at $\beta A_d 1$ the phase shift $\varphi = -180°$ has not yet occurred, the circuit is stable

The angle $\alpha = 180° - \varphi_{(\beta A_d=1)}$ is known as the **phase margin**. It is a measure of the stability of the circuit (Fig. 7.98). If the phase margin is small, then the amplifier with feedback reacts to any change in the input voltage with damped oscillations. If $\alpha = 90°$ then this is the critically damped case, and for $\alpha = 65°$ there are overshoots of about 4%, which is often used in practice.

The **gain margin** is another measure of the stability of a system with feedback, as well as the phase margin (Fig. 7.98).

Internally frequency-compensated op-amps have a frequency compensation, which for a feedback network with $\beta = 1$ shows a phase margin of 65°. This ensures that the amplifier with feedback is always stable. It has the disadvantage that it is very slow for small amounts of feedback ($\beta \ll 1$, $A_{CL} \gg 1$), which happens when a high closed-loop gain is desired.

7.6.3 Comparators

Comparators are operational amplifiers that are operated without feedback. They are used to compare voltages. Therefore the output voltage can only have two states, high or low, depending on the sign of the input voltage V_d. The output is usually an open collector, which is connected to a pull-up resistor.

7.6.4 Circuits with Operational Amplifiers

Operational amplifier circuits can have positive or negative feedback. Circuits with positive feedback show two-state behaviour (e.g. Schmitt trigger) or can oscillate (Wien–Robinson oscillator). Circuits with negative feedback are stable, and the output voltage is proportional to the input voltage with linear feedback. Because of the high gain of the operational amplifier, the voltage difference between the input terminals is practically zero, when negative feedback is used. Calculations using Kirchhoff's laws are not used in the following. Block diagrams are occasionally used to display the circuit principle.

7.6.4.1 Impedance Converter (follower)

The impedance converter is an operational amplifier operated in series–parallel feedback with $\beta = 1$ (Fig. 7.99).

The amount of feedback is $(1 + \beta A_d) \approx A_d$. The transfer function is

$$\frac{V_{\text{out}}}{V_{\text{in}}} = \frac{A_d}{1 + \beta A_d} \approx 1,$$

So, in general, it is assumed that:

$$\boxed{V_{\text{out}} = V_{\text{in}}} \tag{7.124}$$

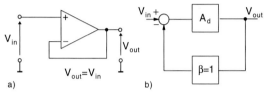

Fig. 7.99. Impedance converter: **a** circuit diagram; **b** block diagram

The input has an extremely high impedance:

$$\boxed{r_{\text{in}} = r_d(1 + A_d) \approx \rightarrow \infty} \tag{7.125}$$

where r_d is the differential input impedance of the operational amplifier.

NOTE: The input bias current is unaffected in this analysis! It loads the input voltage source independently of the amount of feedback. An op-amp with a FET input provides some relief from this problem.

The output has an extremely low impedance:

$$\boxed{r_{\text{out}} = \frac{r'_{\text{out}}}{1 + A_d} \approx 0} \tag{7.126}$$

where r'_{out} is the output impedance of the operational amplifier.

7.6.4.2 Noninverting Amplifier

The noninverting amplifier (Fig. 7.100) is an operational amplifier that is used in series–parallel feedback with $\beta = \dfrac{R_2}{R_1 + R_2}$. The amount of feedback is

$$1 + \frac{R_2}{R_1 + R_2} A_d$$

The transfer function is:

$$\frac{V_{\text{out}}}{V_{\text{in}}} = \frac{A_d}{1 + \beta A_d} \approx 1 + \frac{R_1}{R_2}$$

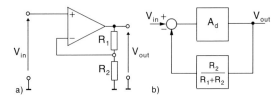

Fig. 7.100. Noninverting amplifier: **a** circuit diagram; **b** block diagram

So, in general, it is assumed that:

$$\boxed{\frac{V_{\text{out}}}{V_{\text{in}}} = 1 + \frac{R_1}{R_2}} \tag{7.127}$$

The input has an extremely high impedance:

$$\boxed{r_{\text{in}} = r_{\text{d}}\left(1 + \frac{R_2}{R_1 + R_2} A_{\text{d}}\right) \approx \to \infty} \tag{7.128}$$

where r_{d} is the differential input impedance of the operational amplifier.

NOTE: The input bias current is not affected by this consideration! The current drained from the input voltage source is independent of the feedback. An op-amp with a FET input provides some relief from this problem.

The output has an extremely low impedance:

$$\boxed{r_{\text{out}} = r'_{\text{out}} \frac{1}{1 + \dfrac{R_2}{R_1 + R_2} A_{\text{d}}} \approx 0} \tag{7.129}$$

where r'_{out} is the output impedance of the operational amplifier.

7.6.4.3 Inverting Amplifier

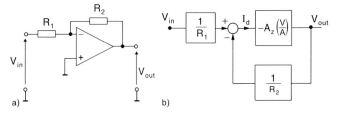

Fig. 7.101. Inverting amplifier: **a** circuit diagram; **b** block diagram

The inverting amplifier (Fig. 7.101) is an op-amp that uses parallel–parallel feedback with $\beta = 1/R_2$. The amount of feedback is

$$1 + \frac{1}{R_2} A_z$$

The input current is determined by R_1. The transfer function is given by:

$$\frac{V_{out}}{V_{in}} = \frac{1}{R_1} \frac{-A_z}{1 + \frac{1}{R_2}(-A_z)} \approx -\frac{R_2}{R_1},$$

So the transfer function is, in general, defined as:

$$\boxed{\frac{V_{out}}{V_{in}} = -\frac{R_2}{R_1}} \qquad (7.130)$$

Its input impedance is:

$$\boxed{r_{in} = R_1}$$

The output has an extremely low impedance:

$$\boxed{r_{out} = \frac{r'_{out}}{1 + \frac{1}{R_2} A_z} \approx 0} \qquad (7.131)$$

where r'_{out} is the output impedance of the operational amplifier.

NOTE: The op-amp is employed here as a transimpedance amplifier, i.e. the transfer function A_z of the amplifier has the qualities of an impedance. Referring to the equivalent circuit in Sect. 7.6.1.15 yields:

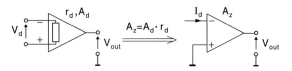

NOTE: The **input bias current** I_{B-} causes an offset voltage. This amounts to $I_{B-} \cdot R_1$. It can be compensated for by connecting a resistor $R = (R_1 || R_2)$ to ground from the noninverting input (Fig. 7.102).

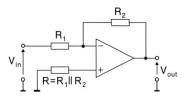

Fig. 7.102. Compensation of the input bias current

7.6.4.4 Summing Amplifier

The summing amplifier, like the inverting amplifier, employs parallel–parallel feedback (Fig. 7.103). The input currents V_i/R_i are summed at the inverting input of the operational

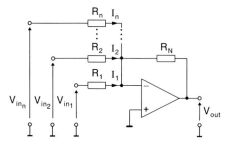

Fig. 7.103. Summing amplifier

amplifier. The output voltage is:

$$V_{\text{out}} = -\sum_{i=1}^{n} I_i \cdot R_N \tag{7.132}$$

Alternatively, V_{out} can be expressed as a function of the input voltages $V_{\text{in}n}$:

$$V_{\text{out}} = -\left(V_{\text{in}1}\frac{R_N}{R_1} + V_{\text{in}2}\frac{R_N}{R_2} + \cdots + V_{\text{in}n}\frac{R_N}{R_n}\right) \tag{7.133}$$

In the case where all resistors are equal it holds that:

$$V_{\text{out}} = -\sum_{i=1}^{n} V_i \tag{7.134}$$

7.6.4.5 Difference Amplifier

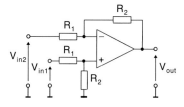

Fig. 7.104. Difference amplifier

The difference amplifier amplifies the difference of two input voltages (Fig. 7.104). Its gain is R_2/R_1.

$$V_{\text{out}} = (V_{\text{in}1} - V_{\text{in}2})\frac{R_2}{R_1} \tag{7.135}$$

The input impedance is $r_{\text{in}} = 2R_1$.

If $V_{in1} = 0$, then the circuit is the same as the inverting amplifier with input-bias current compensation. Input-bias current compensation is automatic in the difference amplifier.

Fig. 7.105 shows a difference amplifier with high input impedance.

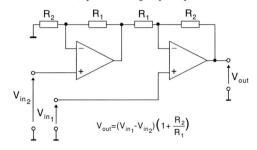

Fig. 7.105. Difference amplifier with high input impedance

Its transfer function is:

$$V_{out} = (V_{in1} - V_{in2})\left(1 + \frac{R_2}{R_1}\right) \tag{7.136}$$

7.6.4.6 Instrumentation Amplifier

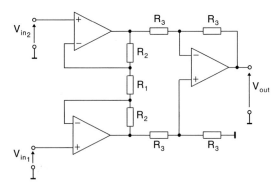

Fig. 7.106. Instrumentation amplifier

The instrumentation amplifier measures the difference between the input voltages V_{in1} and V_{in2} (Fig. 7.106). Its gain is:

$$V_{out} = (V_{in1} - V_{in2}) \cdot \left(1 + 2\frac{R_2}{R_1}\right) \tag{7.137}$$

It has an extremely high input impedance (see Sect. 7.6.4.2).

7.6.4.7 Voltage-Controlled Current Source

The voltage-controlled current source employs series–series feedback as shown in Fig. 7.107a. The input voltage is equated to the voltage drop across the current-sensing resistor R, so it holds that

$$V_{in} = R I_{out} \tag{7.138}$$

A transistor connected in series after the circuit permits higher output currents and has the advantage with the open drain (or open collector) that the choice of output potential is free (Fig. 7.107b).

For earthed loads the current source shown in Fig. 7.108 is suitable. The relationship $I_{out} = V_{in}/R$ is all the more valid the larger R_1 is compared to R.

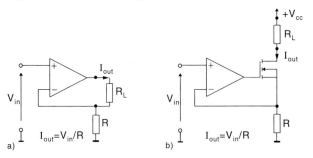

Fig. 7.107. a,b Voltage-controlled current sources employing series–series feedback

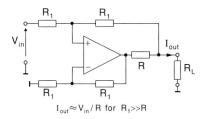

Fig. 7.108. Voltage-controlled current source for earthed/grounded loads

7.6.4.8 Integrator

The integrator works like the inverting amplifier. The input current $I_{in} = V_{in}/R$ charges the capacitor C (Fig. 7.109). Therefore the output voltage is the integral of the input signal:

$$V_{out} = -\frac{1}{RC} \int V_{in}\, dt \tag{7.139}$$

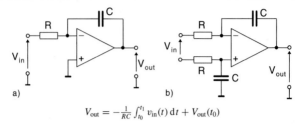

$$V_{\text{out}} = -\frac{1}{RC} \int_{t_0}^{t_1} v_{\text{in}}(t)\,\mathrm{d}t + V_{\text{out}}(t_0)$$

Fig. 7.109. Integrators: **a** simple integrator, **b** differential integrator

For a sinusoidal input the voltage gain is:

$$A_v = -\frac{1}{j\omega RC} \tag{7.140}$$

The input bias current is no longer negligible for large time constants. Relief can be provided either by using input-bias current compensation or – even simpler – by inserting an op-amp with a FET input stage. The input-bias current compensation is carried out in a similar way to the inverting amplifier, except that the noninverting input is connected to ground via a parallel combination of a resistor R and a capacitor C (see Sect. 7.6.4.3).

Integrators are mainly used as I-controllers in negative-feedback systems.

In systems without feedback the output voltage saturates to the positive or negative power supply rails, as the offset voltage and input bias current are integrated as well as the applied voltage signal V_{in}. Integrators must therefore be reset to zero at suitable time intervals, in order to be able to achieve a defined output state for the integration. This can be achieved with a relay or a FET in parallel with the capacitor (Fig. 7.110). In the use of a MOSFET, the internal reverse-current diode limits the output voltage range to positive voltages.

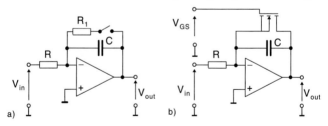

Fig. 7.110. Integrators: reset circuits for $V_{a(t0)} = 0$

7.6.4.9 Differentiator

The input current of the differentiator in Fig. 7.111 is

$$I_{\text{in}} = C \frac{\mathrm{d}V_{\text{in}}}{\mathrm{d}t}$$

This current flows through R, so the output voltage is

$$V_{\text{out}} = -RC\frac{dV_{\text{in}}}{dt} \tag{7.141}$$

For a sinusoidal input the voltage gain is

$$A_{\text{v}} = -j\omega RC \tag{7.142}$$

The differentiator is mainly used as the D-stage in PID controllers.

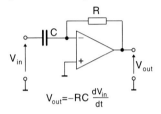

Fig. 7.111. Differentiator

7.6.4.10 AC Voltage Amplifier with Single-Rail Supply

Sometimes amplifiers are operated with only one supply voltage. In that case the reference voltage at the inverting input is set to $V_{CC}/2$ using a voltage divider (Fig. 7.112).

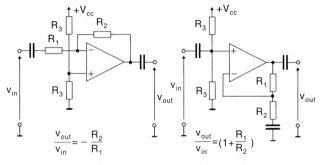

Fig. 7.112. AC voltage amplifier with single-rail supply

7.6.4.11 Voltage Setting with Defined Slew Rate

The output voltage of the circuit in Fig. 7.113 can only change at a rate given by

$$\frac{dV_{\text{out}}}{dt} = \pm V_{\text{out1 max}}\frac{1}{RC} \approx \pm V_{CC}\frac{1}{RC}$$

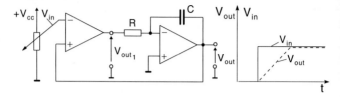

Fig. 7.113. Voltage setting with a defined slew rate

For $V_{out} \neq V_{in}$ the first op-amp's output voltage V_{out1} jumps to one of its supply rails of $\pm V_{out1\,max}$. The output voltage V_{out} therefore changes at a defined slew rate to the value given by $V_{out} = V_{in}$.

7.6.4.12 Schmitt Trigger

Schmitt triggers (comparators with hysteresis) are bistable circuits that use positive feedback. Thus the output voltage can only jump between the output voltage limits of $\pm V_{out\,max}$. By using feedback two thresholds exist for the input voltage, which are defined by the switching of the output voltage. Once a threshold has been exceeded, the other threshold must be exceeded in order to change to a new state.

Schmitt triggers are employed in bistable controllers. They are also used instead of comparators to avoid multiple switching if the input signal is noisy.

Inverting Schmitt Trigger

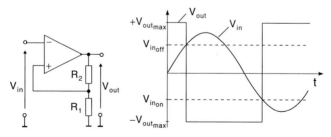

Fig. 7.114. Inverting Schmitt trigger

The trigger levels of the inverting Schmitt trigger (Fig. 7.114) are

$$V_{in\,on} = -\frac{R_1}{R_1 + R_2}V_{out\,max}, \qquad \text{and} \qquad V_{in\,off} = +\frac{R_1}{R_1 + R_2}V_{out\,max} \tag{7.143}$$

Noninverting Schmitt Trigger

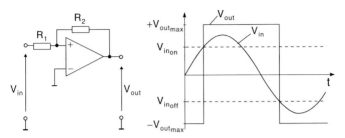

Fig. 7.115. Noninverting Schmitt trigger

The trigger levels of the noninverting Schmitt trigger (Fig.7.115) are

$$V_{\text{in on}} = +\frac{R_1}{R_2} V_{\text{out max}}, \quad \text{and} \quad V_{\text{in off}} = -\frac{R_1}{R_2} V_{\text{out max}} \tag{7.144}$$

7.6.4.13 Triangle- and Square-Wave Generator

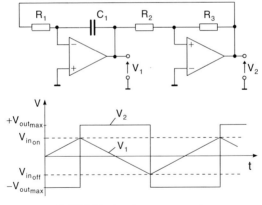

Fig. 7.116. Triangle- and square-wave generator

The triangle- and square-wave generator is a free-running circuit, consisting of an integrator and a noninverting Schmitt trigger (Fig. 7.116). The amplitude of the triangular wave is equal to the Schmitt trigger threshold value. The frequency of the output voltage waveforms is

$$f = \frac{R_3}{R_2} \frac{1}{4 R_1 C_1} \tag{7.145}$$

for a symmetrical output voltage swing of $\pm V_2$.

7.6.4.14 Multivibrator

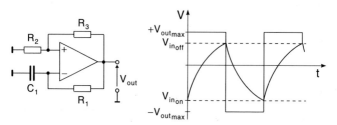

Fig. 7.117. Multivibrator

The multivibrator or square wave generator is a free-running circuit (Fig. 7.117). The switching frequency is

$$f = \frac{1}{2R_1 C_1 \ln\left(1 + \dfrac{2R_2}{R_3}\right)} \quad (7.146)$$

For small hysteresis, i.e. $R_2 \ll R_3$, it holds that

$$f \approx \frac{1}{2R_1 C_1} \frac{R_3}{2R_2} \quad (7.147)$$

7.6.4.15 Sawtooth Generator

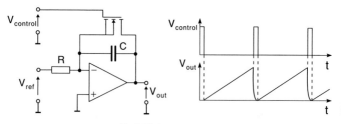

Fig. 7.118. Sawtooth generator

A sawtooth voltage has the form of a ramp (Fig. 7.118). It is generated by charging a capacitor with a constant current and then discharging it in a very short time.

The discharge occurs over a clock pulse, which can be generated externally or internally.

7.6.4.16 Pulse-Width Modulator

Pulse width-modulators (PWM) are mainly used in measurement technology or in switched-mode power supplies. They convert an analogue signal into a digital signal, where the **duty cycle** t_1/T is proportional to the analogue input voltage. Pulse-width modulators are a simple means to prepare analogue signals for digital systems.

Pulse width modulator with fixed pulse frequency:

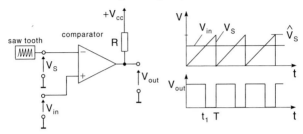

Fig. 7.119. Pulse-width modulator with fixed pulse frequency

The duty cycle is:

$$\boxed{\frac{t_1}{T} = \frac{V_{\text{in}}}{\hat{V}_S}}$$

A pulse-width modulated voltage can be generated by comparing a sawtooth voltage with an analogue voltage (Fig. 7.119). For measurement purposes the sawtooth can be triggered by a digital system and the time t_1 measured. This results in a simple analogue-to-digital converter. This does have the disadvantage, however, that the peak value of the sawtooth must be known, i.e. if necessary must be adjusted.

Precision Pulse-Width Modulator

The accuracy of the pulse-width modulation can be enhanced significantly with the use of an I-controller (Fig. 7.120). The comparison between the desired and the actual value is carried out at the integrator. The integrator output voltage V'_{in} changes value, so that $V_{\text{ref}} \cdot \frac{t_1}{T} = V_{\text{in}}$. The sawtooth amplitude and nonlinearities are not included in the result. The disadvantage of the circuit is that the integrator time constant must be large compared with the periodic time of the sawtooth.

Fig. 7.121 shows a free-running pulse width modulator. The duty cycle of the output voltage is $t_1/T = 0.5$ for $V_{\text{in}} = 0$ V. The accuracy of the modulator is dependent on the symmetry of the bidirectional reference voltage source V_z. The disadvantage of this circuit is that the switching frequency is dependent on the input voltage V_{in}. The switching frequency for $V_{\text{in}} = 0$ V is:

$$f_{(V_{\text{in}}=0)} = \frac{1}{4R_1C_1}. \tag{7.148}$$

The switching frequency decreases with increasing input voltage V_{in}, and $f = 0$ for $V_{\text{in}} = V_z$.

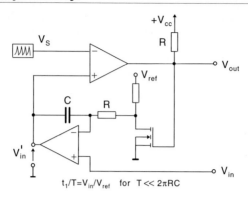

Fig. 7.120. Precision pulse-width modulator

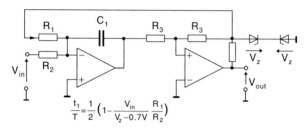

Fig. 7.121. Precision pulse-width modulator

7.7 Active Filters

Filters are circuits with a frequency-dependent transfer function. A distinction is made between **low-pass, high-pass** and **bandpass filters** and **band-stop** or **notch filters**. All of these filters have in common that their transfer function is divided into stop-bands and passbands. The border between the pass- and stop-bands is called the corner or critical frequency. The **critical frequency** (or -3 dB point) is the frequency where the magnitude of the transfer function is -3 dB (i.e. $1/\sqrt{2}$) lower than the pass-band magnitude. This frequency-dependent attenuation of the signal in the stop-band depends on the **order** of the filter. The higher the order the steeper is the frequency-dependent rejection.

Another member of the filter family is the **all-pass filter**. It does not alter the signal amplitude, but changes the phase of the signal depending on its frequency. Band-stop filters and all-pass filter are not covered further in this section.

A further distinction is made between active and passive filters. **Active filters** are filters that contain active components. The active components are used as impedance converters, so that higher-order filters can be combined from series-connected filters of second order, with no extra feedback requirement. This simplifies the design and the calibration of those filters compared to passive filters. Furthermore, the use of active components means that

7.7.1 Low-Pass Filters

7.7.1.1 Theory of Low-Pass Filters

The transfer function of a low-pass filter is explained by the example of a second-order RLC low-pass filter (Fig. 7.122a:

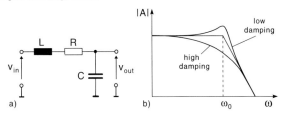

Fig. 7.122. Second-order RLC low-pass filter; **a** circuit, **b** frequency response

$$\frac{v_{out}}{v_{in}} = F(j\omega) = \frac{\frac{1}{j\omega C}}{j\omega L + R + \frac{1}{j\omega C}} = \frac{1}{1 + j\omega RC + (j\omega)^2 LC} \quad (7.149)$$

- For small values of ω the expression for $F(j\omega)$ is approximately equal to 1.
- For large values of ω the quadratic term in the denominator dominates: $F(j\omega)$ drops at a rate of 40 dB/decade.
- The attenuation around the natural frequency defines the transition from passband to stop-band (see Sect. 1.2.6). For smaller attenuation there is a rise due to resonance, whereas for larger attenuation $F(j\omega)$ begins to fall away even before the natural frequency. The attenuation has no influence in the regions of very high or very low values of ω (Fig. 7.122 b).

Normalisation:

Substituting $j\omega$ with the complex frequency s and normalising this with respect to the critical frequency ω_c with $s = \omega_c S$ yields:

$$F(s) = \frac{1}{1 + RC\,s + LC\,s^2}, \qquad \text{with } j\omega = s \quad (7.150)$$

and

$$F(S) = \frac{1}{1 + \omega_c RCS + \omega_c^2 LCS^2}, \qquad \text{with } S = \frac{s}{\omega_c} \quad (7.151)$$

If the S coefficients are replaced by general real coefficients a_1 and b_1, then an independent general function of a second-order low-pass filter can be derived from the circuit:

$$\boxed{F(S) = \frac{1}{1 + a_1 S + b_1 S^2}} \quad (7.152)$$

A low-pass filter of higher order, i.e. a low-pass filter with a steeper drop from the critical frequency, can be realised by connecting in series several low-pass filters, of first and second order.

The general transfer function of a $2n$th-order low-pass filter is then:

$$F(S) = \frac{F_0}{(1 + a_1 S + b_1 S^2) \cdot (1 + a_2 S + b_2 S^2) \cdot \ldots \cdot (1 + a_n S + b_n S^2)} \quad (7.153)$$

- The second-order low-pass filter is the basic building block used to build low-pass filters of higher order.

- A steeper fall-off about the critical frequency can be achieved if several low-pass filters of first and second order are connected in series. First-order low-pass filters can be considered as special versions of second order low-pass filters, for which the coefficient b is equal to zero (Eq. (7.153)). The factor F_0 in the expression takes into consideration any frequency-independent amplification in the low-pass filter.

- The highest power in the denominator polynomial defines the **filter order** of the low-pass filter. It defines the fall-off in the expression 7.153 around the critical frequency. Each power of two in the order produces a fall-off of 20 dB/decade (see Sect. 5.3.2).

- The roots of the denominator polynomial are called the **poles** of $F(S)$. They can be real or complex conjugates, depending on the values of the coefficients a_i and b_i. Complex conjugate poles cause a resonant bump in the region between the stop- and passbands.

- The number of poles is equal to the filter order.

- The **coefficients a_i and b_i** decide the form of the region between the stop- and passbands. Functions with complex conjugate poles have a higher critical frequency compared with functions that have real poles (Fig. 7.123) and therefore cause a steeper transition from the pass- to the stop-band. For this reason most filters (almost without exception) are built with complex conjugate poles.

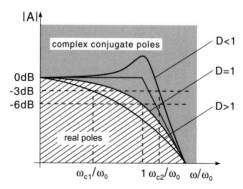

Fig. 7.123. Frequency response of a 2nd order low-pass filter with complex conjugate and real poles

Different filter characteristics are defined, depending on the choice of the coefficients a_i and b_i (Fig. 7.124):

Butterworth: the magnitude response $F(S)$ is flat at the level F_0 almost as far as the critical frequency.

Bessel: below the critical frequency this filter has an optimum square-wave transfer characteristic.

Chebyshev: the magnitude response has a defined amount of ripple in the passband (resonance effect). The fall-off after the critical frequency is particularly steep.

Critical Damping: filters with real poles. All poles have the same value. The filter has no resonance effects.

- Filters of the same order and the same critical frequency, but with different characteristics are distinguished by their a_i and b_i coefficients. This means that filters with different characteristics may be realised by the same circuits using different values for the constituent components.

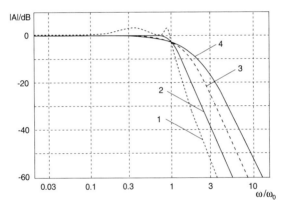

Fig. 7.124. Comparison of fourth-order low-pass filters: 1. Chebyshev, 2. Butterworth, 3. Bessel, 4. Filter with critical damping

The coefficients for different filter characteristics are given in Tables 7.2 to 7.5 up to the sixth order. The magnitude responses of the corresponding transfer functions are shown in Figs. 7.125 to 7.128. The fifth and sixth table columns give the normalised critical frequency of the individual second-order filters and their Q-factors. These parameters are useful in confirming the performance of the individual second-order filters by measurement.

Butterworth:

Table 7.2. Butterworth-filter

Order n	i	a_i	b_i	f_{ci}/f_c	Q_i
1	1	1.0000	0.0000	1.000	–
2	1	1.4142	1.0000	1.000	0.71
3	1	1.0000	0.0000	1.000	–
	2	1.0000	1.0000	1.272	1.00
4	1	1.8478	1.0000	0.719	0.54
	2	0.7654	1.0000	1.390	1.31
5	1	1.0000	0.0000	1.000	–
	2	1.6180	1.0000	0.859	0.62
	3	0.6180	1.0000	1.448	1.62
6	1	1.9319	1.0000	0.676	0.52
	2	1.4142	1.0000	1.000	0.71
	3	0.5176	1.0000	1.479	1.93

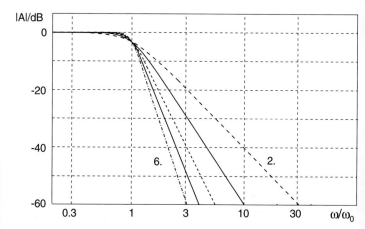

Fig. 7.125. Butterworth low-pass filters second to sixth order

Bessel:

Table 7.3. Bessel-filter

Order n	i	a_i	b_i	f_{ci}/f_c	Q_i
1	1	1.0000	0.0000	1.000	–
2	1	1.3617	0.6180	1.000	0.58
3	1	0.7560	0.0000	1.323	–
	2	0.9996	0.4772	1.414	0.69
4	1	1.3397	0.4889	0.978	0.52
	2	0.7743	0.3890	1.797	0.81
5	1	0.6656	0.0000	1.502	–
	2	1.1402	0.4128	1.184	0.56
	3	0.6216	0.3245	2.138	0.92
6	1	1.2217	0.3887	1.063	0.51
	2	0.9686	0.3505	1.431	0.61
	3	0.5131	0.2756	2.447	1.02

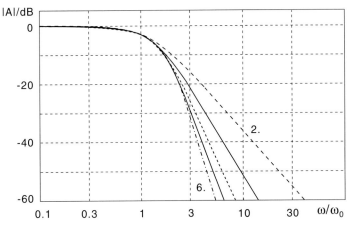

Fig. 7.126. Bessel low-pass filters second to sixth order

Chebyshev with 0.5 dB ripple:

Table 7.4. Chebyshev-filter with 0.5 dB ripple

Order n	i	a_i	b_i	f_{ci}/f_c	Q_i
1	1	1.0000	0.0000	1.0000	–
2	1	1.3614	1.3827	1.0000	0.86
3	1	1.8636	0.0000	0.537	–
	2	0.6402	1.1931	1.335	1.71
4	1	2.6282	3.4341	0.538	0.71
	2	0.3648	1.1509	1.419	2.94
5	1	2.9235	0.0000	0.342	–
	2	1.3025	2.3534	0.881	1.18
	3	0.2290	1.0833	1.480	4.54
6	1	3.8645	6.9797	0.366	0.68
	2	0.7528	1.8573	1.078	1.81
	3	0.1589	1.0711	1.495	6.51

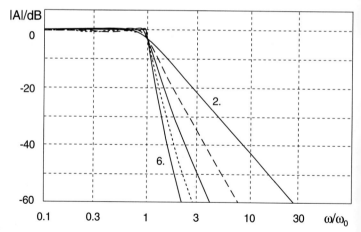

Fig. 7.127. Chebyshev low-pass filters second to sixth order with 0.5 dB ripple

Chebyshev with 3 dB ripple:

Table 7.5. Chebyshev-filter with 3 dB ripple

Order n	i	a_i	b_i	f_{ci}/f_c	Q_i
1	1	1.0000	0.0000	1.000	–
2	1	1.0650	1.9305	1.000	1.30
3	1	3.3496	0.0000	0.299	–
	2	0.3559	1.1923	1.396	3.07
4	1	2.1853	5.5339	0.557	1.08
	2	0.1964	1.2009	1.410	5.58
5	1	5.6334	0.0000	0.178	–
	2	0.7620	2.6530	0.917	2.14
	3	0.1172	1.0686	1.500	8.82
6	1	3.2721	11.6773	0.379	1.04
	2	0.4077	1.9873	1.086	3.46
	3	0.0815	1.0861	1.489	12.78

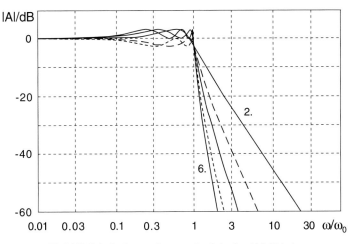

Fig. 7.128. Chebyshev low-pass filters second to sixth order with 3 dB ripple

7.7.1.2 Low-Pass Filter Calculations

Low-pass filter calculations are carried out in the following steps:

- Choice of the filter type, of the critical frequency and of the order.
- Choice of a filter circuit (see also Sect. 7.7.1.3).
- Calculation of the transfer function $F(s)$ and normalisation with $S = s/\omega_c$.
- Conversion of the normalised transfer function according to Eq. (7.153).
- Determination of the component values by comparing the coefficients with the coefficients a_i and b_i (the number of equations is smaller than the number of variables, so that some components can be chosen freely).
- If some component values are unsuitable, the values can be changed without repeating the complete calculation:

> A change of C into C' changes R and L to: $R' = R\dfrac{C}{C'}$, and $L' = L\dfrac{C}{C'}$
>
> A change of R into R' changes C and L to: $C' = C\dfrac{R}{R'}$, and $L' = L\dfrac{R'}{R}$
>
> A change of L into L' changes R and C to: $C' = C\dfrac{L}{L'}$, and $R' = R\dfrac{L'}{L}$

- The fifth and sixth columns of the table, f_{ci}/f_c and Q_i, are useful for confirming the performance of the individual filters of first or second order by measurement.

EXAMPLE: Calculation of a third-order Chebyshev low-pass filter with 3 dB ripple with a critical frequency $f_c = 10$ kHz.

For the realisation, the circuit in Fig. 7.129 was chosen

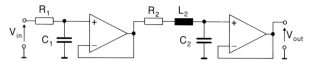

Fig. 7.129. Low-pass filter of third order

The transfer function of the circuit is:

$$F(s) = \frac{1}{1 + R_1 C_1 s} \cdot \frac{1}{1 + R_2 C_2 s + L_2 C_2 s^2}$$

with $S = s/w_c$ follows

$$F(S) = \frac{1}{1 + \underbrace{\omega_c R_1 C_1}_{a_1} S} \cdot \frac{1}{1 + \underbrace{\omega_c R_2 C_2}_{a_2} S + \underbrace{\omega_c^2 L_2 C_2}_{b_2} S^2}$$

Table 7.5 yields:

$$a_1 = 3.3496, \quad b_1 = 0.0000, \quad a_2 = 0.3559, \quad b_2 = 1.1923.$$

C_1 and C_2 are chosen in advance: $C_1 = 10$ nF, and $C_2 = 10$ nF. This yields for R_1, R_2 and L_2:

$$R_1 = \frac{a_1}{\omega_c \cdot C_1} = \frac{3.3496}{2\pi \cdot 10\text{kHz} \cdot 10 \text{ nF}} = 5.334 \text{ k}\Omega$$

$$R_2 = \frac{a_2}{\omega_c \cdot C_2} = \frac{0.3559}{2\pi \cdot 10\text{kHz} \cdot 10 \text{ nF}} = 567 \text{ }\Omega$$

$$L_2 = \frac{b_2}{\omega_c^2 \cdot C_2} = \frac{1.1923}{(2\pi)^2 \cdot 100 \cdot 10^6 \text{Hz}^2 \cdot 10 \text{ nF}} = 30 \text{ mH}$$

7.7.1.3 Low-Pass Filter Circuits

Noninverting First-Order Low-Pass Filter

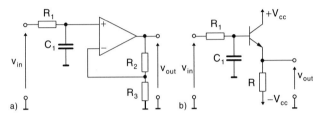

Fig. 7.130. Noninverting first-order low-pass filter **a** with operational amplifier; **b** with emitter follower as impedance converter

The transfer function is:

$$F(\text{S}) = \frac{F_0}{1 + a\text{S}} = \frac{1 + R_2/R_3}{1 + \underbrace{\omega_c R_1 C_1}_{a} \text{S}} \qquad (7.154)$$

Inverting First-Order Low-Pass Filter

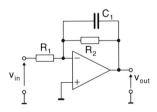

Fig. 7.131. Inverting first-order low-pass filter

The transfer function is:

$$F(\text{S}) = \frac{F_0}{1 + a\text{S}} = \frac{-R_2/R_1}{1 + \underbrace{\omega_c R_2 C_1}_{a} \text{S}} \qquad (7.155)$$

Inverting Second-Order Low-Pass Filter

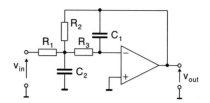

Fig. 7.132. Inverting second-order low-pass filter

The transfer function is:

$$F(S) = \frac{F_0}{1+aS+bS^2} = \frac{-R_2/R_1}{1+\underbrace{\omega_c C_1(R_2+R_3+R_2R_3/R_1)}_{a} S + \underbrace{\omega_c^2 C_1 C_2 R_2 R_3}_{b} S^2} \quad (7.156)$$

C_1 and C_2 are chosen in advance. Then:

$$R_2 = \frac{aC_2 - \sqrt{a^2 C_2^2 - 4C_1 C_2 b(1-F_0)}}{2\omega_c C_1 C_2}, \qquad R_1 = \frac{-R_2}{F_0}, \qquad R_3 = \frac{b}{\omega_c^2 C_1 C_2 R_2}$$

In order to obtain a real value for R_2 it must hold that:

$$\frac{C_2}{C_1} \geq \frac{4b(1-F_0)}{a^2}$$

Noninverting Second-Order Low-Pass Filter

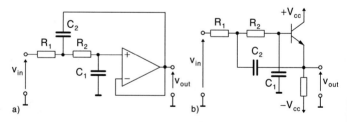

Fig. 7.133. Noninverting second-order low-pass filter **a** with operational amplifier; **b** with emitter follower as impedance converter

The transfer function is:

$$F(S) = \frac{F_0}{1+aS+bS^2} = \frac{1}{1+\underbrace{\omega_c C_1(R_1+R_2)}_{a} S + \underbrace{\omega_c^2 C_1 C_2 R_1 R_2}_{b} S^2} \quad (7.157)$$

C_1 and C_2 are chosen in advance. Then:

$$R_2, R_1 = \frac{aC_2 \pm \sqrt{a^2 C_2^2 - 4bC_1 C_2}}{2\omega_c C_1 C_2}$$

In order to obtain a real value for R_2 it must hold that:

$$\frac{C_2}{C_1} \geq \frac{4b}{a^2}$$

7.7.2 High-Pass Filters

7.7.2.1 Theory of High-Pass Filters

See also Sect. 7.7.1.1: Theory of low-pass filters.

The general transfer function of an nth-order high-pass filter is:

$$F(S) = \frac{F_\infty}{\left(1 + \dfrac{a_1}{S} + \dfrac{b_1}{S^2}\right) \cdot \left(1 + \dfrac{a_2}{S} + \dfrac{b_2}{S^2}\right) \cdot \ \ldots \ \cdot \left(1 + \dfrac{a_n}{S} + \dfrac{b_n}{S^2}\right)} \quad (7.158)$$

As in the case for low-pass filters, a distinction is made between Chebyshev, Butterworth and Bessel high-pass filters. Tables 7.2 to 7.5 are valid for the coefficients a_i and b_i, and F_∞ gives the gain for very high frequencies ($f \to \infty$).

7.7.2.2 High-Pass Filter Circuits

First-Order Noninverting High-Pass Filter

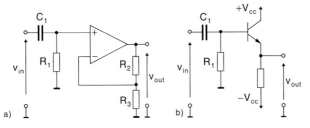

Fig. 7.134. First-order noninverting high-pass filter **a** with operational amplifier; **b** with emitter follower as impedance converter

The transfer function is:

$$F(S) = \frac{F_\infty}{1 + \dfrac{a}{S}} = \frac{1 + \dfrac{R_2}{R_3}}{1 + \dfrac{1}{\omega_c R_1 C_1} \cdot \dfrac{1}{S}}, \qquad a = \frac{1}{\omega_c R_1 C_1} \quad (7.159)$$

First-Order Inverting High-Pass Filter

The transfer function is:

$$F(S) = \frac{F_\infty}{1 + \dfrac{a}{S}} = \frac{-\dfrac{R_2}{R_1}}{1 + \dfrac{1}{\omega_c R_1 C_1} \cdot \dfrac{1}{S}}, \qquad a = \frac{1}{\omega_c R_1 C_1} \tag{7.160}$$

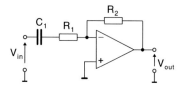

Fig. 7.135. First-order inverting high-pass filter

Second-Order Noninverting High-Pass Filter

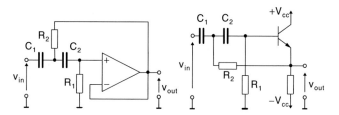

Fig. 7.136. Second-order noninverting high-pass filter

The transfer function is:

$$F(S) = \frac{F_\infty}{1 + \dfrac{a}{S} + \dfrac{b}{S^2}} = \frac{-\dfrac{R_2}{R_1}}{1 + \dfrac{C_1 + C_2}{\omega_c R_1 C_1 C_2} \cdot \dfrac{1}{S} + \dfrac{1}{\omega_c^2 R_1 R_2 C_1 C_2} \dfrac{1}{S^2}} \tag{7.161}$$

with

$$a = \frac{1}{\omega_c R_1 C_1 C_2}, \qquad b = \frac{1}{\omega_c^2 R_1 R_2 C_1 C_2}$$

If $C_1 = C_2 = C$, it then holds that:

$$R_1 = \frac{2}{a \cdot \omega_c C}, \qquad \text{and} \qquad R_2 = \frac{a}{2b \cdot \omega_c C}$$

7.7.3 Bandpass Filters

7.7.3.1 Second-Order Bandpass Filter

The transfer function of a bandpass filter of second order is similar to that of an RLC bandpass filter (Fig. 7.137):

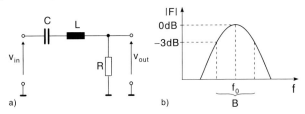

Fig. 7.137. Second-order RLC bandpass filter **a** circuit diagram; **b** amplitude response

- The magnitude response $|F(f)|$ has the value of 1 at the resonant frequency f_0.
- The **centre frequency** of the bandpass filter is equal to the resonant frequency f_0.
- The **bandwidth B** is the frequency range between the -3 dB-points.

The transfer function of the bandpass filter according to Fig. 7.137 is:

$$F(s) = \frac{sRC}{1 + s\underbrace{RC}_{2D/\omega_0} + s^2\underbrace{LC}_{1/\omega_0^2}} \tag{7.162}$$

where D is the damping, ω_0 is the resonant frequency, see also Sect. 1.2.6.1. If the damping D is substituted for by the bandwidth B, and the resonant frequency ω_0 by the resonant frequency f_0, then

$$B = 2D\frac{\omega_0}{2\pi} \quad \text{and} \quad f_0 = \frac{\omega_0}{2\pi}$$

a general expression for a bandpass filter of second order:

$$\boxed{F(s) = \frac{F_0 \cdot s\dfrac{B}{2\pi f_0^2}}{1 + sB/2\pi f_0^2 + s^2 1/(2\pi f_0)^2}} \tag{7.163}$$

The factor F_0 takes into account a frequency-independent amplification, so that the transfer function does not have to have the value of 1 at the resonant frequency.

- The centre frequency and the bandwidth of each bandpass filter of second order can be determined by coefficient comparison with this general transfer function.

EXAMPLE: For the RLC bandpass filter given above:

$$RC = \frac{B}{2\pi f_0^2}, \quad \text{and} \quad LC = \frac{1}{(2\pi f_0)^2},$$

and it follows that

$$f_0 = \frac{1}{2\pi\sqrt{LC}}, \quad \text{and} \quad B = \frac{1}{2\pi} \cdot \frac{R}{L}$$

7.7.3.2 Second-Order Bandpass Filter Circuit

The transfer function of the bandpass filter shown in Fig. 7.138 is:

$$F(s) = \frac{s \cdot \beta \cdot RC}{1 + s \underbrace{RC(3-\beta)}_{B/2\pi f_0^2} + s^2 \underbrace{R^2C^2}_{1/(2\pi f_0)^2}}, \quad \text{with} \quad \beta = 1 + \frac{R_1}{R_2} \quad (7.164)$$

Resonant frequency: $f_0 = \dfrac{1}{2\pi RC}$

Bandwidth: $B = \dfrac{3-\beta}{2\pi RC}$

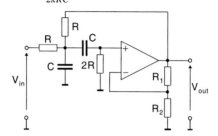

Fig. 7.138. Second-order bandpass filter

The resonant frequency and the bandwidth can be chosen independently of each other. The amplification at the resonant frequency is *not* equal to 1. It is given by $|F(f_0)| = \dfrac{\beta}{3-\beta}$.

To be able to freely select the amplification, a corresponding amplifier must be connected in after the filter circuit.

7.7.3.3 Fourth- and Higher-Order Bandpass Filters

Fourth Order Bandpass Filter

Fourth-order bandpass filters fall off at 40 dB/decade. They can be realised by connecting in series two second-order bandpass filters whose centre frequencies are slightly different. The bandwidth $B = (1/2)\sqrt{2} f_0$ yields a maximally flat passband. The gain in the passband decreases in proportion to the mismatch of the resonant frequencies.

Higher-Order Bandpass Filters with Larger Bandwidth

Bandpass filters of higher orders with a larger bandwidth can be realised by connecting in series a low-pass filter and a high-pass filter with the same characteristics. The bandpass filter fall-off is equal to the low-pass filter and high-pass filter fall-offs. The filter characteristics agree even more for larger values of bandwidth. For Butterworth and Bessel filters this is approximately the case if the critical frequencies from the low- and high-pass filters are separated by a factor of 10. For Chebyshev filters the bandwidth must increase as the order of filter increases (Figs. 7.127 and 7.128).

7.7.4 Universal Filter

A filter can be realised using integrators with feedback (Fig. 7.139).

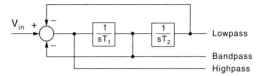

Fig. 7.139. Block diagram of a universal filter

The circuit shown in Fig. 7.139 offers three different output possibilities, depending on whether a high-pass, a low-pass or a bandpass filter is required. The transfer functions are given by:

Low-pass filter:

$$F_{\mathrm{LPF}} = \frac{1}{1 + sT_2 + s^2 T_1 T_2} = \frac{1}{1 + \underbrace{\omega_g T_2}_{a} S + \underbrace{\omega_g^2 T_1 T_2}_{b} S^2} \tag{7.165}$$

Bandpass filter:

$$F_{\mathrm{BPF}} = \frac{sT_2}{1 + s \underbrace{T_2}_{B/2\pi f_0^2} + s^2 \underbrace{T_1 T_2}_{1/(2\pi f_0)^2}}, \qquad f_0 = \frac{1}{2\pi\sqrt{T_1 T_2}}, \qquad B = \frac{1}{2\pi T_1} \tag{7.166}$$

High-pass filter:

$$F_{\mathrm{HPF}} = \frac{1}{1 + \dfrac{1}{T_1}\cdot\dfrac{1}{s} + \dfrac{1}{T_1 T_2}\cdot\dfrac{1}{s^2}} = \frac{1}{1 + \underbrace{\dfrac{1}{\omega_g T_1}}_{a}\cdot\dfrac{1}{S} + \underbrace{\dfrac{1}{\omega_g^2 T_1 T_2}}_{b}\cdot\dfrac{1}{S^2}} \tag{7.167}$$

7.7.5 Switched-Capacitor Filter

In Sect. 7.7.4 it was shown how an active filter can be realised using integrators with feedback. Fig. 7.140 shows an integrator created with a switched capacitor.

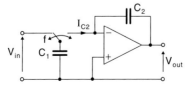

Fig. 7.140. Integrator with a switched capacitor

The capacitor C_1 charges and discharges at a frequency f. The average integration current is proportional to the value of f. Thus the integrator time constant T can be controlled by the frequency f.

The output voltage is:

$$V_{\text{out}} = -\frac{1}{C_2}\int \bar{i}_{C2}\,dt$$

With $\bar{i}_{C2} = f \cdot Q = f \cdot V_{\text{in}} \cdot C_1$ it follows that

$$F(s) = -\frac{C_1}{C_2} f \cdot \frac{1}{s} \tag{7.168}$$

The integrator time constant is: $T = \dfrac{C_2}{f \cdot C_1}$

If a universal filter is built using such integrators, then the critical frequency as well as the filter characteristics can be controlled and determined by using various switching frequencies for the individual integrators.

7.8 Oscillators

Systems with feedback can only oscillate, provided the loop gain is

$$\beta(j\omega) \cdot A(j\omega) \geq 1 \quad \textbf{(Barkhausen criterion)} \tag{7.169}$$

(see also Sect. 7.5.4). An independent oscillation occurs exactly at the frequency at which the phase shift for the feedback loop is

$$(\varphi_\beta + \varphi_A) = 0, 2\pi, 4\pi, \ldots \quad \textbf{(phase criterion)} \tag{7.170}$$

and the gain for the feedback loop is

$$|\beta| \cdot |A| \geq 1 \quad \textbf{(gain criterion)} \tag{7.171}$$

Fig. 7.141. System with positive feedback

- An independent oscillation occurs in systems with feedback, if the phase shift is 0° or integer multiples of 360° and the loop gain is greater than 1.

- Oscillators oscillate with exponentially growing amplitude if the phase criterion is met and the loop gain is *larger* than 1. The amplitude remains constant when the loop gain is *equal* to 1.

Frequency-selective networks are used in the feedback arm if oscillations are to be produced at a definite frequency value. This means that the phase-shift criterion in particular is valid at *only one* frequency. Such frequency-selective networks are usually RC stages, resonant circuits or quartz crystals.

EXAMPLE: The feedback loop gain of the circuit in Fig. 7.142 is:

$$\beta \cdot A = \frac{V_1}{V_{\text{out}}} \frac{V_{\text{out}}}{V_1} = \underbrace{\frac{R_1}{R_1 + R + \mathrm{j}(\omega L - 1/\omega C)}}_{\beta} \cdot A \qquad (7.172)$$

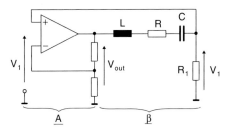

Fig. 7.142. Simple oscillator with series resonant circuit

The *phase-shift criterion* is fulfilled when the imaginary part of β is equal to zero. This is true at the resonant frequency $\omega_0 = \sqrt{1/LC}$. The *feedback-loop gain criterion* is fulfilled when $A \cdot R_1/(R_1 + R) \geq 1$, i.e. the noninverting amplifier gain is chosen to be greater than $(R + R_1)/R_1$.

Figure 7.143 shows the oscillator from Fig. 7.142 with amplitude stabilisation. The JFET increases its impedance with increasing amplitude and thus reduces the noninverting amplifier gain until the loop gain is 1.

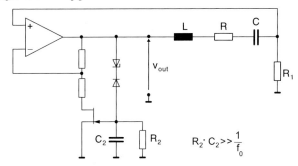

Fig. 7.143. Oscillator with amplitude stabilisation

7.8.1 RC Oscillators

7.8.1.1 Phase-Shift Oscillator

The phase-shift circuit β shifts the phase by at most 270°. The phase shift criterion is fulfilled at $\varphi_\beta = -180°$, as a further 180° shift occurs at the inverting amplifier (Fig. 7.144).

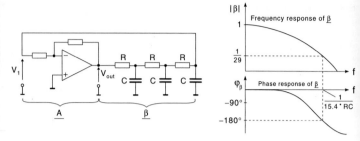

Fig. 7.144. Phase-shift oscillator

The phase-shift criterion is fulfilled by

$$f_0 = \frac{1}{2\pi RC\sqrt{6}} = \frac{1}{15.4 \cdot RC} \qquad (7.173)$$

The feedback loop gain criterion is fulfilled by

$$|A| \geq 29 \qquad (7.174)$$

(Bode plot in Fig. 7.144).

7.8.1.2 Wien Bridge Oscillator

The Wien bridge oscillator feedback circuit is a Wien bandpass filter. The feedback loop transfer function is:

$$\beta \cdot A = \frac{1}{3 + j(\omega RC - 1/\omega RC)} \cdot A \qquad (7.175)$$

The *phase-shift criterion* is fulfilled when the imaginary part of the loop gain is zero, i.e. at $\omega = 1/RC$. The feedback loop gain is then one-third. The *feedback-loop gain criterion* is fulfilled when $A \geq 3$.

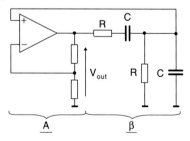

Fig. 7.145. Wien bridge oscillator

Amplitude stabilisation of the Wien bridge oscillator is possible (Fig. 7.143).

7.8.2 LC Tuned Oscillators

LC oscillators use tuned circuits for frequency selection. These can either be series or parallel tuned circuits. They are more stable than RC oscillators, as the phase shift is very large in the range of resonance.

7.8.2.1 Meissner Oscillator

The loop gain is:

$$\beta \cdot A = \frac{1}{1/R_P + j(\omega C - 1/\omega L)} \cdot \frac{N_1}{N_2} \frac{\beta_{AC}}{r_{BE}} \qquad (7.176)$$

The resistor R_P represents the tuned circuit damping. The equivalent resistance r_{BE} represents the input impedance of the common-emitter circuit and can be assumed to be given by $r_{BE} \approx V_T/I_B \approx V_T \cdot (\beta_{DC}/I_C)$ where V_T is the thermal voltage (approximately 25mV), and I_B is the DC base current (Fig. 7.146).

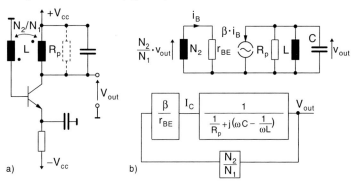

Fig. 7.146. Meissner oscillator in a common-emitter circuit: **a** circuit, **b** equivalent circuit, **c** block diagram

The *phase-shift criterion* is fulfilled when $\omega = 1/\sqrt{LC}$, i.e. at the tuned-circuit resonant frequency.

The *feedback-loop gain criterion* is fulfilled when $\dfrac{N_1}{N_2} \geq \dfrac{r_{BE}}{\beta_{AC} R_P}$.

- The transformer ratio can be chosen to be very small as in practice the base winding only requires a few turns.

7.8.2.2 Hartley Oscillator

The Hartley oscillator uses an inductive voltage divider for feedback. The *phase-shift criterion* is fulfilled at the tuned circuit resonant frequency. The *feedback-loop gain criterion* is approximately fulfilled for the circuit shown in Fig. 7.147a for $L_1 \geq \dfrac{L r_{BE}}{\beta_{AC} R_P}$, or expressed in terms of the turns: $N_1 \geq N \sqrt{\dfrac{r_{BE}}{\beta_{AC} R_P}}$.

- In practice, one or a small number of turns are sufficient for N_1 (r_{BE}, β_{AC} and R_P, see Sect. 7.8.2.1).

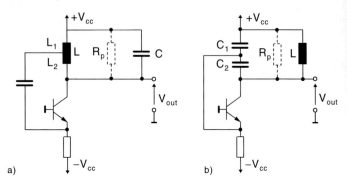

Fig. 7.147. a Hartley oscillator; b Colpitts oscillator

7.8.2.3 Colpitts Oscillator

The Colpitts oscillator uses a capacitive voltage divider for feedback. The *phase-shift criterion* is fulfilled at the resonant frequency for the tuned circuit. The *feedback-loop gain criterion* is approximately fulfilled for the circuit shown in Fig. 7.147b for $C_1 \leq C_2 \dfrac{\beta_{AC} R_P}{r_{BE}}$.

- C_1 is chosen to be very large with respect to C_2 (r_{BE}, β_{AC} and R_P, see Sect. 7.8.2.1).

7.8.3 Quartz/Crystal Oscillators

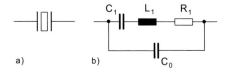

Fig. 7.148. Quartz: a symbol; b equivalent circuit

A quartz is a crystal; it is also an electrically excitable, mechanical device capable of oscillating. It can be represented by an electrical equivalent circuit. The L_1, C_1 and R_1 values are the electrical equivalent values for the mechanical oscillator. The capacitor C_0 represents the capacitance between the electrical terminals of the quartz; its value depends on the circuit layout. Typical values, for example, for a 1 MHz quartz are: $L_1 = 2.53$ H, $C_1 = 0.01$ pF, $R_1 = 50\,\Omega$, $C_0 = 5$ pF.

It is worth noting in particular that such oscillators can achieve Q-factors that are not realisable with electrical circuits. Values between 10^6 and 10^{10} can be achieved.

The equivalent circuit shown in Fig. 7.148b has both a series and a parallel resonance frequency.

The **series resonance** occurs at $\quad \omega_{0S} = \frac{1}{\sqrt{L_1 C_1}}$

The **parallel resonance** occurs at $\quad \omega_{0P} = \frac{1}{\sqrt{L_1 C_1}} \sqrt{1 + \frac{C_1}{C_0}}$

- The series resonance depends only on the mechanical properties of the quartz, which can be very accurate as quartz can be manufactured very precisely. The frequency stability lies between $\Delta f / f_0 = 10^{-4} \ldots 10^{-10}$.

- The series and parallel resonances are very close to each other, as C_0 is much greater than C_1 (Fig. 7.149).

If a capacitor C_S is connected in series with the quartz, then a resonant frequency exists between the series resonance ω_{0S} and the parallel resonance ω_{0P}:

$$\omega_0 = \omega_{0S} \sqrt{1 + \frac{C_1}{C_0 + C_S}} \qquad (7.177)$$

The capacitor C_S permits a high-precision tuning of the oscillation frequency.

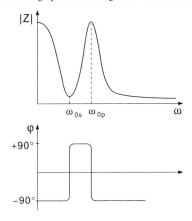

Fig. 7.149. Frequency response of the quartz impedance

7.8.3.1 Pierce Oscillator

Figure 7.150a shows a Pierce oscillator with a CMOS inverter as a driver. This circuit is usually used for CMOS microprocessor clock generation. The oscillator works like the Colpitts oscillator. The current i_1 flows in a closed loop, which is formed by the quartz crystal and the capacitors C_1 and C_2. Because of this, the voltages across C_1 and C_2 are in antiphase. Then $v_{\text{out}} \approx -v_{\text{in}}$. The quartz crystal has a large inductive impedance in this configuration and oscillates close to its parallel resonance. The phase-shift criterion is

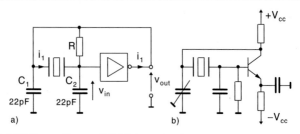

Fig. 7.150. Pierce oscillator: **a** with CMOS inverter; **b** with common-emitter stage as an amplifier

fulfilled in this case, as a further 180° is supplied by the inverter. The feedback loop gain criterion is fulfilled by the very high gain of the inverter. The amplitude of oscillation is limited by the supply voltage, so that the output voltage is approximately a square wave. The resistor R ensures that the circuit will start oscillating by initially charging C_2. It can have a very large value (10 MΩ).

7.8.3.2 Quartz Oscillator with TTL Gates

Here the quartz crystal oscillates at its series resonance frequency. The TTL gates are used as linear amplifiers (Fig. 7.151). The phase-shift criterion is fulfilled if the Quartz impedance is real. The loop gain is greatest when the quartz impedance is at its minimum (feedback-loop gain criterion).

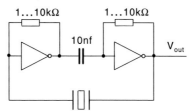

Fig. 7.151. Quartz oscillator with TTL gates

7.8.4 Multivibrators

Multivibrators are self-oscillating digital circuits (Fig. 7.152). The feedback contains a time delay that determines the oscillation frequency. It is unusual to describe multivibrators using the criterion for oscillation.

7.9 Heating and Cooling

Electronic circuits produce power losses, which must be given off as heat into the local environment. The power dissipation is usually given, for example, in the choice of the operating point for a transistor ($P_V = V_{CE}, I_C$). The temperature on the component depends

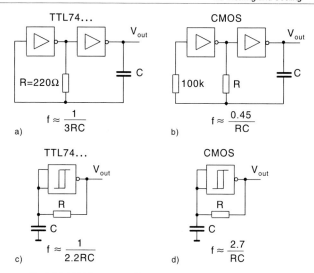

Fig. 7.152. Multivibrators: **a** and **b** with inverters; **c** and **d** with Schmitt triggers

on the geometry of its construction, on the heat-conducting material and the air flow. A large surface area and a good air flow facilitate the heat dissipation. A component with a small surface area and bad air flow will reach a higher temperature.

A **heatsink** and a **fan** are suitable means to keep the component temperature low. A **thermal paste** or **compound** used between the mounting area of the semiconductor and the heatsink, improves the heat conductivity.

7.9.1 Reliability and Lifetime

In electronics the reliability of a component is its ability to function without failure over an acceptable time span. In order to quantify this property the **failure rate** λ is defined:

$$\lambda = \frac{\text{failures}}{\text{total number} \cdot \text{time}} = \frac{\Delta N}{N \cdot \Delta t}$$

ΔN: Number of failures;
N: Number of components;
Δt: Test time.

The **failure rate** defines the average number of failures for usage of the component and over time. The failure rate is measured in **fit (failure in time)**, 1 fit $= 10^{-9} \frac{1}{\text{h}}$.

In addition to the failure rate, the **mean time between failures (MTBF)** is defined:

$$T_{\text{m}} = \frac{1}{\lambda}$$

The MTBF defines the average amount of time before failure. For a large number of equal test components, then after a time T_m probably 63% of the components will have failed.

For a group of n_i electronic devices, containing i different devices and for a device failure rate of λ_i, there is an overall failure rate λ_total and an average MTBF $T_\mathrm{m\,tot}$ of the group of:

$$\lambda_\mathrm{total} = \sum_i n_i \cdot \lambda_i, \quad \text{and} \quad T_\mathrm{m\,total} = \frac{1}{\lambda_\mathrm{total}}$$

The reliability, i.e. the failure rate and the lifetime, of electronic components is mainly dependent on the temperature. The **Arrhenius law** defines this relationship. The failure rate λ is:

$$\lambda = \frac{\text{failures}}{\text{total number} \cdot \text{time}} = \frac{\mathrm{d}N}{N \cdot \mathrm{d}t} = \mathrm{e}^{-\frac{V_\mathrm{a}}{kT}} \tag{7.178}$$

with N: Number of components;
 V_a: Activation energy (eV), 1 eV$= 1.602 \cdot 10^{-19}$ J;
 k: Boltzmann's constant, $1.38 \cdot 10^{-23}$ J/K;
 T: Absolute temperature.

- The **reliability** and **lifetime** of an electronic circuit is mainly dependent on the temperature of the components. The failure rate increases exponentially with the temperature.

The activation energy lies between 0.3 and 1.3 eV, with a typical value of 0.5 eV.

If for a temperature T_1 the failure rate λ_1 is known, then the failure rate for a temperature T_2 is:

$$\lambda_2 = \lambda_1 \mathrm{e}^{-\frac{V_\mathrm{a}}{k}\left(\frac{1}{T_1} - \frac{1}{T_2}\right)} \tag{7.179}$$

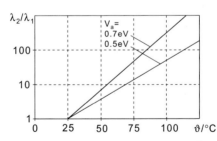

Fig. 7.153. Typical increase in the failure rate as a function of the temperature for two different activation energies V_a

Figure 7.153 shows a typical relationship between the failure rate and the device temperature. The aim of a heatsink calculation therefore should not be to stay directly below the temperature limits given in the data sheets, but rather to keep the temperature as low as possible – in an economical manner.

- The temperature of an electronic circuit should be kept as low as possible – taking financial restraints into account.

7.9.2 Temperature Calculation

Heat dissipation can be modelled and calculated using an electrical equivalent circuit.

7.9.2.1 Thermal Resistance

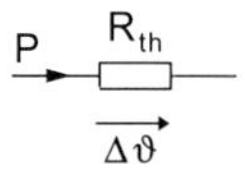

Fig. 7.154. The thermal resistance

A **heat power** P, travelling through a spatial path, causes a **temperature difference** $\Delta\vartheta$. Such a spatial path could be, for example, the path from a p–n junction, where the heat power occurs, to the surrounding environment where the heat is given off. The geometry, material qualities and the air flow of the heat path define the temperature difference.

Thermal resistance R_{th} is defined in a similar manner to electrical resistance using Ohm's law. The thermal resistance replaces the electrical resistance, the heat power P replaces the electrical current and the temperature difference $\Delta\vartheta$ replaces the voltage drop.

Ohm's law for heat conduction is given by:

$$\Delta\vartheta = R_{th} \cdot P \tag{7.180}$$

- The **thermal resistance** R_{th} is given in $\dfrac{K}{W}$ (degrees kelvin per watt).

The thermal resistances R_{th} for individual heat junctions are given in the corresponding data sheets. Thus, for example, the thermal resistance R_{thJC} (JC: junction-case) is given in the transistor data sheets or the thermal resistance of a heatsink is given in the specifications of a heatsink manufacturer. The thermal resistance of a heatsink is given if necessary for forced and natural convection. Figure 7.155 shows the relative change in thermal resistance with forced convection as a function of the air-flow rate.

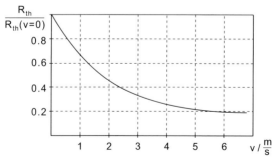

Fig. 7.155. Relative change in thermal resistance with forced convection as a function of the air-flow rate

EXAMPLE: The power dissipation in the p–n junction is passed on from there to the transistor housing, from there via the insulation (e.g. a mica wafer or an aluminium oxide wafer) to the heatsink and from there to the surrounding environment.

Each of these heat junctions has a thermal resistance. Thus the thermal resistance between the p–n junction and the housing R_{thJC}, (JC: junction-case), the thermal resistance of the insulation R_{thINS} and the thermal resistance between the heatsink and the surrounding environment R_{thHS} (Fig. 7.156). The electrical equivalent circuit for these heat junctions in the static case is shown in Fig. 7.157.

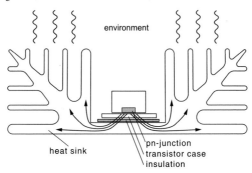

Fig. 7.156. Heat transfer for a transistor p–n junction to the environment

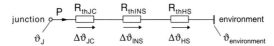

Fig. 7.157. Equivalent circuit for static heat junctions used for the construction of the previous diagram

The junction temperature ϑ_J is:

$$\vartheta_J = \Delta\vartheta_{JC} + \Delta\vartheta_{INS} + \Delta\vartheta_{HS} + \vartheta_{env} = P(R_{thJC} + R_{thINS} + R_{thHS}) + \vartheta_{env} \tag{7.181}$$

7.9.2.2 Thermal Capacity

In addition to the heat conductivity of the material carrying the heat, the **thermal capacity** must also be taken into consideration (Fig. 7.158). This can absorb heat energy. Thus a device will not heat up in an abrupt fashion, but rather will heat up slowly depending on the thermal capacity and amount of heat power.

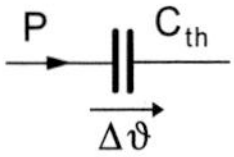

Fig. 7.158. The thermal capacity

The relationship between power and temperature across the thermal capacity corresponds to the analogy of the electrical capacitance:

$$\boxed{P = C_{th} \cdot \frac{d\vartheta}{dt}, \quad \text{or} \quad \Delta\vartheta = \frac{1}{C_{th}} \int P \, dt + \Delta\vartheta_0} \tag{7.182}$$

The **thermal capacity** C_{th} is given in Ws/K (watt-seconds per degree kelvin).

It is calculated from the **specific thermal capacity** c_{th} (Ws/kg K) and the mass m of the material.

$$C_{th} = c_{th} \cdot m \qquad (7.183)$$

The specific thermal capacity is

for copper: $\quad c_{thCu} \approx 400 \dfrac{\text{Ws}}{\text{kg} \cdot \text{K}}$

for aluminium: $c_{thAl} \approx 900 \dfrac{\text{Ws}}{\text{kg} \cdot \text{K}}$

The thermal resistance R_{th} and the thermal capacity C_{th} together form the **thermal time constant** τ_{th}. This is given by: $\tau_{th} = R_{th} \cdot C_{th}$. For transistors it lies between a few hundredths to a few seconds, for heatsinks between minutes and hours.

- For pulsating power dissipation the device temperature can be calculated using the average power, if the thermal time constant is large compared with the periodic time of the power pulses.

EXAMPLE: If the thermal capacities are considered in the construction shown in Fig. 7.156, then this yields the equivalent circuit shown in Fig. 7.159. The transistor case and the heatsink have a thermal capacity. The thermal capacity of the insulation was neglected in this equivalent circuit. The thermal resistance and the thermal capacity together form the thermal time constants $\tau_{JC} = R_{thJC} C_{thJC}$ and $\tau_{HS} = R_{thHS} C_{thHS}$.

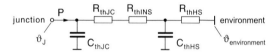

Fig. 7.159. Equivalent circuit for the transient heat junction in the construction of the example in Fig. 7.157

7.9.2.3 Transient Thermal Impedance

Semiconductors can support very large power dissipation for a short time. For impulsive power dissipation the thermal capacities in the vicinity of the junction store the dissipated energy.

For very high frequency power pulses calculations can be performed using the average power. If the power pulses' periodic time is in the range of the thermal time constants, the transient thermal responses are not negligible in the calculation of the junction temperature. The semiconductor manufacturer therefore gives the transient thermal impedance Z_{th}.

The **transient thermal impedance** Z_{th} is given as a function of the impulse duration and the duty cycle D (impulse duration/impulse repetition period, Fig. 7.160).

The temperature difference between the p–n junction and case can be then calculated as:

$$\Delta \vartheta_{JC} = \hat{P} \cdot Z_{thJC}(t_P, T) \qquad (7.184)$$

- The temperature difference is calculated using the amplitude of the dissipated power. The power pulses and their duty cycle are considered in the different curves in the transient thermal impedance diagram with $D = t_p / T$.

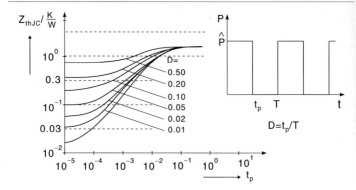

Fig. 7.160. Transient thermal impedance

- The transient thermal resistance is important in the frequency range from a few hertz to a few kilohertz (especially rectifiers and thyristors at 50/60 Hz mains). For higher frequency power pulses the average power and the thermal resistance are usually used for calculations: $\Delta\vartheta = P_{\mathrm{avg.}} \cdot R_{\mathrm{th}}$.

7.10 Power Amplifiers

Power amplifiers offer a large power output with a reasonably good efficiency. The output can usually be regarded as a voltage source with a low source resistance. The gain is usually about 1. Good linearity is obtained as the power amplifier is operated in a feedback system with a large open loop gain.

7.10.1 Emitter Follower

Fig. 7.161 shows the **emitter follower**.

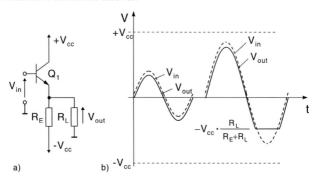

Fig. 7.161. The emitter follower: **a** circuit; **b** voltage waveform

Gain

The gain $A = V_{\text{out}}/V_{\text{in}}$ is:

$$\boxed{A \approx 1} \tag{7.185}$$

Input and Output Impedance

The input impedance r_{in} is:

$$\boxed{r_{\text{in}} \approx \beta \cdot (R_E \| R_L)} \qquad \text{where } \beta \text{ is the small-signal current gain} \tag{7.186}$$

The output impedance r_{out} is:

$$\boxed{r_{\text{out}} \approx \frac{R_{\text{int}} + r_{\text{BE}}}{\beta}} \tag{7.187}$$

with R_{int}: internal resistance of the input voltage source
r_{BE}: dynamic input resistance of the base–emitter junction
$r_{\text{BE}} = h_{11E} \approx \dfrac{V_T}{I_B}$, V_T: thermal voltage 25 mV at $T = 25°C$
I_B: DC base current

Operating Limits

Positive operating limit:

$$\hat{V}_{\text{out max}} \approx +V_{\text{CC}}$$

Negative operating limit:

$$\boxed{\hat{V}_{\text{out min}} \approx -V_{\text{CC}} \cdot \frac{R_L}{R_E + R_L}} \tag{7.188}$$

Maximum Output Power

The maximum available output power is calculated for the case where the peak value of the output voltage is equal to the negative output voltage limit, i.e. that a sinusoidal output is still just about possible.

$$\hat{V}_{\text{out}} = \hat{V}_{\text{out min}} \approx V_{\text{CC}} \cdot \frac{R_L}{(R_E + R_L)}, \qquad P_{\text{out}} = \frac{1}{2} \cdot \frac{\hat{V}_{\text{out}}^2}{R_L} = \frac{1}{2} \cdot V_{\text{CC}}^2 \cdot \frac{R_L^2}{(R_E + R_L)^2 R_L}$$

The derivative dP_{out}/dR_L is calculated and made equal to zero, to discover which load resistance R_L causes the maximum power transfer.

$$\frac{dP_{\text{out}}}{dR_L} = \frac{1}{2} \cdot V_{\text{CC}}^2 \cdot \frac{(R_E + R_L)^2 - R_L \cdot 2 \cdot (R_E + R_L)}{(R_E + R_L)^4} = 0$$

It follows that:

$$R_L = R_E \tag{7.189}$$

The maximum output voltage for $R_L = R_E$ is:

$$\hat{V}_{out} = \frac{V_{CC}}{2} \tag{7.190}$$

- The maximum power is delivered to the load resistance R_L, when R_L is equal to the emitter resistance R_E and the amplitude of the output voltage is $V_{out} = V_{CC}/2$.

The Transistor Power Dissipation

The transistor power dissipation P_{T1} for a sinusoidal output voltage is given by:

$$P_{T1} = \frac{1}{T} \int_0^T v_{CE}(t) \cdot i_C(t) \, dt$$

$$= \frac{1}{T} \int_0^T \left(V_{CC} - \hat{V}_{out} \sin \omega t \right) \cdot \left(\frac{V_{CC} + \hat{V}_{out} \sin \omega t}{R_E} + \frac{\hat{V}_{out} \sin \omega t}{R_L} \right) dt$$

$$P_{T1} = \frac{V_{CC}^2}{R_E} - \frac{1}{2} \frac{\hat{V}_{out}^2}{R_L} - \frac{1}{2} \frac{\hat{V}_{out}^2}{R_E} \tag{7.191}$$

$$P_{T1\,max} = \frac{V_{CC}^2}{R_E} \tag{7.192}$$

- The transistor power dissipation is a maximum when the output is zero.

- The maximum transistor power dissipation is given by $\frac{V_{CC}^2}{R_E}$

The Total Input Power

The total input power is given by:

$$P_{tot} = P_{out} + P_{T1} + P_{RE} \tag{7.193}$$

P_{out}: output power
P_{T1}: transistor power dissipation
P_{RE}: power loss in the emitter resistor R_E

$$P_{out} = \frac{1}{2} \cdot \frac{\hat{V}_{out}^2}{R_L}, \qquad P_{RE} = \frac{1}{T} \int_0^T \frac{v_{RE}^2}{R_E} dt = \frac{1}{T} \int_0^T \frac{(V_{CC} + \hat{V}_{out} \sin \omega t)^2}{R_E} dt = \frac{V_{CC}^2}{R_E} + \frac{1}{2} \frac{\hat{V}_{out}^2}{R_E}$$

$$P_{T1} = \frac{V_{CC}^2}{R_E} - \frac{1}{2} \frac{\hat{V}_{out}^2}{R_L} - \frac{1}{2} \frac{\hat{V}_{out}^2}{R_E}$$

$$\boxed{P_{\text{out}} + P_{T1} + P_{R_E} = P_{\text{total}} = 2\frac{V_{CC}^2}{R_E}} \qquad (7.194)$$

- The total input power P_{total} of the emitter follower is $2\dfrac{V_{CC}^2}{R_E}$ and is *independent* of the load R_L and *independent* of the output voltage V_{out}.

Efficiency

The efficiency is defined as

$$\eta = \frac{\text{delivered power}}{\text{total power}} = \frac{\text{output power}}{\text{input power}} = \frac{P_{\text{out}}}{P_{\text{total}}}$$

The efficiency η reaches its maximum value when the output power is highest, i.e. if $R_E = R_L$ *and* $V_{\text{out}} = V_{CC}/2$.
The efficiency is:

$$\boxed{\eta_{\max} = \frac{P_{\text{out max}}}{P_{\text{total}}} = \frac{V_{CC}^2/8R_L}{2V_{CC}^2/R_E} = \frac{1}{16} = 6.25\%} \qquad (7.195)$$

- The maximum efficiency of the emitter follower is 6.25%.

Class A Operation

Class A operation of an amplifier is defined by:
- the total input power is constant and independent of the load and the output voltage, and
- the transistor current is never zero.

- The emitter follower is an amplifier in class A operation.

7.10.2 Complementary Emitter Follower in Class B Operation

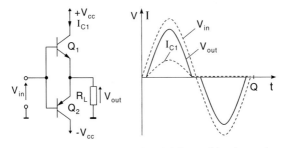

Fig. 7.162. Complementary emitter follower: **a** circuit; **b** diagram of the voltages and currents

In the complementary emitter follower only one transistor conducts at any one time. For a positive input voltage transistor Q_1 conducts, and for a negative input voltage Q_2 conducts. Neither transistor conducts at the zero-crossing point of the input voltage (-0.7 V $< V_{in} <$ $+0.7$ V), the gain in this case being approximately zero. As the transfer characteristic is nonlinear in this region, this is known as **crossover distortion**.

Gain

The gain is:

$$A \approx 1 \tag{7.196}$$

Output Voltage Limit

The output voltage limit is:

$$\hat{V}_{out} \approx \pm V_{CC} \tag{7.197}$$

Input and Output Impedance

The input impedance r_{in} is:

$$r_{in} = \beta \cdot R_L, \quad \text{where } \beta \text{ is the small-signal current gain} \tag{7.198}$$

The output impedance r_{out} is:

$$r_{out} \approx \frac{R_{int} + r_{BE}}{\beta} \tag{7.199}$$

with R_{int}: source resistance of the input voltage source
r_{BE}: dynamic input resistance of the base–emitter junction
$r_{BE} \approx \dfrac{V_T}{I_B}$, V_T: thermal voltage 25 mV at $T = 25°C$
I_B: DC base current

Maximum Output Power

The maximum output power is:

$$P_{out} = \frac{1}{2} \frac{\hat{V}_{out}^2}{R_L} \approx \frac{1}{2} \frac{V_{CC}^2}{R_L} \tag{7.200}$$

Transistor Power Dissipation

The transistor power dissipation for a sinusoidal output voltage per transistor is:

$$P_{T1} = P_{T2} = \frac{1}{T} \int_0^{T/2} \underbrace{\left(V_{CC} - \hat{V}_{out} \sin \omega t\right)}_{V_{CE1}} \cdot \underbrace{\left(\frac{\hat{V}_{out} \sin \omega t}{R_L}\right)}_{I_{C1}} dt = \frac{V_{CC} \cdot \hat{V}_{out}}{\pi \cdot R_L} - \frac{\hat{V}_{out}^2}{4 R_L}$$

In the calculation of the maximum transistor power dissipation the derivative $\dfrac{d P_T}{d \hat{V}_{out}}$ is set to zero. This yields the output voltage at which the maximum transistor power dissipation occurs:

$$\frac{d P_T}{d \hat{V}_{out}} = \frac{V_{CC}}{\pi \cdot R_L} - 2 \frac{\hat{V}_{out}}{4 R_L} = 0 \quad \Rightarrow \quad \hat{V}_{out} = \frac{2}{\pi} V_{CC} = 0.64 \cdot V_{CC}$$

- The maximum transistor power dissipation occurs at an output voltage swing of 64% of the supply voltage (Fig. 7.163).

The maximum transistor power dissipation per transistor is:

$$\boxed{P_{T1\,\max} = P_{T2\,\max} = \frac{V_{CC}^2}{\pi^2 R_L}} \tag{7.201}$$

Input Power

The total input power is given by (Fig. 7.163)

$$P_{\text{total}} = P_{\text{out}} + P_{T1} + P_{T2} \quad \Rightarrow \quad \boxed{P_{\text{total}} = \frac{2 V_{CC} \hat{V}_{out}}{\pi R_L}}$$

and has its maximum in $\hat{V}_{out} = V_{CC}$

Efficiency

$$\eta = \frac{P_{\text{out}}}{P_{\text{total}}} = \frac{\dfrac{1}{2} \dfrac{\hat{V}_{out}^2}{R_L}}{\dfrac{2 V_{CC} \cdot \hat{V}_{out}}{\pi \cdot R_L} - \dfrac{1}{2} \dfrac{\hat{V}_{out}^2}{R_L} + \dfrac{1}{2} \dfrac{\hat{V}_{out}^2}{R_L}} = \frac{\hat{V}_{out}}{V_{CC}} \cdot \frac{\pi}{4} = 0.785 \cdot \frac{\hat{V}_{out}}{V_{CC}}$$

η has its maximum for $\hat{V}_{out} = V_{CC}$:

$$\boxed{\eta_{\max} = 78.5\%} \tag{7.202}$$

Class B operation

Class B operation of an amplifier is defined by:

- the total input power increases proportionally to $\hat{V}_{out}$, and
- each transistor conducts only for half a period.
- The complimentary emitter follower is a class B amplifier.

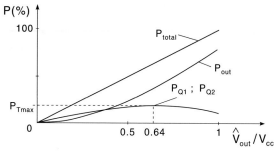

Fig. 7.163. Output power, power dissipated and total input power as function of the level of output voltage swing

7.10.3 Complementary Emitter Follower in Class C Operation

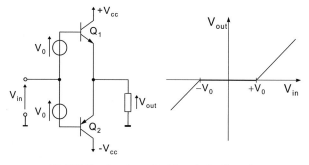

Fig. 7.164. Complementary emitter follower in class C operation

In class C operation the complementary transistors do not conduct in the range of $-V_0 < V_{in} < +V_0$. This causes the amplifier efficiency to improve compared to the class B operation. This is important if the amplifier is activated by a constant amplitude and where the crossover distortion is unimportant, e.g. radio transmitter amplifiers.

7.10.4 The Characteristic Curves of the Operation Classes

Fig. 7.165a shows the operational point in the transistor output characteristics for different operation classes of power amplifiers. Fig. 7.165b shows the corresponding transistor current.

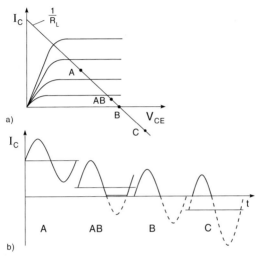

Fig. 7.165. a The characteristic curves of the operation classes (for the transistor T_1); **b** time variation of the collector current with a sinusoidal activation

7.10.5 Complementary Emitter Follower in Class AB Operation

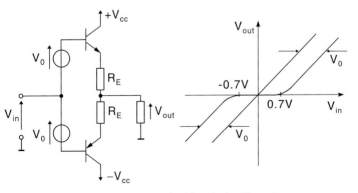

Fig. 7.166. Complementary emitter follower in class AB operation

In class AB operation equal-value bias voltages V_0 are applied to the complementary transistors. V_0 is chosen so that a small quiescent current flows in the transistors at the zero-crossing of the input voltage. The transfer characteristic is thus linearised, and the crossover distortion is reduced (Fig. 7.166). The small quiescent current is chosen so that the heat dissipation in the transistors for no input voltage is small (the power dissipated should be about 10–30 % of the maximum power dissipation). The small quiescent current usually lies between 1–5 % of the peak output current. The small quiescent current is limited by the feedback resistor R_E. This is important particularly in heating of the transistors and the related drop in the base–emitter voltage. If R_E is too small thermal runaway can occur: The transistors heat up, the base–emitter voltage decreases, the small quiescent current increases, the transistor power dissipation also increases causing the transistors to heat up further, the base–emitter voltage further decreases and so on.

7.10.5.1 Biasing for Class AB Operation

Figure 7.167a: The bias voltage is produced by two diode stages. The resistors R_q are chosen so that at maximum-output voltage swing sufficient base current is supplied to the power transistors. This often means that R_q has to be very low, as the voltage drop across them will be very small at maximum-output voltage swing. This also leads in the quiescent case (when $V_{in} = 0$ V) to a large power loss in the R_q resistors as about $V_{CC}/2$ drops across them. The circuits shown in Fig. 7.167d provides some relief by replacing the R_q resistors by current sources.

The R_E feedback resistors prevent the small quiescent current of the power transistors from increasing uncontrollably. The base–emitter voltage of the power transistors drops in the event of heating occurring!. The R_E resistors are chosen so that the small quiescent current is about 1–5 % of the peak output current or, alternatively, when the amplifier is operating at full-output voltage swing that 0.7–2 V are dropped across them. The feedback resistors can be bypassed with diodes, so that the power dissipated in them at full-output voltage swing is not too high.

Figure 7.167b: The bias voltage diodes are replaced by the transistors Q_3 and Q_4. The amplifier input signal power is thus decreased.

Figure 7.167c: The bias voltage diodes are replaced by a transistor circuit Q_3. The transistor circuit appears like a voltage source. The bias voltage is given by: $2V_0 = 0.7\,\text{V}\dfrac{R_1 + R_2}{R_2}$. The resistors R_1 and R_2 can be a potentiometer for a precise quiescent current adjustment. This is particularly important if the power transistors are Darlingtons. In that case, the bias voltage is chosen as: $2V_0 \approx 2.8$ V.

Figure 7.167d: The R_q resistors are replaced by current sources. In the quiescent case ($V_{in} = 0$ V) the power dissipation is therefore clearly reduced or, alternatively, the maximum amplifier output swing increases. The current source is chosen so that at peak-output voltage swing the current requirement of the power transistors is guaranteed.

Figure 7.167e: The amplifier input comes from a common-emitter stage for high voltage gain. This common-emitter stage can be driven directly by a differential amplifier.

Figure 7.167f: The R_q resistor is replaced by the resistors R_{q1} and R_{q2}, with $R_{q1} \ll R_{q2}$. In the quiescent case a voltage of about V_{CC} drops across the **bootstrap** capacitor C. For the output voltage swing the bootstrap capacitor shifts the positive voltage half-cycle of the voltage between R_{q1} and R_{q2} to values higher than the supply voltage V_{CC}. Therefore sufficient voltage remains across R_{q2}, even at full-output voltage swing, to guarantee the base current requirement. R_{q2} is chosen so that a little more than the maximum required base current flows under quiescent conditions. The bootstrap-capacitor is chosen so that an approximately steady DC voltage appears across it (it is a short circuit for AC voltages). The critical frequency of the bootstrap circuit is: $f_c \approx 1/2\pi R_{q1}C$ for $R_{q1} \ll R_{q2}$.

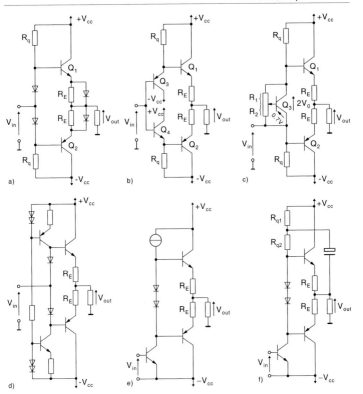

Fig. 7.167. Bias voltage production for AB amplifiers

7.10.5.2 Complementary Emitter Follower with Darlington Transistors

For amplifiers with a large output power or, alternatively, with a large output current, power transistors are created using Darlington or pseudo-Darlington circuits. The transistors Q_1 and Q_2 are power transistors. The transistors Q'_1 and Q'_2 are driver transistors.

Figure 7.168a: The power transistors Q_1 and Q_2 do not conduct in the quiescent state ($V_{in} = 0$ V). The bias voltage V_0 is chosen so that the drop across the feedback resistors R_E is about 0.4 V in the quiescent state (i.e. $V_0 = 2.2$ V). A good linearity is therefore achieved in the crossover region. For larger output voltage swing the power transistors take over the output current.

Figure 7.168b: The bias voltage is chosen at about $V_0 = 2.8$ V. The quiescent current usually lies around 1–5 % of the peak output current.

Figure 7.168c: The pseudo-Darlington circuit uses identical transistor types as the power transistors. The bias voltage V_0 is chosen so that the voltage drop across the feedback resistors R_E in the quiescent state is about 0.4 V (i.e. $V_0 = 1.8$ V).

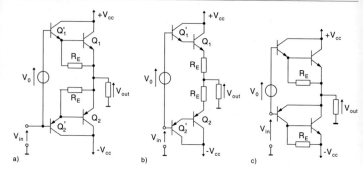

Fig. 7.168. Complementary emitter follower in **a** and **b** Darlington pair, **c** pseudo-Darlington circuit

7.10.5.3 Current-Limiting Complementary Emitter Follower

The current-limiting circuit shown in Fig. 7.169 measures the output current using the resistor R_M (which can be identical to the feedback resistor R_E). If a critical voltage is exceeded, then the base current flows away through the current-limiter circuit, i.e. the output current cannot increase further.

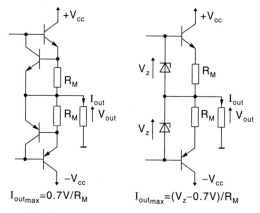

Fig. 7.169. Current-limiting complementary emitter follower

7.10.6 Input Signal Injection to Power Amplifiers

7.10.6.1 Input Signal Injection using a Differential Amplifier

A high linearity and a broad independence from the semiconductor parameters can be achieved by using feedback principles. The differential amplifier uses the difference between the output and the input signals to provide the input signal to a power amplifier. The overall open-loop gain is given by the product of the differential amplifier gain A_1 and the common-emitter stage gain A_2. If the open-loop gain is very large, then the closed-loop gain depends only on the feedback (Fig. 7.170).

The feedback also makes the amplifier input impedance very large and the output impedance very small.

For the **AC voltage gain** the feedback acts like a voltage divider $\dfrac{R_2}{R_1 + R_2}$, and C_2 acts like a short circuit. The gain is therefore:

$$V_\sim = \dfrac{V_1 V_2}{1 + V_1 V_2 \dfrac{R_2}{R_1 + R_2}} \approx \dfrac{R_1 + R_2}{R_2} \qquad (7.203)$$

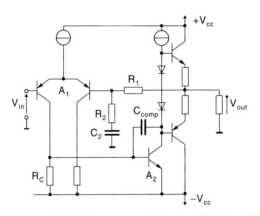

Fig. 7.170. Input signal injection to a power amplifier using a differential amplifier

The **feedback capacitor** C_2 ensures that for DC input voltages the amplifier has complete feedback. This provides a particularly good output voltage zero stability. The DC gain is therefore 1 (Fig. 7.171).

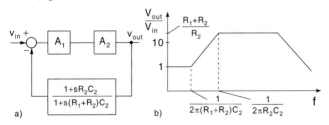

Fig. 7.171. a Block diagram; **b** frequency response of the amplifier with feedback

The **compensation capacitor** C_{comp} decreases the loop gain at high frequencies (see Sect. 7.6.2) in order to reduce the risk of oscillations. C_{comp} should be determined experimentally.

7.10.6.2 Input Signal Injection Using an Op-Amp

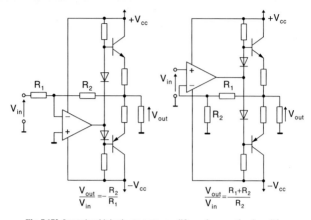

Fig. 7.172. Input signal injection to power amplifiers using operational amplifiers

The power amplifier is included in the op-amp feedback path. The op-amp open-loop gain acts in the feedback loop to produce good linearity.

7.10.7 Switched-Mode Amplifiers

In switched-mode amplifiers the transistors Q_1 and Q_2 work like switches that are alternately switched. The transistor control voltages are created by a pulse-width modulator (PWM). The output voltages of the transistors can only have the values $+V_{CC}$ or $-V_{CC}$. This voltage contains, on the one hand, the pulse-width modulator clock frequency and, on the other hand, the input signal (Fig. 7.173). The clock frequency is suppressed by a second-order LC low-pass filter. Therefore the output signal is a true representation of the input signal.

The switching transistors are usually MOSFETs because of their low losses and short switching time. The switching frequencies are usually in a range of some tens to several hundred kilohertz. In theory the switched mode amplifier is loss free. In practice the efficiency lies between 80 and 90%.

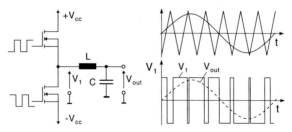

Fig. 7.173. Switched-mode amplifier

7.11 Notation Index

a, a_i	filter coefficient, see Tables 7.2 to 7.5
A	gain
A_{CL}	closed-loop gain of a feedback system
A_d	differential mode gain, open-loop gain of an operational amplifier
A_{OL}	open-loop gain of a feedback system
b, b_i	filter coefficient, see Tables 7.2 to 7.5
B	DC gain of bipolar transistors
B	bandwidth
B	as index: base
C	as index: collector
C	capacitance
C_{th}, c_{th}	thermal capacity (Ws/K), specific thermal capacity (Ws/kg K)
CM	as index: common mode
$CMRR$	common-mode rejection ratio
d	as index: difference
D	damping ratio
D	as index: drain
E	as index: emitter
f	frequency
f_c	critical frequency
f_T	transit frequency
F	transfer function (in the Laplace frequency domain)
g_m	transconductance
G	as index: gate
i	time-varying current, AC current
$i_{s/c}$	short-circuit current
in	as index: input quantity
I	DC current, RMS value of an AC current
I_F	diode forward current
L	inductance
out	as index: output quantity
P	power
r	differential-mode resistance, AC resistance
r_{BE}	differential resistance base–emitter path, V_T/I_B
r_{CE}	differential output resistance of the collector current source
r_{DS}	differential output resistance the drain current source
r_{int}	differential source resistance of a current source
R	resistance
R_{int}	internal resistance of the source, source resistance
R_L	load resistance
R_{th}	thermal resistance (K/W)
s	complex frequency

S	as index: source
S	normalised complex frequency, $S = s/\omega_g$
t	time
T	absolute temperature
T	time constant
T_m	mean time between failure, MTBF
v	time-varying voltage, AC voltage
$v_{o/c}$	open circuit voltage
$v_{o/l}$	open loop voltage
$\hat{V}$	peak magnitude of an AC voltage
V	DC voltage, RMS value of an AC voltage
V_0	offset voltage
V_0, I_0	operating point
ΔV_{BE}	thermal voltage drift of the base–emitter path
V_{CC}	supply voltage
V_F	diode forward voltage
V_T	thermal voltage, about 25 mV at room temperature
$\underline{z}$	differential impedance
$\underline{Z}$	impedance
Z_{th}	transient thermal resistance (K/W)
β	feedback factor, transfer function of the feedback circuit
β	differential mode current gain
β_0	differential mode short-circuit current gain
$\Delta\vartheta$	temperature difference
η	efficiency
φ	phase angle between output and input of a circuit
λ	failure rate
ϑ	temperature
ω	angular frequency
ω_0	resonant angular frequency, centre frequency
ω_c	critical or cutoff angular frequency

7.12 Further Reading

BIRD, J. O.: *Electrical Circuit Theory and Technology*
Butterworth/Heinemann (1999)

BOYLESTAD, R. L.; NASHELSKY, L.: *Electronic Devices and Circuit Theory, 6th Edition*
Prentice Hall (2000)

CRECRAFT, D. I.; GORHAM, D. A.; SPARKES, J. J.: *Electronics*
Chapman & Hall (1993)

FLOYD, T. L.: *Principles of Electric Circuits, 6th Edition*
Prentice Hall (2000)

FLOYD, T. L.: *Electric Circuits Fundamentals, 5th Edition*
Prentice Hall (2001)

FLOYD, T. L.: *Electronic Devices, 5th Edition*
Prentice Hall (1999)

FLOYD, T. L.: *Electronics Fundamentals: Circuits, Devices, and Applications*
Prentice Hall (1997)

GROB, B.: *Basic Electronics, 8th Edition*
McGraw-Hill (1996)

HARPER, C. A.: *Active Electronic Component Handbook, 2nd Edition*
McGraw-Hill (1996)

HOROWITZ, P.; HILL, W.: *The Art of Electronics, 2nd Edition*
Cambridge University Press (1989)

HOROWITZ, P.; HAYES, T. C.: *Student Manual for The Art of Electronics*
Cambridge University Press (1989)

SINGH, J.: *Semiconductor Devices: Basic Principles, 1st Edition*
John Wiley & Sons (2000)

ZVEREV, A. I.: *Handbook of Filter Synthesis*
John Wiley & Sons (1967)

8 Digital Electronics

8.1 Logic Algebra

8.1.1 Logic Variables and Logic Gates

For many signals in electronics only two distinct signal conditions are of interest. For example:

$$\text{current flowing / current not flowing}$$
$$\text{voltage is positive / voltage is negative}$$
$$\text{short circuit / open circuit}$$

A mathematical model for such a system would be a **logic variable**, which can only have two distinct values: usually either zero or one.

$$x = 0, \quad \text{or} \quad x = 1$$

Logic functions can translate a logic variable into a new logic variable. In mathematics systems of logic variables that are related by logic functions are known as Boolean Algebra.

8.1.1.1 Inversion

The **inversion** of a variable x is written as $\bar{x}$.

$$q = \bar{x}$$

The logic variable q assumes the opposite value from x. Any logic function can be represented by a **truth table**.

x	$\bar{x}$
0	1
1	0

The following holds

$$\bar{0} = 1, \quad \bar{1} = 0$$

as well as

$$\bar{\bar{x}} = x$$

- If a logic variable is inverted twice, then it assumes its original value.

8.1.1.2 AND Function

The AND function combines *two* logic variables.

$$q = x \cdot y$$

Spoken as: *x and y*.

The truth table for the AND function is

x	y	$x \cdot y$
0	0	0
0	1	0
1	0	0
1	1	1

Two logic variables can each independently assume either of two values. The four possible combinations are represented in the truth table.

The following then holds for the AND function

$$x \cdot 0 = 0, \qquad x \cdot x = x \tag{8.1}$$
$$x \cdot 1 = x, \qquad x \cdot \overline{x} = 0 \tag{8.2}$$

8.1.1.3 OR Function

$$q = x + y \qquad \text{(not to be confused with the arithmetic 'plus')}$$

Spoken as: *x or y*.

The truth table of the OR function

x	y	$x + y$
0	0	0
0	1	1
1	0	1
1	1	1

The following then holds for the OR function

$$x + 0 = x, \qquad x + x = x \tag{8.3}$$
$$x + 1 = 1, \qquad x + \overline{x} = 1 \tag{8.4}$$

8.1.2 Logic Functions and their Symbols

Logic variables describe electronic signals, while logic functions explain their relationship. The basic elements used to realise these functions are known as **gates**. Special symbols are used as standard for logic gates.

NOTE: The logic symbols used in this chapter are according to EN 60617-12 (formerly IEC 617) and IEEE/ANSI standards. The IEEE standard provides two different types of symbols, *distinctive-shape symbols* and *rectangular-shape symbols*. The first distinguishes the function from the form of the symbol, while the latter consists of a rectangle with a label describing the logic function. In this book the rectangular-shape convention is followed.

8.1.2.1 Inverter (NOT)

$$q = \overline{x}$$

An inverter **inverts** the input signal. The circle at the output side is used to symbolise the inversion (Figs. 8.1 and 8.2).

x	NOT
0	1
1	0

Fig. 8.1. Truth table and symbol for the inverter

- The output of the inverter is 1 only if the input variable has the value of zero.

Fig. 8.2. Distinctive-shape symbol for the inverter

8.1.2.2 AND Gate

$$q = x \cdot y$$

x	y	AND
0	0	0
0	1	0
1	0	0
1	1	1

Fig. 8.3. Truth table and symbol for the AND gate

- The AND function output is 1 only if *both* input variables have the value of 1 (Fig. 8.3), or:

- The AND function output is zero if *at least one* of the input variables has the value of zero.

Fig. 8.4. Distinctive-shape symbol for the AND gate

8.1.2.3 OR Gate

$$q = x + y$$

- The OR function output is 1 if *at least one* of the input variables has the value of 1, or:

x	y	Or
0	0	0
0	1	1
1	0	1
1	1	1

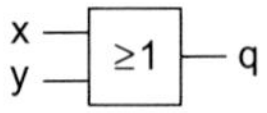

Fig. 8.5. Truth table and symbol of the Or gate

- The Or function output is zero only if *both* input variables have the value of zero (Fig. 8.5).

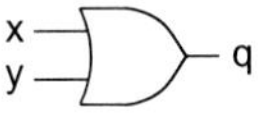

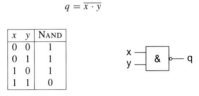

Fig. 8.6. Distinctive-shape symbol for the Or gate

8.1.2.4 Nand Gate

$$q = \overline{x \cdot y}$$

x	y	Nand
0	0	1
0	1	1
1	0	1
1	1	0

Fig. 8.7. Truth table and symbol of the Nand gate

The Nand gate is an And gate followed by an inversion. The circle on the output represents the inversion (Figs. 8.7 and 8.8).

- The Nand function is zero if *both* input variables have the value of 1.

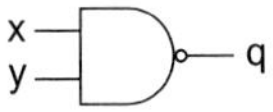

Fig. 8.8. Distinctive-shape symbol of the Nand gate

8.1.2.5 Nor Gate

$$q = \overline{x + y}$$

The Nor gate is an Or gate followed by an inversion. The circle on the output represents the inversion (Figs. 8.9 and 8.10).

- The Nor function is 1 if *both* input variables have the value of zero,

or:

- The Nor function is zero if *at least one* of the input variables has the value of 1.

x	y	NOR
0	0	1
0	1	0
1	0	0
1	1	0

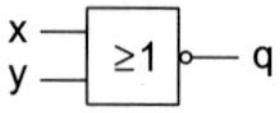

Fig. 8.9. Truth table and symbol for the NOR gate

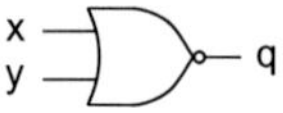

Fig. 8.10. Distinctive-shape symbol of the NOR gate

8.1.2.6 XOR Gate, Exclusive OR

$$q = x \cdot \overline{y} + \overline{x} \cdot y$$

$q = x \oplus y$.

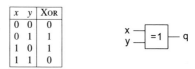

x	y	XOR
0	0	0
0	1	1
1	0	1
1	1	0

Fig. 8.11. Truth table and symbol for the XOR gate

- The XOR function is 1 if *exactly one* of the two input variables has the value of 1,

or:

- The XOR function outputs a 1 only if both input variables are different from each other (Fig. 8.11),

or:

- The XOR function outputs a zero if both input variables are the same.

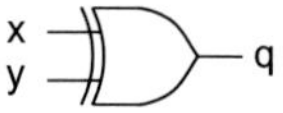

Fig. 8.12. Distinctive-shape symbol of the XOR gate

An XOR gate can also be regarded as a **controlled inverter** (Fig. 8.13). If the second input S is used as the controlling input, then for $S = 0$ the gate is noninverting, and for $S = 1$ it inverts.

8.1.3 Logic Transformations

8.1.3.1 Commutative Laws

$$x \cdot y = y \cdot x , \; x + y = y + x \tag{8.5}$$

Fig. 8.13. XOR gate as a controlled inverter

Variables are interchangeable. In circuit terms: the inputs from AND gates or OR gates can be interchanged.

8.1.3.2 Associative Laws

$$(x \cdot y) \cdot z = x \cdot (y \cdot z) = x \cdot y \cdot z \tag{8.6}$$
$$(x + y) + z = x + (y + z) = x + y + z \tag{8.7}$$

The evaluation of the expressions is the same in each case. In circuit terms: the order of the combination of any two inputs is arbitrary (Fig. 8.14).

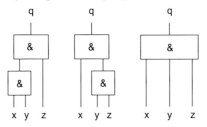

Fig. 8.14. All three circuits are equal according to the associative law; the same is true for OR gates

8.1.3.3 Distributive Laws

$$(x \cdot y) + (x \cdot z) = x \cdot (y + z) \tag{8.8}$$
$$(x + y) \cdot (x + z) = x + (y \cdot z) \tag{8.9}$$

Any variable that is common to two logic expressions can be taken out of the parentheses (Fig. 8.15). there is no equivalent rule in algebra to the one expressed in Eq. 8.9.

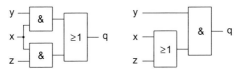

Fig. 8.15. Both circuits are identical according to the distributive law; swapping the AND and OR gates demonstrates an application of the second distributive law

8.1.3.4 Inversion Laws (DeMorgan's Rules)

The inversion laws, also known as DeMorgan's rules, are given by Eqs. 8.10 and 8.11 and are shown in Figs. 8.16 and 8.17.

$$\overline{x} \cdot \overline{y} = \overline{x + y} \tag{8.10}$$
$$\overline{x} + \overline{y} = \overline{x \cdot y} \tag{8.11}$$

Fig. 8.16. The inversion can be shifted from the input to the output; the AND gate is then changed to NOR

Fig. 8.17. The inversion can be shifted from the input to the output; the OR gate is then changed to NAND

Evaluation Rules

The inversion of a variable is always carried out first. All other logic expressions are evaluated from left to right. Any deviation from that order must use suitable parentheses to separate the relevant expressions.

NOTE: When +-signs are used for the OR expression and '·' for the AND expression the same algebra is employed: AND expressions take precedence over OR expressions. In this notation the '·' can be also left out.

EXAMPLE:

$$(x \cdot \overline{y}) + (\overline{x} \cdot y) = x \cdot \overline{y} + \overline{x} \cdot y = x\overline{y} + \overline{x}y$$

8.1.4 Overview: Logic Transformations

EXAMPLE: The following logic term should be simplified using the rules in Table 8.1:

$$\begin{aligned}
q &= \overline{(x \cdot \overline{y}) \cdot (x + y)} & & \text{DeMorgan's rule (14)} \\
&= \overline{x \cdot \overline{y}} + \overline{x + y} & & \text{DeMorgan's rules (14) and (15)} \\
&= (\overline{x} + y) + (\overline{x} \cdot \overline{y}) & & \text{Distributive law (27)} \\
&= \underbrace{[\overline{x} + (\overline{x} \cdot \overline{y})]}_{\overline{x}} + \underbrace{[y + (\overline{x} \cdot \overline{y})]}_{y + \overline{x}} & & \text{From rules (19) and (21)} \\
&= \overline{x} + y + \overline{x} = \overline{x} + y
\end{aligned}$$

Table 8.1. Summary of logic transformations

One variable			
(1)	$\bar{\bar{x}} = x$		
(2)	$x \cdot x = x$	(3)	$x + x = x$
(4)	$x \cdot \bar{x} = 0$	(5)	$x + \bar{x} = 1$
One variable and constants			
(6)	$x \cdot 0 = 0$	(7)	$x + 0 = x$
(8)	$x \cdot 1 = x$	(9)	$x + 1 = 1$
Two variables			
(10)	$x \cdot y = y \cdot x$	(11)	$x + y = y + x$
(12)	$\bar{x} \cdot \bar{y} = \overline{x + y}$	(13)	$\bar{x} + \bar{y} = \overline{x \cdot y}$
(14)	$\overline{x \cdot y} = \bar{x} + \bar{y}$	(15)	$\overline{x + y} = \bar{x} \cdot \bar{y}$
(16)	$\overline{\bar{x} \cdot \bar{y}} = x + y$	(17)	$\overline{\bar{x} + \bar{y}} = x \cdot y$
(18)	$x \cdot (x + y) = x$	(19)	$x + (x \cdot y) = x$
(20)	$x \cdot (\bar{x} + y) = x \cdot y$	(21)	$x + (\bar{x} \cdot y) = x + y$
(22)	$(x \cdot y) + (\bar{x} \cdot y) = y$	(23)	$(x + y) \cdot (\bar{x} + y) = y$
(24)	$(\bar{x} \cdot y) + (x \cdot y) = y$	(25)	$(\bar{x} + y) \cdot (x + y) = y$
Three variables			
(26)	$x \cdot (y + z) =$ $(x \cdot y) + (x \cdot z)$	(27)	$x + (y \cdot z) =$ $(x + y) \cdot (x + z)$

8.1.5 Analysis of Logic Circuits

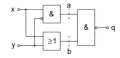

Fig. 8.18. Analysis of logic circuits by segmentation and introduction of variables

To calculate the truth table of a complex logic circuit, the circuit should be broken up at suitable points and the logic value at that point assigned a new variable name. Therefore in the example shown in Fig. 8.18 the temporary variable a is introduced for $x \cdot \bar{y}$. The circle at the input of the AND gate indicates an inversion. The temporary variable b is introduced for $x + y$. The output variable q is the result of a and b passing through a NAND gate. The following truth table shows each of the variables:

x	y	$a = x \cdot \bar{y}$	$b = x + y$	$q = \overline{a \cdot b}$
0	0	0	0	1
0	1	0	1	1
1	0	1	1	0
1	1	0	1	1

The final column of the table shows that the entire circuit expression can be written as $q = \bar{x} + y$. For circuits with several output variables each variable is represented by its own truth table.

8.1.6 Sum of Products and Product of Sums

Solving a problem in digital logic design usually implies using a truth table, which represents the logical relationship between the input and output values. This yields a logical expression for each output variable, and thus the design of the logic circuit. The case where **each input variable** appears in either inverted or noninverted form in each partial term of the output expression is of particular interest.

8.1.6.1 Sum of Products

The **sum of products** (SOP, also canonical sum of products) may be obtained as follows:

- Only rows in the truth table in which the output variable is a logic 1 are considered as partial terms.
- In each of these rows the input variables are operated on by the AND function. A variable in the term is represented by its inverted form if it is 0 in the relevant row, otherwise it is not inverted.
- All partial terms are operated on together by the OR function.

EXAMPLE: In the example the output variables Q and R result from the input variables A, B, C.

A	B	C	Q	R
0	0	0	0	1
0	0	1	1	1
0	1	0	0	1
0	1	1	0	1
1	0	0	1	0
1	0	1	0	1
1	1	0	0	1
1	1	1	1	0

The output variable Q in the truth table has the value of 1 in three cases. This yields the following partial terms:

$\overline{A} \cdot \overline{B} \cdot C$ from the second row,
$A \cdot \overline{B} \cdot \overline{C}$ from the fifth row,
$A \cdot B \cdot C$ from the last row.

The output variable Q is then given by:

$$Q = (\overline{A} \cdot \overline{B} \cdot C) + (A \cdot \overline{B} \cdot \overline{C}) + (A \cdot B \cdot C)$$

The sum of products yields a two-layer combinational circuit as shown in Fig. 8.19.

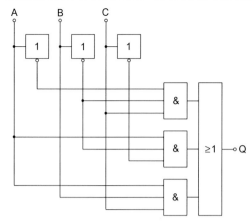

Fig. 8.19. Two-layer combinational circuit using the sum of products

8.1.6.2 Product of Sums

The **product of sums** (POS, also canonical product of sums) may be obtained as follows:

- Only the rows in the truth table in which the output variable is a logic 0, are considered for the partial terms.
- In each of these rows the input variables are operated on together by the OR function. A variable in the partial term is inverted if it is 1 in the relevant row, otherwise it is not inverted.
- All partial terms are operated on together by the AND function.

EXAMPLE: The output variable R in the previous truth table has the value 0 in two cases. This yields the following partial terms:

$$\overline{A} + B + C \quad \text{from the fifth row,}$$
$$\overline{A} + \overline{B} + \overline{C} \quad \text{from the last row.}$$

The output variable R is given by:

$$R = (\overline{A} + B + C) \cdot (\overline{A} + \overline{B} + \overline{C})$$

The product of sums yields a two-layer combinational circuit as shown in Fig. 8.20.

- Both POS and SOP solutions may still contain redundancies, i.e. they can be further simplified. The sum of products yields short expressions for variables that have the value 1 in a few cases. For the opposite case the product of sums yields the more compact solution.

NOTE: The sum of products is preferred in TTL design. The product of sums is preferred in the realisation of logic functions using programmable logic devices (PLDs).

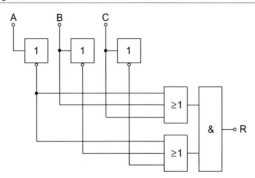

Fig. 8.20. Two-layer combinational circuit using the product of sums

8.1.7 Systematic Reduction of a Logic Function

Both of the following techniques are methods to find the reduced logic functions of a given truth table. Aim: to have the lowest possible number of logic gates in the electronic realisation.

- *Karnaugh map:* Graphic technique limited to a few input variables;
- *Quine–McCluskey technique*: For any number of variables; more sophisticated technique, nevertheless easy to program for computer-aided engineering.

8.1.7.1 Karnaugh Map

The Karnaugh map is a representation of the truth table in rows and columns, in a manner that from one entry to the next *only* one input variable changes. The Karnaugh map for four input variables A, B, C and D is shown in Table 8.2. The configuration used in the sum of products is shown first, followed by the product of sums representation. The input variables are shown in each entry. The number in the upper right of each cell in the table is the decimal value of the combined input bits.

For each table entry the output level is entered that corresponds to that input variable combination. In moving from one table entry to the next in either the horizontal or vertical direction only one input variable changes. Table cells on the edge of the table are not hemmed in, however, as they are considered as also being neighbours of the corresponding extreme cell in the same row or column. Therefore, the far-right cell in the first row is adjacent to the far-left cell in the same row, and also to the far-right cell in the last row. The table can be considered as being overlaid on a toroidal surface. For three input variables the table size is reduced by two rows. It is not possible to visibly represent more than four input variables.

The following steps are taken to arrive at the **reduced sum of products**:

- Adjacent cells with 1 as an entry are grouped together. The group size can only be in powers of two, that is, 1, 2, 4 or 8 cells. The largest possible groupings should be made.
- All cells must be in at least one group. Each cell can ,however, be in more than one group.

8.1 Logic Algebra

Table 8.2. Karnaugh map for four input variables

SOP	$\overline{A} \cdot \overline{B}$	$\overline{A} \cdot B$	$A \cdot B$	$A \cdot \overline{B}$
$\overline{C} \cdot \overline{D}$	0000 0	0100 4	1100 12	1000 8
$\overline{C} \cdot D$	0001 1	0101 5	1101 13	1001 9
$C \cdot D$	0011 3	0111 7	1111 15	1011 11
$C \cdot \overline{D}$	0010 2	0110 6	1110 14	1010 10

POS	$A + B$	$A + \overline{B}$	$\overline{A} + \overline{B}$	$\overline{A} + B$
$C + D$	0000 0	0100 4	1100 12	1000 8
$C + \overline{D}$	0001 1	0101 5	1101 13	1001 9
$\overline{C} + \overline{D}$	0011 3	0111 7	1111 15	1011 11
$\overline{C} + D$	0010 2	0110 6	1110 14	1010 10

- An AND-ed term should be noted for each group that represents *only* the variables contained within the group. For groups with two cells one variable can be discarded, for groups with four cells two etc.
- The resultant terms are OR-ed.

EXAMPLE: The reduced sum of product of the logic variable Q is found using the following truth table (Table 8.3).

Table 8.3. Truth table

	A	B	C	D	Q
0	0	0	0	0	1
1	0	0	0	1	1
2	0	0	1	0	1
3	0	0	1	1	0
4	0	1	0	0	1
5	0	1	0	1	1
6	0	1	1	0	0
7	0	1	1	1	0
8	1	0	0	0	0
9	1	0	0	1	0
10	1	0	1	0	1
11	1	0	1	1	1
12	1	1	0	0	0
13	1	1	0	1	0
14	1	1	1	0	0
15	1	1	1	1	1

Q	$\overline{A} \cdot \overline{B}$	$\overline{A} \cdot B$	$A \cdot B$	$A \cdot \overline{B}$
$\overline{C} \cdot \overline{D}$	1 0	1 4	12	8
$\overline{C} \cdot D$	1 1	1 5	13	9
$C \cdot D$	3	7	1 15	1 11
$C \cdot \overline{D}$	1 2	6	14	1 10

This results in the Karnaugh map shown above. For clarity only the cells containing a 1 have been represented in the map. In the upper left corner a group of four 1s can be formed. Two cells can be grouped in the third row. It is pointless to form a group with 1 in the lower right corner with the 1 directly over it. It is far better to group it with the 1 in the lower left corner. This yields the following terms:

Group (0, 4, 1, 5): $\overline{A} \cdot \overline{C}$
Group (15, 11): $A \cdot C \cdot D$
Group (2, 10): $\overline{B} \cdot C \cdot \overline{D}$

Complete expression:

$$Q = (\overline{A} \cdot \overline{C}) + (A \cdot C \cdot D) + (\overline{B} \cdot C \cdot \overline{D})$$

Instead of the original eight terms each with four variables, only one term with two and two terms with three variables remain after the reduction.

The following steps are taken to arrive at the **reduced products of sums**:

- Adjacent cells with 0 as an entry are grouped together. The group size can only be in powers of two, that is, 1, 2, 4 or 8 cells. The largest possible groupings should be made.
- All cells must be in at least one group. Each cell can, however, be in more than one group.
- An OR-ed term should be noted for each group, that represents *only the* variables contained within the group. For groups with two cells one variable can be discarded, and for groups with four cells two, etc.
- The resultant terms are AND-ed.

EXAMPLE: The reduced product of sums of the logic variable R is found using the following truth table (Table 8.4).

Table 8.4. Truth table

	A	B	C	D	R
0	0	0	0	0	0
1	0	0	0	1	1
2	0	0	1	0	0
3	0	0	1	1	1
4	0	1	0	0	1
5	0	1	0	1	0
6	0	1	1	0	1
7	0	1	1	1	1
8	1	0	0	0	0
9	1	0	0	1	1
10	1	0	1	0	0
11	1	0	1	1	1
12	1	1	0	0	1
13	1	1	0	1	0
14	1	1	1	0	1
15	1	1	1	1	0

R	$A+B$	$A+\overline{B}$	$\overline{A}+\overline{B}$	$\overline{A}+B$
$C+D$	0 0	4	12	0 8
$C+\overline{D}$	1	0 5	0 13	9
$\overline{C}+\overline{D}$	3	7	0 15	11
$\overline{C}+D$	0 2	6	14	0 10

These results are shown in the Karnaugh map alongside. For clarity only the cells containing a 0 have been represented in the map. The four cells in the corners can form a group together by crossing over the edges. Two cells in the middle form a horizontal group. A further vertical group can be formed in the middle right. This yields the following terms:

Group (0, 8, 2, 10): $\quad B + D$
Group (5, 13): $\quad\quad\overline{B} + C + \overline{D}$
Group (13, 15): $\quad\;\overline{A} + B + C$

The overall expression is then:

$$R = (B + D) \cdot (\overline{B} + C + \overline{D}) \cdot (\overline{A} + B + C)$$

Instead of the original seven terms each with four variables, only one term with two and two terms with three variables remain after the reduction.

Consideration of Undefined States

Occasionally, the state of the output variables for a certain combination of input variables is not defined or is not relevant. Such states are denoted by an ×. The states are said to be undefined or *don't care* states. In the Karnaugh map undefined states can be organised into groups at will. The grouping is chosen for the best simplification of the output expression.

EXAMPLE: The reduced form of the logic variable S is found using the following truth table (Table 8.5).

Table 8.5. Truth table

	A	B	C	D	S
0	0	0	0	0	1
1	0	0	0	1	1
2	0	0	1	0	0
3	0	0	1	1	×
4	0	1	0	0	×
5	0	1	0	1	×
6	0	1	1	0	×
7	0	1	1	1	×
8	1	0	0	0	0
9	1	0	0	1	0
10	1	0	1	0	0
11	1	0	1	1	×
12	1	1	0	0	1
13	1	1	0	1	0
14	1	1	1	0	1
15	1	1	1	1	×

S	$\overline{A}\cdot\overline{B}$	$\overline{A}\cdot B$	$A\cdot B$	$A\cdot\overline{B}$
$\overline{C}\cdot\overline{D}$	1 (0)	× (4)	1 (12)	(8)
$\overline{C}\cdot D$	1 (1)	× (5)	(13)	(9)
$C\cdot D$	× (3)	× (7)	× (15)	× (11)
$C\cdot\overline{D}$	(2)	× (6)	1 (14)	(10)

S	$A+B$	$A+\overline{B}$	$\overline{A}+\overline{B}$	$\overline{A}+B$
$C+D$	(0)	× (4)	0 (12)	0 (8)
$\overline{C}+\overline{D}$	(1)	× (5)	0 (13)	0 (9)
$C+D$	× (3)	× (7)	× (15)	× (11)
$\overline{C}+D$	0 (2)	× (6)	(14)	0 (10)

To calculate the **sum of products** the undefined states in the first and second rows should be defined as 1. This yields a group of four in the upper left corner of the Karnaugh map. Equally, the two 1 s in the third column can be grouped together by traversing the table boundary. This yields the following terms:

$$S_1 = (\overline{A}\cdot\overline{C}) + (A\cdot B\cdot\overline{D})$$

For the **product of sums** the 0 s in the last column can be grouped together with the undefined states in the third row to form a group of four.

The 0 in lower left corner can be made into another group of four by combining with the three undefined states. The 0 in the third column has not yet been grouped. The undefined states to the left and below allow it to form another group of four. This yields the following terms:

$$S_2 = (\overline{A} + B) \cdot (A + D) \cdot (\overline{B} + C)$$

Both functions S_1 and S_2 correctly represent the truth table, although they lead to *different* logic expressions.

8.1.7.2 The Quine–McCluskey Technique

The Quine–McCluskey minimisation technique proceeds from the sum of products representation of the function to be minimised. For the sum of products, the product terms in which *each* variable appears are known as **minterms**.

EXAMPLE: The first term of the logic variable $Q = A\overline{B}C\overline{D} + ABD$ is a minterm, but the second is not, as it does not contain the variable C.

NOTE: Sum terms in which each variable appears once are known as **maxterms**.

A product term P is called **implicant** of Q if for $P = 1, \quad Q = 1$ holds.

EXAMPLE: The minterm $A\overline{B}C\overline{D}$ is an implicant of Q, since if $A\overline{B}C\overline{D} = 1$, then also $Q = 1$. The same is true for the product term ABD.

A **prime implicant** is a term that is no longer an implicant if one of the variables is omitted.

EXAMPLE: For $Q = A\overline{B}C\overline{D} + ABC + \overline{A}\,\overline{B}C\overline{D}$ the second term is a prime implicant. The first and third terms are not prime implicants, as for $\overline{B}C\overline{D} = 1$ it follows that $Q = 1$.

The Quine–McCluskey technique works in two steps:

1. Define the prime implicants;
2. Define the minimum overlap.

Defining the Prime Implicants

For the term

$$Q = AB\overline{C}D + ABC\overline{D} + \overline{A}BC\overline{D} + AB\overline{C}\,\overline{D} + \overline{A}B\overline{C}\,\overline{D} + A\overline{B}\,\overline{C}\,\overline{D}$$

the prime implicants should be defined. To that end all of the product terms are entered into a list. For each variable the value entered is the one required to make it equal to 1.

$ABCD$	$ABCD$	$ABCD$
(1) 1 1 0 1 *	(14) 1 1 0 -	(2435) - 1 - 0
(2) 1 1 1 0 *	(23) - 1 1 0 *	(2345) - 1 - 0
(3) 0 1 1 0 *	(24) 1 1 - 0 *	
(4) 1 1 0 0 *	(35) 0 1 - 0 *	
(5) 0 1 0 0 *	(45) - 1 0 0 *	
(6) 1 0 0 0 *	(46) 1 - 0 0	

Terms that differ only in the value of *one* variable are said to be **adjacent**. This is true, for example for the terms in the first and fourth row (1101 and 1100). The related product terms are $AB\overline{C}D$ and $AB\overline{C}\,\overline{D}$. Such terms can be **shortened** by the variable that appears as its own complement in the expressions.

$$AB\overline{C}D + AB\overline{C}\,\overline{D} = AB\overline{C}(D + \overline{D}) = AB\overline{C}$$

Such adjacent terms are searched for in each table entry. These are marked (in this example by a star), and the shortened form is entered into a new table (the second column). The variable to be removed is marked by a dash. The numbers in parentheses refer to rows in the previous column that gave rise to the shortening.

In the new table the search for similar terms is resumed. The process ends when no further terms can be shortened. Identical terms, like the last two in the third column, are only taken into consideration once.

All terms that have no marking in a table could not be shortened, and are therefore prime entries. In the example shown, these are the terms (14), (46) and (2435). The prime entries for the function Q are therefore

$$AB\overline{C}, \quad A\overline{C}\,\overline{D} \quad \text{and} \quad B\overline{D}$$

Some of the prime entries possibly still contain redundancies. These can be minimised as follows.

Defining the Minimum Overlap

All of the product terms of the original expression that was to be minimised are entered into a table. Then those prime implicants are marked for which the product terms in the row are implicants. (Or all possible prime implicants that are fully contained within a product term mean that that term must be marked.)

	$AB\overline{C}$	$A\overline{C}\,\overline{D}$	$B\overline{D}$
$AB\overline{C}D$	×		
$\overline{A}BC\overline{D}$			×
$ABC\overline{D}$			×
$AB\overline{C}\,\overline{D}$	×	×	×
$\overline{A}B\overline{C}\,\overline{D}$			×
$A\overline{B}\,\overline{C}\,\overline{D}$		×	

Starting with the longest term, the prime implicants are discard as long as *at least one* X remains *in each row*.

In the example, no prime implicant contains redundancy, so the minimum expression is given by

$$Q = AB\overline{C} + A\overline{C}\,\overline{D} + B\overline{D}$$

Computer programmes can easily carry out the searching, ordering and marking of the tables and can also handle larger numbers of variables.

8.1.8 Synthesis of Combinational Circuits

A **combinational circuit** is a logic circuit whose output variable depends only on the values of the applied inputs. Combinational circuits have no internal memory. Opposite: sequential circuit (Sect. 8.3).

The product of sums and sum of products expressions make it possible to build combinational circuits using AND, OR and inverter gates. AND and OR gates as well as inverters can also be represented by NAND or NOR gates (Fig. 8.21). Each combinational circuit can be realised using various combinations of NAND or NOR gates.

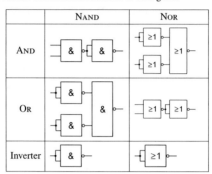

Fig. 8.21. Representation of the basic logic functions using only NAND or NOR gates

8.1.8.1 Implementation Using only NAND Gates

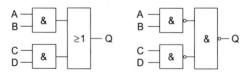

Fig. 8.22. Implementation of a combinational circuit using only NAND gates

Starting with the *sum of products* expression, both AND as well as OR gates can be replaced by NAND gates (Fig. 8.22). The equivalence of both circuits can be seen from DeMorgan's rule:

$$(A \cdot B) + (C \cdot D) = \overline{\overline{A \cdot B} \cdot \overline{C \cdot D}}$$

8.1.8.2 Implementation Using only NOR Gates

Starting with the *products of sums*, both AND as well as OR gates can be replaced by NORs (Fig. 8.23). The equivalence of both circuits can be seen from DeMorgan's rule:

$$(A + B) \cdot (C + D) = \overline{\overline{A + B} + \overline{C + D}}$$

NOTE: There are two alternatives to this approach: using multiplexers (Sect. 8.4.2) or using programmable logic devices (Sect. 8.6.5).

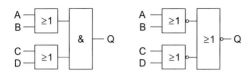

Fig. 8.23. Implementation of a combinational circuit using only NOR gates

8.2 Electronic Realisation of Logic Circuits

8.2.1 Electrical Specification

8.2.1.1 Voltage Levels

Logic states are represented by voltage levels in digital circuits. There is a defined voltage range for each logic level.

$$\begin{aligned}&\textbf{Logic high voltage level range (H):} \quad \text{Range closer to } +\infty\\&\textbf{Logic low voltage level range (L):} \quad \text{Range closer to } -\infty\end{aligned}$$

For any logic family the voltage level ranges are standardised. The actual output voltage of a logic gate depends on the loading, on the temperature and on the supply voltage. Moreover, the output voltage level may vary from device to device, for the same operating conditions (sample variations). **Typical output voltage levels** specify the midvalues of the defined range of voltages. Voltage values in between the ranges for H and L do occur for a short time in logic gates, but they are **undefined** states.

The mapping of voltage level ranges to logic values is an arbitrary process.

$$\begin{aligned}&\textbf{Positive logic:} \quad H \cong 1, \quad L \cong 0\\&\textbf{Negative logic:} \quad L \cong 1, \quad H \cong 0\end{aligned}$$

Unless otherwise specified, positive logic is presumed to apply. It is by far the most commonly employed.

8.2.1.2 Transfer Characteristic

The transfer characteristic shows the relationship between the output and input voltage of a logic gate. Figure 8.24 shows the transfer characteristic of an inverter. The ideal shape is in the form of a step. The actual shape of the transfer characteristic depends on the temperature.

Any circuit must be suitably designed, in order to connect the outputs of logic gates directly to the inputs of the following logic gates. This is guaranteed within any given *logic family*.

The **threshold voltage** is the input voltage for which both input and output voltages are the same. It is at the intersection of the transfer characteristic with the straight line of slope 1 emanating from the origin (note the different scales on the axes).

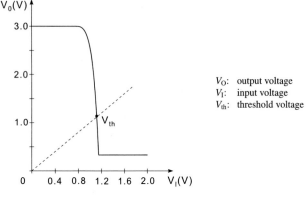

Fig. 8.24. Transfer characteristic of an inverter

8.2.1.3 Loading

For any logic family the loading data can be used to estimate the number of inputs in subsequent logic gates a given logic-circuit output can drive.

Fan-in: is a measure of the input current expressed in multiples of the standard input current of that logic family.

Fan-out: number of standard inputs a given output is able to drive.

- The sum of the fan-ins of all gate inputs driven from the the same output must not exceed its stated fan-out.

NOTE: Fan-outs may be different for logic high and low voltage levels. The smaller of the two values must be observed for the logic design.

8.2.1.4 Noise Margin

For serially connected logic gates it must be ensured that the output signal of the first gate is recognised correctly by the second gate. Manufacturers give the so-called **guaranteed static noise margin** for their gates, which hold even for the worst operating conditions (temperature, load, supply voltage).

- The **static noise margin** of the logic high state is the difference between the lowest output voltage V_{OHmin} and the lowest allowable input voltage V_{IHmin} of the following gates that will still be accepted as a logic high voltage level.
- The **static noise margin** of the logic low state is the difference between the highest output voltage V_{OLmax} and the highest allowable input voltage V_{ILmax} of the following gates that will still be accepted as a logic low voltage level.

The noise margin gives the maximum value a noise voltage may have without causing an error to occur at the gates (see Fig. 8.25). In the example these are 0.7 V for the H state and 0.4 V for the L state. This holds for noise signals that last longer than the gates' propagation delay (tens of nanoseconds).

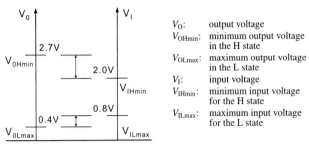

Fig. 8.25. Definition of the noise margin (sample data are examples from TTL-LS Gates)

The behaviour for very short spikes is described by the **dynamic noise margin**. This depends on the spike duration. For very short spikes a higher voltage is allowed before the device will produce an error.

The **typical noise margin** is the difference between typical output voltage and the threshold voltage V_{th}.

8.2.1.5 Propagation Delay Time

The propagation delay time is the time difference between the edges of the input signal and the resulting change in the output signal. Edges are defined by input or output voltage crossing the threshold voltage, respectively (Fig. 8.26).

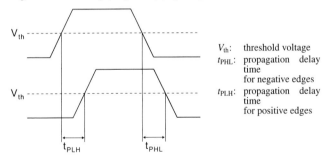

V_{th}: threshold voltage
t_{PHL}: propagation delay time for negative edges
t_{PLH}: propagation delay time for positive edges

Fig. 8.26. Definition of the propagation delay time

- For a state change from H to L the propagation delay time t_{PHL} applies.
- For a state change from L to H the propagation delay time t_{PLH} applies.

8.2.1.6 Rise Times

Transition times are defined as follows:

Rise time: t_{LH} for a (positive) logic low–high transition

Fall time: t_{HL} for a (negative) logic high–low transition

- The time between 10% and 90% of the steady-state value is measured.

NOTE: In some data sheets other reference points are used.

NOTE: Rise and fall times for a given logic element can differ significantly. For normal gates they lie in the range of a few nanoseconds. This is why the oscilloscope rise time cannot be neglected when measuring their rise time. The measured time is given by

$$t_{meas} = \sqrt{t_{LH}^2 + t_{oscil}^2}$$

and must be corrected accordingly.

- Rise and fall times depend on the load. In particular, the load capacitance is important, because it must be discharged by the output current.

8.2.1.7 Power Loss

The power loss in a digital logic circuit is consists of a **static** component, caused by the quiescent currents, and a **dynamic** component, which depends on the discharge currents of the internal and external capacitances.

The power loss depends therefore on the load and the frequency. It also depends fundamentally on the fabrication process. See also the sections from 8.2.3.

8.2.1.8 Minimum Slew Rate

Logic circuit inputs require input signals with steep slopes, otherwise the output signals will be unstable. The minimum required **slew rate** is usually given in the data sheets in V/µs.

Output signals from digital logic circuits have the minimum slew rate for the permissible load. External signals with slow transits can be a problem. Application of Schmitt triggers solve this problem (see Sect. 8.2.6.4).

8.2.1.9 Integration

Logic gates are nowadays almost exclusively realised by integrated circuits. The integration results in a space savings as well as a reduction of propagation times, power requirements and cost. However, integrated logic circuits for normal use are only produced in standardised function elements. For the design of logic circuits the availability of the desired logic combination must always be checked.

Integrated logic circuits are realised in two completely different processes. This results in two *logic families*, the TTL and the CMOS families. The former is based on the application of bipolar technology,;the latter on the integrated field effect transistor technology.

8.2.2 Overview: Notation in Data Sheets

f_{max} (*maximum clock frequency*): Maximum clock frequency at the input of a bistable circuit for which the operation of the device according to the data sheet is still guaranteed.

I_{CC} (*supply current*): The average current drawn from the voltage supply by the circuit.

I_{CCPD} (*power-down supply current*): The current drawn by the circuit when in power-down mode (caused by a *power-down* signal).

I_{IH} (*high-level input current*): The current flowing into the input of a circuit when a logic high voltage level has been applied.

I_{IL} (*low-level input current*): The current flowing into the input of a circuit when a logic low voltage level has been applied.

I_{OH} (*high-level output current*): The current flowing *into* the output of a circuit for an output logic high voltage level.

NOTE: This value is usually negative, as the current flows from the output.

I_{OL} (*low-level output current*): The current flowing into the output of a circuit for an output logic low voltage level.

I_{OS} (*short-circuit output current*): The current flowing into the output of a circuit when the output is connected with ground. This is usually given for the output voltage level H.

NOTE: This value is negative.

I_{OZH} (*high-impedance state output current with high-level voltage applied*): The maximum current that flows into the *three-state* output of a circuit, where the output is in a high-impedance state, and an external logic high voltage level is applied at the output.

I_{OZL} (*high-impedance state output current with low-level voltage applied*): The maximum current that flows into the *three-state* output of a circuit, where the output is in a high-impedance state, and an external logic low voltage level is applied at the output.

NOTE: This value is negative.

V_{IH} (*high-level input voltage*): The input voltage that corresponds to the voltage level H. Mostly given as the minimum allowable applied voltage that the circuit element will accept as a logic high voltage level.

V_{IL} (*low-level input voltage*): The input voltage that corresponds to the logic low voltage level. Mostly given as the maximum allowable applied voltage, that the circuit element will accept as a logic low voltage level.

V_{OH} (*high-level output voltage*): The output voltage that appears when the logic device is excited so that a logic high voltage level appears at the output. Mostly given as the minimum guaranteed value.

NOTE: V_{OH} depends strongly on load and temperature.

V_{OL} (*low-level output voltage*): The output voltage that appears when the logic device is excited so that a logic low voltage level appears at the output. Mostly given as a maximum guaranteed value.

NOTE: V_{OL} depends strongly on load and temperature.

t_{dis} (*disable time*): Valid for *three-state outputs*. This is the propagation delay measured between reference points of the switch-off signal and the output signal, where the output switches from a defined voltage level to a high-impedance state.

t_h	*(hold time)*: Minimum time necessary for a signal to be applied, to achieve the desired reaction.	
t_w	*(pulse width)*: Time interval between the defined reference points on the first and the second edges of an impulse.	
t_{pd}	*(propagation delay time)*: Propagation delay time of a logic element. Time between the reference points of an input signal and of the resulting output signals.	

NOTE: Sometimes there are differences between t_{PLZ} and t_{PHZ} depending on the active output voltage level.

t_h *(hold time)*: Minimum time necessary for a signal to be applied, to achieve the desired reaction.

t_w *(pulse width)*: Time interval between the defined reference points on the first and the second edges of an impulse.

t_{pd} *(propagation delay time)*: Propagation delay time of a logic element. Time between the reference points of an input signal and of the resulting output signals.

NOTE: Sometimes there are differences between t_{pLH} and t_{pHL} depending on the edges chosen.

t_r *(rise time)*: Time interval between the signal passing through 10% and 90% of its steady state for a rising edge.

t_f *(fall time)*: Time interval between the signal passing through 90% and 10% of its steady state for a falling edge.

t_{pxz} see t_{dis} for both logic high and low voltage level.

Table 8.6. Signal representation in data sheets

Signal	Input	Output
———	Must be constant	Is constant
＼＼	May change from high to low	Changes from high to low
／／	May change from low to high	Changes from low to high
✕✕✕	Either change is allowed	Unpredictable state
⟩⟨	—	Centreline represents the high-impedance state (for *three-state* outputs)

8.2.3 TTL Family

The transistor–transistor logic (TTL) devices are produced in different series. The following holds for each of them:

- +5 V supply voltage;
- arbitrary connection of components because of compatible input and output signals;
- pin compatibility for devices of the same name even if they are different TTL series.

8.2.3.1 TTL Devices

The essential qualities of the different devices are given below and in Table 8.7 (in parentheses the notation is given for a four-NAND gate device):

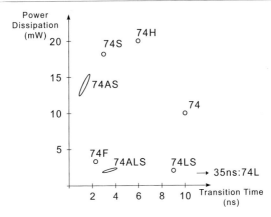

Fig. 8.27. Switching times and power dissipation of different TTL devices

Table 8.7. Electrical specifications for TTL devices

$V_{CC} = 5$ V $\vartheta = 25°C$		**TTL devices**		
		74LS00	74ALS00	74F00
Input voltage	V_{ILmax}	0.8 V	0.8 V	0.8 V
	V_{IHmin}	2.0 V	2.0 V	2.0 V
Output voltage	V_{OLmax}	0.5 V	0.5 V	0.5 V
	V_{OHmin}	2.7 V	2.7 V	2.7 V
Threshold voltage	V_{th}	1.3 V	1.5 V	1.5 V
Fan-out		20	20	33
Output current	I_{OLsink}	8 mA	8 mA	20 mA
Propagation delay time typ./max.	t_{PLH}	9/15 ns	4 ns	4/5 ns
	t_{PHL}	10/15 ns	5 ns	3/4 ns
Rise time	t_{LH}	10 ns	5 ns	3 ns
Fall time	t_{HL}	6 ns	5 ns	3 ns
(for a 15 pF load)				
Minimum slew rate (of the input voltage)		1V/μs	5V/μs	
Power dissipation (per gate)		2 mW	1.2 mW	4 mW

Standard TTL series (7400). Historically the first device; very low-priced; was the industry standard for decades.

High-speed TTL series (74H00). Slightly faster than the standard series because lower resistance in the layout; sales are no longer significant.

Low-power TTL series (74L00). Much slower than the standard series; lower power dissipation; sales are no longer significant.

Schottky TTL series (74S00). Insertion of Schottky transistors and diodes gave rise to much shorter switching times; small number of different types.

Low-power Schottky TTL series (74LS00). Insertion of Schottky transistors; lower power dissipation; device with up to now the greatest number of different types; industry standard.

Advanced low-power Schottky TTL series (74ALS00). Shorter switching times than the LS series; lower power dissipation; great number of different types; very complex circuits for microprocessor applications can occasionally be found exclusively as ALS types.

FAST series (74F00). Fast devices; only a few manufacturers.

Advanced Schottky TTL series (74AS00). Extremely short switching times between 1 and 2 ns, however, moderate power loss; can be used to replace high-speed ECL circuits.

High-speed CMOS series (74HC00). Not a TTL device, but this CMOS series is pin and function compatible with the TTL series; for qualities and comparison see Sect. 8.2.4.

8.2.3.2 Basic TTL Gate Circuit

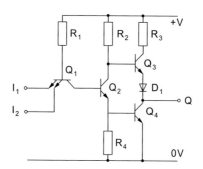

Fig. 8.28. Basic circuit of a TTL–NAND gate

The basic structure of one (of four) NAND gates in the 7400 device is shown in Fig. 8.28. If one of the emitters (I_1 or I_2) is connected to ground, then the transistor Q_1 is turned on. This turns Q_2 off, and Q_4 is then also turned off. The base of Q_3 is connected via R_2 to the supply voltage. Q_3 is turned on. The output Q is connected to the logic H potential via R_3, Q_3, D_1. R_3 (150–500 Ω) limits the current.

For a positive input voltage on I_1 *and* I_2 the current no longer flows through R_1 and out through the emitter, but rather flows over the Q_2 base–emitter junction. Q_2 turns on and then turns Q_4 on. Q_4 bypasses the resistance R_4, which provides feedback for Q_2. This rapidly forces the amplification of Q_2 upwards. Q_4 is completely on and can sink current from the output Q.

The characteristic output stage with the three-semiconductor structure of Q_3, D_1, Q_4 on top of each other like the faces of a totem pole leads to the notation *totem pole*.

The basic structure of the TTL circuit has the following characteristics:

- For a logic low voltage level at the input, the driver circuit must **sink current**.

- For a logic high voltage level at the input, the driver circuit must **source a small current**.

- The *totem pole* has a fairly high impedance for the logic high voltage level. The output functions as a **current source**. The output current is limited by the resistor. The logic high potential drops below the allowable limit if the output current is too large.
- The output stage has a low impedance for the logic low potential when it operates as a **current sink**. The dynamic resistance of the lower output transistor and thermal loading are the limits on this operation.
- The *totem-pole* configuration means that TTL outputs cannot be connected in parallel. (For other output stages see Sect. 8.2.6.)

NOTE: Some manufacturers permit the parallel connection of the outputs of two gates if the same logic signal is applied within the **same device**.

Fig. 8.29. Input and output currents for TTL gates; the values given apply to LS gates

A fan-out of 20 for logic high and logic low voltage levels can be calculated from the values given.

8.2.4 CMOS Family

Complementary metal-oxide semiconductor (CMOS) devices are manufactured in different series. The following holds for each of them:

- +5 V–15 V supply voltage (also 3 V–18 V);
- extremely low input currents;
- very small power dissipation in static operation and for low frequencies;
- output currents in the logic high and low states are equally large.

The essential qualities of the different series are (the notation for a four-NAND gate device is given in parentheses):

CMOS series A (CD4011A). Historically the first device; has been superseded since then.
CMOS series B (CD4011B). Industry-standard; largest spectrum of different types; standardised, manufacturer-independent static specification data.
LOCMOS series (HEF4011B). Higher switching times than the CMOS series B; transfer characteristic step-like.
High-speed CMOS series (74HC00). Pin and function compatible to the equivalently numbered TTL devices; tenfold faster switching times and higher output currents than the CMOS series B; for frequencies below 20 MHz lower power dissipation than the LS TTL series; differs from the TTL supply voltage 2 V–6 V.
High-speed CMOS series (74HCT00). Offshoot of the HC series with a more limited input voltage range 4.5 V–5.5 V; the input side is TTL voltage level compatible.
Advanced high-speed CMOS series (74AC00 or 74ACT00). Even faster than the HC series; very high output currents of 24 mA in the logic high and low states; input side is TTL voltage level compatible.

The electrical specification data for the CMOS series, which are dependent on the supply voltage, are given in Table 8.8.

Table 8.8. Electrical specification for CMOS devices

$\vartheta = 25\,°C$		CMOS device				
		HEF4011B			74HC00	74AC00
Supply voltage	V_{CC}	5V	10V	15V	4.5V	
Input voltage	V_{ILmax}	1.5 V	3 V	4 V	0.9 V	1.35 V
	V_{IHmin}	3.5 V	7 V	11 V	3.2 V	2.0 V
Output voltage	V_{OLmax}	50 mV			100 mV	
(for $I_O \leq 1\,\mu A$)	V_{OHmin}	$V_{CC} - 50$mV			4.9 V	
Output current	I_{OLmax}	0.4 mA	1.1 mA	3.0 mA	20 mA	24 mA
Propagation delay time	t_{PLH}	35 ns	16 ns	13 ns	8 ns	5 ns
	t_{PHL}	16 ns	13 ns	12 ns	8 ns	4 ns
Rise/fall time	t_{LH}	25 ns	15 ns	11 ns	6 ns	1.5 ns
(with a 15 pF load)						
Input current		$\leq 0.3\,\mu A$			0.3 μA	0.1 μA

Propagation delay times and the output rise/fall times depend strongly on the load capacitance. For a load of 50 pF the times for the HEF series approximately double.

8.2.5 Comparison of TTL and CMOS

Because of the low input currents in the 74HC CMOS series, the number of gates that can be connected is not defined by the resistive load. The maximum allowable load capacitance is much more of a limit (typically, 5 pF per gate input).

Table 8.9. Fan-outs of TTL and CMOS devices

↓Input	TTL devices					CMOS devices	
	74xx	74LSxx	74Sxx	74ALSxx	74Fxx	74HCxx	74ACxx
74xx	10	5	12	5	12	2	15
74LSxx	20	20	50	20	50	10	60
74Sxx	8	4	10	4	10	2	12
74ALSxx	20	20	50	20	50	20	120
74Fxx	20	13	33	13	33	6	40
74HCxx	> 50						
74ACxx	>50						

8.2.5.1 Other Logic Families

Other logic families apart from the successful CMOS and TTL logic families are also in use:

ECL (*emitter-coupled logic*): Achieves switching times below 1 ns, as the transistors are not operated as saturated switches; technical data: high-impedance differential-inputs, low-impedance outputs; high power dissipation of about 50 mW per gate; ECL circuits always offer Q and $\overline{Q}$ outputs; currents are not switched off, but are rerouted, so there

Table 8.10. Comparison of TTL and CMOS data

	LS–TTL	**CMOS**
Switching speed	10 ns	40 ns (5 V) – 15 ns (15 V)
Power dissipation	Up to about 3 MHz constant, then rising	Linearly with frequency; above 5 MHz (at 5 V) higher than LS–TTL
Fan-out	20	> 50
Output impedance	25 Ω (for low)	250 Ω (high and low)

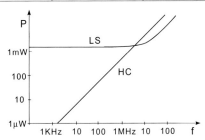

Fig. 8.30. Power dissipation for TTL and CMOS devices

are lower noise voltages on the supply lines; operating voltage is -5.2 V, and very fast circuits require an additional supply voltage of -2.0 V.

APPLICATION: Computers, high-speed signal processing.

LSL (*low speed logic, with high noise immunity*): Logic circuits with high immunity can be realised by raising the voltage and increasing the switching times; the switching times can be further increased by using external capacitors; internal Zener diodes at the input lift the threshold voltage up to about 6 V; for a 12 V supply voltage the noise margin amounts to 5 V; the switching times are 150 ns and more.

APPLICATION: Industrial control in noisy environments.

RTL (*resistor–transistor logic*): Predecessor of the TTL circuits, which was replaced by DTL; a simplification was achieved using resistors; great limitations due to the influence of adjacent gates.

DTL (*diode–transistor logic*): Predecessor of TTL; fan-in using diodes.

GaAs : This is not a new logic family, but rather a new kind of transistor manufacturing technology using gallium-arsenide; very short switching times in the range of 10 ps (=0.01 ns!); optoelectronic components are prepared using the same technology; the development of a combined opto-electronic logic is anticipated (circuits with light or current).

8.2.6 Special Circuit Variations

8.2.6.1 Outputs with Open Collector

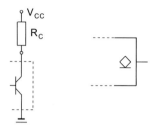

Fig. 8.31. Open collector-output and circuit symbol

Devices with *open collector output* connect the collector of the output transistor to the output, without connecting it to the supply voltage via a transistor as in the *totem-pole* output (Fig. 8.31). For CMOS circuits this implies an *open drain* output.

- An *open-collector* output must always be connected via a resistor to the positive supply voltage.

Advantages:

- *Open-collector* outputs can be connected in parallel without causing problems (see also Sect. 8.2.6.2).
- The load can be connected to a voltage that is higher than the supply voltage of the logic device. The only limit is the maximum allowable breakdown voltage of the output transistor.

8.2.6.2 Wired AND/OR

A 'wired AND' circuit is made by connecting together two outputs with open collectors (Fig. 8.32). It is sufficient that *one* of the output transistors conducts for a logic low voltage level to be present on the common output. The truth tables show the logic levels at the collectors of the transistors that would be present, if each transistor were present alone:

Y	X	Q
Low	Low	Low
Low	High	Low
High	Low	Low
High	High	High

Y	X	Q
0	0	0
0	1	0
1	0	0
1	1	1

Y	X	Q
1	1	1
1	0	1
0	1	1
0	0	0

For positive logic the circuit behaves like a logic AND gate (middle table), and for negative logic like an OR gate (right table). Hence the notation wired AND or OR.

In this manner several outputs can be connected together creating a bus.

To make sure valid logic levels appear during operation, the common collector (pull-up) resistor R_C must be suitably selected (Fig. 8.33).

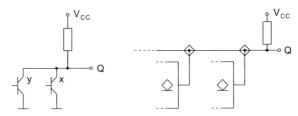

Fig. 8.32. Principle of a wired-AND-circuit and related circuit symbol

For a logic high voltage level the currents I_{QH} the of the output transistors and the input currents I_{IH} flow through the resistor. It must be sufficiently small so that the output voltage level does not drop below the allowable logic high voltage input level of subsequent circuits.

For a logic low voltage level in the worst case only one transistor is switched on. The resistance must be at least large enough so that the maximum collector current I_{OLmax} is not exceeded. In addition, the input currents I_{IL} of the connected inputs flow through it.

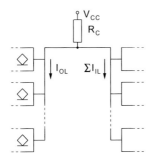

Fig. 8.33. Selection of the size of the pull-up resistor

$$R_{\max} = \frac{V_{CC} - 2.4 \text{ V}}{K \cdot I_{QH} + N \cdot I_{IH}}$$

$$R_{\min} = \frac{V_{CC} - 0.4 \text{ V}}{I_{OLmax} - N \cdot |I_{IL}|}$$

K: Number of outputs connected in parallel

N: Number of inputs the parallel connected in parallel (each fan-in = 1)

NOTE: In practice the smallest allowable value is chosen, to achieve the maximum switching speed.

EXAMPLE: For the 74LS TTL family the output leakage current amounts to
$I_{QH} < 250$ µA, the input current $|I_{IL}| < 0.4$ mA per input and the maximum collector current $I_{OLmax} = 8$ mA.

For LS–TTL: $\quad R_C = \dfrac{5 \text{ V} - 0.4 \text{ V}}{8 \text{ mA} - N \cdot 0.4 \text{ mA}} = \dfrac{4.6 \text{ V}}{(20 - N) \cdot 0.4 \text{ mA}}$

8.2.6.3 Tri-State Outputs

In a circuit with **tri-state**[*] output *both* transistors of the final push–pull stage can be switched into the high-impedance state by an enable signal. Such devices are suitable for bus systems (Fig. 8.34). In the high-impedance state the device acts as if it were not present.

- The three output states are denoted by H, L and Z.

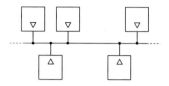

Fig. 8.34. Connection of several *tri-state* devices to a bus

8.2.6.4 Schmitt Trigger Inputs

Devices with Schmitt trigger inputs have *two* different threshold voltages, depending on whether the output state is high or low. The transfer characteristic of a Schmitt trigger is therefore different for turning on from turning off.

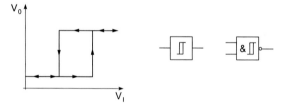

Fig. 8.35. Transfer characteristic of a Schmitt trigger; circuit symbol of gates with Schmitt trigger inputs

The difference between turn-on and turn-off is known as **hysteresis** (Fig. 8.35). For TTL circuits this typically amounts to about 0.8 V, for CMOS circuits it depends on the applied voltage:

$V_H = 0.27 \cdot V_{CC} - 0.55$ V

Application:

- Inputs with very slow edges can be used with the Schmitt trigger, which leads to reduction of transition time.
- In conjunction with RC gates, they can be used for pulse stretching or to build an oscillator (Fig. 8.36).

---

[*] The notation tri-state was originally a trade name. It is more widely used than the notation *three-state output*

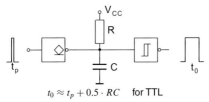

Fig. 8.36. Pulse stretching using Schmitt trigger gates; the falling edges are delayed

8.3 Combinational Circuits and Sequential Logic

- A **combinational circuit** is a logic circuit whose output states only depend on the signal state applied at its inputs. This is known as **combinational logic**.
- A **sequential circuit** employs internal memory. The output states depend not only on the present input states but also on previous states. This is known as **sequential logic**.

8.3.1 Dependency Notation

The dependency notation is based on the DIN 40 900 Norm (Part 12). It gives a representation of the effects of external signals in complex digital circuits (Fig. 8.37). A distinction is made between controlling and controlled connections. The following rules apply:

- Each input is labelled by an identifying symbol. This is noted within the circuit symbols.
- Inputs affecting other inputs are identified by a letter that denotes the kind of influences. The identifying symbol of the affected input will also be denoted.

The dependency notation is different for the following cases (cited from the DIN 40 900) and is summarised in Table 8.11:

G-dependency: This represents an AND gate with its dependent connections. A Gx-input in state 0 internally drives the connections controlled by it to 0, otherwise they remain unchanged.

V-dependency: This represents an OR gate with its dependent connections. A Vx-input in state 1 internally drives the connections controlled by it to 1, otherwise they remain unchanged.

N-dependency: This represents XOR gate with its dependent connections. A controllable inversion is thereby realised. A Nx-connection in state 1 inverts the controlled connections. Otherwise it leaves its state unaffected.

Z-dependency: This function acts like an internal connection. Z-dependent connections copy their logic value. Z-dependency often is combined with other dependencies.

C-dependency: This realises a control function. A Cx-connection in state 0 causes all dependent connections to be ineffective. Else it can exercise its intended function.

S-dependency: Connections that are dependent on an Sx-input assume the state that they would assume for the combination of $S = 1$, $R = 0$. This happens independently of the actual state at the R-input. In state 0 the controlling connection is ineffective.

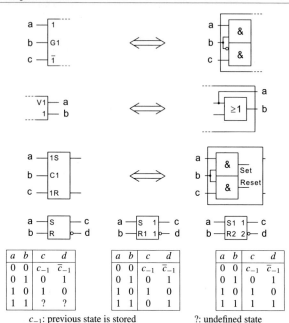

a	b	c	d
0	0	c_{-1}	$\bar{c}_{-1}$
0	1	0	1
1	0	1	0
1	1	?	?

a	b	c	d
0	0	c_{-1}	$\bar{c}_{-1}$
0	1	0	1
1	0	1	0
1	1	0	1

a	b	c	d
0	0	c_{-1}	$\bar{c}_{-1}$
0	1	0	1
1	0	1	0
1	1	1	1

c_{-1}: previous state is stored   ?: undefined state

Fig. 8.37. Explanation of the dependency notation

R-dependency: Connections that are dependent on an Rx-input assume the state that they would assume for the combination of $R = 1$, $S = 0$. This happens independently of the actual state at the S-input. In state 0 the controlling connection is ineffective.

EN-dependency: This describes an enable dependency (*enable*). EN-controlled inputs become only effective when they are in the state 1. The EN-dependency can often be found for *open-collector* and *tri-state* outputs. The outputs are set to the high-impedance state by the 0 state of the controlling inputs.

A-dependency: This denotes the choice of an address, in particular for memory. The controlling inputs are weighted to the power of 2. The Ax-inputs have the same effect as an enable signal on the resulting address.

M-dependency: This denotes a switching in different operating conditions (modes), e.g. up/down counting.

T-dependency: Tx-controlled connections change their state as soon as the controlling input has the value 1.

CT-dependency: Denotes that for a certain counter state or register contents an action will be carried out, e.g. a carry signal.

8.3.1.1 Overview: Dependency Notation

Table 8.11. Dependency notation

Symbol	Dependency	Action for 1/0
A	Address	Address selected/not selected
C	Clock, control	Allows/inhibits action
CT	Contents	Permitted action/inputs blocked
EN	Enable	Permitted action/outputs high impedance
G	AND	Unaltered state/state = 0
M	Mode	Mode selected/not selected
N	Controlled inversion	Inverted state/noninverted state
R	Reset	Reaction as for $R = 1$, $S = 0$/no reaction
S	Set	Reaction as for $S = 1$, $R = 0$/no reaction
T	Toggle	State changes/stays the same
V	OR	State = 1/unaltered state
Z	Connection	State = 1/state = 0

The notation 'action' means that controlled inputs have their normally defined effect on the function of the circuit elements and that controlled outputs assume the internal logic state that is given by the function of the circuit elements.

8.3.2 Circuit Symbols for Combinational and Sequential Logic

Figure 8.38 shows some examples of circuits illustrating the the use of the dependency notation. The first example shows a buffer whose output signal can be inverted by choice. Multiple dependencies can be combined, as can be seen in the second example. The logic sequence is given by the numbers on the affected connection. The third example shows a bidirectional buffer whose tri-state outputs can be driven into the high-impedance state, depending on the state of the inputs c. Thereby the direction of data transmission is defined.

A 2-to-1-multiplexer is shown in the next example. The variable c is fed into the control block and selects, through the AND gate, which of the inputs to connect to the output. This is an additional example for the notation that the controlling signal influences the connection inverted ($\bar{1}$).

A ROM with 32×4-bit memory capacity is shown in the following example. The five address lines $a_0 \ldots a_4$ select the addresses 0 to 31. The four outputs of the ROM are controlled by an enable input.

The last example shows a counter that counts from 0 to 7. The edge-triggered clock input influences the counter state, whose binary value in each position is entered in parentheses. The output in the upper right will be active synchronous to the clock only for counter state 7 (CT = 7).

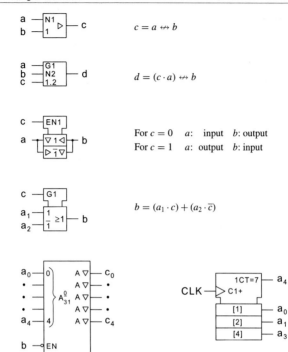

Fig. 8.38. Examples of circuit symbols with the dependency notation

8.4 Examples of Combinational Circuits

8.4.1 1-to-n Decoder

A **decoder** activates exactly *one* of n possible outputs. The selection is made using control signal inputs. The active state is often a logic low state. One-to-ten-decoders are also known as **BCD-decimal decoders**.

EXAMPLE: Truth table of a 1-to-4 decoder (Fig. 8.39):

APPLICATION: Code conversion, selection of memory elements in microprocessor systems.

8.4.2 Multiplexer and Demultiplexer

Multiplexers are electronically controlled selection circuits.

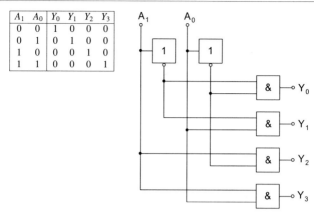

A_1	A_0	Y_0	Y_1	Y_2	Y_3
0	0	1	0	0	0
0	1	0	1	0	0
1	0	0	0	1	0
1	1	0	0	0	1

Fig. 8.39. Truth table and circuit of a 1-to-4 decoder

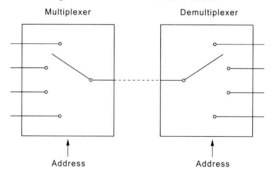

Fig. 8.40. Multiplexer and demultiplexer

A **multiplexer** connects one of n input signals to *a single* output line. The selection is carried out using an address (Fig. 8.40). Multiplexers are also known as **data selectors**.

The opposite operation holds for the **demultiplexer**, as it connects a signal from *a single* input to one of n outputs using an address.

The demultiplexer follows from the 1-to-n decoder. The addressed output does not go high, but rather passes on the voltage level of the input signal.

- Multiplexers are also suitable for realising arbitrary logic functions.

EXAMPLE: The circuit of a logic function with four input variables is to be found. The logic function is described in a truth table. The voltage level of the input variables is fed to the four address lines of a multiplexer with 16 inputs. Each of the 16 inputs is fixed high or low, depending on the truth table. Any logic function with four variables can be realised in this way.

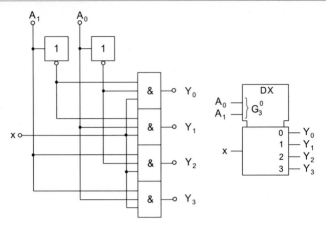

Fig. 8.41. Circuit of a demultiplexer and its circuit symbol

8.4.2.1 Overview of Circuits

Multiplexer		
CMOS	TTL	Inputs
4515	74150	16
4512	74151	8
4539	74153	2×4

Demultiplexer		
Outputs	CMOS	TTL
16	4514	74154
8	74 HCT 138	74138
2×4	74 HCT 139	74139

8.5 Latches and Flip-Flops

Flip-flops are **bistable** triggered switches. A flip-flop is said to be **set** when its output is high, otherwise it is said to be **reset**.

8.5.1 Flip-Flop Applications

Flip-flops are used in:

- registers (see Sect. 8.7);
- shift registers (see Sect. 8.7);
- memories (see Sect. 8.6);
- counters (see Sect. 8.8);
- frequency dividers;
- state memories (see Sect. 8.9).

8.5.2 SR Flip-Flop

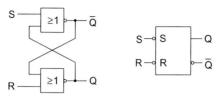

Fig. 8.42. SR flip-flop and its circuit symbol

A feature of this kind of flip-flop are the cross-coupled inverting gates. An SR flip-flop, shown in Fig. 8.42, is composed of NOR gates. The inputs are known as set or reset, respectively. The truth table is given by:

S	R	Q	$\overline{Q}$
0	0	Q_{-1}	$\overline{Q}_{-1}$
0	1	0	1
1	0	1	0
1	1	0	0

Q_{-1}: previous state

The output state will not change if the inputs are $S = 0$, $R = 0$. For inputs of $S = R = 1$ the output is $Q = \overline{Q} = 0$, which is *logically* impossible. By changing both input signals to $S = R = 0$ the output state cannot be specified without other information being supplied. This should therefore be avoided.

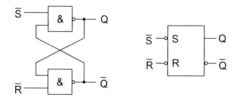

Fig. 8.43. $\overline{SR}$-latch with NAND gates and its circuit symbol

A flip-flop that is set or reset by a logic low level is shown in Fig. 8.43. The truth table is then:

$\overline{S}$	$\overline{R}$	Q	$\overline{Q}$
0	0	1	1
0	1	1	0
1	0	0	1
1	1	Q_{-1}	$\overline{Q}_{-1}$

Q_{-1}: previous state

8.5.2.1 SR Flip-Flop with Clock Input

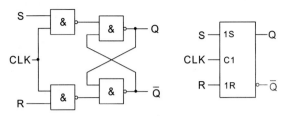

Fig. 8.44. SR flip-flop with clock input and its circuit symbol

The SR flip-flop can be expanded to become a **gated SR flip-flop** (Fig. 8.44). Only while CLK is in the high state, can the output states be changed by the RS-inputs. In state $CLK =$ low the previous state remains as it was, independent of the RS inputs. The truth table is:

CLK	S	R	Q	$\overline{Q}$	
1	0	0	Q_{-1}	$\overline{Q}_1$	⎫
1	0	1	0	1	⎬ as SR flip-flop
1	1	0	1	0	⎬
1	1	1	?	?	⎭
0	X	X	Q_{-1}	$\overline{Q}_{-1}$	memory state

8.5.3 D Flip-Flop

With a D flip-flop the illegal input combinations are avoided by arranging the circuit elements suitably. The truth table is:

CLK	D	Q	$\overline{Q}$	
0	X	Q_{-1}	$\overline{Q}_{-1}$	memory
1	0	0	1	⎫ transparent
1	1	1	0	⎭

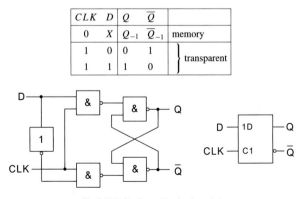

Fig. 8.45. D flip-flop and its circuit symbol

As long as the clock signal $CLK = 1$, the flip-flop is **transparent** for the data signal D, i.e. the output signal follows the data signal. If the gate signal assumes the value 0, then the present state of the data line is stored and is independent of further changes on D (Fig. 8.45). The D flip-flop is also known as the *D-latch*.

8.5.4 Master–Slave Flip-Flop

The transparency of the D flip-flop is lost when two are connected in series. The flip-flops are controlled by two complementary clock signals. This configuration is known as a *master–slave* flip-flop. The Q' output of the *master* follows the D-signal as long as $CLK = 1$. The *slave* flip-flop remains locked. If the clock signal drops to 0, then the *master* flip-flop locks up, and the subsequent following slave flip-flop copies the logic state of the *master's* Q output.

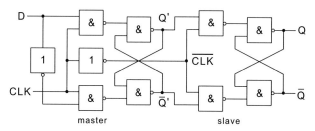

Fig. 8.46. Edge-triggered D flip-flop

Figure 8.47 shows a master–slave flip-flop, composed of two SR flip-flops in series.

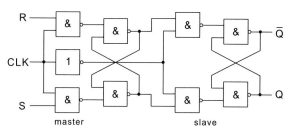

Fig. 8.47. SR master–slave flip-flop

The two flip-flops are alternatively blocked by the complementary clock signal. When the clock signal $CLK = 1$ the state of the first flip-flop is given by the RS input signals. If the clock signal drops to 0, the *master* flip-flop is blocked and stores its state that was present before the clock transition. The *slave* flip-flop receives the complement signal $\overline{T} = 1$ and thus becomes transparent. The state of the first flip-flop appears at the output. This *master–slave* flip-flop is not transparent at any moment. Input states of $R = S = 1$ cause undefined output states as for the simple SR flip-flop.

8.5.5 JK Flip-Flop

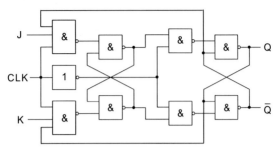

Fig. 8.48. JK *master–slave* flip-flop

Undefined output states are avoided by the JK flip-flop by coupling back the complementary output states Q and $\overline{Q}$. The flip-flop inputs are **preparatory inputs** and are denoted by J and K. The information read in on the positive edges appears at the output only on the following negative edges. This is known as **delayed outputs**. They are denoted by ¬ at the output. The truth table for an applied clock signal $CLK = 010$ is shown in Fig. 8.49:

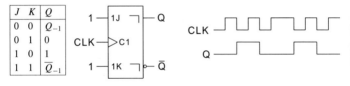

Fig. 8.49. Truth table of the JK flip-flop and circuit as a binary divider

The states $J/K = 0/1$ and $J/K = 1/0$ set the flip-flop to the respective state of the J-input *synchronously* with the negative edge of the clock signal.

A special case applies for $J = K = 1$. The JK flip-flop inverts its previous state. The flip-flop operates as a frequency divider or *scaler* (Fig. 8.49). This is also known as a *toggle* flip-flop.

Most flip-flops have additional *asynchronous* set or reset inputs. These have priority over the JK inputs.

NOTE: As long as the clock $CLK = 1$, the states at the JK inputs may not change. For flip-flops with JK data lockout this limitation is not valid.

8.5.6 Flip-Flop Triggering

Different kinds of *triggering* are used with flip-flops.

Unclocked flip-flops: Their state depends only on the set/reset inputs.
Clocked flip-flops: The actual time when the information is passed on is defined by a clock signal.
Level-triggered flip-flops: The information transfer is defined by the voltage level of the control signal.

Edge-triggered flip-flops: The information transfer is defined by the state transition of the control signal.

8.5.7 Notation for Flip-Flop Circuit Symbols

The circuit symbols denoted by DIN 40 900 (Part 12) are as follows:

Dynamic input: The (transient) internal 1-state corresponds to the transition from the external 0-state to the 1-state. Otherwise the internal logic state is 0.

Dynamic input with inversion: The (transient) internal 1-state corresponds to the transition from the external 1-state to the 0-state. Otherwise the internal logic state is 0.

Dynamic input with polarity indicator: The (transient) internal 1-state corresponds to the transition from the external high state to the low state. Otherwise the internal logic state is 0.

Delayed output: The state change at this output is postponed until the triggering signal returns to its original state.

NOTE: The internal logic state of inputs affecting the output states must not change as long as the input causing the change is still in the internal 1-state.

D-input: The internal logic state of the D-input is stored by the element.

NOTE: The internal logic state of this input is always dependent on a gating input or output.

J-input: If this input assumes the internal state 1, a 1 is stored in the element. In the internal state 0 it has no effect on the element.

K-input: If this input assumes the internal state 1, a 0 is stored in the element. In the internal state 0 it has no effect on the element.

NOTE: The combination $J = K = 1$ causes a change of the internal logic state into its complementary state.

R-input (*reset*): If this input assumes the internal state 1, a 1 is stored in the element. In the internal state 0 it has no effect on the element.

S-input (*set*): If this input assumes the internal state 1, a 0 is stored in the element. In the internal state 0 it has no effect on the element.

NOTE: The effect of the combination $R = S = 1$ is not defined by the symbol.

T-input (*toggle*): If this input assumes the internal state 1, the internal state of the output changes to its complementary state. In the internal state 0 it has no effect on the element.

8.5.8 Overview: Flip-Flops

The most popular flip-flop types are listed in Table 8.12:

Table 8.12. Types of flip-flops

Circuit symbol	Flip-flop	Triggering
▷T	T flip-flop	Edge-triggered, clocked
S, R		Not clocked, level-triggered
1S, C1, 1R	SR flip-flop	Clocked, one-level triggered
1S, ▷C1, 1R		Clocked, edge-triggered
1J, C1, 1K	JK flip-flop	Clocked, two-level triggered
1J, ▷C1, 1K		Clocked, two-edge triggered
1D, C1	D flip-flop	Clocked, one-level triggered
1D, ▷C1		Clocked, edge-triggered

8.5.9 Overview: Edge-Triggered Flip-Flops

Edge-triggered flip-flops make the design of sequential circuits very clear and therefore are also frequently used in programmable logic devices (PLD). Figure 8.50 shows the waveform diagram for the four types of edge-triggered flip-flops. All flip-flops shown are **positive edge-triggered**.

The **SR flip-flop** is set by the positive edges of the clock signal if the *set* input is high. Repeated set levels do *not* change the output state. The flip-flop is reset if the *reset* input is high at the time of the positive clock edge. $R = S = 1$ at the time of the positive clock edge leads to an undefined state. Otherwise the combination is allowed.

The **D flip-flop** assumes the value at the data input with the positive clock edge. The sloped edges in the waveform diagram indicate that the precise timing of the transients is irrelevant for the circuit's function.

The **T flip-flop** divides the clock signal by 2. For an approximately constant clock frequency this is known as a **frequency divider**.

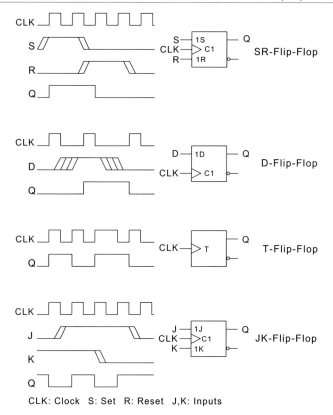

Fig. 8.50. Waveform diagram of the four edge-triggered flip-flop types

For the edge-triggered **JK flip-flop** the output signals depend on the asynchronous inputs J and K. For the combination $J = K = 1$, the flip-flop operates as a T flip-flop, and for the combination $J = K = 0$ it stores the previous state.

Flip-flops are **storage** elements. The truth table descriptions must therefore take into account the state before the triggering clock edge arrived. This is denoted by Q_{-1}. The truth table in sequential logic leads to the **flip-flop transition table**. This is the synthesis table required to define any output state transition from Q_{-1} to Q for a given input signal. The output states *after* the relevant clock edge can be given as a function of the state *before* the edges and other control signals. This is known as the **characteristic expression** for the given logic element.

8.5.10 Synthesis of Edge-Triggered Flip-Flops

When building a circuit using programmable logic devices (PLD), it is often necessary to realise various edge-triggered flip-flops from simple logic elements. The following sections show the necessary steps in a successful design.

SR Flip-Flop (Edge-Triggered)

As there can be several combinations of signals that can cause the same state transition $Q_{-1} \to Q$, the transition table can contain several entries in a single row.

- Characteristic expression: $Q = S + (\overline{R} \cdot Q_{-1}) = S + \overline{R} \cdot Q_{-1}$.

  The condition $S \cdot R = 0$ must be maintained to avoid the undefined state.

SR flip-flop			
R	S	Q_{-1}	Q
0	0	0	0
0	0	1	1
0	1	×	1
1	0	×	0
1	1	×	?

×: don't care
?: undefined state

Synthesis table			
Q_{-1}	Q	R	S
0	0	0	0
		1	0
0	1	0	1
1	0	1	0
1	1	0	0
		0	1

Compact Form			
Q_{-1}	Q	R	S
0	0	×	0
0	1	0	1
1	0	1	0
1	1	0	×

D Flip-Flop (Edge-Triggered)

D flip-flop		
D	Q_{-1}	Q
0	0	0
0	1	0
1	0	1
1	1	1

Synthesis table		
Q_{-1}	Q	D
0	0	0
0	1	1
1	0	0
1	1	1

- Characteristic expression: $Q = D$

| D flip-flop ||||||
Clear	Preset	D	Q_{-1}	Q
0	0	0	0	0
		0	1	0
		1	0	1
		1	1	1
0	1	×	×	1
1	0	×	×	0
1	1	×	×	?

×: don't care
?: undefined state

Sometimes the *preset* and *clear* inputs govern the operation of these flip-flops. They have priority over the data inputs. Asynchronous *preset* and *clear* inputs operate immediately on the output signal, synchronous only at the next relevant clock edge.

T Flip-Flop (Edge-Triggered)

A T flip-flop with *preset* and *clear* inputs has the following truth table:

| T flip-flop ||||
Clear	Preset	Q_{-1}	Q
0	0	0	1
0	0	1	0
0	1	×	1
1	0	×	0
1	1	×	?

×: don't care
?: undefined state

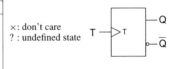

| Synthesis table ||||
Q_{-1}	Q	Clr	Pre
0	0	1	0
0	1	0	0
		0	1
1	0	0	0
		1	0
1	1	0	1

| Compact Form ||||
Q_{-1}	Q	Clear	Preset
0	0	1	0
0	1	0	x
1	0	×	0
1	1	0	1

- Characteristic expression:

$$Q = (\overline{Q_{-1}} \cdot \overline{\text{Clear}}) + \text{Preset} = \overline{Q_{-1} \cdot \text{Clear}} + \text{Preset}$$

with the condition that $\text{Clear} \cdot \text{Preset} = 0$

JK Flip-Flop (Edge-Triggered)

JK flip-flop			
J	K	Q_{-1}	Q
0	0	0	0
0	0	1	1
0	1	×	0
1	0	×	1
1	1	0	1
1	1	1	0

×: don't care

By including the *preset* and *clear* inputs the transition table can be expanded to:

JK flip-flop					
Clear	Preset	J	K	Q_{-1}	Q
0	0	0	0	0	0
		0	0	1	1
		0	1	×	0
		1	0	×	1
		1	1	0	1
		1	1	1	0
0	1	×	×	×	1
1	0	×	×	×	0
1	1	×	×	×	?

Synthesis table			
Q_{-1}	Q	J	K
0	0	0	0
		0	1
0	1	1	0
		1	1
1	0	0	1
		1	1
1	1	0	0
		1	0

Compact Form			
Q_{-1}	Q	J	K
0	0	0	×
0	1	1	×
1	0	×	1
1	1	×	0

×: don't care
? : undefined state

- Characteristic expression:

$$Q = (J \cdot \overline{Q}_{-1}) + (\overline{K} \cdot Q_{-1}) = J \cdot \overline{Q}_{-1} + \overline{K} \cdot Q_{-1}$$

8.5.11 Overview: Flip-Flop Circuits

TTL	Function	CMOS
74118	Six SR flip-flops	4042[a]
7474[b]	Two D flip-flops, edge-triggered	4013
7475[b]	Four D flip-flops	4042
7473[b]	Two JK flip-flops	
74107[b]	Two JK flip-flops	
7476[b]	Two JK *master–slave* flip-flops	4027
74111	Two JK *master–slave* flip-flops with data lockout	

[a] Only four flip-flops.
[b] Also available as 74 HCxxx or 74 HCTxxx *high-speed* CMOS series.

8.6 Memory

Strictly speaking, semiconductor memory can be divided into

- addressable memory;
- programmable logic devices.

Addressable memory are used for data, programs, etc. These are what is usually meant when referring to **memory**. Programmable logic devices memorise logic function connections. These are described in Sect. 8.6.5.

Addressable memory can be categorised according to its access points:

- ROM (*read-only memory*) permanent memory;
- RAM (*random access memory*) read–write memory.

ROM-storage is non-volatile, that is, the memory contents are not lost on removal of the supply voltage. Also, the memory contents cannot be altered.

RAM-storage is volatile, that is, the memory contents are lost on removal of the supply voltage. Also, the memory contents can be both written and read.

The notation *random access memory* (unconstrained memory access) is for historical reasons. Both memory types can be accessed as the user chooses. Semiconductor memories are organised in a way that the memory location is freely accessible for reading or writing if its **address** has been specified correctly. The memory capacity is always to powers of 2, as the addresses are encoded in binary.

- **Bit-oriented** memories store a single bit at each address.
- **Word-oriented** memories store 4, 8, 16 or 32 bits at each address.

8.6.1 Memory Construction

Memory elements are organised in matrix form. The address is divided internally into row and column addresses. Each is decoded by a row or column decoder. The memory element at the intersection of the selected row and column is selected by AND-ing. It is connected to the data bus. The R/$\overline{\text{W}}$ (*read/write*) signal selects whether the memory element is to be written to or read from (Fig. 8.51).

In addition to the read/write signal R/$\overline{\text{W}}$, a CS (*chip select*) signal is employed. This selects the overall memory device. For $CS = 0$ the data output is in a high-impedance state. This permits the use of several memory elements on a bus system.

The CS and R/$\overline{\text{W}}$ gate signals generate a write enable WE. This enables the D flip-flop in the addressed memory location (Fig. 8.52). For memories that store entire words several of these memory locations are situated in parallel. For any given address an entire **memory cell** can be accessed. For permanent memory (ROM) the R/$\overline{\text{W}}$ line can be eliminated. The data lines D_{in} and D_{out} are connected internally, the R/$\overline{\text{W}}$ signal switches the output gates into a high-impedance state in the write mode.

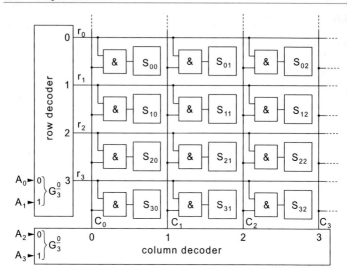

Fig. 8.51. Principle of construction of a memory element

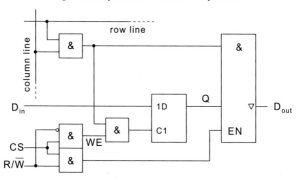

Fig. 8.52. Equivalent circuit for a memory element

8.6.2 Memory Access

All signals must meet certain conditions for proper memory access.

Reading:

- A certain amount of time t_{AA} must pass after applying the address, because of the internal propagation delays before the data is valid at the output. This is the **address access time** t_{AA} or simply **access time**.

Writing:

- A certain amount of time t_{AS} must pass after applying the address, before the write enable R/$\overline{W}$ goes low (*address setup time*).
- The write enable R/$\overline{W}$ must stay low for a minimum amount of time t_{W_p} (*write pulse width*).
- The data is read in at the positive edges of the R/$\overline{W}$ write-enable signal. Also the data must be applied in a stable manner for a minimum amount of time t_{DW} (*data valid to end of write*).
- After the change of the R/$\overline{W}$ write-enable signal the data and address lines must maintain their values for a minimum amount of time t_H (*hold time*).

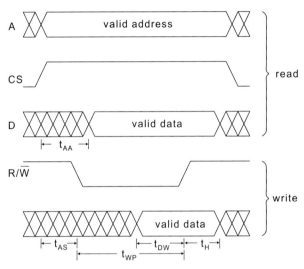

Fig. 8.53. Waveform diagram of a read and a write operation

The minimum total time for a write operation is

$$t_W = t_{AS} + t_{W_p} + t_H$$

where t_W is the **write cycle time**.

8.6.3 Static and Dynamic RAMs

- **Static RAMs** maintain their memory contents in the presence of a supply voltage without requiring extra external circuitry (**SRAM**).
- **Dynamic RAMs** have to be periodically refreshed, or else the memory contents are lost (**DRAM**).

In static RAM each memory element is realised using a flip-flop. For CMOS RAM six transistors are required per bit.

In the effort to use the smallest chip area per bit possible, memory elements have been realised using a single MOSFET transistor. The memorisation is achieved using charge packets in the transistor's gate–source capacitor. The charge is held for a relatively short time, so the memory must be refreshed every few milliseconds.

During a read access the entire memory row is refreshed. If the application does not automatically access each row in the memory to refresh it, this must be realised with separate circuitry. It is worth the extra outlay as DRAM have roughly 4 times higher integration density. For higher memory capacity many address lines are required. This implies a large housing for the IC. Column and row addresses are multiplexed to reduce the number of external pins, as shown in Fig. 8.54. The address conversion occurs in the internal interim memory using the *column address strobe* (CAS) and *row address strobe* (RAS) signals.

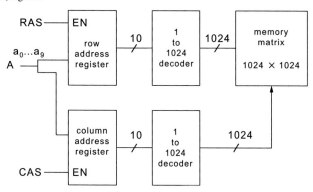

Fig. 8.54. Address multiplexing and interim memories in a 1 MBit DRAM

8.6.3.1 Variations of RAM

Dynamic RAM controller: Logic taking care of automatic refreshing of DRAM memory contents.

Pseudostatic RAMs: Dynamic RAM where the *refresh* logic is already integrated.

Multiport RAM: Frequent design where one port maybe written to, while the other can only be read.

EXAMPLE: Video memory: each port has separate address and data lines.

Arbiter: Priority logic that resolves access conflicts in multiport memories. In small memory devices this is integrated in the multiport memory chip.

FIFO: (*first in, first out*) memory realising a buffer (Fig. 8.55). The memory is equipped with input and output ports. Addressing is automatically performed internally. The data are output in the order in which they were originally input. A FIFO uses two address registers that point to the first and last entries in the buffer queue. Addressing is arranged in cyclic fashion, hence the name **ring memory**.

ECC memory: (*error-correcting code*) is memory that stores redundant bits for error control purposes. Individual bit errors can be detected and corrected (EDC *error detection and correction*). Popular combinations of information/parity bits are 8/5, 16/6 and 32/7.

EDC controller: Logic circuit realising the ECC memory error detection and correction.

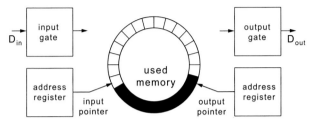

Fig. 8.55. Logic model of a FIFO memory

8.6.4 Read-Only Memory

Read-only memory (ROM) is read-only during normal operation. It is nonvolatile, i.e. the memory contents remain intact even after removal of supply voltage. The basic structure is a diode array. Diodes are located at the intersection of row and column conductors. Actually, the memory contents are not realised by the presence of a diode, but by their electrical connection to the column conductor (Fig. 8.56).

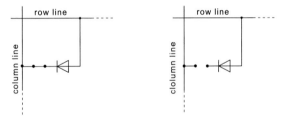

Fig. 8.56. Principle of memorisation of a bit in a ROM

There are different types of ROM:

- **ROM** (*read-only memory*): Data contents are burnt in during the last step of the manufacturing process in the form of a metallisation mask (**mask-programmable** ROM). The lead time is high. This is economical only for large production quantities.
- **PROM** (*programmable read-only memory*): Can be irreversibly programmed by the user. A precisely specified overcurrent/overvoltage pulse stream either burns through a link (*fusible link*) or pn junctions of the coupling elements are shorted (*avalanche-induced migration*).
- **EPROM** (*erasable programmable read-only memory*): Can be completely erased by the user using intensive exposure to ultraviolet light. Distinguishing feature: quartz window on the top of the IC housing. Coupling elements are FETs with 'floating gates'. This creates a highly isolated capacitor whose charge, influenced by the FET's threshold voltage, represents the information storage. EPROMs are generally slower than PROMs.

- **EEPROM** (*electrically erasable read-only memory*), also **EAROM** (*electrically alterable read-only memory*): Memory cells can be selectively electrically programmed *and* erased. The total number of programme/erase cycles is limited to about 10^4. Because of lower prices EEPROMs replace EPROMs.

  EEPROMs are also combined with RAM (known as *flash* EEPROM) in a single device, to gain the advantages of RAM (fast and frequent access) and of EEPROM (nonvolatile).

8.6.5 Programmable Logic Devices

Programmable logic devices (PLD) store logic connections. Their structure is oriented along the normal representation of logic functions. Each has an array of AND and OR connections that can be programmed by the user to make or break, i.e. the user can modify the array to achieve the desired function.

8.6.5.1 Principle of Operation

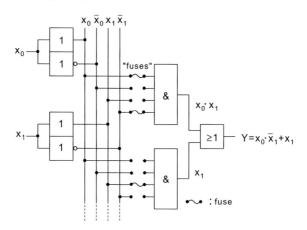

Fig. 8.57. Principle of a programmable logic device

Figure 8.57 shows the principle of a **PLD**. Two input signals are stored/appear on column conductors in both inverted and noninverted form. These are then connected to several AND gates whose outputs connect to an OR gate. Programmability here simply means making the proper breaks in the connections. A simpler representation of the configuration in Fig. 8.57 is shown in Fig. 8.58. The crosses represent connections.

The programming process therefore is the same as that of a PROM. Figure 8.59 shows a comparison of three PLD structures.

- **PROM**s consists of a fixed AND array that provides the address decoding. The OR array is programmable and holds the the memory contents.
- **PAL**s on the other hand consist of a programmable AND array, whereas the OR array is fixed.

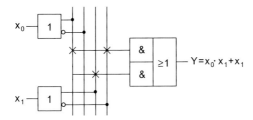

Fig. 8.58. Compact representation of the PLD circuit

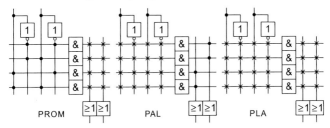

• : fixed connection   × : fused (programmable) connection

Fig. 8.59. Principle structure of PROM, PAL and PLA

- **PLAs** offer both programmable AND as well as OR arrays. They are therefore more flexible than PROMs or PALs; however, the propagation time through the array is higher.

8.6.5.2 PLD Types

The basic structure of the PLD architecture means that there are several variations of programmable logic array elements. The differences lie in the method used to program the array (using fuses, diodes or FETs), and in the programmability of the AND and OR arrays as well as in their ability to be subsequently reprogrammed.

PROM: Fixed AND array that provides the address decoding and a programmable OR array. The connections are metallic and behave like fuses that can be burnt.

EPROM: (*erasable programmable read-only memory*) PROM version. Fixed AND array. Programmable OR array. The coupling elements are FETs with isolated gates. The information is stored as charge in the gate capacitor. Erasure is achieved by removing the charge.

PAL: (*programmable array logic*) Fixed OR array. The AND array is programmable.

HAL: (*hardware array logic*) A mask-programmed version of the PAL produced by the manufacturer.

PLA: (*programmable logic array*) Both the AND as well as the OR array are programmable. PLAs are therefore more flexible, but also require more design effort. Will be replaced by LCA (logic cell array).

EPLD: (*erasable programmable logic device*) This has the same structure as the PAL. The coupling elements are the same as for EPROMs. In this manner EPLDs are UV-erasable and reusable.

IFL: (*integrated fuse logic*) General expression for different kinds of programmable logic devices.

FPGA: (*field-programmable gate array*)
FPLA: (*field-programmable logic array*)
FPLS: (*field-programmable logic sequencer*)

LCA: (*logic cell array*) offers reconfigurable logic blocks. The kind of connection is stored in a nonvolatile memory. This is how the LCA is programmed (trademark of XILINX).

AGA: (*alterable gate-array logic*) alterable gate array.

GAL: (*generic array logic*) electrically erasable gate array, with a PAL structure and programmable output configurations. Can replace many PAL types.

Table 8.13 shows an overview of the different kinds of PLD. The ROM is included for comparison:

Table 8.13. Properties of PDAs

PLD Type	AND array	OR array	Memory
ROM	Fixed	Mask	Mask
PROM	Fixed	Programmable	Fuse
EPROM	Fixed	Programmable	Stored charge
PAL	Programmable	Fixed	Fuse
HAL	Mask	Fixed	Mask
PLA	Programmable	Programmable	Fuse
EPLD	Programmable	Fixed	Stored charge
LCA	Programmable	Programmable	Stored charge
AGA	Programmable	Programmable	Stored charge
GAL	Programmable	Fixed	Stored charge

A characteristic of the PLDs is a 'last fuse'. If these are burnt through then the contents of the programmed array are no longer electrically reachable. Therefore there is a certain amount of security against unauthorised copying of the internal structure.

8.6.5.3 Output Circuits

PALs are equipped with various output circuits. These are shown in Fig. 8.60.

The following types are realised:

High-H output: The signal is available after the OR gate.

Low-L output: The signal is inverted.

Complement-C output: The signal and its complement are both output. This is a rare and uneconomical solution since many output pins are required.

Programmable-P output: The polarity of the output can be selected by using an XOR gate as a controlled inverter. The controlling input of the XOR gate is connected to ground by a fuse.

XOR-X output: Two OR outputs are XOR-ed. This structure is applied almost exclusively in arithmetic units.

Sharing-S output: Creates a 'poor man's' FPLA out of a PAL. This version of the PLD offers a small programmable output OR array.

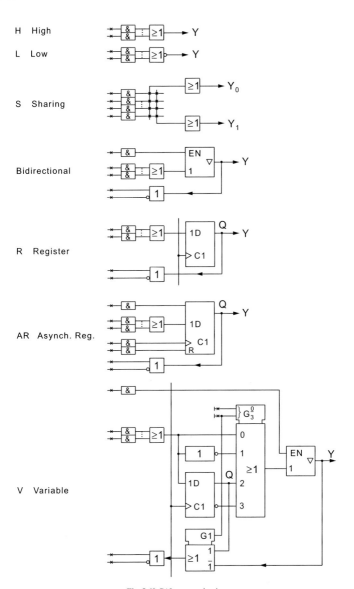

Fig. 8.60. PAL output circuits

Bidirectional-B outputs: Can be programmed as inputs or as feedback of interim results (multiple use of a partial term). The output circuit is *tri-state* capable. The *enable* signal can be derived of a logic combination of the input signals.

Register-R outputs: At defined point in time the outputs states are transferred into D flip-flops. The clock line is common to all gates. This structure is suitable for the synthesis of sequential logic circuits.

Asynchronously registered AR output: Set, reset and clock signal are obtained as logic terms.

Variable-V output: New variation of PAL (or GAL) are equipped with output macro cells (OLMC: *output logic macro cell*) that can be programmed using control bits to realise one of the output versions H, L or R.

8.7 Registers and Shift Registers

Registers are flip-flop configurations for the interim storage of signal states. The 4-, 8- or 16-bit register (*latch*) is a parallel configuration of D flip-flops, which have a common clock.

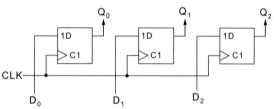

Fig. 8.61. 3 bit register using D flip-flops

Shift registers are flip-flops in a ring circuit, i.e. the output of one flip-flop is connected to the input of the next flip-flop.

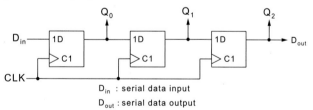

Fig. 8.62. 3 bit shift register

All flip-flops are clocked by the same clock signal. The delayed input signal appears at the output.

The connection between the inputs and outputs can be separated, and an external signal can be input by using a multiplexer. This is known as **loadable shift register with parallel access**. The signal *load* controls the acceptance of the data. Such a shift register is known as a **parallel in serial out** (PISO) and as a **serial in parallel out** (SIPO).

Shift registers are available as 4, 8, 16 and more flip-flop suites.

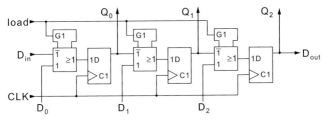

Fig. 8.63. Loadable shift register

8.8 Counters

Counters are sequential logic that have a defined series of internal flip-flop states dependent on the applied clock signals. The internal states do not necessarily correspond to a common number representation. The type of control is used to differentiate between them:

- **Synchronous counters:** All flip-flops are clocked in parallel (simultaneously).
- **Asynchronous counters:** At least one of the flip-flops receives a clock signal that has been generated within the circuit.
- **Semisynchronous counters:** Synchronous counter elements are connected in series. Such counters are synchronous in sections, but overall are asynchronous.

Forms of representation of counter states:

- **Binary counter:** The counter state is represented in binary form.
- **BCD counter:** Each decimal place of the counter state is individually represented in binary form.
- **Others:** The counter state represents other codes (1 of 10, biquinary (e.g. 74 393), etc.).

Forms of counting direction:

- **Up counter,**
- **Down counter,**
- **Up/Down counter,**
- Counters with **separate** up and down count **inputs**.

Forms of flip-flop configuration:

- **Walking-ring counter**: Consist of a shift register whose contents are cyclically shifted.
- **Johnson counter** (*switch tail ring counter*): A special form of the walking-ring counter.

Forms of control options:

- **Programmable counter** (up counter): This allows a defined counter state to be loaded in parallel and the count to proceed from this new state.

8.8.1 Asynchronous Counters

8.8.1.1 Binary Counter

The following relationship can be read from the truth table of a binary counter:

- An output variable z_i changes value, if the next lowest variable z_{i-1} changes state from 1 to 0. This rule is highlighted by the horizontal lines in the following table.

Counter state	z_2 2^2	z_1 2^1	z_0 2^0
0	0	0	0
1	0	0	1
2	0	1	0
3	0	1	1
4	1	0	0
5	1	0	1
6	1	1	0
7	1	1	1

The realisation of an asynchronous binary counter, as shown in Fig. 8.64, follows from this table. The complementary output of the D flip-flop is fed back to its input. In this manner each flip-flop, at each relevant clock edge, takes a complementary signal at its D-input (*toggle* flip-flop). Each flip-flop functions as a 1 : 2 frequency divider. It can be seen from the waveform diagram that the counter state can be directly read from the output variables.

The counter shown in the diagram returns to the start state of 0 after the counter state 7. It runs through a total of 8 states. It is therefore called a **modulo-8** counter. Each extra flip-flop extends the range of the counter by a power of 2.

The transition of the counter state 7 to the counter state 0 shows an essential disadvantage of the asynchronous counter: The positive clock edges at the first flip-flop's input cause it to toggle. The complementary output changes from $0 \rightarrow 1$ and changes the following flip-flop and so on. However, each flip-flop can only toggle once the previous flip-flops has toggled. The clock input positive edge is delayed at each flip-flop by the propagation delay time t_{PH}. The counter state is correct only after all flip-flops have settled. In the meantime, incorrect output states are on the output lines. Because of this carryover delay this counter is known as a *ripple-through counter*.

Figure 8.65 shows an asynchronous binary counter created from JK flip-flops. The flip-flops are edge-triggered on the positive edge.

Figure 8.66 shows the circuit symbol for an asynchronous binary counter. The reset input 11 forms part of the control block for all of the flip-flops. The input 10 is negative edge-triggered. The output 9 influences the input 1 internally (Z-dependency). For a transition from 0 to 1 its state changes (T-dependency). The other outputs work in a corresponding manner. The simplified circuit symbol is shown next to it. This is used if the asynchronous operation does not have to be explicitly identified.

8.8.1.2 Decimal Counter

Decimal counters are often used in applications where the counter state is shown in decimal form. In order to keep the decoder requirement low, a counter is used for each decimal

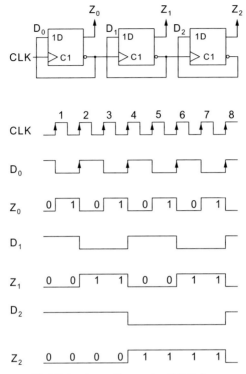

Fig. 8.64. Asynchronous binary counter with D flip-flops

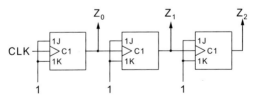

Fig. 8.65. Asynchronous binary counter with JK flip-flops

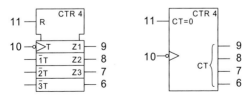

Fig. 8.66. Circuit symbols for the asynchronous binary counter

place that counts from 0 to 9. Normally the counter states are represented in binary form so the counters are known as **BCD** counters (*binary coded decimal*).

The truth table shows the internal representation of the counter state for a BCD-counter. As each entry has a weighting of 8, 4, 2 and 1, this is known as an **8421-code**.

The decimal counter in Fig. 8.67 derives from a 4-bit binary counter. The NAND gate resets all of the flip-flops the moment both z_1 and z_3 assume the value 1. That occurs for the (irregular) counter state 10. This state lasts only for the duration of the signal delays in the counter. The reset signal is therefore a spike.

Counter state	z_3 2^3	z_2 2^2	z_1 2^1	z_0 2^0
0	0	0	0	0
1	0	0	0	1
2	0	0	1	0
3	0	0	1	1
4	0	1	0	0
5	0	1	0	1
6	0	1	1	0
7	0	1	1	1
8	1	0	0	0
9	1	0	0	1
10[†]	1	0	1	0

[†]transient irregular state

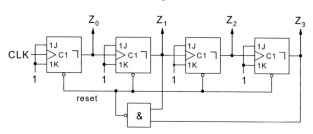

Fig. 8.67. Asynchronous decimal counter

In practice such circuits are avoided, as their correct operation depends critically on the propagation delays. The configuration shown in Fig. 8.68 avoids this problem by blocking the counter stages via preparation inputs. This means that the counter goes directly to state 0 at the following clock edge after counter state 9.

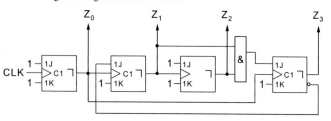

Fig. 8.68. Asynchronous decimal counter

Figure 8.69 shows the circuit symbol of a decimal counter. The input 1 is negative edge-triggered and causes the counter to count upwards (plus-sign). The counter divides by 10 (CTR DIV 10). The counter state is available at the connections 3, 5, 6, 7 encoded in binary form. The braces gather the outputs their power to the base two is written as [0..3]. Input 2 causes a reset, recognisable by a signal $CT = 0$. Resetting is carried out asynchronously, as no C-dependency is given.

Fig. 8.69. Circuit symbol of a decimal counter

Any arbitrary number of digits can be realised by connecting several decade counters serially (Fig. 8.70).

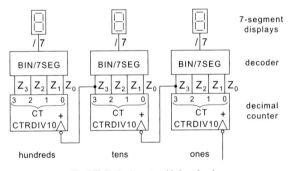

Fig. 8.70. Decimal counter with three decades

The counter state of each decade is transformed through a BCD/7 segment decoder for display purposes. The Z_3 output signal is carried over to the next-higher decade. This signal has a negative edge only when resetting of the counters occurs, which triggers the following counter. The carryover signal of the highest decade can be used to set an SR flip-flop and thereby indicate the overflow of the count.

8.8.1.3 Down Counter

A down counter *decreases* the counter state at each input pulse.

The following relationship can be seen from the truth table of a down counter:

- An output variable Z_i changes its value if the next lowest variable Z_{i-1} changes state from 0 to 1. This rule is highlighted at the horizontal lines.

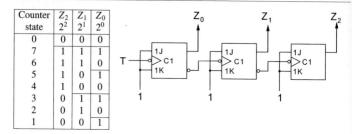

Counter state	Z_2 2^2	Z_1 2^1	Z_0 2^0
0	0	0	0
7	1	1	1
6	1	1	0
5	1	0	1
4	1	0	0
3	0	1	1
2	0	1	0
1	0	0	1

Fig. 8.71. Asynchronous 3-bit down counter

Unlike the up counter, the flip-flop *complementary* outputs are connected to the subsequent clock inputs.

8.8.1.4 Up/Down Counter

A counter with programmable counting direction is obtained if the outputs of the flip-flops are fed into XOR gates (Fig. 8.72). This allows the toggling of the output polarity over a common control line and thus determines either up or down counting.

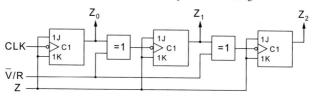

Fig. 8.72. Asynchronous switchable up/down counter

NOTE: Changing the count direction should not occur during a count process, as this causes a polarity change at the flip-flop inputs and causes uncontrolled counting. The Z-input blocks the J-K inputs during the switch.

8.8.1.5 Programmable Counter

Programmable counters can be (pre)loaded with a defined counter state.

Figure 8.73 shows a programmable 4-bit counter. A logic high voltage level at the *load* input causes all flip-flops to be set or reset, depending on the applied signals at the parallel inputs.

Figure 8.74 shows the circuit symbol for a programmable 4-bit counter on the left. The loading process is triggered by the input marked with load (C-dependency). The value of the individual stages is shown in the parentheses. The output in the control block on the right for counter state 15 ($CT = 15$) assumes the state 1. It supplies a carryover signal for further expansion of the counter.

Programmable down counters that stop on reaching the zero state or begin a new loaded count are of particular importance, especially in microprocessor systems. Such counters are called **presettable counters**. Figure 8.74 shows on the right side a down counter

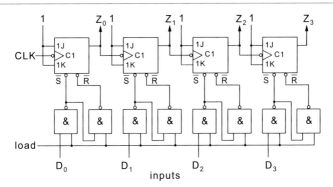

Fig. 8.73. Programmable 4-bit counter

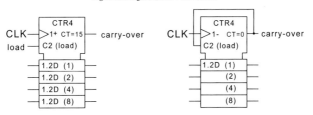

Fig. 8.74. Circuit symbols for the programmable counter

that loads a new count state when it reaches count state zero. It can be seen from the dependency notation for the load and clock input that the load signal is effective only after the subsequent clock signal. If the counter is also loaded with the number m, then it runs through $m + 1$ cycles. It then functions as a **modulo-$(m + 1)$ counter**.

APPLICATION: Such counters are employed in programmable frequency dividers or timers.

8.8.2 Synchronous Counters

Up counter				Down counter			
counter-state	z_2 2^2	z_1 2^1	z_0 2^0	counter-state	z_2 2^2	z_1 2^1	z_0 2^0
0	0	0	0	7	1	1	1
1	0	0	1	6	1	1	0
2	0	1	0	5	1	0	1
3	0	1	1	4	1	0	0
4	1	0	0	3	0	1	1
5	1	0	1	2	0	1	0
6	1	1	0	1	0	0	1
7	1	1	1	0	0	0	0

The table shows the count states for a binary counter. The following rule can be derived from the truth table for an up counter:

- An output variable z_i changes its value when all lower value variables have a value of 1 *and a new count impulse arrives.*

This rule holds for counting downwards:

- An output variable z_i changes its value when all lower value variables have a value of 0 *and a new count impulse arrives.*

These rules are taken into consideration in the design of synchronous counters (Fig. 8.75). The characteristic feature of synchronous counters is that the clock signal is simultaneously (synchronously) fed to all flip-flops. In order that the flip-flops toggle only at the permitted states, the set-inputs of the flip-flops must be fed by a suitable combinational preparation circuit.

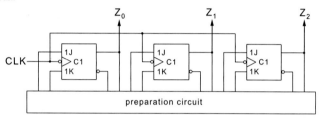

Fig. 8.75. Principle of a synchronous counter

The nature of the input circuit follows from the rules just mentioned. For an up counter, a flip-flop may only toggle at the clock edge when all lower value stages have the state 1. It follows that:

$$S_0 = 1, \qquad S_1 = Z_0, \qquad S_2 = Z_0 \cdot Z_1, \qquad S_3 = Z_0 \cdot Z_1 \cdot Z_2$$

The AND gates in Fig. 8.76 implement these logic expressions.

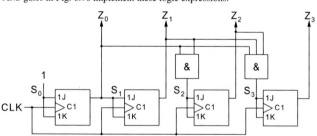

Fig. 8.76. Synchronous binary counter

8.8.2.1 Cascading Synchronous Counters

It is often a problem to design synchronous counters whose count capacity exceeds the capacity of an individual counter device. This is explained with the example of the 71 191 4-bit synchronous counter. The 71 191 is positive edge-triggered and is equipped with two suitable outputs to extend the count range.

Min/max: The output min/max goes low when either the up counter reaches the maximum counter state (15), or the down counter reaches zero.

RCE (*ripple count enable*): This output is logic 0 when the *enable* input and the min/max input are low and count input is at logic 0.

Figure 8.77 shows the obvious circuit to extend the count range. The RCE output of each counter stage is connected to the clock input of the following device. This can be described as being **semisynchronous** or **partially synchronous**. The clock is fed in parallel only to the flip-flops in the first counter device. The maximum count speed decreases with the length of the counter.

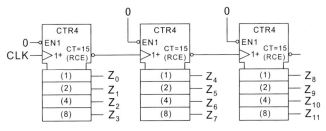

Fig. 8.77. Semisynchronous binary counter

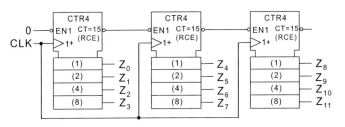

Fig. 8.78. Synchronous binary counter with serial carryover

In the circuit in Fig. 8.78 the entire multistage counter operates synchronously, but the carryovers are produced serially. Each counter stage is equipped with an *enable* input that blocks the counter and the carryover generation. The first counter is continuously enabled. The enable input of the next stage is fed with the carryover signal of the preceding counter stage. So, for example, the second counter device can only continue counting as long as the first counter outputs the carryover signal. This is the case for exactly one clock period.

The circuit in Fig. 8.79 permits the fastest operation, as the carryover is output in parallel. The output min/max goes low, when the maximum is reached in counting upwards, or zero is reached when counting downwards. All counter stages are supplied with the same clock signal in parallel.

NOTE: The carryover signal gates are already integrated in some counters (e.g. in the 74 163). Figure 8.80 shows a circuit example (see also the manufacturer's application notes).

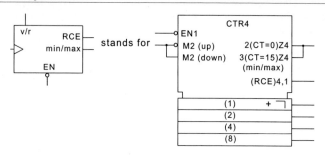

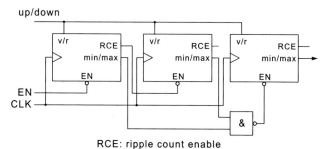

Fig. 8.79. Synchronous binary counter with parallel carryover

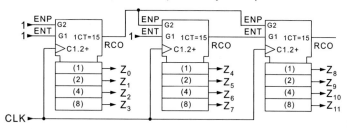

Fig. 8.80. Synchronous binary counter with parallel carryover without external gates

8.8.3 Overview: TTL and CMOS Counters

Tables 8.14 and 8.15 give an overview of TTL and CMOS counter properties.

Explanation of table entries:

A – asynchronous counter	BCD	– BCD-counter
S – synchronous counter	B	– binary code
± – up/down counter	1/10	– 1-to-10 code
↑ – counter triggers on positive edges	7-segment	– seven-segment code
↓ – counter triggers on negative edges	J	– Johnson counter

The range of the counter is given by the number of bits. If it is not a base 2 number, then the amount of counter states is also given.

AC – asynchronous clear		SL	– synchronous load
SC – synchronous clear		OC	– open collector
AS – asynchronous set		ENT, ENP	– inputs for parallel carryover generation without external gates
AL – asynchronous load			
		P	– programmability

The clock frequencies that are given are guaranteed values. Typical values are around 50% higher. Many of the listed TTL counters are also available as ALS devices with higher clock frequencies.

8.8.3.1 TTL Counters

Table 8.14. Properties of TTL counters

Type	A/S	Edge	Range [Bit]/Number	Code	P	Reset	Frequency (guaranteed) [MHz]	Observations
LS 90	A	↓	4/10	BCD	AS	AC	32	Can be set to 9
LS 92	A	↓	4/12	B	–	AC	32	
LS 93	A	↓	4	B	–	AC	32	Succeeded by LS 293
LS 142	A	↑	4/10	1/10	–	AC	20	With latch, decoder, OC driver 60 V
LS 143	A	↑	4/10	7-Seg	–	AC	12	As LS 142 with 7-segment decoder, LED constant-current outputs
LS 144								as LS 143 with 15 V OC driver
LS 160	S	↑	4/10	BCD	AL	AC	25	
LS 161	S	↑	4	B	SL	AC	25	As LS 163 with AC
LS 162	S	↑	4/10	BCD	SL	SC	25	
LS 163	S	↑	4	B	SL	SC	25	As LS 161 with SC
LS 168	S±	↑	4/10	BCD	SL	–	25	ENT, ENP inputs
LS 169	S±	↑	4	B	SL	–	25	
LS 176	A	↓	4/10	BCD/5-2	AL	AC	35	Depending on external circuitry: BCD or biquinary code
LS 177	A	↓	4	B	AL	AC	35	
LS 190	S±	↑	4/10	BCD	AL	–	20	
LS 191	S±	↑	4	B	AL	–	20	
LS 192	S±	↑	4/10	BCD	AL	AC	25	Separate clock inputs for up/down counter
LS 193	S±	↑	4	B	AL	AC	25	Separate clock inputs for up/down counter
LS 196	A	↓	4/10	BCD	AL	AC	30	
LS 197	A	↓	4	B	AL	AC	30	
LS 290	A	↓	4/10	BCD	AS	AC	32	
LS 293	A	↓	4	B	–	AC	32	As LS 93 with supply pins at corners
LS 390	A	↓	8/100	BCD	–	AC	25	Two LS 290 in a single housing
LS 393	A	↓	8	B	–	AC	25	Two LS 293 in a single housing

8.8.3.2 CMOS Counters

The frequencies given are for a 50 pF load at 5/10/15 V.

Table 8.15. Properties of CMOS counters

Type	A/S	Edge	Range [Bit]/Number	Code	P	Reset	Frequency (guaranteed) [MHz]	Observations
4017	S	↑/↓	5/10	1/10	–	AC	3/8/12	Johnson counter
4018	S	↑	5/2..10	J	AL	AC	2/6/8	Johnson counter
4020	A	↓	14	B	–	AC	5/13/18	
4022	S	↑	4/8	1/8	–	AC	3/8/12	Johnson counter
4024	A	↓	7	B	–	AC	5/13/18	
4029	S±	↑	4 or 4/10	B/BCD	AL	–	4/12/18	Switchable binary/decimal counter
4040	A	↓	12	B	–	AC	5/13/18	
4060	A	↓	14	B	–	AC	4/10/15	Gates for oscillator
4510	S±	↑	4/10	BCD	AL	AC	5/12/17	
4516	S±	↑	4	B	AL	AC	5/12/17	
4518	S	↑/↓	2×4/100	BCD	–	AC	3/7/10	Two decimal counters
4520	S	↑/↓	2×4	B	–	AC	3/7/10	Two binary counters
4522	A-	↑/↓	4/10	BCD	AL	AC	6/12/16	Down counter
4526	A-	↑/↓	4	B	AL	AC	6/12/16	Down counter
4534	A	↓	20/10^5	BCD	–	AC	2.5/6/8	BCD multiplex output
4737	A	↑	16/20000	BCD	AS	AC	3/8/10	Multiplex output
40160	S	↑	4/10	BCD	SL	AC	5/12/17	
40161	S	↑	4	B	SL	AC	5/12/17	
40162	S	↑	4/10	BCD	SL	SC	5/12/17	
40163	S	↑	4	B	SL	SC	5/12/17	
40192	S±	↑	4/10	BCD	AL	AC	3/9/13	Separate clock inputs for up/down counter
40193	S±	↑	4	B	AL	AC	3/9/13	Separate clock inputs for up/down counter

Some of the devices are also available as HCT-CMOS devices with significantly higher allowable clock frequencies.

8.9 Design and Synthesis of Sequential Logic

Two design methods for sequential logic are presented aiming for different implementations, namely

- Sequential logic realised with programmable logic devices (PLD);
- Sequential logic realised with addressable memory (ROM).

Example A

Implementation of a programmable 3-bit counter. All data are in positive logic.

Requirements:
Reset: Reset counter states to zero.
Load: Load parallel applied data into counter.
Mode: L counts up, H counts down.

Inputs:
$D_0 \ldots D_2$: data inputs

Outputs:
$z_0 \ldots z_2$: counter state encoded in binary

Carry/borrow: not used in this example for clarity reasons.

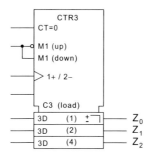

Fig. 8.81. Circuit symbol of the sequential logic to be realised

The following state transition table describes the operation of the control signals *reset*, *load*, *mode* as well as the data inputs D_i ($i = 0 \ldots 2$) and the outputs z_i ($i = 0 \ldots 2$). Counter state z_{-i}^* means the counter state *before* the triggering edge.

reset	load	D_i	z_i	$\overline{z_i}$
0	0	×	$z_i^{*\ddagger}$	$\overline{z_i^*}$
0	1	0	0	1
0	1	1	1	0
1	×	×	0	1

The following logic expression holds for any counter output z_i:

$$z_i = \overline{reset} \cdot \overline{load} \cdot z_i^* + \overline{reset} \cdot load \cdot D_i \tag{8.12}$$

NOTE: For PLDs with inverting outputs an expression for $\overline{z_i}$ can be derived. It can be seen from the table that:

$$\overline{z_i} = \overline{reset} \cdot \overline{load} \cdot \overline{z_i^*} + \overline{reset} \cdot load \cdot \overline{D_i} + reset$$

The same applies to the following expressions.

A further state transition table must be created for the actual count process. The count state order is influenced by the count direction or *mode* signal. ZS is a quantity that gives the counter state. The quantities with an asterisk denote the new states *after* the triggering clock signal.

mode	ZS	z_2	z_1	z_0	ZS^*	z_2^*	z_1^*	z_0^*
	0	0	0	0	1	0	0	1
	1	0	0	1	2	0	1	0
	2	0	1	0	3	0	1	1
	3	0	1	1	4	1	0	0
0	4	1	0	0	5	1	0	1
	5	1	0	1	6	1	1	0
	6	1	1	0	7	1	1	1
	7	1	1	1	0	0	0	0
	0	0	0	0	7	1	1	1
	1	0	0	1	0	0	0	0
	2	0	1	0	1	0	0	1
	3	0	1	1	2	0	1	0
1	4	1	0	0	3	0	1	1
	5	1	0	1	4	1	0	0
	6	1	1	0	5	1	0	1
	7	1	1	1	6	1	1	0

Therefore each counter position can be represented by a truth table, in which the *only* signal combinations that are entered are the ones that lead to $z_i^* = 1$. A synthesis table can be derived from this to describe the transition $z_i \rightarrow z_i^*$.

Least significant bit (LSB):

z_0	z_0^*	mode	z_2	z_1	z_0
0	0	-	-	-	-
0	1	0	×	×	0
		1	×	×	0
1	0	0	×	×	1
		1	×	×	1
1	1	-	-	-	-

This yields the expression

$$z_0^* = \overline{z_0} \tag{8.13}$$

Middle counter digit:

$z_1^* = 1$ **for**					$z_1^* = 1$ **for**			
mode	z_2	z_1	z_0		mode	z_2	z_1	z_0
0	0	0	1		0	×	0	1
0	0	1	0		0	×	1	0
0	1	0	1		1	×	0	0
0	1	1	0		1	×	1	1
1	0	0	0					
1	0	1	1					
1	1	0	0					
1	1	1	1					

The table on the right is a summary of the table on the left. This yields the expression for z_1^*

$$z_1^* = \overline{mode} \cdot \overline{z_1} \cdot z_0 + \overline{mode} \cdot z_1 \cdot \overline{z_0} + mode \cdot \overline{z_1} \cdot \overline{z_0} + mode \cdot z_1 \cdot z_0 \tag{8.14}$$

Most significant bit (MSB):

$z_2^* = 1$	**for**		
mode	z_2	z_1	z_0
0	0	1	1
0	1	0	0
0	1	0	1
0	1	1	0
1	0	0	0
1	1	0	1
1	1	1	0
1	1	1	1

$z_2^* = 1$	**for**		
mode	z_2	z_1	z_0
0	0	1	1
0	1	0	×
×	1	1	0
1	0	0	0
1	1	×	1

The table on the right is a summary of the table on the left. This yields the expression for z_2^*

$$z_2^* = \overline{mode} \cdot \overline{z_2} \cdot z_1 \cdot z_0 + \overline{mode} \cdot z_2 \cdot \overline{z_1} + z_2 \cdot z_1 \cdot \overline{z_0} + mode \cdot \overline{z_2} \cdot \overline{z_1} \cdot \overline{z_0} \\ + mode \cdot z_2 \cdot z_0 \tag{8.15}$$

The following expressions for the individual counter bits can be derived by inserting expressions (8.13) to (8.15) into expression (8.12):

$$z_0^* = \overline{reset} \cdot load \cdot D_0 + \overline{reset} \cdot \overline{load} \cdot \overline{z_0}$$

$$z_1^* = \overline{reset} \cdot load \cdot D_1 \\ + \overline{reset} \cdot \overline{load} \cdot \overline{mode} \cdot \overline{z_1} \cdot z_0 + \overline{reset} \cdot \overline{load} \cdot \overline{mode} \cdot z_1 \cdot \overline{z_0} \\ + \overline{reset} \cdot \overline{load} \cdot mode \cdot \overline{z_1} \cdot \overline{z_0} + \overline{reset} \cdot \overline{load} \cdot mode \cdot z_1 \cdot z_0$$

$$z_2^* = \overline{reset} \cdot load \cdot D_2 \\ + \overline{reset} \cdot \overline{load} \cdot \overline{mode} \cdot \overline{z_2} \cdot z_1 \cdot z_0 + \overline{reset} \cdot \overline{load} \cdot \overline{mode} \cdot z_2 \cdot \overline{z_1} \\ + \overline{reset} \cdot \overline{load} \cdot z_2 \cdot z_1 \cdot \overline{z_0} + \overline{reset} \cdot \overline{load} \cdot mode \cdot \overline{z_2} \cdot \overline{z_1} \cdot \overline{z_0} \\ + \overline{reset} \cdot \overline{load} \cdot mode \cdot z_2 \cdot z_0$$

These expressions in the sum of products or product of sums form can be directly realised in a suitable PLD with output registers. In practice, the logic expressions are produced using computer-aided engineering software, which can generate the required layout (**PAL assembler**).

A method that is less focussed on the logic expressions for the combinational circuit but more on the states of the circuit to be designed is shown in the next example.

Example B

A circuit is to be designed that controls a traffic light at a pedestrian crossing. The sequence of the individual traffic lights, to which the circuit states should correspond, are represented in a state diagram (Fig. 8.82). Each state of the circuit is represented by a circle, and possible transitions from one state to another are represented by arrows. If the transitions can only occur under certain conditions, then the conditions are written beside the arrow.

State	Car	Pedestrian	Next state
0	Green	r	1
1	Amber	r	2
2	Red	r	3
3	Red	g	4
4	Red	r	5
5	Red/Amber	r	0

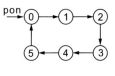

Fig. 8.82. State diagram of the traffic light control

For **synchronous circuits** a transition must only happen at the relevant clock edge. An arrow pointing back to the same circle means that the state is unchanged. Systems that can be described by a number of states and their transitions are known as **finite state machines**.

The table in Fig. 8.82 shows the individual states of the traffic light controller. The traffic light colours are denoted by their capitalised names for cars and by lower case letters for pedestrians.

Figure 8.82 shows the state diagram related to the table. The system should go to state 1 after being powered on. This is shown by the arrow with the notation *pon* (*power on*). Although states 2 and 4 activate the same traffic light colour, they are defined by different states as they have different subsequent states. The six states are passed through cyclically. The circuit can be very easily realised using a modulo-6 counter, whose outputs control a small memory element (ROM). This would translate the counter state in the table above into its corresponding traffic light colour (Fig. 8.83).

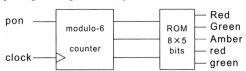

Fig. 8.83. Circuit realisation with a counter and ROM

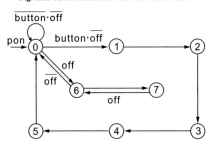

Fig. 8.84. Expanded traffic light control state diagram

8.9 Design and Synthesis of Sequential Logic

A real pedestrian crossing traffic light responds to the pressing of a button. The state diagram should therefore be expanded to include a *button*. In addition, the amber light should flash if an external *off* signal is received, and the pedestrian signals should deactivate (Fig. 8.84).

The state transition table then becomes:

State	Car	Pedestrian	Condition	Next state
0	Gr	r	Button · $\overline{\text{off}}$	1
			off	6
			$\overline{\text{Button}}$ · off	0
1	A	r		2
2	R	r		3
3	R	g		4
4	R	r		5
5	R/A	r		0
6	A	-	off	7
			$\overline{\text{off}}$	0
7	-	-		6

The circuit shown in Fig. 8.85 is suitable to realise the required control.

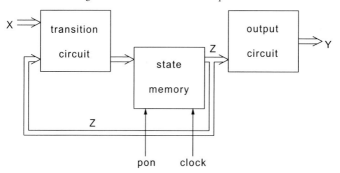

Fig. 8.85. Circuit with memory states, transition combinational and output circuits

The counter in Fig. 8.83 is replaced by a **state memory**. It stores the current state that is encoded into the **state vector** $z(t_n)$ as a series of binary digits. The subsequent state $z(t_{n+1})$ is given by the actual state and any possible input quantities, i.e. the **input vector** x (*qualifier*). The processing of the input vector x and state vector z is performed in the **transition logic** (usually a ROM). The state vector z is processed in the **output logic**. The result is the **output vector** y. In the case of the traffic light controller these are the signals for the traffic light colours. The state memory is also influenced by the clock and the power-on signal *pon*.

The traffic light controller passes through eight states, for which three flip-flops are sufficient. The state vector width is therefore 3. It is useful to employ a (P)ROM for the transition logic. Part of the address of the ROM consists of the state vector, and the rest is the input vector. The input signals can modify the next states under certain conditions and are therefore known as *qualifiers*. The ROM addresses are formed by the *off* and *button* qualifiers as well as the state vector.

$$\underbrace{\overbrace{x_1 \quad x_0}^{\text{Qualifier}} \quad \overbrace{z_2 \quad z_1 \quad z_0}^{\text{State}}}_{\text{ROM address}}$$

The ROM contents are therefore:

Table 8.16. Memory content of transition circuit

ROM address	State	*off*	*Button*	Next state
0	0	0	0	0
1		0	1	1
2		1	0	6
3		1	1	6
4				2
⋮	1	×		⋮
7				2
8				3
⋮	2	×		⋮
11				3
12				4
⋮	3	×		⋮
15				4
16				5
⋮	4	×		⋮
19				5
20				0
⋮	5	×		⋮
23				0
24	6	0	0	0
25		0	1	0
26		1	0	7
27		1	1	7
28				6
⋮	7	×		⋮
31				6

For the traffic light controller in the example above a 32 × 3-bit ROM s required for the transition logic. The output combinational logic requires an 8 × 5-bit ROM. For such a small amount of memory it makes sense to give the combinational logic circuit the structure shown in Fig. 8.86.

The transition logic here unites both functions of the previous combinational logic circuits. A row of this 32 × 8-bit memory has the contents shown in Table 8.17. The table shows a section of this memory.

NOTE: Traffic light colours usually last for different durations. That can be achieved by splitting up one traffic light phase over several states of the circuit or by influencing the clock generator with additional output signals.

Table 8.17. Memory section of the 32 × 8-bit memory

ROM address	State	*off*	*Button*	Next state	Light Colour					Contents (dec.)
					R	Ge	Gr	r	g	
⋮	⋮			⋮			⋮			⋮
3	0	1	1	6	0	0	1	1	0	198
⋮	⋮			⋮			⋮			⋮

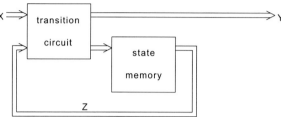

Fig. 8.86. Circuit with memory states and transition logic

For the coding of 9 states, 4 flip-flops are required that can assume in total 16 states. It is good practice to code out *all* 7 **illegal states** so that every state yields a defined output state for a transition. Thus it can be avoided that a circuit is left stuck in unreachable states after a noise glitch.

The transition logic in the circuit in Fig. 8.85 contains the transitions from one state to the following, where the input variables can modify the target state. For processors this is known as a conditional jump. For this reason this memory is also known as **program ROM**. The ROM used to decode the states into output signals is called the **output ROM**.

Extensive state diagrams with many transition conditions are often more readily realised using microprocessors, which moreover offer a greater amount of flexibility.

8.10 Further Reading

ALMAINI, A. E. A.: *Electronic Logic Systems, 3rd Edition*
Prentice Hall (1994)

DEMASSA, T. A.; CICCONE, Z.: *Digital Integrated Circuits, 1st Edition*
John Wiley & Sons (1995)

DORF, R. C.: *The Electrical Engineering Handbook, Section VIII*
CRC press (1993)

FLOYD, T. L.: *Digital Fundamentals, 7th Edition*
Prentice Hall (2000)

FLOYD, T. L.: *Electronics Fundamentals: Circuits, Devices, and Applications, 5th Edition*
Prentice Hall (2000)

FLOYD, T. L.: *Electronic Devices, 5th Edition*
Prentice Hall (1998)

KATZ, R. H.: *Contemporary Logic Design*
Benjamin/Cummings (1994)

MANO, M. M.: *Digital Design, 2nd Edition*
Prentice Hall (1995)

MANO, M. M.; KIME, C. R.: *Logic and Computer Design Fundamentals, 2nd Edition*
Prentice Hall (2000)

WAKERLY, J. F.: *Digital Design, 3rd Edition*
Prentice Hall (2000)

WILKINSON, B.: *Digital System Design, 2nd Edition*
Prentice Hall (1992)

ZWOLIŃSKI, M.: *Digital System Design with VHDL*
Prentice Hall (2000)

9 Power Supplies

Power supplies are electronic circuits that are designed to supply other electronic circuits or applications in a suitable way with electric energy. For example they can convert the mains voltage into a stabilised DC voltage for a microcontroller. So-called **uninterruptable power supplies** (UPS) convert the DC voltage of a battery to 230 V/50 Hz AC voltage, for instance, to supply a computer.

The most common application is the conversion of mains voltage into a smaller voltage, which is suitable for the connected circuits.

To achieve this it is necessary

- to isolate the mains from the electronics for the protection of the user, and
- to provide a stabilised DC voltage, i.e. the DC voltage has to be independent from variations in the mains voltage and also in the load.

The isolation is always achieved with transformers. These can either be operated at the mains frequency, or at high frequencies in switched-mode power supplies. High frequency allows the use of smaller components at similar ratings.

The stabilisation of the voltage can be done with a transistor operating in its active region. It can also be done by using switched-mode techniques, which optimise the efficiency of the power supply and reduce the physical dimensions.

9.1 Power Transformers

Transformers convert the mains voltage to a lower level and realise the **electrical isolation** between the mains and the low voltage. It is important that transformers are safely constructed components and therefore have to be approved according to national standards. The **national signs of approval** are printed on transformers (Fig. 9.1).

Fig. 9.1. National approval signs

In the European Community the **EU sign of conformity**, or in short the **CE sign**, has replaced the individual approval signs (Fig. 9.2). The CE sign states that for a component (in this case, a transformer) all relevant EU standards have been maintained. The manufacturer is *responsible* for the tests and has to confirm this in the so-called EU conformity declaration. The tests themselves can be done by a certified tester. The approval of the test only becomes relevant in the case of a dispute.

Fig. 9.2. EU sign of conformity

- The **primary** winding of the transformer is the mains winding, and the **secondary** winding is the electrically isolated low-voltage winding.
- The **rated power** is the product of the secondary rated voltage and the RMS value of the maximum secondary current. The value of the rated power is given in VA.
- The **rated voltage** is the mains voltage for the primary side, and for the secondary side the voltage at the rated current, i.e. the voltage when supplying the rated power.
- The **loss factor** is the ratio of no-load voltage to the rated voltage. Common values are between 1.35 and 1.15 for transformers with a rated power between 3 and 20 VA.

The internal resistance of the transformer can be calculated from the no-load voltage and the rated voltage.

$$R_{\text{int}} = \frac{\text{no-load voltage} - \text{rated voltage}}{\text{rated current}} \tag{9.1}$$

NOTE: Very small transformers are sometimes designed with a high internal resistance in order to make them short-circuit proof. This is done to avoid the need for fuses.

- On the primary side the **protection** of the transformer against overload is achieved with a fuse. In the case where the load is unevenly distributed between the secondary windings, they each have to be protected additionally.
- In order to make the transformer **short-circuit proof**, the manufacturer inserts a positive temperature coefficient element (PTC) or a heat-sensitive switch in the primary winding. In this case the use of a fuse is not necessary.

9.2 Rectification and Filtering

Usually the secondary voltage is rectified and filtered, i.e. the pulsating DC voltage after the rectifier is smoothed with a capacitor.

The filter capacitor is charged by a pulsating current for a period defined by the angle φ (Fig. 9.3). It is dependent on the internal resistance of the transformer and the capacitance of the filtering capacitor. Common values are between 30° and 50°.

The **output current** I_{out} equals the average value of the diode current I_F. The RMS value of the diode current can reach values up to twice the output current. The peak value of the diode current lies between 4 and 6 times the output current (Fig. 9.3).

$$\bar{I}_F = I_{\text{out}}, \qquad I_{F\,\text{RMS}} \approx 1.5\ldots 2 \cdot I_{\text{out}}, \qquad \hat{I}_F \approx 4\ldots 6 \cdot I_{\text{out}}$$

The RMS value of the diode current equals the RMS value of the secondary transformer current. This must be considered when choosing the apparent power of the transformer.

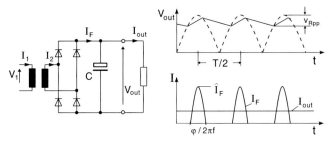

Fig. 9.3. Rectification and filtering

- The apparent power of the transformer S_N must be approximately twice the value of the output power $V_{out} \cdot I_{out}$.

The **filter capacitor** is usually chosen so that the **peak-to-peak ripple voltage** V_{Rpp} is approximately 20% of the output voltage V_{out}. The discharge time of the capacitor is approximately half of the periodic time of the mains frequency. Using the capacitor formula $i = C \dfrac{dv}{dt}$, the required capacitance can be determined as:

$$C \approx \frac{I_{out} \cdot T/2}{V_{Rpp}} = \frac{I_{out} \cdot T/2}{V_{out} \cdot 0.2} \tag{9.2}$$

For the 50 Hz mains the capacitor C is chosen thus:

$$C \ (\mu F) \approx \frac{I_{out} \ (mA)}{V_{out} \ (V)} \cdot 50 \tag{9.3}$$

Assuming the mains voltage is 10% under its rated value, the ripple voltage is 20%. If the diode voltage drops are not considered, then it holds for the **minimum output voltage** $V_{out\,min}$:

$$V_{out\,min} \approx 0.9 \cdot V_N \cdot \sqrt{2} \cdot 0.8 \tag{9.4}$$

Therefore for the required **transformer rated voltage** V_N:

$$V_N \geq V_{out\,min} \tag{9.5}$$

NOTE: In most power supplies a voltage regulator follows the filtering capacitor. Usually voltage regulators require a voltage drop of approximately 3 V. For this reason the minimum output voltage of the filtering circuit is very important: its value must be approximately 3 V higher than the regulated voltage.

9.2.1 Different Rectifier Circuits

Half-Wave Rectifier

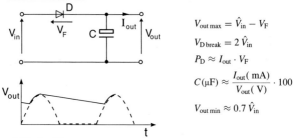

$$V_{\text{out max}} = \hat{V}_{\text{in}} - V_F$$
$$V_{D\,\text{break}} = 2\,\hat{V}_{\text{in}}$$
$$P_D \approx I_{\text{out}} \cdot V_F$$
$$C(\mu F) \approx \frac{I_{\text{out}}(\text{mA})}{V_{\text{out}}(\text{V})} \cdot 100$$
$$V_{\text{out min}} \approx 0.7\,\hat{V}_{\text{in}}$$

Bridge Rectifier

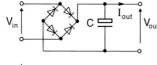

$$V_{\text{out max}} = \hat{V}_{\text{in}} - 2\,V_F$$
$$V_{D\,\text{break}} = \hat{V}_{\text{in}}$$
$$P_{D\,\text{tot}} = 2 I_{\text{out}} \cdot V_F$$
$$C(\mu F) \approx \frac{I_{\text{out}}(\text{mA})}{V_{\text{out}}(\text{V})} \cdot 50$$
$$V_{\text{out min}} \approx 0.7\,\hat{V}_{\text{in}}$$

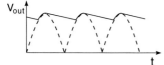

Full-Wave Rectifier

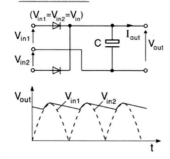

$$V_{\text{out max}} = \hat{V}_{\text{in}} - V_F$$
$$V_{D\,\text{break}} = 2\,\hat{V}_{\text{in}}$$
$$P_{D\,\text{tot}} \approx I_{\text{out}} \cdot V_F$$
$$C(\mu F) \approx \frac{I_{\text{out}}(\text{mA})}{V_{\text{out}}(\text{V})} \cdot 50$$
$$V_{\text{out min}} \approx 0.7\,\hat{V}_{\text{in}}$$

Fig. 9.4. A variety of rectifying circuits (value for C is valid for 50 Hz mains frequency)

Full-Wave Dual-Supply Rectifier

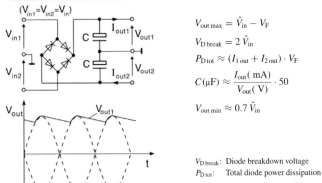

$$V_{out\,max} = \hat{V}_{in} - V_F$$

$$V_{D\,break} = 2\,\hat{V}_{in}$$

$$P_{D\,tot} \approx (I_{1\,out} + I_{2\,out}) \cdot V_F$$

$$C(\mu F) \approx \frac{I_{out}(\text{mA})}{V_{out}(\text{V})} \cdot 50$$

$$V_{out\,min} \approx 0.7\,\hat{V}_{in}$$

$V_{D\,break}$: Diode breakdown voltage
$P_{D\,tot}$: Total diode power dissipation

Fig. 9.5. A variety of rectifying circuits (value for C is valid for 50 Hz mains frequency)

Table 9.1. Comparison of the rectifier circuits

	Advantages	Disadvantages
Half-wave rectifier	Simple circuit	Large capacitor, high current RMS value
Bridge rectifier	One secondary winding, breakdown voltage $V_{D\,break} = \hat{V}_{in}$	High diode losses
Full-wave rectifier	Low diode losses (suitable for high currents)	Two secondary windings, breakdown voltage $V_{D\,break} = 2\hat{V}_{in}$
Full-wave dual-supply rectifier	One bridge rectifier for two output voltages, similar load on both secondary windings	Breakdown voltage $V_{D\,break} = 2\hat{V}_{in}$

9.3 Analogue Voltage Stabilisation

Voltage regulators are used to maintain a voltage at a constant level, independent of **voltage variation** in the mains and the **load variation**.

9.3.1 Voltage Stabilisation with Zener Diode

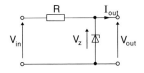

Fig. 9.6. Voltage stabilisation with zener diode

The **output voltage** is equal to the zener voltage (Fig. 9.6):

$$V_{\text{out}} = V_z \tag{9.6}$$

The **maximum power loss** P_{Lz} in the zener diode occurs when no load is connected to the circuit ($I_{\text{out}} = 0$):

$$P_{Lz} = \frac{V_{\text{in max}} - V_z}{R} \cdot V_z \tag{9.7}$$

The **maximum available output current** is given by:

$$I_{\text{out max}} = \frac{V_{\text{in min}} - V_z}{R} \tag{9.8}$$

If the output current becomes larger than $I_{\text{out max}}$ no current will flow through the zener diode and V_{out} drops below V_z. The **maximum short-circuit current** is:

$$I_{s/c} = \frac{V_{\text{in max}}}{R}$$

9.3.2 Analogue Stabilisation with Transistor

The **output voltage** is:

$$V_{\text{out}} = V_z - V_{\text{BE}} \approx V_z - 0.7 \text{ V} \tag{9.9}$$

The transistor Q_2 is configured as a current source with a current of $I_s = 0.7 \text{ V}/R_1$. The source current is chosen so that at the rated load the transistor Q_1 receives the required base current, and a small current flows through the zener diode(Fig. 9.7). Hence, the output voltage is kept at $V_{\text{out}} = V_z - 0.7 \text{ V}$ for all different loads, between no-load and the rated load. If there is an overload the transistor Q_3 opens, thus reducing the base current of Q_1 so that the **maximum output current** is limited to $I_{\text{out max}} = 0.7 \text{ V}/R_M$.

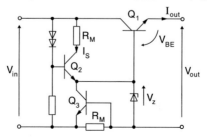

Fig. 9.7. Analogue regulation with transistors

9.3.3 Voltage Regulation

The **output voltage** is:

$$V_{out} = V_{ref} \cdot \frac{R_1 + R_2}{R_2} \tag{9.10}$$

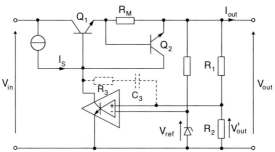

Fig. 9.8. Voltage regulation

The operational amplifier amplifies the difference $(V_{ref} - V'_{out})$, which is the difference between the desired value and the actual value of the output voltage. With its open collector output it controls the base current of Q_1 by sinking more or less of the source current I_S. For example, if V_{out} is too high, the operational amplifier takes the base current from transistor Q_1, turning the transistor down and thus lowering the output voltage. The transistor Q_2 takes the current I_S in the case of an overload, i.e. when $I_{out} > \dfrac{V_{BE}}{R_M} \approx \dfrac{0.7\,V}{R_M}$. If the loop gain was too large causing the circuit to oscillate, then a PI controller may be used as the amplifier (R_3, C_3 in Fig. 9.8).

To obtain a **variable output voltage** it is possible to create the reference voltage with a potentiometer and to feed this voltage into the inverting input of the operational amplifier. Then the desired value is adjustable. It is always better to vary the reference value than to change R_1 and R_2, since this does not affect the control loop and therefore the stability of the system. The range of the adjustment via R_1/R_2 is only appropriate for fine tuning of output voltage.

9.3.3.1 Integrated Voltage Regulators

There is a large variety of integrated voltage regulators available. Usually they are short-circuit proof, no-load proof and have temperature protection. The 78xx series for positive voltages and the 79xx series for negative voltages are well known for fixed voltages. These fixed voltage regulators are available for different current ratings.

EXAMPLE: Figure 9.9 shows an example of a ± 12 V voltage supply. In addition to the previously described circuits there are some ceramic capacitors of 100 nF close to the voltage regulator. Their purpose is to reduce possible oscillations in the regulator.

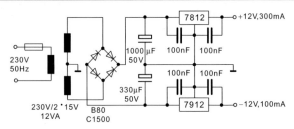

Fig. 9.9. Example: voltage supply with fixed voltage regulators

9.4 Switched Mode Power Supplies

Switched-mode power supplies (SMPS) are used in nearly all electronic systems. Every television set and computer is powered by an SMPS, as is most state-of-the-art industrial equipment. Battery-powered equipment also uses SMPS to provide a constant internal supply voltage independent of the charge state of the battery. SMPS are also used to achieve a higher supply voltage than that of the powering battery voltage. This is normally required for tape recorders, CD players, notebooks, mobile phones and cameras. SMPS have remarkable advantages when compared to linear regulated power supplies. Theoretically, SMPS work loss-free, and in practice efficiencies of about 70% to 95% are achieved. This results in low-temperature operation and consequently high reliability. The other major advantage is that SMPS operate at high frequencies, which results in small low-weight components. Compared to linear power supplies SMPS are therefore inherently more efficient, smaller, lighter and cheaper to manufacture.

In general, all SMPS have the same principle of operation. Small quantities of energy are taken from an input voltage by an electronic switch (transistor), which switches at high frequency. The switching frequencies are normally in the range of 20 kHz to 300 kHz, depending on the required performance. The ratio between turn-on and turn-off time of the switch determines the average energy flow. A low-pass filter is placed at the output of all SMPS to smooth the discontinuous energy flow. The high efficiency of SMPS is a direct result of the theoretically loss-free switching component and low-pass filter.

There are a number of different types of SMPS, as described below. Although similar in principle, the manner of operation differs greatly between topology types.

SMPS can be configured as **secondary** or as **primary switched power supplies**. Secondary power supplies have no isolation between the input and output. They are used in applications where isolation (in respect to mains) already exists or where isolation is not required, for example, in battery-powered devices. Primary switched power supplies offer an isolation between input and output. Their switching transistors operate on the primary side of a transformer. The energy is transferred to the secondary side at a high-frequency via a high frequency transformer. Because of the high operating frequency the transformer can be relatively small.

There are three basic SMPS configurations. These are **flyback**, **forward** and **resonant converter**. Flyback converters transfer their energy during the off-time of the transistors. Forward converters transfer their energy during the on-time of the transistors. Resonant converters use a resonant circuit for switching the transistors when they are at the zero-current or zero-voltage point, resulting in reduced stress on the switching transistors.

9.4.1 Single-Ended Converters, Secondary Switched SMPS

9.4.1.1 Buck Converter

The **buck converter** converts an input voltage into a lower output voltage. It is also called a **step-down converter**.

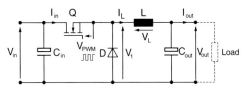

Fig. 9.10. Buck converter

Figure 9.10 shows the circuit diagram of a buck converter. The transistor Q operates as the switch, which is turned on and off by a pulse-width modulated control voltage V_{PWM} operating at high frequency. The ratio $\dfrac{t_1}{T}$, where t_1 is the on-time and T the periodic time, is called the **duty cycle**.

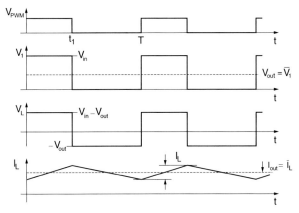

Fig. 9.11. Voltages and currents of the buck converter

In the following analysis it is assumed that the conducting voltage drops of the transistor and the diode are zero.

During the on-time of the transistor the voltage V_1 is equal to V_{in}. When the transistor switches off (blocking phase), the inductor L continues to drive the current through the load in parallel with C_{out} and back through the diode. Consequently, the voltage V_1 is zero. The voltage V_1 stays at zero during the off-time of the transistor provided that the current I_L does not reduce to zero. This is called **continuous-mode operation**. In this mode V_1

is a voltage that changes between V_{in} and zero, corresponding to the duty cycle of V_{PWM}, (Fig. 9.11). The low-pass filter, formed by L and C_{out}, produces an average value of V_1, i.e. $V_{out} = \overline{V_1}$. Therefore for continuous mode:

$$V_{out} = \frac{t_1}{T} V_{in} \qquad (9.11)$$

- For the continuous mode the output voltage is a function of the duty cycle and the input voltage, and is independent of the load.

The inductor current I_L has a triangular shape, and its average value is determined by the load. The peak-to-peak current ripple ΔI_L is dependent on L and can be calculated with the help of Faraday's law:

$$V = L \frac{di}{dt} \quad \rightarrow \quad \Delta i = \frac{1}{L} \cdot V \cdot \Delta t$$

$$\rightarrow \quad \Delta I_L = \frac{1}{L} (V_{in} - V_{out}) \cdot t_1 = \frac{1}{L} V_{out} (T - t_1) \qquad (9.12)$$

For $V_{out} = \frac{t_1}{T} V_{in}$ and a switching frequency f it follows for the continuous mode:

$$\Delta I_L = \frac{1}{L} (V_{in} - V_{out}) \cdot \frac{V_{out}}{V_{in}} \cdot \frac{1}{f} \qquad (9.13)$$

- The current ripple ΔI_L is independent of the load. The average value of the current I_L is equal to the output current I_{out}.

At low load current, where $I_{out} \leq \frac{\Delta I_L}{2}$, the current I_L reduces to zero during every switching cycle. This is called **discontinuous-mode**, and for this mode the calculations above are not valid.

Calculation of L and C_{out}

In order to calculate the value of L a realistic value of ΔI_L must be selected. The problem is as follows: If ΔI_L is selected at a very low value, the value of L has to be relatively high, which would require a very heavy and expensive inductor. If ΔI_L is assigned a high level the switch-off current of the transistor would be very high (this would result in high losses in the transistor). A good compromise is to design for: $\Delta I_L \approx 0.2\, I_{out}$

For L follows:

$$L = \frac{1}{\Delta I_L} (V_{in} - V_{out}) \cdot \frac{V_{out}}{V_{in}} \cdot \frac{1}{f} \qquad (9.14)$$

The maximum value of the inductor current is:

$$\hat{I}_L = I_{out} + \tfrac{1}{2} \Delta I_L \qquad (9.15)$$

Assuming that the inductor ripple current is small compared to its DC current, the RMS value of the current flowing through the inductor is given by:

$$I_{L(RMS)} \approx I_{out} \qquad (9.16)$$

The capacitor C_{out} is chosen usually for the cutoff frequency of the LC_{out}-low-pass filter, which is approximately 100 to 1000 times lower than the switching frequency. An exact calculation of the capacitor depends on its maximum AC current rating and its serial equivalent impedance Z_{max}. Both values can be verified from the relevant data sheet.

The current ripple ΔI_L causes a voltage ripple ΔV_{out} at the output capacitor C_{out}. For normal switching frequencies this voltage ripple is determined by the equivalent impedance Z_{max}.

The output voltage ripple is given by Ohm's law:

$$\boxed{\Delta V_{out} \approx \Delta I_L \cdot Z_{max}} \qquad (9.17)$$

The choice of the output capacitor depends not on its capacitance, but on its series equivalent impedance Z_{max} at the switching frequency, which can be verified from the capacitor data sheet.

9.4.1.2 Boost Converter

The **boost converter** converts an input voltage to a higher output voltage. The boost converter is also called a **step-up converter**.

Boost converters are used in battery-powered devices, where the electronic circuit requires a higher operating voltage than that supplied by the battery, e.g. notebooks, mobile phones and camera flashes.

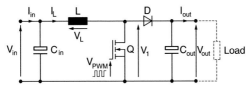

Fig. 9.12. Boost converter

Figure 9.12 shows the basic circuit diagram of the boost converter. The transistor Q operates as a switch, which is turned on and off by a pulse-width modulated control voltage V_{PWM}.

In the following analysis is be assumed that the conducting voltage drops of the transistor and the diode are zero (during switching).

During the on-time of the transistor, the voltage across L is equal to V_{in} and the current I_L increases linearly. When the transistor is turned off, the current I_L flows through the diode and charges the output capacitor.

The function of the boost converter can also be described in terms of energy balance. During the on-time of the transistor the inductance is charged with energy, and during the off-time of the transistor this energy is transferred from the inductor through the diode to the output capacitor.

If the transistor is not turned on and off by the clock pulse, the output capacitor charges via L and D to the level $V_{out} = V_{in}$. When the transistor is switched the output voltage will increase to higher levels than the input voltage.

In a similar manner to the buck converter (Sect. 9.4.1.1) a distinction is made between the discontinuous and continuous mode, depending upon whether the inductor current I_L reduces to zero during the off-time of the transistor.

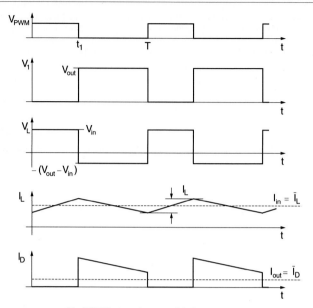

Fig. 9.13. Voltages and currents of the boost converter

With the help of Faraday's law the continuous mode and steady-state conditions (see also Fig. 9.13) can be established: $\Delta I_L = \frac{1}{L} V_{in} \cdot t_1 = \frac{1}{L} (V_{out} - V_{in}) \cdot (T - t_1)$. This yields:

$$V_{out} = V_{in} \frac{T}{T - t_1} \qquad (9.18)$$

- For the continuous mode the output voltage is a function of the duty cycle and the input voltage, and is independent of the load.
- The boost converter is *not* short-circuit proof, because there is inherently no switch-off device in the short-circuit path.

NOTE: If the boost converter is not regulated in a closed loop but is controlled by a fixed duty cycle of a pulse generator (this could be the case for a laboratory set-up), the boost converter is *not* no-load proof. This is because each switching cycle results in energy in the choking coil being transferred to the output capacitor. This will result in the output voltage continuously increasing until the devices are eventually destroyed.

Calculation of L and C_{out}

As with the buck converter, the starting point for calculating L is to select a value of current ripple ΔI_L of about 20% that of the input current: $\Delta I_L \approx 0.2\, I_{in}$. The input current

I_{in} can be calculated by assuming zero losses (input power = output power), therefore:
$V_{in} \cdot I_{in} = V_{out} \cdot I_{out} \rightarrow I_{in} = I_{out} \dfrac{V_{out}}{V_{in}}$

L can be calculated as follows:

$$L = \frac{1}{\Delta I_L}(V_{out} - V_{in})\frac{V_{in}}{V_{out}} \cdot \frac{1}{f} \qquad (9.19)$$

The peak value of the inductor current is (Fig. 9.13):

$$\hat{I}_L = I_{in} + \frac{1}{2}\Delta I_L \qquad (9.20)$$

Assuming that the inductor ripple current is small compared to its DC current, the RMS value of the current flowing through the inductor is given by:

$$I_{L(RMS)} \approx I_{in} \qquad (9.21)$$

The output capacitor is charged by pulses (Fig. 9.13). The ripple ΔV_{out} of the output voltage results from the pulsating charge current I_D and is mainly determined by the impedance Z_{max} at the switching frequency of capacitor C_{out}. The value of Z_{max} can be verified from the capacitor data sheet.

The output voltage ripple is given by Ohm's law:

$$\Delta V_{out} \approx I_D \cdot Z_{max} \qquad (9.22)$$

9.4.1.3 Buck-Boost Converter

The **buck-boost converter** converts a positive input voltage to a negative output voltage.

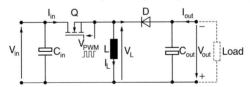

Fig. 9.14. Buck-boost converter

Figure 9.14 shows the basic circuit of the buck-boost converter. The transistor Q works as a switch, which is turned on and off by the pulse-width-modulated voltage V_{PWM}. During the on-time of the transistor, the inductor current I_L increases linearly. During the off-time the current I_L is continuous and charges the output capacitor C_{out}. Note the polarity of the output voltage in Fig. 9.14.

For the continuous mode and with steady-state conditions the output voltage is given by:

$$V_{out} = V_{in}\frac{t_1}{T - t_1} \qquad (9.23)$$

The inductor current I_L is given by (Fig. 9.15):

$$\overline{I_L} = I_{out}\frac{T}{T-t_1} = I_{out}\left(\frac{V_{out}}{V_{in}}+1\right), \quad \text{and} \quad \Delta I_L = \frac{1}{L}V_{in}t_1 = \frac{1}{L}\cdot\frac{V_{in}V_{out}}{V_{in}+V_{out}}\cdot\frac{1}{f}$$

(9.24)

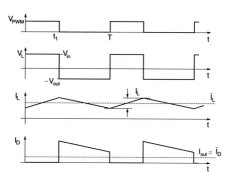

Fig. 9.15. Voltages and currents of the buck-boost converter

9.4.2 Primary Switched SMPS

9.4.2.1 Flyback Converter

The **flyback converter** belongs to the primary switched converter family, which means there is isolation between input and output. Flyback converters are used in nearly all mains-supplied electronic equipment for low power consumption, up to approximately 300 W, examples of which include televisions, personal computers, printers, etc.

Flyback converters have a remarkably low number of components when compared to other SMPS. They also have the advantage that several isolated output voltages can be regulated by one control circuit.

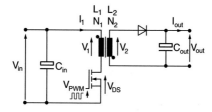

Fig. 9.16. Flyback converter

Figure 9.16 shows the basic circuit of a flyback converter. The transistor works as a switch, which is turned on and off by the pulse-width modulated control voltage V_{PWM}. During the on-time of the transistor the primary voltage of the transformer V_1 is equal to the input voltage V_{in}, which results in the current I_1 increasing linearly. During this phase, energy is stored in the transformer core. During the on-phase the secondary current is zero, because the diode is blocking. When the transistor is turned off the primary current I_1 is interrupted, and the voltages at the transformer invert (due to Faraday's law $v = L\frac{di}{dt}$), the diode conducts and the energy moves from the transformer core via the diode to the output capacitor C_{out}.

During the on-phase of the transistor the drain–source voltage V_{DS} is equal to zero (Fig. 9.17). During the off-time of the transistor, the output voltage V_{out} will be transformed back to the primary side, and the drain–source voltage theoretically steps up to $V_{DS} = V_{in} + V_{out} \cdot \frac{N_1}{N_2}$. If a mains voltage of 230 V/50 Hz is used, V_{DS} will jump to approximately 700 V. In practice, this voltage will be even higher due to the self-induction of the leakage inductance of the transformer. To allow for this effect the minimum rated drain–source breakdown voltage of the transistor must be 800 V.

The transformer is not a 'normal' one, in that its function is to store energy during the on-time of the transistor and to deliver this energy during the off-time via the diode to the output capacitor. In fact, the transformer is a storage inductor (often called a choke) with a primary and secondary winding. To store energy the transformer core needs an air gap (normal transformers do not have an air gap). An important consideration for this transformer is that primary and secondary windings are closely coupled in order to achieve a minimum leakage inductance. It should be noted that the energy of leakage inductance cannot be transferred to the secondary side and is therefore dissipated as heat on the primary side.

Design of the Flyback Converter

Regarding the primary voltage of the transformer V_1, its average value $\overline{V_1}$ must be equal to zero for steady-state conditions (if not, the current would increase to infinity).

This yields: $V_{in} \cdot t_1 = V_{out} \cdot \frac{N_1}{N_2} \cdot (T - t_1)$, and:

$$\boxed{V_{out} = V_{in} \cdot \frac{N_2}{N_1} \cdot \frac{t_1}{T - t_1}} \tag{9.25}$$

The turns ratio of the transformer should be chosen so that for the rated output power the on-time (energy charge time) t_1 is equal to the off-time (energy discharge time) $T - t_1$. This leads to the turns ratio:

$$\boxed{\frac{N_1}{N_2} = \frac{V_{in}}{V_{out}}} \tag{9.26}$$

In this case, the breakdown voltage of the transistor and the reverse voltage of the diode must be:

$$\boxed{\text{Transistor:} \quad V_{DS} = V_{in} + V_{out} \cdot \frac{N_1}{N_2} \approx 2V_{in}} \tag{9.27}$$

$$\boxed{\text{Diode:} \quad V_R = V_{out} + V_{in} \cdot \frac{N_2}{N_1} \approx 2V_{out}} \tag{9.28}$$

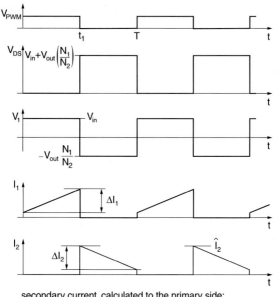

secondary current calculated to the primary side:

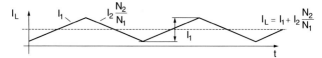

Fig. 9.17. Voltages and currents at the flyback converter

It should be noted that the rated breakdown voltage of the transistor must be chosen significantly higher, because at the turn-off instant the energy of the leakage inductance L_{leak} will not be taken over by the secondary winding. To keep the overvoltage within an acceptable range a **snubber circuit** is required (Fig. 9.18). At the instant of turn-off the current of the leakage inductance L_{leak} is diverted through the diode D and charges the capacitor C. The power is dissipated in resistor R.

If R and C are required to operate at 230 V AC, the value of R has to be determined experimentally to ensure that the DC voltage across C falls within the region of 350 V to 400 V.

In order to design the transformer, the primary inductance L_1 has to be calculated first(Fig. 9.16). L_1 has to store energy during the on-time of the transistor, which is the energy required at the output. This energy is given by $W = P_{\text{out}} \cdot T$, where T is the period of the switching frequency, and P_{out} is the rated power. This energy is stored in the primary inductance during the first half of the period time and is transferred to the output capacitor during the second half of the switching period. As before, the switching period is divided into two equal parts, one part to store the energy and the other part to transfer the energy.

During the on-time of the transistor, the voltage across the primary inductance is equal to V_{in}, and the current I_1 is a ramp waveform (Fig. 9.18). For every cycle of the input energy

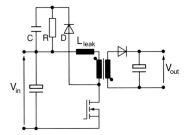

Fig. 9.18. Snubber circuit to limit the peak voltage across the transistor

it follows that:
$$W = V_{in} \frac{\hat{I}_1}{2} \frac{T}{2} \quad \text{(Fig. 9.19)}$$

This energy is stored in L_1 and can be calculated as:
$$W = \frac{1}{2} L_1 \hat{I}_1^2$$

For the size of the primary inductance this leads to:
$$L_1 \approx \frac{V_{in}^2}{8 \, P_{out} \cdot f}.$$

The calculation above assumes an efficiency of 100%. If we consider an efficiency of η, it means that we have to store more energy in L_1 and not all of this energy is delivered to the output. Then L_1 can be calculated as follows:

$$\boxed{L_1 \approx \frac{V_{in}^2}{8 \, P_{out} \cdot f} \cdot \eta} \qquad (9.29)$$

Efficiency η has to be estimated because its value is not known at this point in time. However, $\eta \approx 0.75$ is normally a good estimate.

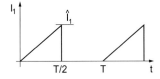

Fig. 9.19. Shape of the input current I_1 for rated power

The peak value of the current I_1 is: $\hat{I}_1 = \dfrac{4 \cdot P_{out}}{V_{in} \cdot \eta}$

The RMS-value of the current I_1 is: $I_{1\,RMS} = \dfrac{\hat{I}_1}{\sqrt{6}}$

The core of the transformer and the windings can now be calculated with the help of Sect. 9.4.5.

NOTE: The core of the transformer must have a sufficiently large gap, in which the major part of the magnetic energy can be stored (refer to Sect. 9.4.5).

The output capacitor C_{out} is charged by pulses (refer to Fig. 9.17). The ripple ΔV_{out} of the output voltage results from the pulsating charge current I_2 and is mainly determined by the impedance Z_{max} of the capacitor. The value of Z_{max} can be verified from the capacitor data sheet.

The magnitude of the ripple voltage is given as follows:

$$\Delta V_{out} \approx \hat{I}_2 \cdot Z_{max}$$

The input capacitor C_{in} can be calculated for 230 V/50 Hz as follows:

$$C_{in} \approx 1 \frac{\mu F}{W} \cdot P_{in}$$

A special feature of the flyback converter is the possibility of controlling several isolated output voltages with only one control circuit (Fig. 9.20).

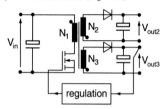

Fig. 9.20. Flyback converter for several output voltages

One output voltage $V_{out\,3}$ is regulated (Fig. 9.20). Voltage $V_{out\,2}$ is coupled to $V_{out\,3}$ via the turns ratio: $\dfrac{V_{out2}}{V_{out3}} = \dfrac{N_2}{N_3}$. The energy that is stored in L_1 (N_1) during the on-time of the transistor moves during the off-time to the outputs. These output voltages maintain their values in relationship to the turns ratio. The output voltages, when viewed in relation to the turns ratio from the primary side, appear to be in parallel.

9.4.2.2 Single-Transistor Forward Converter

The **single-transistor forward converter** belongs to the primary switched converter family as there is isolation between input and output. It is suitable for output powers of up to 1 kW. The single-transistor forward converter is also called a single-ended forward converter (Fig. 9.21).

The forward converter transfers the energy during the on-time of the transistor. During this time the voltage V_1 is equal to the input voltage. The winding N_2 is in the same direction as N_1. When the transistor is on, the voltage V_2 at N_2 is given by $V_2 = V_{in} \dfrac{N_2}{N_1}$. The voltage V_2 drives the current I_2 through the diode D_2, which during this time is equal to I_3, through L, which charges the output capacitor C_{out}.

During the off-time of the transistor, N_1 and N_2 are without current. The inductor L draws its current through diode D_3. The value of voltage V_3 is equal to zero (neglecting the forward voltage drop of D_3).

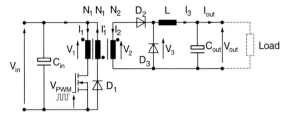

Fig. 9.21. Single transistor forward converter

During the off-time of the transistor, the magnetic flux of the transformer has to decrease to zero. The core is demagnetised with N_1' via D_1 to V_{in}. Since N_1' has the same number of turns as N_1, the demagnetisation needs an equal time interval as the on-time. For this reason the minimum off-time has to be as long as the on-time. This causes a maximum duty cycle t_1/T of 0.5 for the single-transistor forward converter.

During the off-time, the voltage at N_1' is equal to the input voltage V_{in}. This voltage will be transformed back to the primary winding N_1 and for V_1 follows: $V_1 = -V_{in}$. Because of this the drain–source voltage steps up to $V_{DS} \geq 2V_{in}$ when the transistor is turned off (Fig. 9.22).

In comparison to the transformer of the flyback converter, the transformer in this forward converter is a 'normal' transformer. Its job is not to store energy but to transfer energy. For this reason the core has no air gap.

- The breakdown voltage of the transistor has to be $V_{DS} > 2V_{in}$.
- The windings N_1 and N_1' must be closely coupled. However, a snubber circuit (as shown in Fig. 9.18, Sect. 9.4.2.1) is necessary.
- In comparison to the flyback converter, the forward converter can only have one regulated output voltage.
- The maximum duty cycle is $\dfrac{t_1}{T} = 0.5$.

Design of the Single-Transistor Forward Converter

The output voltage V_{out} is equal to the average value of V_3. The maximum duty cycle is 0.5. This leads to (see also Sect. 9.4.1.1):

$$V_{out} = V_{in} \cdot \frac{N_2}{N_1} \cdot \frac{t_1}{T} \tag{9.30}$$

For the turns ratio it follows that:

$$\boxed{\frac{N_2}{N_1} = 2 \cdot \frac{V_{out}}{V_{in}}, \quad \text{and} \quad N_1 = N_1'} \tag{9.31}$$

For further calculation of the transformer see Sect. 9.4.5.

To calculate L the method used for the buck converter is appropriate. Initially the current ripple ΔI_3 of the inductor current I_3 has to be selected. A value of 20% of the output

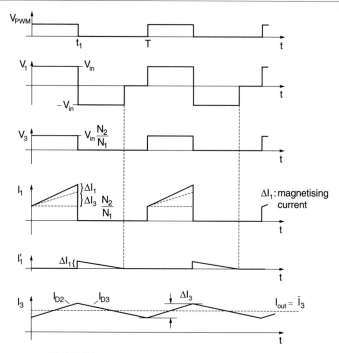

Fig. 9.22. Voltages and currents at the single transistor forward converter

current is normally acceptable: $\Delta I_3 \approx 0.2 \cdot I_{out}$. Assuming a maximum duty cycle of 0.5, this leads to:

$$L = \frac{V_{out} \cdot T/2}{\Delta I_3} \qquad (9.32)$$

The value of C_{out} depends on the acceptable voltage ripple ΔV_{out} of the output voltage. This voltage ripple is mainly determined by the impedance Z_{max} of the output capacitor C_{out}:

$$\Delta V_{out} \approx \Delta I_L \cdot Z_{max}$$

The value of Z_{max} can be verified from the data sheet of C_{out}.

The input capacitor C_{in} for 230 V/50 Hz should be:

$$C_{in} \approx 1 \frac{\mu F}{W} \cdot P_{in}$$

Two-Transistor Forward Converter

The **two-transistor forward converter** is a variant of the single transistor forward converter (Fig. 9.23).

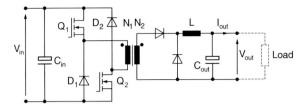

Fig. 9.23. Two-transistor forward converter

The transistors Q_1 and Q_2 switch at the same time. During the on-time of the transistors, the voltage at the primary winding is equal to the input voltage V_{in}. During the off-time of the transistors the transformer is demagnetised via the diodes D_1 and D_2 into the input voltage V_{in}. In comparison to the single-transistor forward converter this converter has the advantage that its transistors have to block the input voltage only and the winding N_1' is not required. In addition, the coupling of the transformer windings is no longer critical. These advantages make this converter type suitable for significantly higher output powers in comparison to the single-transistor forward converter.

The calculation of the components is equivalent to the single-transistor forward converter.

- For the two-transistor forward converter the breakdown voltage of the transistors is only required to be $V_{DS} = V_{in}$.
- The two-transistor forward converter can be used for powers up to a few kilowatts. It is a simple converter, which is not critical with regard to design and operation.

9.4.2.3 Push–Pull Converters

The **push–pull converter** is suitable for high-power design.

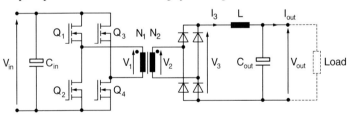

Fig. 9.24. Push–pull converter, here: full-bridge type

The push–pull converter drives the high-frequency transformer with an AC voltage, where the negative as well as the positive half swings transfer energy. The primary voltage V_1 can be $+V_{in}$, $-V_{in}$ or $zero$, depending on which pair of transistors (Q_1, Q_4 or Q_2, Q_3) are turned on or off. At the secondary side the AC voltage is rectified and smoothed by L and C_{out} (Fig. 9.24).

For the continuous mode it follows that (see also Sect. 9.4.1.1):

$$\boxed{V_{out} = V_{in} \cdot \frac{N_2}{N_1} \cdot \frac{t_1}{T}} \qquad (9.33)$$

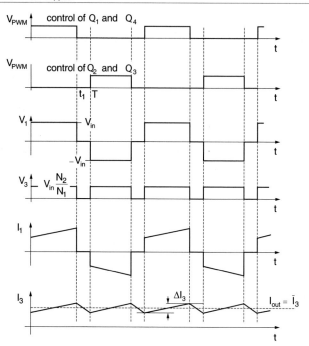

Fig. 9.25. Voltages and currents at the push–pull converter

The duty cycle $\frac{t_1}{T}$ may theoretically increase to 100%. This is not possible in practice because the serial connected transistors Q_1, Q_2 or Q_3, Q_4 have to be switched with a time difference in order to avoid a short-circuit of the input supply. The turns ratio of the transformer must be such that:

$$\boxed{\frac{N_2}{N_1} \geq \frac{V_{\text{out}}}{V_{\text{in}}}} \qquad (9.34)$$

- The transistors of the push–pull converter can be switched with the maximum duty cycle of 0.5. This leads to the maximum duty cycle of $\frac{t_1}{T} = 1$ after rectification.

The calculation of L and C_{out} follows that of the buck converter (Sect. 9.4.1.1).

Half-Bridge Push–Pull Converter

A variant of the push–pull converter is the **half-bridge push–pull converter**. The capacitors C_1 and C_2 divide the input voltage V_{in} into two. Therefore the magnitude of the primary voltage V_1 is $\pm V_{\text{in}}/2$. In comparison to the full-bridge push–pull converter, it follows that for the half-bridge type the turns ratio of the transformer $\frac{N_2}{N_1} \geq 2\frac{V_{\text{out}}}{V_{\text{in}}}$.

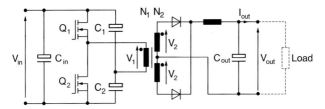

Fig. 9.26. Half-bridge push–pull converter with full-wave rectifier

In Fig. 9.26 a two-diode full-wave rectifier is used instead of a full-wave bridge rectifier. The choice of the rectifier type depends on the output voltage and current. The difference between these two rectifier types is that the current has to pass through two diodes in the bridge type and only one diode in the full-wave type. Consequently, the full-wave type is used for high currents (to reduce the rectifier losses), and the bridge type is used for high-voltage purposes in order to save one secondary winding of the transformer.

9.4.2.4 Resonant Converters

Resonant converters use resonant circuits to switch the transistors when they are at the zero-current or zero-voltage point. This reduces the stress on the switching transistors and the radio interference. A distinction is made between between zero-voltge-switching (ZVS) and zero-current-switching (ZCS) resonant converters.

To control the output voltage, resonant converters are driven by a constant pulse duration at a variable frequency. The pulse duration is required to be equal to half of the resonant period time for switching at the zero-crossing points of current or voltage.

There are many different types of resonant converters. For example, the resonant circuit can be placed at either the primary or secondary side of the transformer. Another alternative is that a serial or parallel resonant circuit can be used, depending on whether it is required to turn off the transistor when the current is zero or the voltage is zero.

The technique of resonant converters is described below, with the ZCS push–pull resonant converter offered as an example.

ZCS Push-Pull Resonant Converter

Figure 9.27 shows the ZCS push–pull resonant converter. The resonant circuit is formed by L and C. Assume an initial condition of the voltage V_C across C equal to zero. If the transistor Q_1 is now turned on, a sinusoidal current half-swing starts through Q_1, L, Tr, C and C_{in}. This half-swing charges the capacitor C from zero to V_{in}. If this first half-sinusoidal swing is completed, Q_1 can be switched off without losses. After a short delay Q_2 can be switched and a next half-sinusoidal swing starts, thus discharging C from V_{in} back to 0 Volts.

Each half-sinusoidal swing transfers a certain amount of energy from the primary to the secondary side of the transformer. The transformer Tr operates on its primary side as a voltage source. For the duration of the current swing through the primary winding, the output voltage V_{out} will be transformed to the primary side: $V'_{out} = V_{out} \dfrac{N_2}{N_1}$. The

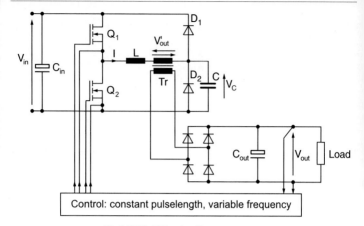

Fig. 9.27. The ZCS push–pull resonant converter

energy that is transferred by every half-swing is equal to $W = V'_{out} \cdot \int i(t) \, dt$. This energy will be transferred twice in each resonant period. This leads to an output power of $P_{out} = W \cdot 2 f_{switching}$ ($f_{switching}$: frequency of the converter). Figure 9.28 shows an equivalent circuit for one half-swing.

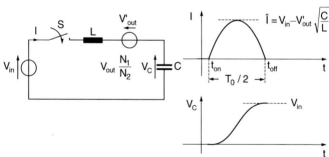

Fig. 9.28. Equivalent circuit for one half sinusoidal swing of the ZCS push–pull resonant converter

The resonant frequency is:

$$f_0 = \frac{1}{2\pi\sqrt{LC}} \tag{9.35}$$

This leads to the minimum on-time of the transistors. The on-time should be a little higher than half of the resonant period time to ensure that the current reduces to zero. For maximum energy transfer, V'_{out} must be half of V_{in}. This leads to the turns ratio of the transformer:

$$V'_{out} = \frac{1}{2} V_{in} \quad \Rightarrow \quad \frac{N_1}{N_2} = \frac{1}{2} \cdot \frac{V_{in}}{V_{out}} \tag{9.36}$$

The maximum output power is achieved if one half-current swing instantly follows the next.

The transferred energy of each half-swing further depends on the value of C and L. The higher the value of C and the lower the value of L, to maintain a certain resonant frequency, the higher the amount of energy transfer (see also the peak value of the current in Figs. 9.28 and 9.29).

To achieve a certain output power P_{out}, considering $V'_{out} = V_{in}/2$, it can be shown that for L and C:

$$\sqrt{\frac{L}{C}} = \frac{\left(\frac{V_{in}}{2}\right)^2 \cdot \frac{2}{\pi} \cdot \frac{f_{Switching}}{f_0}}{P_{out}} \quad \Rightarrow \quad C = \frac{1}{2\pi \cdot \sqrt{\frac{L}{C}} \cdot f_0}, \quad \text{and} \quad L = \left(\sqrt{\frac{L}{C}}\right)^2 \cdot C$$

(9.37)

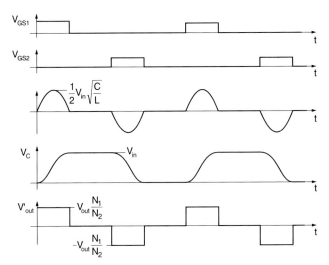

Fig. 9.29. Voltages and currents at the ZCS push–pull resonant converter (V_{GS} : Gate–Source control voltage)

In addition to the general advantages of resonant converters, having lower switching losses and lower radio interference, this particular resonant converter has two more additional advantages:

- The ZCS push–pull resonant converter can regulate several output voltages using one control circuit, as for the flyback converter. This is because several output voltages seem to be connected in parallel when viewed from the primary side. Therefore the energy always passes to that output having the lowest voltage value, taking into consideration the turns ratio.
- The ZCS push–pull resonant converter is both no-load and short-circuit proof, without any additional electronic precautions being required. The output voltage cannot reach

more than twice the nominal value, as then $V'_{out} = V_{in}$. The current cannot reach more than twice the nominal output current, as then $V'_{out} = 0$ and $\hat{I} = V_{in}\sqrt{C/L}$.

- This converter has minor switching losses and EM interference.

9.4.3 Overview: Switched-Mode Power Supplies

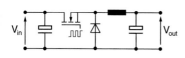

Buck converter

- $V_{out} \leq V_{in}$
- Short-circuit and no-load proof simply achievable
- V_{GS} has to float
- Usage: Replacement for analogue voltage regulators

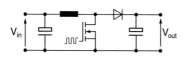

Boost converter

- $V_{out} \geq V_{in}$
- Not short-circuit proof
- Not no-load proof if not operating in a closed loop
- Usage: Battery-supplied devices, such as notebooks, mobile phones, camera flashes

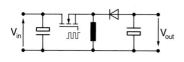

Inverting converter/buck-boost converter

- $V_{out} < 0$ V
- Short-circuit proof easily achievable
- Not no-load proof if not operating in a closed loop
- Usage: Generation of an additional negative voltage from a positive supply voltage

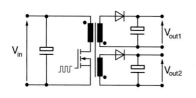

Flyback converter

- Several isolated output voltages can be regulated by one control circuit
- Output power up to several hundred watts
- Wide range of input and output voltages (mains voltage 85–270 VAC achievable)
- Transistor breakdown voltage $V_{DS} \geq 2V_{in}$
- Very good magnetic coupling required
- Big core with air gap necessary

9.4 Switched Mode Power Supplies

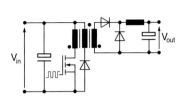

Single-transistor forward converter

- One isolated controllable output voltage
- Output power up to several hundred watts
- Transistor breakdown voltage $V_{DS} \geqq 2V_{in}$
- Duty cycle $\dfrac{t_{on}}{T} \leqq 0.5$
- Very good magnetic coupling required
- Small core without air gap

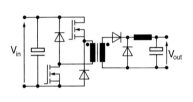

Two-transistor forward converter

- One isolated controllable output voltage
- Output power up to several kilowatts
- Transistor breakdown voltage $V_{DS} = V_{in}$
- Duty cycle $\dfrac{t_{on}}{T} \leqq 0.5$
- No extraordinary magnetic coupling necessary
- Small core without air gap

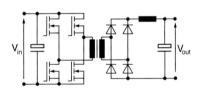

Full-bridge push–pull converter

- One isolated controllable output voltage
- Output power up to many kilowatts
- Transistor breakdown voltage $V_{DS} = V_{in}$
- No extraordinary magnetic coupling necessary
- Small core without air gap
- Balancing problems

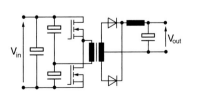

Half-bridge push–pull converter

- One isolated controllable output voltage
- Output power up to several kilowatts
- Transistor breakdown voltage $V_{DS} = V_{in}$
- No extraordinary magnetic coupling necessary
- Small core without air gap
- Balancing problems

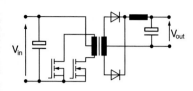

Push–pull converter with common based transistors

- One isolated controllable output voltage
- Output power up to several hundred watts
- Transistor breakdown voltage $V_{DS} \geq 2V_{in}$
- Small core without air gap
- Very good magnetic coupling between the primary coils required
- Balancing problems

ZCS push–pull resonant converter

- Several isolated output voltages achievable
- Output power up to several kilowatts
- Transistor breakdown voltage $V_{DS} = V_{in}$
- No extraordinary magnetic coupling necessary
- Small core without air gap
- Control with fixed pulse duration and variable frequency
- In case the output power is low compared to the rated power the frequency can be audible

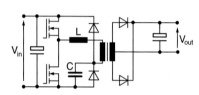

9.4.4 Control of Switched-Mode Power Supplies

The output voltage of a switched-mode power supply is kept constant with the help of closed loop control. The value of the output voltage (actual value) is compared with a reference voltage (nominal voltage). The difference between the actual and nominal values controls the duty cycle of the transistor driver. The control loop regulates the variation of the mains and of the output current change. This is called **line regulation** and **load regulation**.

There are two different methods of regulation: **voltage-mode** and **current-mode** control. The voltage-mode control is the 'traditional' method of regulation. Most modern systems use current-mode control, which is the basis of nearly all IC switched-mode controllers.

Both controller types can be explained using a boost converter as shown in Fig. 9.30:

9.4.4.1 Voltage-Mode Control

The output voltage V_{out} is compared to the reference voltage V_{ref} via a voltage divider R_1, R_2. Then the difference $V_{ref} - V'_{out}$ is amplified by the PI-regulator. A pulse width modulator (PWM, see Sect. 7.6.4.16) converts the output voltage of the PI regulator V_2 into a pulse-width modulated voltage t_1/T. The output of the PWM controls the transistor of the boost converter (see also Sect. 9.4.1.2).

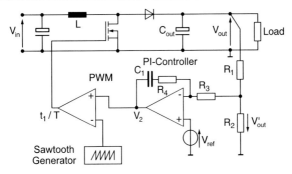

Fig. 9.30. Voltage-modecontrol for a boost converter

The closed loop operates as follows: if the output voltage V_{out} is too low, the voltage V'_{out} will be lower than the reference voltage V_{ref}. This will cause the output voltage V_2 of the PI regulator to increase. In the PWM circuit V_2 is compared with a sawtooth signal, and as V_2 increases the duty cycle t_1/T also increases. This causes the output voltage to increase until $V'_{out} = V_{ref}$.

9.4.4.2 Current-Mode Control

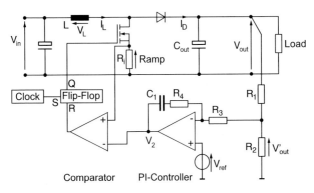

Fig. 9.31. Current-mode control for a boost converter

The output voltage V_{out} is compared to a reference voltage V_{ref} via the voltage divider R_1, R_2 and then amplified by the PI regulator. The output voltage of the PI regulator is compared with the ramp voltage across the current measuring resistor R_i. When the voltage across R_i exceeds V_2 the output of the comparator resets an SR flip-flop and turns the transistor off. The SR flip-flop is preset by the clock. The transistor is turned on by the clock and turned off when the ramp voltage (which means the inductor current) reaches a certain value. In this way the PI regulator directly controls the inductor current.

The closed loop operates as follows: if the output voltage V_{out} is too low, the voltage V'_{out} will be lower than the reference voltage V_{ref}. This causes the output voltage of the PI regulator V_2 to increase. The comparator compares the voltage V_2 with the ramp voltage across R_i. In this way V_2 determines the value to which the ramp voltage across R_i increases (which means the value to which the inductor current I_L increases) before the transistor is turned off. If V_2 increases because the V'_{out} is lower than V_{ref}, the inductor current will increase until V'_{out} is exactly equal to the reference voltage.

9.4.4.3 Comparison: Voltage-Mode vs. Current-Mode Control

The PI regulator of the current-mode control regulates the inductor current directly. This current feeds the output capacitor C_{out} and the load resistance R_L. Together C_{out} and R_L form a first-order system and the step response is an exponential function.

The voltage-mode control regulates the duty cycle t_1/T, which means that the voltage across L is controlled. This voltage operates on a second-order system formed by L, C_{out} and R_L. The step response of such a system is a sinusoidal transient approaching a fixed value.

Current-mode control therefore has a better control response; for this reason most controllers are current-mode types.

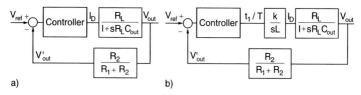

Fig. 9.32. Block-diagrams for **a** current-mode and **b** voltage-mode control

9.4.4.4 Design of the PI Controller

The PI-controlled system tends to oscillate if the capacitance C_1 is selected at too small a value and if the resistor R_4 is too high a value. To alleviate this problem C_1 should initially be selected high (a $1-\mu F$ foil capacitor is a good choice for most control circuits). The value of R_4 should be selected so that the cutoff frequency of the PI controller stays well below the cutoff frequency of L and C_{out}:

$$\frac{1}{2\pi\sqrt{LC_{out}}} \geq 10 \frac{1}{2\pi R_4 C_1} \tag{9.38}$$

The controller should now operate in a stable mode (if not, internal interference or a bad board architecture could be a problem). To improve the reaction of the closed loop, C_1 can be decreased step by step with a corresponding increase of R_4. If the loop starts to oscillate, C_1 can be increased by a factor of 10 while R_4 has to be decreased by the same factor. Using these design guides the loop will operate in a stable mode with sufficient regulation speed for most applications.

NOTE: In many control circuits the operational amplifier (normally called the error amplifier) is a transconductance amplifier. It supplies an output current (very high output impedance), which is proportional to the input voltage. In this case R_4 and C_1 are connected between the output and ground to achieve the PI characteristic of the controller.

9.4.5 Design of Inductors and High-Frequency Transformers

Inductors *store* energy, **transformers** *transfer* energy. This is the main difference. The magnetic cores are significantly different for inductors and high-frequency transformers: inductors need an air gap for storing energy, but transformers do not. Transformers for flyback converters have to store energy, which means they are not high-frequency transformers; they are in fact inductors with primary and secondary windings. The material of the cores is normally **ferrite**. Other materials with high permeability and with a high saturation point are also used.

9.4.5.1 Calculation of Inductors

An inductor with a certain inductance L and a certain peak current $\hat{I}$ can be determined by the following calculation:

Inductors should store energy. The stored energy of an inductor is $W = \frac{1}{2}L\hat{I}^2$. This energy is stored as magnetic field energy within the ferrite core and within the air gap (Fig. 9.33). The core size increases with increasing requirements for stored energy.

- The size of an inductor increases approximately proportionally to the energy to be stored.

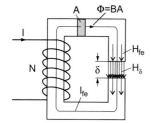

I Inductor current
N Number of turns
A Cross-sectional area of the core
l_{fe} Magnetic length of the core
δ Air gap
Φ Magnetic flux
B Magnetic flux density
    Magnetic field strength within the ferrite
H_δ Magnetic field strength within the air gap

Fig. 9.33. Inductor with its magnetic and mechanical parameters

The field energy in the inductor is given as:

$$W = \frac{1}{2}\int \vec{H} \cdot \vec{B} \, dV \approx \underbrace{\frac{1}{2}\vec{H}_{Fe} \cdot \vec{B}_{Fe} \cdot V_{Fe}}_{\text{Energy in the ferrite}} + \underbrace{\frac{1}{2}\vec{H}_\delta \cdot \vec{B}_\delta \cdot V_\delta}_{\text{Energy in the air gap}} \tag{9.39}$$

The magnetic field density $\vec{B}$ is continuous and is approximately equal within the air gap and the ferrite, i.e. $\vec{B} \approx \vec{B}_{Fe} \approx \vec{B}_\delta$. The magnetic field strength $\vec{H}$ is not continuous. Within the air gap it is increased by a factor μ_r with respect to the field strength within the ferrite. If this is substituted into Eq. (9.39) and considering $\vec{B} = \mu_0\mu_r \cdot \vec{H}$, $V_{Fe} = l_{Fe} \cdot A$ and $V_\delta = \delta \cdot A$ this leads to:

$$\boxed{W \approx \frac{1}{2}\frac{B^2}{\mu_0}\left(\frac{l_{Fe}}{\mu_r} + \delta\right) \cdot A} \tag{9.40}$$

The relative permeability μ_r of the ferrite is 1000–4000. It should be noted that the magnetic length of the ferrite is reduced by μ_r in the above equation. Therefore it can be seen that the energy is mainly stored within the air gap.

This leads to: $W \approx \dfrac{1}{2} \dfrac{B^2 \cdot A \cdot \delta}{\mu_0}$

- Inductors require an air gap, in which the energy is stored.

Because the energy is stored within the air gap, an inductor requires a certain volume for the air gap to store a certain amount of energy. The energy is given by $\dfrac{1}{2} L \hat{I}^2$. The core material has a limit for the maximum magnetic flux density B. This limit is approximately $B_{max} = 0.3$ T for the usual ferrite materials. This leads to a minimum required volume of the air gap:

$$V_\delta = A \cdot \delta \geq \dfrac{L \hat{I}^2 \cdot \mu_0}{B_{max}^2}, \qquad \text{where} \quad B_{max} = 0.3 \text{ T} \tag{9.41}$$

Knowing the required volume of the air gap, a core can be selected from a data book of ferrite cores.

The number of turns N can be calculated with help of the magnetic conductance A_L (often simply called the A_L value):

$$N = \sqrt{\dfrac{L}{A_L}} \qquad A_L\text{: magnetic conductance} \tag{9.42}$$

The A_L value can be verified from the data book of the ferrite cores. The maximum flux density within the ferrite can be calculated using the data of the core data sheet. The maximum flux density must usually not exceed 0.3 T.

$$B = \dfrac{L \cdot \hat{I}}{N \cdot A_{min}} = \dfrac{N \cdot A_L \cdot \hat{I}}{A_{min}} \;\overset{!}{\to}\; \leq 0.3 \text{ T} \tag{9.43}$$

Where A_{min} is the minimum cross-sectional area of the core. The flux density has its maximum at A_{min}. The value of A_{min} can be verified from the data sheet.

Calculation of the Wire

The current density J of the wire can be chosen between 2 and 5 A/mm (depending upon the size and the insulation, which determines the heat transport out of the inductor). This leads to the diameter of the wire d:

$$d = \sqrt{\dfrac{4 \cdot I_{RMS}}{\pi \cdot J}}, \qquad \text{where} \quad J = 2 \ldots \underline{3} \ldots 5 \; \dfrac{\text{A}}{\text{mm}^2} \tag{9.44}$$

9.4.5.2 Calculation of High-Frequency Transformers

High-frequency transformers transfer electric power. Their physical size depends upon the power to be transferred and upon the operating frequency. The higher the frequency the smaller the physical size. Usually frequencies are between 20 and 100 kHz. The material of the core is ferrite.

Data books for appropriate cores provide information about the possible transfer power for various cores.

Therefore the first step in designing a high-frequency transformer is to choose a suitable core with the help of the data book, since the size of the core is dependent on the transferred power and the frequency. The second step is to calculate the number of primary turns. This number determines the magnetic flux density within the core. The number of secondary turns is the ratio of the primary to secondary voltages. Following this, the diameters of the primary and secondary wires can be calculated, appropriate to the required RMS values of the currents.

Calculation of the Minimum Number of Primary Turns

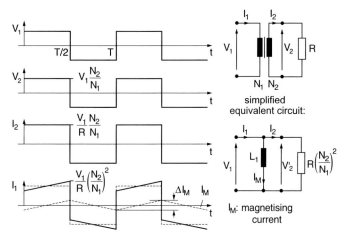

Fig. 9.34. Transformer voltages and currents

The voltage V_1 at the primary side of the transformers has a rectangular shape. This results in an input current I_1, which is the addition of the back-transformed secondary current I_2 and the magnetising current I_M (Fig. 9.34). To keep the magnetising current I_M low, a magnetic core without an air gap is used.

The rectangular voltage V_1 causes a triangular shape for the magnetising current I_M. The magnetising current is approximately independent of the secondary current I_2 (see the simple equivalent circuit in Fig. 9.34). The magnetising current is approximately proportional to the magnetic flux or flux density. The input voltage V_1 determines the magnetic flux. The physical relationships are given by Faraday's law: $V = N \cdot \dfrac{d\Phi}{dt}$.

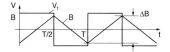

Fig. 9.35. Transformer input voltage and magnetic flux density

For the transformer shown in Fig. 9.34 it follows that:

$$\Delta B = \frac{V_1 \cdot T/2}{N_1 \cdot A} \tag{9.45}$$

- The change ΔB in flux density depends on the frequency $f = 1/T$ and the number of turns N_1. The higher the frequency and the number of turns, the lower the flux density change.

The minimum number of turns $N_{1\,\text{min}}$ can be calculated to ensure that a certain change of flux density ΔB is not exceeded. The saturation flux density of about $\hat{B} \approx 0.3$ T (which means $\Delta \hat{B} \approx 0.6$ T) cannot be used in high-frequency transformers. In push–pull converters, traversing the hysteresis loop with every clock cycle would result in unacceptable losses, i.e. heat generation. If no further information on core losses and thermal resistance is available, ΔB should be limited to $\Delta B \approx 0.3$–0.2 T for operating frequencies from 20 to 100 kHz. The lower the value of ΔB, the lower the core losses.

This leads to a minimum number of turns for N_1:

$$\boxed{N_{1\,\text{min}} \geq \frac{V_1 \cdot T/2}{\Delta B \cdot A_{\text{min}}} \quad \text{where} \quad \Delta B \approx 0.2\text{–}0.3 \text{ T}} \tag{9.46}$$

where A_{min} is the minimum cross-sectional area of the core. This is where the flux density is at a maximum. The value of A_{min} can be checked from the data sheet.

NOTE: In single-ended forward converters the core is magnetised in a unipolar direction only. In push–pull converters the core is dual-polarity magnetised.

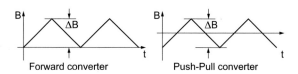

Forward converter   Push-Pull converter

The calculation of the minimum number of turns $N_{1\,\text{min}}$ is equal for these different types of switched-mode power supplies.

Calculation of the Wire Diameter

The diameter of the conductors depends on the RMS value of the current. The current can be calculated from the power.

For the full-bridge push–pull converter:

$$I_{1\,\text{RMS}} \approx \frac{P_{\text{out}}}{V_{\text{in}}} \quad \text{and} \quad I_{2\,\text{RMS}} = \frac{P_{\text{out}}}{V_{\text{out}}}$$

For the half-bridge push–pull converter:

$$I_{1\,\text{RMS}} \approx \frac{2P_{\text{out}}}{V_{\text{in}}} \quad \text{and} \quad I_{2\,\text{RMS}} = \frac{P_{\text{out}}}{V_{\text{out}}}$$

For the single ended forward converter:

$$I_{1\,\text{RMS}} \approx \frac{\sqrt{2}\,P_{\text{out}}}{V_{\text{in}}} \quad \text{and} \quad I_{2\,\text{RMS}} = \frac{\sqrt{2}\,P_{\text{out}}}{V_{\text{out}}}$$

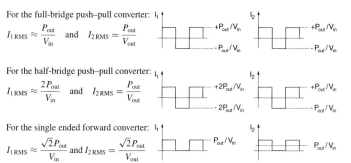

The magnetising current can be neglected in this calculation. The current density can be chosen in a range of 2 to 5 A/mm², depending on the thermal resistance of the choke. The cross section A_{wire} and the diameter d_{wire} can be calculated as follows:

$$A_{\text{wire}} = \frac{I}{J} \quad \text{and} \quad d_{\text{wire}} = \sqrt{\frac{I \cdot 4}{J \cdot \pi}}, \quad \text{where} \quad J = 2 \ldots \underline{3} \ldots 5 \; \frac{\text{A}}{\text{mm}^2} \tag{9.47}$$

Normally cores are designed so that the available winding cross-sectional area is sufficient for this calculation. Primary and secondary windings need the same winding cross-sectional area.

NOTE: If good coupling is important, the primary and secondary windings should be placed on top of each other. Improved coupling is achieved if the windings are interlocked. In the following example the coupling is bad in (a), good in (b) and best in (c).

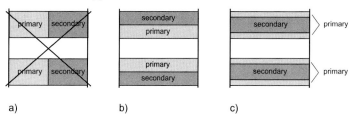

a) b) c)

NOTE: The primary number of turns should not be chosen significantly higher than $N_{1\,\text{min}}$. Otherwise the copper losses of the wire would increase needlessly because of the longer conductor.

Other literature even gives an optimum value ΔB_{opt}, where the sum of the hysteresis and the copper losses are minimised.

NOTE: For high frequencies and large diameters of the wire the skin effect must be considered. For operating frequencies of more than 20 kHz and wire diameters of more than 1 mm², stranded wire or copper foil should be used.

9.4.6 Power Factor Control

The European Standards EN61000-3-2 defines limits for the harmonics of line current. This concerns appliances, which are determined for the domestic market and have an input power of ≥ 75 W (special regulations; see EN61000-3-2). Some limit values from this standard are given in Table 9.2. In practice, this standard means that for many applications a mains rectifier with smoothing is not allowed because of the amount of harmonics (Fig. 9.36).

Table 9.2. RMS limits for the harmonics of the line current

Harmonic order n	Input power 75 to 600 W maximum value of harmonic current per watt (mA/W) / maximum (A)	Input power > 600 W maximum value of harmonic current (A)
3	3.4 / 2.30	2.30
5	1.9 / 1.14	1.14
7	1.0 / 0.77	0.77
9	0.5 / 0.4	0.40
11	0.35 / 0.33	0.33

To keep the line current approximately sinusoidal, a boost converter can be used (Fig. 9.37). In this case the boost converter is called a **power factor preregulator** or **power factor correction** (PFC). In comparison to the simple boost converter, the PFC is controlled in a different way: the output voltage is higher than the input voltage as for the boost converter, but the transistor is turned on and off in such a way that a sinusoidal input current is achieved instead of a constant output voltage. The transistor is driven in such a way that the inductor current $I_{in(t)}$ follows the shape of the rectified mains $V_{in}(t)$. The output voltage of the PFC is controlled to approximately $\overline{V_{out}} \approx 380$ V .

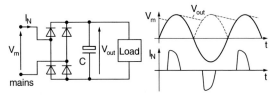

Fig. 9.36. Normal rectifying and smoothing of the mains voltage and the mains current

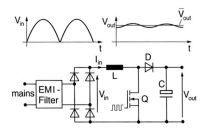

Fig. 9.37. Boost converter as a power factor preregulator

9.4.6.1 Currents, Voltages and Power of the PFC

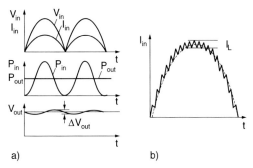

Fig. 9.38. a Currents, voltages and power of the PFC; and **b** the magnified input current with its high-frequency ripple

For the following calculations it is assumed that the output power is constant:

$$P_{\text{out}} = V_{\text{out}} \cdot I_{\text{out}} = \text{const.} \tag{9.48}$$

The input current should be controlled to a sinusoidal shape and should be in phase with the input voltage. The input power is now pulsating and can be calculated as follows:

$$P_{\text{in}}(t) = \frac{\hat{V}_{\text{in}} \cdot \hat{I}_{\text{in}}}{2} \cdot (1 - \cos 2\omega t) \tag{9.49}$$

The input power consists of a DC part, $P_{\text{in}=} = \frac{\hat{V}_{\text{in}} \cdot \hat{I}_{\text{in}}}{2}$, and an AC part, $P_{\text{in}\sim} = \frac{\hat{V}_{\text{in}} \cdot \hat{I}_{\text{in}}}{2} \cdot \cos 2\omega t$. The DC part is equal to the output power P_{out}, provided the PFC is loss-free.

$$P_{\text{in}} = \frac{\hat{V}_{\text{in}} \cdot \hat{I}_{\text{in}}}{2} = V_{\text{out}} \cdot I_{\text{out}} = P_{\text{out}} \tag{9.50}$$

In practice, an efficiency of about $\eta = 95\%$ is realistic, which means that $P_{\text{in}} \approx \frac{P_{\text{out}}}{0.95}$.

The output capacitor C_{out} is charged by the pulsating input power P_{in} and discharged by the constant output power P_{out}. This causes a voltage ripple ΔV_{out} across C_{out}, depending on the value of C_{out}. For 230 V/50 Hz supply, providing $V_{\text{out}} = 380$ V and $\Delta V_{\text{out}}/V_{\text{out}} = 10\%$, C_{out} can be calculated:

$$C_{\text{out}} \approx 0.5 \frac{\mu F}{W} \tag{9.51}$$

The choke L determines the high-frequency ripple of the input current ΔI_L (Fig. 9.38 b). The higher the inductance and the higher the clock frequency f, the lower is the current ripple. If $\Delta I_L = 20\%$ of the peak value of the input current $\hat{I}_{\text{in}}$, and assuming that the mains voltage has a minimum value of $V_{\text{in min}} = 200$ V, it follows that:

$$L \approx \frac{50 \cdot 10^3}{f \cdot P_{\text{in}}} \qquad L(\text{H}), \ f(\text{Hz}), \ P(\text{W}) \tag{9.52}$$

and for the maximum inductor current:

$$I_{L\,max} = \hat{I}_{in\,max} + \frac{1}{2}\Delta I_L = 1.1 \cdot \frac{2P_{in}}{\hat{V}_{in\,min}} \qquad (9.53)$$

9.4.6.2 Controlling the PFC

A variety of integrated PFC controllers are available for the switching transistor. Usually the data sheets and application examples of those ICs are very extensive. Nevertheless, it is very important to understand the working principles of the controller in order to design proper circuits (Fig. 9.39).

In general, two feedback circuits are required:

One controller for the input current in order to keep it sinusoidal (*input current control*), and one controller to keep the average output voltage constant (*output voltage control*).

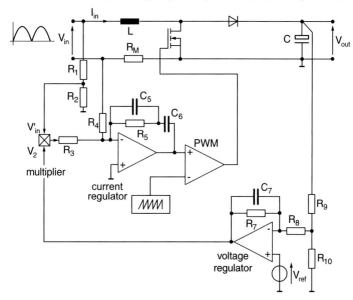

Fig. 9.39. The control loops of the PFC

The input control-loop current is led by the input voltage. In this case, the input current acquires the same shape as the input voltage, and consequently the power factor of the mains current will be unity.

The output voltage is controlled by comparing it to a constant reference voltage.

The multiplier links the two loops. The output of the multiplier is sinusoidal, and its magnitude depends on the output voltage-control loop. If the output voltage decreases from its nominal value, the output voltage of the voltage control amplifier increases, which causes the magnitude of the multiplier output to increase. Consequently, the RMS value of the input current also increases.

For proper operation:

- The low-pass filter R_5, C_5 should have a cutoff frequency of approximately 10% of f_{PWM} in order to suppress the current ripple of f_{PWM} in the current measurement at R_M.
- The cutoff frequency the PI regulator R_5, C_6 should be approximately 10 times higher than the frequency of the mains: $f_{R_5 C_6} \approx 500$ Hz.
- The cutoff frequency of the output voltage regulator R_7, C_7 should be 10% of the output voltage ripple (100 Hz): $f_{R_7 C_7} \approx 10$ Hz.
- The *RMS value* of the input current is controlled by the output voltage-control loop, while the input current control-loop creates a *sinusoidal* input current.

9.4.7 Radio-Frequency Interference Suppression of Switched-Mode Power Supplies

Switched-mode power supplies generate **radio-frequency interference** caused by the high frequency switching. This interference propagates through space by means of electromagnetic fields or via the mains supply in the form of currents and voltages. Legislation limits the levels of permitted interference. These limits are published in the European Standards. Table 9.3 gives some of the most important limits for non-stationary high-frequency equipment (interference class B). High-frequency equipment is that which operates at a frequency in excess of 9 kHz.

Table 9.3. Limits for mobile high frequency equipment class B

Quantity	Frequency range	Limits	Standard
Electromagnetic interference at 10 m distance	30 to 230 MHz	30 dB (µV/m)	EN55022
	230 to 1000 MHz	37 dB (µV/m)	Class B
Current harmonics at the mains	0 to 2 kHz	see Table 9.2 (PFC)	EN61000
Conducted-mode interference voltages at the mains with respect to earth	0.15 to 0.5 MHz**	66 to 56 dB (µV) Q* 56 to 46 dB (µV) M*	EN55022 Class B
	0.5 to 5 MHz	56 dB (µV) Q* 46 dB (µV) M*	
	5 to 30 MHz	60 dB (µV) Q* 50 dB (µV) M*	

* Q: Measured with quasi-peak detector
   M: Measured with average detector
** Linear decrease to the logarithm of the frequency

9.4.7.1 Radio-Frequency Interference Radiation

High-frequency equipment **emission** of radio-frequency interference is measured as the radio noise field strength (µV/m). The amount of radio-frequency interference radiation depends on the rise time of the switched currents and voltages and significantly on the layout of the printed circuit board. To keep the radio-frequency interference radiation low, three principles should be adhered to:

- Meshes, in which switched currents flow, should be designed as small as possible in the (surrounding) area to keep their electromagnetic field low.
- Nodes whose potential with respect to earth step up and down with switching should be as small as possible in volumetric space, to keep parasitic capacitance to earth low.
- The switched-mode power supply should have a metal housing.

NOTE: In addition to reduction of the interference radiation, the first two principles are also beneficial in keeping the conducted interference leaving the power supply via the mains low. It should also be noted that a high interference level results in inaccurate switching of the transistors and problems with the closed loop-control circuit. This often causes audible noise.

9.4.7.2 Mains Input Conducted-Mode Interference

Switched-mode power supplies take high-frequency currents out of the mains. These currents cause a voltage drop at the source impedance of the mains, which can be measured at the mains terminals. According to the European Standards the interference voltages have to be measured between the mains terminals and earth. For this measurement specific radio interference test equipment is needed, which includes a **radio-frequency interference meter** and an **artificial mains network**. This equipment is required to define a specific mains impedance for comparable measurements. To reduce conducted-mode interference special **radio-frequency interference filters** or electromagnetic interference filters (EMI-filters) are employed.

A distinction is made between three different types of radio interference voltage (Fig. 9.40):

- **Unsymmetric radio-frequency interference voltage**: This is the high-frequency voltage between earth and each mains terminal. This is the only voltage measured in accordance with standards. The limits in Table 9.3 are valid for this voltage only.
- **Common-mode radio-frequency interference voltage (asymmetric radio-interference voltage)**: This is the sum of all unsymmetric interference voltages with respect to earth.
- **Differential-mode radio-frequency interference voltage (symmetric radio-frequency interference voltage)**: This is the high-frequency voltage between the mains terminals.

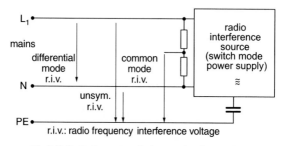

Fig. 9.40. Single-phase mains radio-frequency interference voltages

Although the legislation requires only measurement of the unsymmetric radio-frequency interference voltages, the common-mode *and* differential-mode interferences are decisive for radio-frequency interference suppression. The respective suppression of common-mode and differential-mode interference requires different designs and components.

9.4.7.3 Suppression of Common-Mode Radio-Frequency Interference

Common-mode radio-frequency interference voltages at the mains terminals L_1 and N (for three-phase mains L_1, L_2, L_3 and N) are common-mode voltages with respect to earth potential PE, which means they are equal in magnitude and phase. The interference currents $I_\approx$ which are driven by this common-mode voltage, are also common-mode currents. These flow via earth (earth conductor) and back through the parasitic capacitance C_earth. C_earth is very low. Therefore the common-mode interference voltage has a very high impedance, which means that this interference source acts like a current source. A low-pass filter to suppress the interference voltages at the mains terminals must therefore be arranged as in Fig. 9.41. Looking from the switched-mode power supply, the required low-pass filter must have a shunt capacitor (C_y) and a **current-compensated choke**. Current-compensated chokes are wound so that no magnetic field is generated by the operating current (50 or 60 Hz), see Fig. 9.42. Therefore the choke only acts against the common-mode interference current and does not affect the operating current.

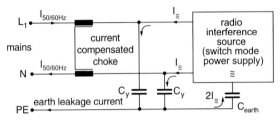

Fig. 9.41. Suppression of asymmetric (common-mode) radio frequency interference voltages

The capacitors are called **y-capacitors**. Y-capacitors have to fulfil special safety requirements because they would connect the mains phase to ground in case of a fault. Y-capacitors may not exceed a certain capacitance to ensure that the permitted maximum **earth leakage current** is not exceeded. The earth leakage current is a 50 Hz current (or 60 Hz in certain countries). The maximum earth leakage current is 3.5 mA (in medical equipment it is a maximum of 0.5 mA). According to the standards for the measurement of earth leakage current, terminals L_1 and N have to be connected, and the maximum mains voltage has to be applied between L_1 & N and PE. This means that the y-capacitors are in parallel. For European 230 V/50 Hz mains it follows that for the maximum y-capacitor:

$$C_y \leq \frac{1}{2} \cdot \frac{230\text{ V} + 10\%}{2\pi 50\text{ Hz} \cdot 3.5\text{ mA}} \approx 22\text{ nF}$$

9.4.7.4 Suppression of Differential-Mode Radio Frequency Interference

Differential-mode radio interference voltages are high-frequency voltages between the mains terminals L_1 and N. To reduce the interference level, an LC low-pass filter has to be inserted between the mains conductors L_1 and N (Fig. 9.43). The differential-mode interference voltage results mainly from the pulsed current, which is taken by the switched-mode power supply from the mains rectifier smoothing capacitor. Because of the impedance

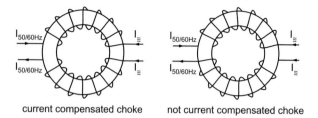

Fig. 9.42. *Left:* current-compensated choke for common-mode interference, *right:* not current-compensated choke (in this case, a ring-core double choke with powder core) for differential-mode interferences

of the smoothing capacitor, a high frequency voltage is generated between L_1 and N. This is a low impedance, which means that the interference source acts as a voltage source. Looking from the switched-mode power supply the interference filter must be arranged using a series choke followed by a shunt capacitor (Fig. 9.43). The choke must not be a compensated choke, because differential-mode interference current and 50 Hz-operating current (which is also a differential type) cause a magnetic field within the core (Fig. 9.42). To avoid saturation these **EMI suppression chokes** require an air gap. In a **powder-core choke** the air gap is not visible, because the air gap is achieved with iron powder, which is glued together. Air gap size can be fixed by the amount of glue used. **Open cores** are also used. With this type the magnetic field loop closes through space. Powder choke cores and other ring cores are preferred because they have a lower magnetic field outside the core.

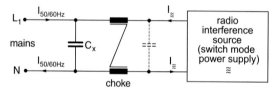

Fig. 9.43. Suppression of differential-mode interference

The capacitors for this purpose are called **x-capacitors**. They have a lower test-voltage than y-capacitors and are not limited in their value. Foil-type capacitors up to 1 µF are normally used.

NOTE: Sometimes the impedance of the differential-mode interference source is approximately equal to the mains impedance. In that case a p-low-pass filter using two x-capacitors are appropriate (in Fig. 9.43 dotted lined).

9.4.7.5 Complete Radio-Frequency Interference Filter

Figure 9.44 shows a complete **radio-frequency interference filter**. The component values can be found iteratively and with the help of experience. With the radio-interference meter only the unsymmetric interference voltages can be measured. Therefore it is not possible to differentiate between common-mode and differential-mode interference. In practice, the operating frequency and several harmonics are differential-mode interference and all high frequencies, say above 5 MHz, are common-mode. Often a powder core choke is not required.

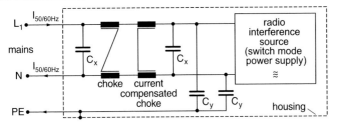

Fig. 9.44. Radio-frequency interference filter for common-mode and differential-mode filtering

9.5 Notation Index

A	cross-sectional area
A_L	magnetic conductance
B_{FE}	magnetic flux density in iron/ferrite
B_δ	magnetic flux density in the air gap
C	capacitor
D	diode
f	frequency
f_0	resonant frequency
ΔV, ΔI	voltage ripple, current ripple
in	as index: input value
H_{FE}	magnetic field strength in iron/ferrite
H_δ	magnetic field strength in the air gap
I	DC current, RMS value of a current
$\hat{I}$	peak value of a current
$\bar{I}$	average value of a current
I_F	current in a diode in forward direction
$I_{s/c}$	short circuit current
J	current density
l_{FE}	magnetic length of the iron/ferrite core
L	inductivity
out	as index: output value
P	power
P_L	power loss
PWM	pulse-width modulated
R	as index: rated value
R_M	current measurement resistor
RMS	root-mean square
t_1	on-time of a transistor
t_1/T	duty cycle
T	period time

V_{BE}	base–emitter voltage
V_F	forward voltage drop at a diode
V_{FE}	magnetic volume of a ferrite core
V_{max}	maximum value of voltage
V_{min}	minimum value of voltage
V_{ref}	reference voltage
V_{Rpp}	peak-to-peak voltage ripple
V_{PWM}	pulse width modulated voltage
V_z	zener voltage
V_δ	volume of the air gap
Z	impedance
Z_{max}	impedance of a capacitor (in the data sheet usually for 10 kHz)
δ	length of the air gap
μ_0	permeability of air/vacuum, $1.257 \cdot 10^{-6}$ Vs/Am
μ_r	relative permeability

9.6 Further Reading

BILLINGS, K.: *Switchmode Power Supply Handbook*
McGraw-Hill (1999)

CHRYSSIS, G.: *High Frequency Switching Power Supplies*
McGrawHill (1984)

RASHID, M. H.: *Power Electronics: Circuits, Devices, and Applications, 2nd Edition*
Prentice Hall (1993)

RASHID, M. H.: *Spice for Circuits and Elecronics using Pspice, 2nd Edition*
Prentice Hall (1995)

TIHANYI, L.: *Electromagnetic Compatibility in Power Electronics*
Butterworth/Heinemann (1995)

A Mathematical Basics

A.1 Trigonometric Functions

A.1.1 Properties

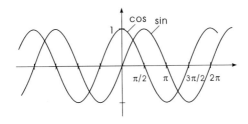

Fig. A.1. Graphs of sine and cosine functions

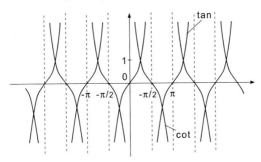

Fig. A.2. Graphs of tangent and cotangent functions

Special values					
$\alpha =$	0	$\pi/6$	$\pi/4$	$\pi/3$	$\pi/2$
	0°	30°	45°	60°	90°
$\sin x =$	0	1/2	$\sqrt{2}/2$	$\sqrt{3}/2$	1
$\cos x =$	1	$\sqrt{3}/2$	$\sqrt{2}/2$	1/2	0
$\tan x =$	0	$\sqrt{3}/3$	1	$\sqrt{3}$	∞

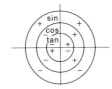

Fig. A.3. Signs of the trigonometric functions

Conversions

	$\cos\alpha$	$\sin\alpha$	$\tan\alpha$
$\cos\alpha =$	–	$\pm\sqrt{1-\sin^2\alpha}$	$\dfrac{1}{\pm\sqrt{1+\tan^2\alpha}}$
$\sin\alpha =$	$\pm\sqrt{1-\cos^2\alpha}$	–	$\dfrac{\tan\alpha}{\pm\sqrt{1+\tan^2\alpha}}$
$\tan\alpha =$	$\dfrac{\pm\sqrt{1-\cos^2\alpha}}{\cos\alpha}$	$\dfrac{\sin\alpha}{\pm\sqrt{1-\sin^2\alpha}}$	–

$$\sin^2\alpha + \cos^2\alpha = 1; \qquad \tan\alpha = \frac{\sin\alpha}{\cos\alpha}; \qquad \cot\alpha = \frac{1}{\tan\alpha}$$

Squares of trigonometric functions

	$\sin^2\alpha$	$\cos^2\alpha$	$\tan^2\alpha$
$\sin^2\alpha =$	–	$1-\cos^2\alpha$	$\dfrac{\tan^2\alpha}{1+\tan^2\alpha}$
$\cos^2\alpha =$	$1-\sin^2\alpha$	–	$\dfrac{1}{1+\tan^2\alpha}$
$\tan^2\alpha =$	$\dfrac{\sin^2\alpha}{1-\sin^2\alpha}$	$\dfrac{1-\cos^2\alpha}{\cos^2\alpha}$	–

Symmetry properties

$\sin(-\alpha)$	$=$	$-\sin\alpha$	odd function
$\cos(-\alpha)$	$=$	$\cos\alpha$	even function
$\tan(-\alpha)$	$=$	$-\tan\alpha$	odd function

Sums and differences with π

$x =$	$(\pi/2 - \alpha)$	$(\pi - \alpha)$	$(\pi + \alpha)$	$(\pi/2 + \alpha)$
$\sin x =$	$\cos\alpha$	$\sin\alpha$	$-\sin\alpha$	$\cos\alpha$
$\cos x =$	$\sin\alpha$	$-\cos\alpha$	$-\cos\alpha$	$-\sin\alpha$
$\tan x =$	$\cot\alpha$	$-\tan\alpha$	$\tan\alpha$	$-\cot\alpha$

A.1.2 Sums and Differences of Trigonometric Functions

$$\sin\alpha + \sin\beta = 2\sin\left(\frac{\alpha+\beta}{2}\right)\cdot\cos\left(\frac{\alpha-\beta}{2}\right) \tag{A.1}$$

$$\sin\alpha - \sin\beta = 2\cos\left(\frac{\alpha+\beta}{2}\right)\cdot\sin\left(\frac{\alpha-\beta}{2}\right) \tag{A.2}$$

$$\cos\alpha + \cos\beta = 2\cos\left(\frac{\alpha+\beta}{2}\right)\cdot\cos\left(\frac{\alpha-\beta}{2}\right) \tag{A.3}$$

$$\cos\alpha - \cos\beta = -2\sin\left(\frac{\alpha+\beta}{2}\right) \cdot \sin\left(\frac{\alpha-\beta}{2}\right) \tag{A.4}$$

$$\tan\alpha \pm \tan\beta = \frac{\sin(\alpha \pm \beta)}{\cos\alpha \cdot \cos\beta} \tag{A.5}$$

A.1.3 Sums and Differences in the Argument

$$\sin(\alpha \pm \beta) = \sin\alpha\cos\beta \pm \cos\alpha\sin\beta \tag{A.6}$$

$$\cos(\alpha \pm \beta) = \cos\alpha\cos\beta \mp \sin\alpha\sin\beta \tag{A.7}$$

$$\tan(\alpha \pm \beta) = \frac{\tan\alpha \pm \tan\beta}{1 \mp \tan\alpha\tan\beta} \tag{A.8}$$

A.1.4 Multiples of the Argument

$$\sin 2\alpha = 2\sin\alpha\cos\alpha \tag{A.9}$$

$$\cos 2\alpha = \cos^2\alpha - \sin^2\alpha \tag{A.10}$$

$$\tan 2\alpha = \frac{2\tan\alpha}{1 - \tan^2\alpha} \tag{A.11}$$

$$\sin 3\alpha = 3\sin\alpha - 4\sin^3\alpha \tag{A.12}$$

$$\cos 3\alpha = 4\cos^3\alpha - 3\cos\alpha \tag{A.13}$$

$$\tan 3\alpha = \frac{3\tan\alpha - \tan^3\alpha}{1 - 3\tan^2\alpha} \tag{A.14}$$

$$\sin 4\alpha = 8\cos^3\alpha\sin\alpha - 4\cos\alpha\sin\alpha \tag{A.15}$$

$$\cos 4\alpha = 8\cos^4\alpha - 8\cos^2\alpha + 1 \tag{A.16}$$

$$\tan 4\alpha = \frac{4\tan\alpha - 4\tan^3\alpha}{1 - 6\tan^2\alpha + \tan^4\alpha} \tag{A.17}$$

$$\sin\frac{\alpha}{2} = \pm\sqrt{\frac{1-\cos\alpha}{2}} \tag{A.18}$$

$$\cos\frac{\alpha}{2} = \pm\sqrt{\frac{1+\cos\alpha}{2}} \tag{A.19}$$

$$\tan\frac{\alpha}{2} = \pm\sqrt{\frac{1-\cos\alpha}{1+\cos\alpha}} \tag{A.20}$$

NOTE: In Eqs. (A.18)–(A.20) the sign of the square root must be equal to the sign of the function on the left side of the equation.

A.1.5 Weighted Sums of Trigonometric Functions

$$a \cdot \cos\alpha + b \cdot \cos\beta = c \cdot \cos\gamma \qquad (A.21)$$

with

$$c = \sqrt{a^2 + b^2 + 2ab \cdot \cos(\alpha - \beta)}; \qquad \tan\gamma = \frac{a \cdot \sin\alpha + b \cdot \sin\beta}{a \cdot \cos\alpha + b \cdot \cos\beta}; \qquad (A.22)$$
$$a \cdot \sin\alpha + b \cdot \sin\beta = c \cdot \sin\gamma; \qquad (A.23)$$

with

$$c \text{ and } \tan\gamma \quad \text{as in Eq. (A.22)}$$

A.1.6 Products of Trigonometric Functions

$$\cos\alpha \cdot \cos\beta = \frac{1}{2}\big[\cos(\alpha-\beta) + \cos(\alpha+\beta)\big] \qquad (A.24)$$

$$\cos\alpha \cdot \sin\beta = \frac{1}{2}\big[\sin(\alpha+\beta) - \sin(\alpha-\beta)\big] \qquad (A.25)$$

$$\sin\alpha \cdot \sin\beta = \frac{1}{2}\big[\cos(\alpha-\beta) - \cos(\alpha+\beta)\big] \qquad (A.26)$$

$$\sin\alpha \cdot \cos\beta = \frac{1}{2}\big[\sin(\alpha-\beta) + \sin(\alpha+\beta)\big] \qquad (A.27)$$

$$\sin(\alpha+\beta) \cdot \sin(\alpha-\beta) = \cos^2\beta - \cos^2\alpha \qquad (A.28)$$

$$\cos(\alpha+\beta) \cdot \cos(\alpha-\beta) = \cos^2\beta - \sin^2\alpha \qquad (A.29)$$

A.1.7 Triple Products

$$\cos\alpha \cdot \cos\beta \cdot \cos\gamma = \frac{1}{4}\big[\cos(\alpha+\beta+\gamma) + \cos(-\alpha+\beta+\gamma) \qquad (A.30)$$
$$+ \cos(\alpha-\beta+\gamma) + \cos(\alpha+\beta-\gamma)\big]$$

$$\cos\alpha \cdot \cos\beta \cdot \sin\gamma = \frac{1}{4}\big[\sin(\alpha+\beta+\gamma) + \sin(-\alpha+\beta+\gamma) \qquad (A.31)$$
$$+ \sin(\alpha-\beta+\gamma) - \sin(\alpha+\beta-\gamma)\big]$$

$$\cos\alpha \cdot \sin\beta \cdot \sin\gamma = \frac{1}{4}\big[-\cos(\alpha+\beta+\gamma) - \cos(-\alpha+\beta+\gamma) \qquad (A.32)$$
$$+ \cos(\alpha-\beta+\gamma) + \cos(\alpha+\beta-\gamma)\big]$$

$$\sin\alpha \cdot \sin\beta \cdot \sin\gamma = \frac{1}{4}\big[-\sin(\alpha+\beta+\gamma) + \sin(-\alpha+\beta+\gamma) \qquad (A.33)$$
$$+ \sin(\alpha-\beta+\gamma) + \sin(\alpha+\beta-\gamma)\big]$$

A.1.8 Powers of Trigonometric Functions

$$\cos^2 \alpha = \frac{1}{2}(1 + \cos 2\alpha) \tag{A.34}$$

$$\sin^2 \alpha = \frac{1}{2}(1 - \cos 2\alpha) \tag{A.35}$$

$$\cos^3 \alpha = \frac{1}{4}(\cos 3\alpha + 3 \cos \alpha) \tag{A.36}$$

$$\sin^3 \alpha = \frac{1}{4}(3 \sin \alpha - \sin 3\alpha) \tag{A.37}$$

$$\cos^4 \alpha = \frac{1}{8}(\cos 4\alpha + 4 \cos 2\alpha + 3) \tag{A.38}$$

$$\sin^4 \alpha = \frac{1}{8}(\cos 4\alpha - 4 \cos 2\alpha + 3) \tag{A.39}$$

A.1.9 Trigonometric Functions with Complex Arguments

$$\cos z = \frac{1}{2}e^{jz} + \frac{1}{2}e^{-jz} \tag{A.40}$$

$$\sin z = \frac{1}{2j}e^{jz} - \frac{1}{2j}e^{-jz} \tag{A.41}$$

A.2 Inverse Trigonometric Functions (Arc Functions)

Arc functions are the inverse functions of trigonometric functions.

$$\arcsin(\sin \alpha) = \alpha \qquad \arccos(\cos \alpha) = \alpha$$
$$\text{arccot}(\cot \alpha) = \alpha \qquad \arctan(\tan \alpha) = \alpha$$

NOTE: These functions are designated on calculators as $\sin^{-1}$, $\cos^{-1}$ and $\tan^{-1}$.

Because of the periodicity of the trigonometric functions their inverse functions are ambiguous. Therefore the **principal values** are defined.

$$-\pi/2 \leq \arcsin x \leq +\pi/2 \tag{A.42}$$

$$0 \leq \arccos x \leq \pi \tag{A.43}$$

$$-\pi/2 \leq \arctan x \leq +\pi/2 \tag{A.44}$$

Within the range of the principal values it holds that

$$\arcsin x = \pi/2 - \arccos x = \arctan(x/\sqrt{1-x^2}) \tag{A.45}$$

$$\arccos x = \pi/2 - \arcsin x = \text{arccot}(x/\sqrt{1-x^2}) \tag{A.46}$$

$$\arctan x = \pi/2 - \text{arccot} x = \arcsin(x/\sqrt{1+x^2}) \tag{A.47}$$

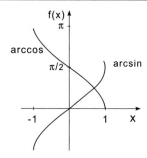

Fig. A.4. Graphs of the arcsine and arccosine functions within the range of the principal values

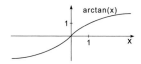

Fig. A.5. Graph of the arctan function

A.3 Hyperbolic Functions

$$\cosh z = \frac{1}{2}e^z + \frac{1}{2}e^{-z} \tag{A.48}$$

$$\sinh z = \frac{1}{2}e^z - \frac{1}{2}e^{-z} \tag{A.49}$$

$$\tanh z = \frac{e^z - e^{-z}}{e^z + e^{-z}} \tag{A.50}$$

The **addition theorems of the hyperbolic functions** are obtained by formally substituting

$$\sin z \to j \sinh z; \qquad \cos z \to \cosh z$$

EXAMPLE: $\cos^2 z + \sin^2 z \to \cosh^2 z + j^2 \sinh^2 z = \cosh^2 z - \sinh^2 z = 1$

A.4 Differential Calculus

A.4.1 Basics of Differential Calculus

While $f(x), u(x), v(x)$ are functions with an existing derivative, a is a purely real constant.

$$\begin{aligned}
a' &= 0 \\
(au)' &= au' \\
(u+v)' &= u' + v' \\
(u \cdot v)' &= uv' + vu' \qquad &\text{Product rule} \\
\left(\frac{u}{v}\right)' &= \frac{u'v - uv'}{v^2} \qquad &\text{Division rule} \\
[f(u(x))]' &= f'(u) \cdot u'(x) \qquad &\text{Chain rule}
\end{aligned}$$

A.4.2 Derivatives of Elementary Functions

$f(x)$	$f'(x)$
a	0
x	1
ax^n	anx^{n-1}
a^x	$a^x \ln a$
e^{ax}	ae^{ax}
x^x	$x^x(1+\ln x)$
$\log_a x$	$\dfrac{1}{x}\log_a e$
$\ln x$	$\dfrac{1}{x}$
$\sin x$	$\cos x$
$\cos x$	$-\sin x$
$\tan x$	$\cos^{-2} x = 1 + \tan^2 x$
$\cot x$	$-\sin^{-2} x = -(1+\cot^2 x)$
$\sinh x$	$\cosh x$
$\cosh x$	$\sinh x$

$f(x)$	$f'(x)$
$\tanh x$	$\cosh^{-2} x = 1 - \tanh^2 x$
$\coth x$	$-\sinh^{-2} x = 1 - \coth^2 x$
$\arcsin x$	$\dfrac{1}{\sqrt{1-x^2}}$
$\arccos x$	$-\dfrac{1}{\sqrt{1-x^2}}$
$\arctan x$	$\dfrac{1}{1+x^2}$
$\text{arccot}\, x$	$-\dfrac{1}{1+x^2}$
$\text{arsinh}\, x$	$\dfrac{1}{\sqrt{x^2+1}}$
$\text{arcosh}\, x$	$\dfrac{1}{\sqrt{x^2-1}}$
$\text{artanh}\, x$	$\dfrac{1}{1-x^2}, \quad x^2 < 1$
$\text{arcoth}\, x$	$\dfrac{1}{1-x^2}, \quad x^2 > 1$

A.5 Integral Calculus

A.5.1 Basics of Integral Calculus

$$\int_a^b f(x)\,dx = -\int_b^a f(x)\,dx \tag{A.51}$$

$$\int_a^b f(x)\,dx = \int_a^c f(x)\,dx + \int_c^b f(x)\,dx \tag{A.52}$$

$$\int_a^b f(x)\,dx - \int_a^c f(x)\,dx = \int_c^b f(x)\,dx \tag{A.53}$$

$$\int_a^b f(x) \pm g(x)\,dx = \int_a^b f(x)\,dx \pm \int_a^b g(x)\,dx \tag{A.54}$$

$$\int_a^b u\,dv = u(b)v(b) - u(a)v(a) - \int_a^b v\,du \tag{A.55}$$

A.5.1.1 Integrals of Elementary Functions

$$\int x^n \, dx = \frac{x^{n+1}}{n+1} \quad \text{for } n \neq -1 \tag{A.56}$$

$$\int \frac{dx}{x} = \ln x \tag{A.57}$$

$$\int f(x) f'(x) \, dx = \frac{1}{2} (f(x))^2 \tag{A.58}$$

$$\int \frac{f'(x)}{f(x)} \, dx = \ln(f(x)) \tag{A.59}$$

$$\int \frac{f'(x)}{2\sqrt{f(x)}} \, dx = \sqrt{f(x)} \tag{A.60}$$

$$\int e^x \, dx = e^x \tag{A.61}$$

$$\int e^{ax} \, dx = \frac{1}{a} e^{ax} \tag{A.62}$$

$$\int \ln x \, dx = x \ln x - x \tag{A.63}$$

$$\int a^x \ln a \, dx = a^x \tag{A.64}$$

$$\int \sin x \, dx = -\cos x \tag{A.65}$$

$$\int \cos x \, dx = \sin x \tag{A.66}$$

$$\int \cot x \, dx = \ln |\sin x| \tag{A.67}$$

$$\int \frac{dx}{\sin^2 x} = -\cot x \tag{A.68}$$

$$\int \frac{dx}{\cos^2 x} = \tan x \tag{A.69}$$

$$\int \sinh x \, dx = \cosh x \tag{A.70}$$

$$\int \cosh x \, dx = \sinh x \tag{A.71}$$

$$\int \frac{dx}{\sinh^2 x} = -\coth x \tag{A.72}$$

$$\int \frac{dx}{\cosh^2 x} = \tanh x \tag{A.73}$$

$$\int \frac{dx}{\sqrt{1-x^2}} = \arcsin x = -\arccos x \tag{A.74}$$

$$\int \frac{\mathrm{d}x}{\sqrt{x^2 - 1}} = \operatorname{arcosh} x = \ln\left(x + \sqrt{x^2 - 1}\right) \tag{A.75}$$

$$\int \frac{\mathrm{d}x}{\sqrt{x^2 + 1}} = \operatorname{arsinh} x = \ln\left(x + \sqrt{x^2 + 1}\right) \tag{A.76}$$

$$\int \frac{\mathrm{d}x}{1 + x^2} = \arctan x = -\operatorname{arccot} x \tag{A.77}$$

$$\int \frac{\mathrm{d}x}{1 - x^2} = \operatorname{artanh} x, \qquad \text{for } x^2 < 1$$

$$= \operatorname{arcoth} x, \qquad \text{for } x^2 > 1 \tag{A.78}$$

$$\int \frac{1}{a^2 + x^2}\, \mathrm{d}x = \frac{1}{a} \arctan\left(\frac{x}{a}\right), \quad a \neq 0 \tag{A.79}$$

$$\int \frac{1}{a^2 - x^2}\, \mathrm{d}x = \frac{1}{2a} \ln\left|\frac{a + x}{a - x}\right| \tag{A.80}$$

$$\int \frac{\sqrt{1 + x}}{\sqrt{1 - x}}\, \mathrm{d}x = \arcsin x - \sqrt{1 - x^2} \tag{A.81}$$

A.5.2 Integrals Involving Trigonometric Functions

$$\int \sin mx \, \mathrm{d}x = -\frac{1}{m} \cos mx \tag{A.82}$$

$$\int \sin^2 x \, \mathrm{d}x = -\frac{1}{2} \sin x \cos x + \frac{x}{2} = -\frac{1}{4} \sin 2x + \frac{x}{2} \tag{A.83}$$

$$\int \sin^3 x \, \mathrm{d}x = -\frac{1}{3}(\sin^2 x + 2) \cdot \cos x = -\frac{3 \cos x}{4} + \frac{\cos^3 x}{12} \tag{A.84}$$

$$\int \sin^n x \, \mathrm{d}x = -\frac{1}{n} \sin^{n-1} x \cos x + \frac{n-1}{n} \int \sin^{n-2} x \, \mathrm{d}x \tag{A.85}$$

$$\int \frac{\mathrm{d}x}{\sin x} = \ln\left|\frac{1}{\sin x} - \cot x\right| = \ln\left|\tan \frac{x}{2}\right|$$

$$= -\frac{1}{2} \ln\left(\frac{1 + \cos x}{1 - \cos x}\right) = \operatorname{artanh}(\cos x) \tag{A.86}$$

$$\int \frac{\mathrm{d}x}{\sin^2 x} = -\cot x \tag{A.87}$$

$$\int \sin(a + bx)\, \mathrm{d}x = -\frac{1}{b} \cos(a + bx) \tag{A.88}$$

$$\int \cos mx \, \mathrm{d}x = \frac{1}{m} \sin mx \tag{A.89}$$

$$\int \cos^2 x \, \mathrm{d}x = \frac{1}{2} \sin x \cos x + \frac{x}{2} = \frac{1}{4} \sin 2x + \frac{x}{2} \tag{A.90}$$

$$\int \cos^3 x \, dx = \frac{1}{3} \sin x (\cos^2 x + 2) = \frac{3}{4} \sin x + \frac{\sin^3 x}{12} \tag{A.91}$$

$$\int \cos^n x \, dx = \frac{1}{n} \sin x \cos^{n-1} x + \frac{n-1}{n} \int \cos^{n-2} x \, dx \tag{A.92}$$

$$\int \frac{dx}{\cos x} = \ln \left| \frac{1}{\cos x} + \tan x \right| = \ln \tan \left(\frac{\pi}{4} + \frac{x}{2} \right)$$

$$= \frac{1}{2} \ln \left(\frac{1 + \sin x}{1 - \sin x} \right) = \operatorname{artanh}(\sin x) \tag{A.93}$$

$$\int \frac{dx}{\cos^2 x} = \tan x \tag{A.94}$$

$$\int \cos(a + bx) \, dx = \frac{1}{b} (\sin a + bx) \tag{A.95}$$

$$\int \sin x \cos x \, dx = \frac{\sin^2 x}{2} \tag{A.96}$$

$$\int \frac{dx}{\sin x \cos x} = \ln(\tan x) \tag{A.97}$$

$$\int \sin mx \sin nx \, dx = \frac{\sin(m-n)x}{2(m-n)} - \frac{\sin(m+n)x}{2(m+n)}, \text{ for } m^2 \neq n^2 \tag{A.98}$$

$$\int \cos mx \cos nx \, dx = \frac{\sin(m-n)x}{2(m-n)} + \frac{\sin(m+n)x}{2(m+n)}, \text{ for } m^2 \neq n^2 \tag{A.99}$$

$$\int \sin mx \cos nx \, dx = -\frac{\cos(m-n)x}{2(m-n)} - \frac{\cos(m+n)x}{2(m+n)}, \text{ for } m^2 \neq n^2 \tag{A.100}$$

$$\int \sin x \cos^n x \, dx = -\frac{1}{n+1} \cos^{n+1} x \tag{A.101}$$

$$\int \sin^n x \cos x \, dx = \frac{1}{n+1} \sin^{n+1} x \tag{A.102}$$

$$\int \tan x \, dx = -\ln |\cos x| \tag{A.103}$$

$$\int \cot x \, dx = \ln |\sin x| \tag{A.104}$$

$$\int \tan^2 x \, dx = \tan x - x \tag{A.105}$$

$$\int \cot^2 x \, dx = -\cot x - x \tag{A.106}$$

$$\int x \sin x \, dx = \sin x - x \cos x \tag{A.107}$$

$$\int x^2 \sin x \, dx = 2x \sin x - (x^2 - 2) \cos x \tag{A.108}$$

$$\int x^3 \sin x \, dx = (3x^2 - 6)\sin x - (x^3 - 6x)\cos x \tag{A.109}$$

$$\int x^n \sin x \, dx = -x^n \cos x + n \int x^{n-1} \cos x \, dx \tag{A.110}$$

$$\int x \sin^2 x \, dx = \frac{x^2}{4} - \frac{x}{4}\sin 2x - \frac{1}{8}\cos 2x \tag{A.111}$$

$$\int x^2 \sin^2 x \, dx = \frac{x^3}{6} - \left(\frac{x^2}{4} - \frac{1}{8}\right)\sin 2x - \frac{x}{4}\cos 2x \tag{A.112}$$

$$\int \frac{\sin x}{x} \, dx = x - \frac{x^3}{3 \cdot 3!} + \frac{x^5}{5 \cdot 5!} - \frac{x^7}{7 \cdot 7!} + - \cdots \tag{A.113}$$

$$\int \frac{\sin x}{x^n} \, dx = -\frac{\sin x}{(n-1)x^{n-1}} + \frac{1}{n-1}\int \frac{\cos x}{x^{n-1}} \, dx, \quad \text{for } n \neq 1 \tag{A.114}$$

$$\int x \cos x \, dx = \cos x + x \sin x \tag{A.115}$$

$$\int x^2 \cos x \, dx = 2x \cos x + (x^2 - 2)\sin x \tag{A.116}$$

$$\int x^3 \cos x \, dx = (3x^2 - 6)\cos x + (x^3 - 6x)\sin x \tag{A.117}$$

$$\int x^n \cos x \, dx = x^n \sin x - n \int x^{n-1} \sin x \, dx \tag{A.118}$$

$$\int x \cos^2 x \, dx = \frac{x^2}{4} + \frac{x}{4}\sin 2x + \frac{1}{8}\cos 2x \tag{A.119}$$

$$\int x^2 \cos^2 x \, dx = \frac{x^3}{6} + \left(\frac{x^2}{4} - \frac{1}{8}\right)\sin 2x + \frac{x}{4}\cos 2x \tag{A.120}$$

$$\int \frac{\cos x}{x} \, dx = \ln|x| - \frac{x^2}{2 \cdot 2!} + \frac{x^4}{4 \cdot 4!} - \frac{x^6}{6 \cdot 6!} + - \cdots \tag{A.121}$$

$$\int \frac{\cos x}{x^n} \, dx = -\frac{\cos x}{(n-1)x^{n-1}} - \frac{1}{n-1}\int \frac{\sin x}{x^{n-1}} \, dx, \quad \text{for } n \neq 1 \tag{A.122}$$

A.5.3 Integrals Involving Exponential Functions

$$\int e^x \, dx = e^x \tag{A.123}$$

$$\int e^{-x} \, dx = -e^{-x} \tag{A.124}$$

$$\int e^{ax} \, dx = \frac{1}{a}e^{ax} \tag{A.125}$$

$$\int e^{-x^2} \, dx = \frac{x}{0! \cdot 1} - \frac{x^3}{1! \cdot 3} + \frac{x^5}{2! \cdot 5} - \ldots \quad \text{Gaussian error integral} \tag{A.126}$$

$$\int x e^{ax}\, dx = \frac{1}{a^2} e^{ax}(ax - 1) \tag{A.127}$$

$$\int x^n e^{ax}\, dx = \frac{x^n}{a} e^{ax} - \frac{n}{a} \int x^{n-1} e^{ax}\, dx$$

$$= e^{ax} \left[\frac{x^n}{a} - \frac{n x^{n-1}}{a^2} + \frac{n(n-1) x^{n-2}}{a^3} - + \cdots \right] \tag{A.128}$$

$$\int \frac{e^{ax}}{x}\, dx = \ln x + ax + \frac{a^2 x^2}{2 \cdot 2!} + \frac{a^3 x^3}{3 \cdot 3!} + \frac{a^4 x^4}{4 \cdot 4!} + \cdots \tag{A.129}$$

$$\int e^{ax+c} \sin(bx + d)\, dx = \frac{e^{ax+c}}{a^2 + b^2} \left[a \sin(bx + d) - b \cos(bx + d) \right] \tag{A.130}$$

$$\int e^{ax+c} \cos(bx + d)\, dx = \frac{e^{ax+c}}{a^2 + b^2} \left[a \cos(bx + d) + b \sin(bx + d) \right] \tag{A.131}$$

A.5.4 Integrals Involving Inverse Trigonometric Functions

$$\int \arcsin \frac{x}{a}\, dx = x \arcsin \frac{x}{a} + \sqrt{a^2 - x^2} \tag{A.132}$$

$$\int \arccos \frac{x}{a}\, dx = x \arccos \frac{x}{a} - \sqrt{a^2 - x^2} \tag{A.133}$$

$$\int \arctan \frac{x}{a}\, dx = x \arctan \frac{x}{a} - a \ln(\sqrt{a^2 + x^2}) \tag{A.134}$$

$$\int \text{arccot} \frac{x}{a}\, dx = x \, \text{arccot} \frac{x}{a} + a \ln(\sqrt{a^2 + x^2}) \tag{A.135}$$

$$\int x \arctan \frac{x}{a}\, dx = -\frac{ax}{2} + \frac{x^2 + a^2}{2} \arctan \frac{x}{a} \tag{A.136}$$

A.5.5 Definite Integrals

$$\int_0^{\frac{\pi}{2}} \sin^n x\, dx = \int_0^{\frac{\pi}{2}} \cos^n x\, dx$$

$$= \frac{1 \cdot 3 \cdot 5 \cdots (n-1)}{2 \cdot 4 \cdot 6 \cdots n} \frac{\pi}{2}, \quad \text{for even } n$$

$$= \frac{2 \cdot 4 \cdot 6 \cdots (n-1)}{1 \cdot 3 \cdot 5 \cdots n}, \quad \text{for odd } n$$

$$= \frac{\sqrt{\pi}}{2} \frac{\Gamma\left(\frac{n}{2} + \frac{1}{2}\right)}{\Gamma\left(\frac{n}{2} + 1\right)}, \quad \text{for } n > -1 \tag{A.137}$$

$$\int_0^\infty \frac{\sin ax}{x}\,\mathrm{d}x = \frac{\pi}{2}, \quad \text{for } a > 0$$

$$\phantom{\int_0^\infty \frac{\sin ax}{x}\,\mathrm{d}x} = 0, \quad \text{for } a = 0$$

$$\phantom{\int_0^\infty \frac{\sin ax}{x}\,\mathrm{d}x} = -\frac{\pi}{2}, \quad \text{for } a < 0 \tag{A.138}$$

$$\int_0^\infty \frac{\cos ax}{x}\,\mathrm{d}x = \infty \tag{A.139}$$

$$\int_{-\infty}^\infty \frac{\cos ax}{x}\,\mathrm{d}x = 0 \tag{A.140}$$

$$\int_0^\pi \sin^2(ax)\,\mathrm{d}x = \frac{\pi}{2}, \quad \text{for } a \neq 0 \tag{A.141}$$

$$\int_0^\pi \cos^2(ax)\,\mathrm{d}x = \frac{\pi}{2}, \quad \text{for } a \neq 0 \tag{A.142}$$

$$\int_0^\pi \sin mx \cdot \sin nx\,\mathrm{d}x = \int_0^\pi \cos mx \cdot \cos nx\,\mathrm{d}x \tag{A.143}$$

$$ = 0, \quad \text{for } m \neq n, \quad \text{with } m, n = 1, 2, 3, \ldots$$

$$ = \frac{\pi}{2}, \quad \text{for } m = n, \quad \text{with } m, n = 1, 2, 3, \ldots$$

$$\int_{-a}^{+a} \sin \frac{m\pi x}{a} \cdot \sin \frac{n\pi x}{a}\,\mathrm{d}x = \int_{-a}^{+a} \cos \frac{m\pi x}{a} \cdot \cos \frac{n\pi x}{a}\,dx \tag{A.144}$$

$$ = 0, \quad \text{for } m \neq n \quad \text{with } m, n = 1, 2, 3, \ldots$$

$$ = a, \quad \text{for } m = n \quad \text{with } m, n = 1, 2, 3, \ldots$$

$$\int_{-a}^{+a} \sin \frac{m\pi x}{a} \cdot \cos \frac{n\pi x}{a}\,\mathrm{d}x = 0, \quad \text{with } m, n = 1, 2, 3, \ldots \tag{A.145}$$

$$\int_0^\infty \frac{\sin mx \cdot \sin nx}{x}\,\mathrm{d}x = \frac{1}{2}\ln\frac{m+n}{m-n}, \quad \text{with } m > n > 0 \tag{A.146}$$

$$\int_0^\infty \frac{\sin mx \cdot \cos nx}{x}\,dx = \frac{\pi}{2}, \quad \text{for } m > n \geq 0 \tag{A.147}$$

$$= \frac{\pi}{4}, \quad \text{for } m = n > 0$$

$$= 0, \quad \text{for } n > m \geq 0$$

$$\int_0^\infty \sin x^2\,dx = \int_0^\infty \cos x^2\,dx = \frac{1}{2}\sqrt{\frac{\pi}{2}} \tag{A.148}$$

$$\int_0^\infty e^{-ax}\,dx = \frac{1}{a}, \quad \text{with } a > 0 \tag{A.149}$$

$$\int_0^\infty e^{-a^2 x^2}\,dx = \frac{1}{2a}\sqrt{\pi}, \quad \text{with } a > 0 \tag{A.150}$$

$$\int_0^\infty x e^{-x^2}\,dx = \frac{1}{2} \tag{A.151}$$

$$\int_0^\infty x^2 e^{-x^2}\,dx = \frac{1}{4}\sqrt{\pi} \tag{A.152}$$

$$\int_0^\infty x^2 e^{-a^2 x^2}\,dx = \frac{\sqrt{\pi}}{4a^3}, \quad a > 0 \tag{A.153}$$

$$\int_0^\infty e^{-ax} \sin(nx)\,dx = \frac{n}{a^2 + n^2}, \quad \text{with } a > 0 \tag{A.154}$$

$$\int_0^\infty e^{-ax} \cos(nx)\,dx = \frac{a}{a^2 + n^2}, \quad \text{with } a > 0 \tag{A.155}$$

$$\int_0^\infty e^{-a^2 x^2} \cos bx\,dx = \frac{\sqrt{\pi}}{2a} e^{-b/4a^2}, \quad a > 0 \tag{A.156}$$

$$\int_0^\infty x^n e^{-ax}\,dx = \frac{n!}{a^{n+1}}, \quad a > 0,\ n > 0,\ \text{integer} \tag{A.157}$$

A.6 The Integral of the Standard Normal Distribution

For a normally distributed quantity x with a mean value μ and a standard deviation σ the **normalised** random quantity $z = (x - \mu)/\sigma$ is distributed according to a **normalised standard (Gaussian) distribution**. The following pages show tables of the integral of this distribution.

$$\Phi(z) = \frac{1}{\sqrt{2\pi}} \int_0^z e^{\frac{-x^2}{2}} \, dx$$

Application		Problem statement
(−z 0 z)	$p = 2 \cdot \Phi(z)$	Probability p that the value does **not** deviate **more** than $|z|$ from the average value (higher or lower).
		Example: Given a set of 100 Ω ± 5% resistors. What is the portion of the components deviating **not more than** ±15 Ω from the nominal value? $z = (115 - 100)/5 = 3.0 \Rightarrow p = 2 \cdot \Phi(z) = 99.7\%$
(−z 0 z)	$p = 1 - 2 \cdot \Phi(z)$	Probability that the value does deviate **more** than $|z|$ from the average value (higher or lower).
		Example: What part of the components have an actual resistance value below 90 Ω or above 110 Ω? $z = 2.0 \Rightarrow p = 1 - 2 \cdot \Phi(z) = 4.55\%$
(0 z)	$p = 0.5 - \Phi(z)$	Probability that the average is exceeded by more than z.
		Example: What percentage of the components have an actual value exceeding 110 Ω? $z = 2.0 \Rightarrow p = 0.5 - \Phi(z) = 2.275\%$
(0 z_1 z_2)	$p = \Phi(z_1) - \Phi(z_2)$	Probability that the value is between z_1 and z_2.
		Example: What is the probability that a resistance value of the set is between 114.5 Ω and 115 Ω? $z_1 = (114.5 - 100)/5 = 2.9$, $z_2 = 3.0 \Rightarrow$ $p = \Phi(3.0) - \Phi(2.9)$ $= 0.498\,650\,0 - 0.498\,134\,1 = 0.05\%$

NOTE: For these kinds of problems it is reasonable to include many digits in the calculations, since many digits will cancel out each other calculating differences of almost equal numbers.

Confidence Intervals

![two-tailed](−z 0 z)	![two-tailed tails](−z 0 z)	![one-tailed](0 z)	z
90.0 %	10.0 %	5.0 %	1.645
95.0 %	5.0 %	2.5 %	1.960
98.0 %	2.0 %	1.0 %	2.326
99.0 %	1.0 %	0.5 %	2.576
99.5 %	0.5 %	0.25 %	2.807
99.8 %	0.2 %	0.1 %	3.091
99.9 %	0.1 %	0.05 %	3.293
99.95 %	0.05 %	0.025 %	3.483

A.6 The Integral of the Standard Normal Distribution

Integral of the Standard Normal Distribution

z		0	1	2	3	4	5	6	7	8	9
0.0	0.0	000	040	080	120	160	199	239	279	319	359
0.1		398	438	478	517	557	596	636	675	714	753
0.2		793	832	871	910	948	987	0.1 026	0.1 064	0.1 103	0.1 141
0.3	0.1	179	217	255	293	331	368	406	443	480	517
0.4		554	591	628	664	700	736	772	808	844	879
0.5		915	950	985	0.2 019	0.2 054	0.2 088	0.2 123	0.2 157	0.2 190	0.2 224
0.6	0.2	257	291	324	357	389	422	454	486	517	549
0.7		580	611	642	673	704	734	764	794	823	852
0.8		881	910	939	967	995	0.3 023	0.3 051	0.3 078	0.3 106	0.3 133
0.9	0.3	159	186	212	238	264	289	315	340	365	389
1.0		413	438	461	485	508	531	554	577	599	621
1.1		643	665	686	708	729	749	770	790	810	830
1.2		849	869	888	907	925	944	962	980	997	0.4 015
1.3	0.4	032	049	066	082	099	115	131	147	162	177
1.4		192	207	222	236	251	265	279	292	306	319
1.5		332	345	357	370	382	394	406	418	429	441
1.6		452	463	474	484	495	505	515	525	535	545
1.7		554	564	573	582	591	599	608	616	625	633
1.8		641	649	656	664	671	678	686	693	699	706
1.9		713	719	726	732	738	744	750	756	761	767
2.0	0.4	772 499	777 845	783 084	788 218	793 249	798 179	803 008	807 739	812 373	816 912
2.1		821 356	825 709	829 970	834 143	838 227	842 224	846 137	849 966	853 713	857 379
2.2		860 966	864 475	867 907	871 263	874 546	877 756	880 894	883 962	886 962	889 894
2.3		892 759	895 559	898 296	900 969	903 582	906 133	908 625	911 060	913 437	915 758
2.4		918 025	920 237	922 397	924 506	926 564	928 572	930 531	932 443	934 309	936 128
2.5		937 903	939 634	941 322	942 969	944 574	946 138	947 664	949 150	950 600	952 012
2.6		953 388	954 729	956 035	957 307	958 547	959 754	960 929	962 074	963 188	964 274
2.7		965 330	966 358	967 359	968 332	969 280	970 202	971 099	971 971	972 820	973 645
2.8		974 448	975 229	975 988	976 725	977 443	978 140	978 817	979 476	980 116	980 737
2.9		981 341	981 928	982 498	983 051	983 589	984 111	984 617	985 109	985 587	986 050
3.0	0.4	986 500	986 937	987 361	987 772	988 170	988 557	988 932	989 296	989 649	989 991
3.1		990 323	990 645	990 957	991 259	991 552	991 836	992 111	992 377	992 636	992 886
3.2		993 128	993 363	993 590	993 810	994 023	994 229	994 429	994 622	994 809	994 990
3.3		995 165	995 335	995 499	995 657	995 811	995 959	996 102	996 241	996 375	996 505
3.4		996 630	996 751	996 868	996 982	997 091	997 197	997 299	997 397	997 492	997 584
3.5		997 673	997 759	997 842	997 922	997 999	998 073	998 145	998 215	998 282	998 346
3.6		998 409	998 469	998 527	998 583	998 636	998 688	998 739	998 787	998 834	998 878
3.7		998 922	998 963	999 004	999 042	999 080	999 116	999 150	999 184	999 216	999 247
3.8		999 276	999 305	999 333	999 359	999 385	999 409	999 433	999 456	999 478	999 499
3.9		999 519	999 538	999 557	999 575	999 592	999 609	999 625	999 640	999 655	999 669
4.0	0.4 999	683	696	709	721	733	744	755	765	775	784
4.1		793	802	810	819	826	834	841	848	854	860
4.2		866	872	878	883	888	893	898	902	906	911
4.3		915	918	922	925	929	932	935	938	941	943
4.4		946	948	951	953	955	957	959	961	963	964
4.5		966	968	969	970	972	973	974	976	977	978
5.0		997 129	997 274	997 412	997 544	997 669	997 787	997 900	998 008	998 110	998 207

B Tables

B.1 The International System of Units (SI)

SI Base Units			
Quantity	Name	Symbol	Definition
Length	Meter	m	1 m is the length that light passes in 1/299 792 458 seconds.
Time	Second	s	1 s is the duration of 9 192 631 770 periods of the radiation corresponding to the transition between the two hyperfine levels of the ground state of the ^{133}Cs atom.
Mass	Kilogram	kg	1 kg is the mass of the international kilogram prototype (a platinum–iridium cylinder).
Electric current	Ampere	A	1 A is that constant current which, if maintained in two straight parallel conductors of infinite length, of negligible circular cross section, and placed 1 m apart in vacuum, would produce a force equal to $2 \cdot 10^{-7}$ N per meter of length.
Temperature	Kelvin	K	1 K is the fraction 1/273.16 of the thermodynamic temperature of the triple point of water.
Amount of substance	Mole	mol	1 mol is the amount of substance of a system that contains as many elementary entities as there are atoms in 0.012 kg of carbon ^{12}C.
Luminous intensity	Candela	cd	1 cd is the luminous intensity, in a defined direction, of a source that emits monochromatic radiation of frequency 540 THz and that has a radiant intensity in that direction of 1/683 W/sr.

The SI system (French: *Système International d'Unités*) consists of

- seven base units (e.g. the ampere),
- derived coherent units (e.g. the Watt second),
- additional noncoherent accepted units (e.g. the hour).

The speed of light is defined as $c_0 = 299\,792\,458$ m/s and thus combines the two basic units length and time. The coherent derived units are products or quotients of the basic units. The noncoherent units include various proportional factors (apart from powers of ten), such as, for example, 3600 for seconds and hours.

B.1.1 Decimal Prefixes

A **quantity** consists of a **value** and a **unit**. One (and only one) decimal prefix may be used before the unit symbol, e.g. kΩ for 10^3 Ω. Usually only powers of 1000 are used for this, e.g. kilo, mega, milli. Some (for historical reasons) exceptions are cm, hPa, decibel and some others.

Decimal prefixes					
Symbol	Prefix	Factor	Symbol	Prefix	Factor
d	deci-	10^{-1}	D	deca-	10^1
c	centi-	10^{-2}	H	hecto-	10^2
m	milli-	10^{-3}	k	kilo-	10^3
μ	micro-	10^{-6}	M	mega-	10^6
n	nano-	10^{-9}	G	giga-	10^9
p	pico-	10^{-12}	T	tera-	10^{12}
f	femto-	10^{-15}	P	peta-	10^{15}
a	atto-	10^{-18}	E	exa-	10^{18}

Expressions in the USA: 10^9, *billion*; 10^{12}, *trillion*; 10^{15}, *quadrillion*; 10^{18}, *quintillion*. In France and Germany a billion is actually defined as 1000 times the American billion.

In circuit diagrams the capacitance and resistance values are often noted in a short form. The decimal prefix then replaces the decimal point.

```
3k3          is 3.3 kΩ
3p3          is 3.3 pF
6M8          is 6.8 MΩ
2n7          is 2.7 nF
2R2          is 2.2 Ω
4μ7 or 4u7   is 4.7 μF
```

B.1.2 SI Units in Electrical Engineering

The units used in Electrical Engineering are derived from the SI basic units. The most common units are listed in the table below.

Units in Electrical Engineering			
Symbol	Unit	Relationship	Unit for
A	Ampere	Base unit	Electric current
C	Coulomb	As	Electric charge
cd	Candela	Base unit	Luminous intensity
F	Farad	As/V	Capacitance
H	Henry	Vs/A	Inductance
Hz*	Hertz	1/s	Frequency
J	Joule	Ws	Energy, work
K	Kelvin	Base unit	Temperature
kg	Kilogram	Base unit	Mass
kWh[†]	Kilowatthour	3.6 MJ	Work
lm	Lumen	cd sr	Luminous flux
lx	Lux	lm/m^2	Illumination
m	Meter	Base unit	Length
N	Newton	kg m/s^2	Force
Ω	Ohm	V/A	Resistance
S	Siemens	1/Ω	Admittance
s	Second	Base unit	Time
T	Tesla	Vs/m^2	Magnetic flux density
V	Volt	J/C = Ws/As	Voltage
W	Watt	AV	Power
Wb	Weber	Vs	Magnetic flux

*The unit hertz (Hz) is only used for frequencies. The unit of angular frequency is s^{-1}.

[†]Noncoherent permitted unit.

The units cancel each other out for quantities that are defined as the ratio of two similar quantities. These are described as **relative quantities**. These could be, for example, the solid angle (in steradians), the efficiency and logarithmic power ratios (decibel).

B.2 Naturally Occurring Constants

Physical Constants	
Permeability of free space (Permeability constant)	$\mu_0 = 4 \cdot \pi \cdot 10^{-7}$ H/m $= 1.256\,637\,06 \cdot 10^{-6}$ Vs/Am
Absolute dielectric constant (Permittivity constant)	$\epsilon_0 = 1/(\mu_0 \cdot c^2)$ $= 8.854\,187\,82 \cdot 10^{-12}$ As/Vm
Speed of light in vacuum	$c = 2.997\,924\,58 \cdot 10^8$ m/s
Elementary charge of the electron	$e = 1.602\,177\,33 \cdot 10^{-19}$ C
Boltzmann-constant	$k = 1.380\,658 \cdot 10^{-23}$ J/K
Electron rest mass	$m_e = 9.109\,389\,7 \cdot 10^{-31}$ kg

NOTE: For most calculations it is sufficient to consider four digits of the constants.

B.3 Symbols of the Greek Alphabet

Letters of the Greek alphabet			
Letter	Name	Letter	Name
α	Alpha	ν	Nu
β	Beta	ξ, Ξ	Xi
γ, Γ	Gamma	o	Omicron
δ, Δ	Delta	π, Π	Pi
ε	Epsilon	ϱ	Rho
ζ	Zeta	σ, Σ	Sigma
η	Eta	τ	Tau
$\theta, \vartheta, \Theta$	Theta	υ, Υ	Upsilon
ι	Iota	ϕ, φ, Φ	Phi
κ	Kappa	χ	Chi
λ, Λ	Lambda	ψ, Ψ	Psi
μ	Mu	ω, Ω	Omega

B.4 Units and Definitions of Technical–Physical Quantities

Quantity	Symbol	Definition	Unit	Name
Length	l, r, s	Base unit	m	Meter
Area	A	$= l^2$	m²	
Volume	V	$= l^3$	m³	
Time	t, T, τ	Base unit	s	Second
Velocity	v	$= ds/dt$	m/s	
Acceleration	a	$= dv/dt$	m/s²	
Frequency	f	$= 1/T$	1/s = Hz	Hertz
Angular frequency	ω	$= 2\pi/T$	1/s	
Mass	m	Base unit	kg	Kilogram
Mass density	ρ	$= m/V$	kg/m³	
Force	F	$= m \cdot a$	kg m/s² = N	Newton
Pressure	p	$= F/A$	N/m² = Pa	Pascal
Momentum	p	$= m \cdot v = \int F \, dt$	kg m/s	
Angular momentum	L	$= J \cdot \omega$	kg m²/s	
Torque	M	$= r \cdot F$	N m	
Moment of inertia	J	$= \int r^2 \, dm$	kg m²	
Current	I	Base unit	A	Ampere
Current density	J	$= dI/dA$	A/m²	
Charge	Q	$= \int I \, dt$	As = C	Coloumb
Voltage	V	$= W/Q$	V	Volt
Electric field strength	E	$= F/Q$	V/m	
Energy, work	W	$= \int P \, dt$	W s = J	Joule
Power	P	$= V \cdot I$	W	Watt
Apparent power	S		VA	
Reactive power	Q		var	
Resistance	R	$= V/I$	Ω	Ohm
Specific resistance	ρ	$= R \cdot A/l$	Ω m	
Admittance	G	$= 1/R$	S	Siemens, Ω^{-1} or ℧
Conductivity	σ	$= 1/\rho$	S/m	
Electric displacement	D	$= dQ/dA$	As/m²	
Capacitance	C	$= Q/U$	F	Farad
Magnetic flux density	B	$= F/Q \cdot v$	Vs/m² = T	Tesla
Magnetic field strength	H		A/m	
Magnetic flux	Φ	$= \int B \, dA$	Vs = Wb	Weber
Inductance	L	$= V/(d\Psi/dt)$	Vs/A = H	Henry

B.5 Imperial and American Units

Unit	Symbol	In SI Units	Conversion factor
Length			
inch	in	25.4 mm	0.0393701 in/mm
mil = 1/1000 in	mil	25.4 µm	0.0393701 mil/µm
foot = 12 in	ft	0.30468 m	3.28084 ft/m
yard = 3 ft	yd	0.9144 m	1.09361 yd/m
(statute) mile = 1760 yd	mi	1.60934 km	0.62137 mi/km
Area			
square inch	sq in	6.4516 cm^2	0.155 sq in/mm^2
square mil	sq mil	$6.4516 \cdot 10^{-4}$ mm^2	1550 sq mil/mm^2
circular mil	CM	$0.5067 \cdot 10^{-3}$ mm^2	1.974 CM/mm^2
M circular mil	MCM	0.5067 mm^2	1.974 MCM/mm^2
Volume			
cubic inch	cu in	16.387 cm^3	0.061024 cu in/cm^3
cubic foot = 1728 cu in	cu ft	28.317 dm^3	0.035315 cu ft/dm^3
cubic yard = 27 cu ft	cu yd	0.76455 m^3	1.30795 cu yd/m^3
fluid ounce (UK)	fl oz	28.413 cm^3	0.035195 fl oz/cm^3
fluid ounce (US)	fl oz	29.574 cm^3	0.033813 fl oz/cm^3
gallon (US) = 128 fl oz	gal	3.78543 dm^3	0.264170 gal/dm^3
Mass			
ounce	oz	28.3459 g	0.0352739 oz/g
pound = 16 oz	lb	0.453592 kg	2.204622 lb/kg
Force			
pound force	lbf	4.445 N	0.225 lbf/N
poundal = 1 lb · ft/s^2	pdl	0.1383 N	7.23 pdl/N
Density			
pound per cubic foot	lb/ft^3	16.02 kg/m^3	$0.0624 \dfrac{\text{lb} \cdot \text{m}^3}{\text{ft}^3 \cdot \text{kg}}$
Work			
British thermal unit	BTU	1.055056 kJ	0.947817 BTU/kJ
horsepower hour	HPhr	2.6845 MJ	0.37251 HPhr/MJ
Power			
BTU per second	BTU/s	1.055056 kW	0.947817 BTU/kWs
BTU per hour	BTU/h	0.293071 W	3.41214 BTU/Wh
horse power	HP	0.74570 kW	1.34102 HP/kW
Wire weights (Mass per unit length)			
pound per foot	lb/ft	1.488 kg/m	$0.672 \dfrac{\text{lb} \cdot \text{m}}{\text{kg} \cdot \text{ft}}$
pound per yard	lb/yd	0.496 kg/m	$2.016 \dfrac{\text{lb} \cdot \text{m}}{\text{kg} \cdot \text{yd}}$
pound per mile	lb/mi	0.2818 kg/km	$3.548 \dfrac{\text{lb} \cdot \text{km}}{\text{kg} \cdot \text{mi}}$

Unit	Symbol	In SI Units	Conversion factor
Electrical conductors (electrical quantities with respect to conductor length)			
ohms per 1000 feet	$\Omega/1000$ ft	$3.28\,\Omega/\text{km}$	0.3047 m/ft
ohms per 1000 yards	$\Omega/1000$ yd	$1.0936\,\Omega/\text{km}$	0.9144 m/yd
megohms per mile	$\text{M}\Omega/\text{mi}$	$0.6214\,\Omega/\text{km}$	1.6093 m/mi
microfarads per mile	$\mu\text{F/mi}$	$0.6214\,\mu\text{F/km}$	1.6093 km/mi
micromicrofarads per foot	$\mu\mu\text{F/ft}$	3.2808 pF/m	0.30468 m/ft
decibel per 100 ft	dB/100 ft	32.75 dB/km	0.305 m/ft
		3.77 Np/km	$0.2653\,\dfrac{\text{dB}\cdot\text{km}}{\text{Np}\cdot\text{ft}}$
decibel per 1000 yd	dB/1000 yd	1.094 dB/km	0.9144 m/yd
		0.126 Np/km	$7.943\,\dfrac{\text{dB}\cdot\text{m}}{\text{Np}\cdot\text{yd}}$
decibel per mile	dB/mi	0.621 dB/km	1.609 m/mi
		0.0715 Np/km	$13.98\,\dfrac{\text{dB}\cdot\text{km}}{\text{Np}\cdot\text{mi}}$
Optical units			
lambert	L	3183 cd/m^2	$\pi\cdot 10^{-4}$ lam^2/cd
foot-lambert	fL	3.42626 cd/m^2	0.291864 ft la m^2/cd
candela per square inch	cd/sq in	1555.0 cd/m^2	$64.308\cdot 10^{-3}$ m^2/sq in
candela per square foot	cd/sq ft	10.7639 cd/m^2	0.092903 m^2/sq ft
foot-candle	fc	10.7639 lx	0.092903 ft cd/lx
Temperature			
degree Fahrenheit	°F	$5/9$ K	$9/5$ °F/K
		for temperature differences	
	°F	$5/9(x\,°\text{F}-32\,°\text{F})\,°\text{C}$	$(9/5\cdot x\,°\text{C}+32\,°\text{C})\,°\text{F}$
		for absolute temperatures	

EXAMPLE: A width of 3/8 in is equivalent to $3/8 \cdot 25.4$ mm ≈ 9.5 mm. On the other hand 10 mm is equal to $10 \cdot 0.0393701$ in ≈ 0.4 in.

B.6 Other Units

Many of these units are rarely used.

Symbol	Name	In SI units
′	arc minute	$1/60°$
′	foot	0.30468 m
″	arc second	$1/3600°$
″	inch	25.4 mm
a, yr	year	
Å	Angstroem	0.1 nm
asb	Apostilb	$1/\pi$ cd/m²
at	atmosphere, technical	98.0665 kPa
atm	atmosphere	101.325 kPa
bar	bar	100 kPa
bbl	barrel (US)	1.59 hl
Bi	biot	10 A
Bq	becquerel	1/s
bu (UK)	bushel	36.37 l
bu (US)	bushel	35.24 l
c	Neuminute	$\pi/2 \cdot 10^4$ rad
cal	calorie	4.1868 J
cbm	cubic meter	1 m³
cc	Neusekunde	$\pi/2 \cdot 10^6$ rad
ccm	cubic centimeter	1 cm³
Ci	curie	$3.7 \cdot 10^{10}$ Bq
Cic	cicero	$12\,p \approx 4.5$ mm
CM	circular mil	$5.06707 \cdot 10^{-4}$ mm²
cmm	cubic millimeter	1 mm³
cwt (UK)	hundred weight	50.80 kg
cwt (US)	long hundred weight	50.80 kg
d	day	86 400 s
Dez	dez	$\pi/18$ rad
dr av	dram	1.772 g
dry pt (US)	dry pint	0.5506 l
dyn	dyne	10^{-5} N
erg	erg	10^{-7} J
eV	electron volt	$1.602 \cdot 10^{-19}$ J
fL	foot-lambert	3.426 cd/m²
F	Fermi	1 fm
Fr	franklin	$\approx 1/3 \cdot 10^{-9}$ C

Other Units		
Symbol	Name	In SI units
G	gauss	10^{-4} T
g	gon	$1.1111°$
γ	gamma	$1\,\mu g$
gal (UK)	gallon	4.5466 l
gal (US)	gallon	3.7854 l
Gb	Gilbert	$10/4\pi$ A
Gon	gon	$1.1111°$
gr	grain	64.8 mg
grd	grad	1 K
Gy	gray	1 J/kg
h	hour	3 600 s
hl	hectoliter	100 l
hp	horsepower	745.7 W
k	karat (metric)	0.200 g
Kal	kilocalorie	4.1868 kJ
kcal	kilocalorie	4.1868 kJ
kp	kilopound	9.806 65 N
kWh	kilowatt hour	$3.6 \cdot 10^6$ J
L	lambert	$1/\pi \cdot 10^4$ cd/m^2
lbf	pound-force	4.448 N
lb wt	pound weight	4.48 N
M	maxwell	10^{-8} Wb
μ	micron	$1\,\mu m$
MCM	1000 circular mils	0.5067 mm^2
ml	milliliter	1 cm^3
mm Hg	millimeter of mercury	133.322 Pa
mm Q	see mmHg	
mrad	millirad	1/1000 rad
Np	neper	8.686 dB
nt	nit	1 cd/m^2
nx	nox	10^{-3} lx
Oe	oersted	$1000/4\pi$ A/m
p	pond	$9.806\,65 \cdot 10^{-3}$ N
p	point, typographic	0.376 065 mm
pdl	poundal	0.1383 N
ph	phot	10^4 lm/m^2
PS (German)	horsepower	735.498 75 W
psf	pound (weight) per square foot	47.88 Pa
psi	pound (weight) per square inch	6895 Pa
pt (UK)	pint	0.5683 l
pt (US)	pint	0.4731 l

Other Units		
Symbol	Name	In SI units
q	quarter (mass)	12.7 kg
qmm	square millimetre	1 mm^2
qt (US)	quart	0.94631
R	roentgen	$258 \cdot 10^{-6}$ C/kg
rad	radian	57.29578°
rem	rem	0.01 J/kg
sb	stilb	10^4 cd/m^2
sh cwt	short hundredweight	45.36 kg
sh tn	short ton	907.2 kg
sm	nautical mile	1852 m
sr	steradian	solid angle
Sv	sievert	1 J/kg
t	metric ton	1000 kg
t (UK)	ton	1016 kg
Torr, torr	torr	133.322 Pa

B.7 Charge and Discharge Curves

Function $e^{-t/\tau}$										
t/τ	0	1	2	3	4	5	6	7	8	9
0	1.0000	0.9048	.8187	.7408	.6703	.6065	.5488	.4966	.4493	.4066
1	.3679	.3329	.3012	.2725	.2466	.2231	.2019	.1827	.1653	.1496
2	.1353	.1225	.1108	.1003	.0907	.0821	.0743	.0672	.0608	.0550
3	.0498	.0450	.0408	.0369	.0334	.0302	.0273	.0247	.0224	.0202
4	.0183	.0166	.0150	.0136	.0123	.0111	.0101	.0091	.0082	.0074
5	.0067	.0061	.0055	.0050	.0045	.0041	.0037	.0033	.0030	.0027

EXAMPLE: A 4.7 µF capacitor is discharged via a 1 kΩ resistor. What is the voltage across the capacitor after 10 ms?

The time constant of the RC combination is 4.7 ms, therefore 10 ms is equivalent to approximately 2.3 time constants τ. For this value the table yields 0.1003. This means that the voltage across the capacitor has decreased to 10%.

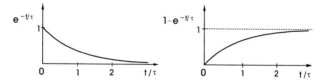

Fig. B.1. Discharge and charge characteristics of an RC combination

Function $1 - e^{-t/\tau}$										
t/τ	0	1	2	3	4	5	6	7	8	9
0	0.0000	.0952	.1813	.2592	.3297	.3935	.4512	.5034	.5507	.5934
1	.6321	.6671	.6988	.7275	.7534	.7769	.7981	.8173	.8347	.8504
2	.8647	.8775	.8892	.8997	.9093	.9179	.9257	.9328	.9392	.9450
3	.9502	.9550	.9592	.9631	.9666	.9698	.9727	.9753	.9776	.9798
4	.9817	.9834	.9850	.9864	.9877	.9889	.9899	.9909	.9918	.9926
5	.9933	.9939	.9945	.9950	.9955	.9959	.9963	.9967	.9970	.9973

EXAMPLE: A discharged 4.7 µF capacitor is charged from a 5 V voltage source via a 1 kΩ resistor. What is the voltage across the capacitor after 10 ms?

The time constant of the RC combination is 4.7 ms, therefore 10 ms is equivalent to approximately 2.3 time constants τ. For this value the table yields 0.8997. The voltage across the capacitor is therefore $5 \text{ V} \cdot 0.8997 \approx 4.5 \text{ V}$.

B.8　IEC Standard Series

E96 ±1%	E48 ±2%	E24 ±5%	E12 ±10%	E6 ±20%
1.00	1.00	1.0↑	1.0↑	1.0↑
1.02				
1.05	1.05			
1.07				
1.10	1.10	1.1		
1.13				
1.15	1.15			
1.18				
1.21	1.21	1.2	1.2	
1.24				
1.27	1.27			
1.30				
1.33	1.33	1.3		
1.37				
1.40	1.40			
1.43				
1.47	1.47	1.5	1.5	1.5
1.50				
1.54	1.54			
1.58				
1.62	1.62	1.6		
1.65				
1.69	1.69			
1.74				
1.78	1.78	1.8	1.8	
1.82				
1.87	1.87			
1.91				
1.96	1.96	2.0		
2.00				
2.05	2.05			
2.10				
2.15	2.15	2.2	2.2	2.2
2.21				
2.26	2.26			
2.32				
2.37	2.37	2.4		
2.43				
2.49	2.49			
2.55				
2.61	2.61			
2.67				
2.74	2.74	2.7	2.7	
2.80				
2.87	2.87			
2.94				
3.01	3.01	3.0↓		
3.09				
3.16	3.16↓		3.3↓	3.3↓

E96 ±1%	E48 ±2%	E24 ±5%	E12 ±10%	E6 ±20%
3.24				
3.32	3.32	3.3↑	3.3↑	3.3↑
3.40				
3.48	3.48			
3.57				
3.65	3.65	3.6		
3.74				
3.83	3.83			
3.92				
4.02	4.02	3.9	3.9	
4.12				
4.22	4.22			
4.32		4.3		
4.42	4.42			
4.53				
4.64	4.64	4.7	4.7	4.7
4.75				
4.87	4.87			
4.99				
5.11	5.11	5.1		
5.23				
5.36	5.36			
5.49				
5.62	5.62	5.6	5.6	
5.76				
5.90	5.90			
6.04				
6.19	6.19	6.2		
6.34				
6.49	6.49			
6.65				
6.81	6.81	6.8	6.8	6.8
6.98				
7.15	7.15			
7.32				
7.50	7.50	7.5		
7.68				
7.87	7.87			
8.06				
8.25	8.25	8.2	8.2	
8.45				
8.66	8.66			
8.87				
9.09	9.09	9.1		
9.31				
9.53	9.53			
9.76				
10.0	10.0↓	10↓	10↓	10↓

The horizontal lines mark the approximate intervals that are covered by the given tolerances.

EXAMPLE: For a calculated resistance value of 1.17 kΩ a resistance of 1.15 kΩ is chosen from the E48 series or a 1.2 kΩ from the E24 series. If only E6 series resisitors are available, then a 1.0 kΩ resistance is chosen.

The values of the IEC series form a harmonic series. Each value has the same ratio to its preceding value. This ratio is $\sqrt[6]{10}$ for the E6 series, $\sqrt[12]{10}$ for the E12 series, etc. The values are chosen so that for the given tolerances a minimum number of resistors have to be kept in stock.

B.9 Resistor Colour Code

E96, E48, E24				
1. Ring	2./3. Ring	4. Ring	5. Ring	6. Ring
Colour / 1. Digit	2./3. Digit	Factor	Tolerance	Temp. coeff.
Silver		0.01 Ω	±10%	
Gold		0.1 Ω	±5%	
Black	0	1.0 Ω		±250 · 10⁻⁶/K
Brown	1 / 1	10 Ω	±1%	±100 · 10⁻⁶/K
Red	2 / 2	100 Ω	±2%	±50 · 10⁻⁶/K
Orange	3 / 3	1 kΩ		±15 · 10⁻⁶/K
Yellow	4 / 4	10 kΩ		±25 · 10⁻⁶/K
Green	5 / 5	100 kΩ	±5%*	±20 · 10⁻⁶/K
Blue	6 / 6	1 MΩ		±10 · 10⁻⁶/K
Purple	7 / 7	10 MΩ		±5 · 10⁻⁶/K
Grey	8 / 8	100 MΩ*		±1 · 10⁻⁶/K
White	9 / 9	0.1 Ω*	±10%*	
Colour / 1. Digit	2. Digit	Factor	Tolerance	—
1. Ring	2. Ring	3. Ring	4. Ring	—
E6, E12, E24				

*In case the conductivity of gold and silver varnish cannot be tolerated, the following replacements can be made:
Gold   is replaced by   white for   0.1 Ω, and by   green for   ±5%.
Silver   is replaced by   grey for   0.01 Ω, and by white for   ±10%.

EXAMPLE: A resistor with the colour rings grey, red, red, gold has a resistance of 8.2 kΩ with a tolerance ±5 %.

Tolerances and temperature coefficients may be marked by letters.

Tolerance								
Letter	B	C	D	F	G	J	K	M
%	±0.1	±0.25	±0.5	±1	±2	±5	±10	±20

Temperature coefficient							
Letter	T	E	C	K	J	L	D
10⁻⁶/K	±10	±25	±50	±100	±150	±200	+200/ − 500

EXAMPLE: A reference resistor with the five colour rings green, blue, red, brown, red and with the letter E has a resistance value of 5620 Ω with a tolerance of ±2% and a temperature coefficient of $\pm 25 \cdot 10^{-6}$ K.

B.10 Parallel Combination of Resistors

High-precision resistors are not always available in all 96 or 192 values per decade of the E series. Many manufacturers produce the values of the E12 series with a tolerance of 1% or better. The values of the finer series can be approximated with parallel combinations of those resistors. The following table lists those values that are closest to the values of E96.

Value	$R_1 \| R_2$	Value	$R_1 \| R_2$	Value	$R_1 \| R_2$
100	100	210	220||4700	464	560||2700
102	120||680	214	220||8200	470	470
105	120||820	220	270||1200	479	560||3300
107	120||1000	229	270||1500	487	820||1200
110	220||220	230	390||560	500	1000||1000
112	120||1800	235	330||820	509	560||5600
115	120||2700	245	270||2700	524	560||8200
118	120||6800	250	270||3300	530	820||1500
120	120	253	270||3900	545	1000||1200
121	220||270	261	270||8200	560	560
123	180||390	264	390||820	563	820||1800
127	150||820	270	270	579	680||3900
130	180||470	280	560||560	594	680||4700
133	150||1200	287	330||2200	606	680||5600
136	150||1500	294	330||2700	618	680||6800
140	150||2200	300	330||3300	629	820||2700
143	150||3300	310	390||1500	643	1000||1800
147	150||6800	317	330||8200	667	1200||1500
150	150	321	390||1800	680	680
153	180||1000	331	390||2200	698	820||4700
158	220||560	340	680||680	715	820||5600
161	180||1500	349	390||3300	732	820||6800
165	330||330	358	470||1500	750	1500||1500
169	180||2700	365	390||5600	767	1000||3300
174	180||5600	373	470||1800	796	1000||3900
179	330||390	382	560||1200	820	820
180	180	390	390	825	1000||4700
182	270||560	400	470||2700	848	1000||5600
186	220||1200	411	470||3300	872	1000||6800
192	220||1500	419	470||3900	891	1000||8200
196	220||1800	434	470||5600	918	1200||3900
200	220||2200	440	470||6800	956	1200||4700
206	220||3300	451	820||1000	964	1500||2700

B.11 Selecting Track Dimensions for Current Flow

Copper tracks on printed circuit boards heat up when a current flows through them. The graph shown in Fig. B.2 permits the selection of the required track width as a function of the temperature increase and the cross-sectional area of the copper. The values are for orientation for an ambient temperature of 20°C without external cooling.

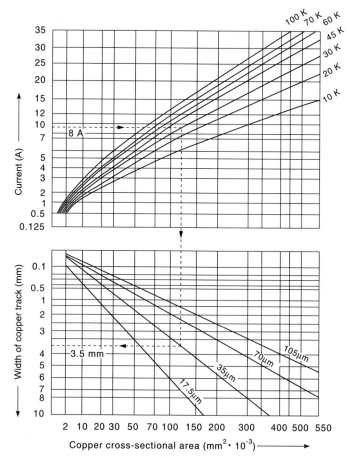

Fig. B.2. Copper track selection

EXAMPLE: A copper track should carry 8 A, while not increasing the temperature by more than 30 K. A cross-sectional area of about 0.12 mm² is therefore required. For a track depth of 35 µm, this implies a conductor width of 3.5 mm.

B.12 American Wire Gauge

In the US, gauges are given in American wire gauge numbers (AWG). They are derived from the steps of the manufacturing process of copper wire.

AWG	Diameter (in)	Cross section (MCM)	Diameter (mm)	Cross-sectional area (mm^2)
0000	0.4600	211	11.7	107
000	0.4100	168	10.4	84.9
00	0.3650	133	9.27	67.5
0	0.3250	105	8.25	53.5
1	0.2890	83.7	7.35	42.4
2	0.2580	66.3	6.54	33.6
4	0.2040	41.8	5.19	21.2
6	0.1620	26.3	4.12	13.3
8	0.1280	16.5	3.26	8.35
10	0.1020	10.4	2.59	5.27
12	0.0810	6.51	2.05	3.30
14	0.0640	4.12	1.63	2.09
16	0.0510	2.58	1.29	1.31
18	0.0400	1.63	1.024	0.824
20	0.0320	1.02	0.813	0.519
22	0.0253	0.641	0.643	0.325
24	0.0210	0.405	0.511	0.205
26	0.0159	0.254	0.405	0.129
28	0.0126	0.159	0.320	0.0804
30	0.0100	0.101	0.255	0.0511
32	0.0080	0.0639	0.203	0.0324
34	0.0063	0.0397	0.160	0.0201
36	0.0050	0.0250	0.127	0.0127
38	0.0040	0.0161	0.102	0.00817
40	0.0031	0.0097	0.079	0.00490
4/0	see 0000	–	–	–
3/0	see 000	etc.	–	–

MCM: 1000 circular mils
1 MCM = 0.5067 mm^2

The 0000, 000, etc. in AWG are also denoted by 4/0, 3/0, etc.

The following rules hold for AWG numbers:

- An AWG 10 wire has a diameter of close to 0.1 in, a cross-sectional area of about 10 MCM and (for copper wire) a resistance of 1 Ω/1000 ft.

- An increase of 3 AWG numbers doubles the cross-sectional area and the wire weight, and decreases the wire resistance by a factor of 2.

- An increase of 6 AWG numbers doubles the cross-sectional diameter.

- An increase of 10 AWG numbers increases the cross-sectional area 10 times.

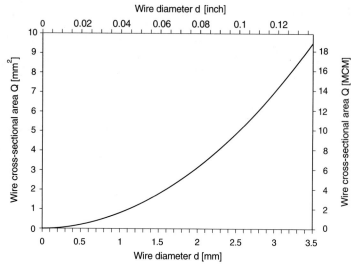

Fig. B.3. Cross-sectional area $Q = \left(\dfrac{d}{2}\right)^2 \cdot \pi$

B.13 Dry Cell Batteries

Coding of the cells		
IEC notation	Open-circuit voltage	Chemical system
R	1.5 V	Zinc–carbon
CR	3.3 V	Manganese dioxide–lithium
ER	3.8 V	Chromium–lithium
LR	1.45 V	Zinc–Alkali metal-manganese
MR	1.35 V	Zinc–Mercury oxide
NR	1.40 V	Zinc–manganese dioxide–mercury oxide
PR	1.40 V	Zinc–air
SR	1.55 V	Zinc–silver oxide

B.13 Dry Cell Batteries

Dimensions	Format			
	Zinc–Carbon	Alkali–manganese	Ni–Cd battery	NiMH–battery
	Capacity	Capacity	Capacity	Capacity
Mono cell 33 mm ⌀ × 60 mm	R20 7.3 Ah	LR20 18 Ah	KR35/62 4 Ah	5 Ah
Baby cell 26 mm ⌀ × 50 mm	R14 3.1 Ah	LR14 7 Ah	KR27/50 2 Ah	2.6 Ah
Mignon cell 14.5 mm ⌀ × 50 mm	R6 1.1 Ah	LR6 2.3 Ah	KR15/51 0.75 Ah	1.1 Ah
Micro cell 10.5 mm ⌀ × 44.5 mm	R03 0.5 Ah	LR03 1.2 Ah	KR10/44 0.2 Ah	0.45 Ah
Microdyn cell 12 mm ⌀ × 30 mm	R1 0.6 Ah	LR1 0.8 Ah	KR12/30 0.15 Ah	–
9 V pack 15.5 mm × 25 mm × 48 mm	6F22 0.4 Ah	6LF22 0.6 Ah	TR7/8 0.15 Ah	0.12 Ah
4.5 V flat battery 22 mm × 62 mm × 65 mm	3R12 2 Ah	–	–	–

The cell capacity values give an orientation. The cell capacity depends greatly on the kind of discharge and the operating temperature. Notation according to IEC.

International notation of batteries							
	Mono cell	Baby cell	Mignon cell	Micro cell	Microdyn cell	Transistor battery	Flat battery
IEC	R20	R14	R6	R03	R1	6F22	3R12
USA	D	C	AA	AAA	N	6AM6	–
Japan	UM1	UM2	UM3	UM4	UM5	–	UM10

The notations are valid for zinc–carbon batteries.

B.14 Notation of Radio-Frequency Ranges

Frequency range	Wavelength	Description	Acronym
30–300 Hz	10 000–1000 km	Extremely low frequency	ELF
300Hz–3 kHz	1000–100 km	Infralow frequency	ILF
3–30 kHz	100–10 km	Very low frequency	VLF
30–300 kHz	10–1 km	Low frequency	LF
		Long wave	LW
300–3000 kHz	1000–100 m	Medium wave	MW
3–30 MHz	100–10 m	High frequency	HF
		Short wave	SW
30–300 MHz	10–1 m	Very high frequency	VHF
		Ultrashort wave	USW
300–3000 MHz	100–10 cm	Ultrahigh frequency	UHF
3–30 GHz	10–1 cm	Super high frequency	SHF
30–300 GHz	10–1 mm	Extremely high frequency	EHF
300–3000 GHz	1–0.1 mm	Hyperhigh frequency	HHF

Range	Meaning	CCIR band	CCITT notation
ELF	Eextremely low frequency		
ILF	Infralow frequency		
VLF	Very low frequency	4	Miriametric
LF	Low frequency	5	Kilometric
MF	Middle frequency	6	Hectometric
HF	High frequency	7	Decametric
VHF	Very high frequency	8	Metric
UHF	Ultrahigh frequency	9	Decimetric
SHF	Super high frequency	10	Centimetric
EHF	Extremely high frequency	11	Millimetric
HHF	Hyperhigh frequency	12	Submillimetric

B.15 Ratios

In measurement logarithmic ratios of values are often used. Measured values and reference values must have the same dimensions (e.g. power, current). For complex values the ratio of the absolute values (e.g. apparent power) is considered.

A destinction is made between **power** and **field values**. Power values are proportional to the power, whereas field values to the power of 2 are proportional to the power.

The decibel (dB) is used as the unit for logarithmic ratios to the base 10. The neper (Np) is used for natural logarithmic ratios, although this occurs less frequently.

$$1 \text{ Np} = 8.685889 \text{ dB} \qquad 1 \text{ dB} = 0.115129 \text{ Np} \tag{B.1}$$

Power attenuation (log of the power value ratio):

$$a_\text{P} = 10 \cdot \lg \frac{P_1}{P_2} \text{ dB} \tag{B.2}$$

Voltage attenuation (log of the field value ratio):

$$a_\text{V} = 20 \cdot \lg \frac{V_1}{V_2} \text{ dB}, \quad V_1, V_2 \text{ at the same source resistance} \tag{B.3}$$

B.15.1 Absolute Voltage Levels

Absolute levels perform the ratio calculation with respect to a defined reference value. The absolute power level is given by:

$$P_\text{L} = 10 \cdot \lg \frac{P}{1 \text{ mW}} \text{ dB(mW)} \quad \text{or dBm} \tag{B.4}$$

The dBm is used frequently, for example, for laser diode power output.

The **absolute voltage level** is defined as

$$P_\text{SP1} = 20 \cdot \lg \frac{V}{0.775 \text{ V}} \text{ dB(0.775 V)} \tag{B.5}$$

So a voltage level of 0 dB(0.775 V) corresponds to a voltage of 0.775 V. A power of 1 mW will be therefore dissipated by a 600 Ω resistance. The voltage level is often given with respect to 1 V.

$$P_\text{SP2} = 20 \cdot \lg \frac{V}{1 \text{ V}} \text{ dB(V)} \quad \text{or dBV} \tag{B.6}$$

Other reference values are also used.

Reference Values: Level Ratios

$R_{in} = R_{out}$ (Ω)	P_{ref} (mW)	V_0 (V)	dB(mW)	Application
600	1	0.77459	0	Standard
75	1	0.27386	0	RF
60	1	0.24494	0	Measurement
50	1	0.22360	0	Measurement
150	1	0.389	0	Telephony
500	6	1.73205	7.78	USA telephony
600	6	1.1898	7.78	USA telephony
600	12.5	2.739	10.97	USA telephony

There is no uniform usage of reference values, so care must be taken in applying them!

Reference values: voltage levels

Notation	Reference value	dB(0.775 V)
dBV	0 dBV=1 V	2.2
dBmV	0 dBmV=1 mV	−57.8
dBµV	0 dBµV=1 µV	−117.8

B.15.1.1 Conversion of Power and Voltage Level Ratios

The power level ratio corresponds to the voltage level ratio only for a resistance of 600 Ω. For absolute level ratios, measured across a resistance, R,

$$\text{power level ratio} = \text{voltage level ratio} + \text{correction factor } \Delta \qquad (B.7)$$

$$P_L = P_{SPl} + \underbrace{10 \cdot \lg \frac{600\ \Omega}{R}}_{\Delta} \qquad (B.8)$$

The unit of the correction factor is the dB. Depending on the resistor R where the level is measured, it follows that

Correction factors

R (Ω)	50	60	75	150	500	600	1200
Δ (dB)	10.79	10.00	9.03	6.02	0.79	0	-3.01

B.15.2 Relative Levels

Relative level	Gain	Attenuation	Relative level	Gain	Attenuation
0.0	1.0000	1.0000	0.5	1.0593	0.9441
0.1	1.0116	0.9886	0.6	1.0715	0.9333
0.2	1.0233	0.9772	0.7	1.0839	0.9226
0.3	1.0351	0.9661	0.8	1.0965	0.9120
0.4	1.0471	0.9550	0.9	1.1092	0.9016
1.0	1.1220	0.8913	11.0	3.5481	0.2818
1.5	1.1885	0.8414	11.5	3.7584	0.2661
2.0	1.2589	0.7943	12.0	3.9811	0.2512
2.5	1.3335	0.7499	12.5	4.2170	0.2371
3.0	1.4125	0.7079	13.0	4.4668	0.2239
3.5	1.4962	0.6683	13.5	4.7315	0.2113
4.0	1.5849	0.6310	14.0	5.0119	0.1995
4.5	1.6788	0.5957	14.5	5.3088	0.1884
5.0	1.7783	0.5623	15.0	5.6234	0.1778
5.5	1.8836	0.5309	15.5	5.9566	0.1679
6.0	1.9953	0.5012	16.0	6.3096	0.1585
6.5	2.1135	0.4732	16.5	6.6834	0.1496
7.0	2.2387	0.4467	17.0	7.0795	0.1413
7.5	2.3714	0.4217	17.5	7.4989	0.1334
8.0	2.5119	0.3981	18.0	7.9433	0.1259
8.5	2.6607	0.3758	18.5	8.4140	0.1189
9.0	2.8184	0.3548	19.0	8.9125	0.1122
9.5	2.9854	0.3350	19.5	9.4406	0.1059
10.0	3.1623	0.3162	20.0	10.0000	0.1000
40	10^2	10^{-2}	100	10^5	10^{-5}
60	10^3	10^{-3}	120	10^6	10^{-6}
80	10^4	10^{-4}	140	10^7	10^{-7}

EXAMPLE: The power amplification for a 48.5 dB amplifier is required; 48.5 dB = 8.5 dB + 40 dB. From the table it can be seen that 8.5 dB corresponds to a gain of 2.6607, while 40 dB means an amplification of 100. The product of the two yields a voltage amplification of 266. If a more exact result is required, e.g. for 48.7 dB, then the values for 0.7 dB + 8 dB + 40 dB should be taken from the table. This yields $1.0839 \cdot 2.5119 \cdot 100 = 272.26$.

B.16 V.24 Interface

The interface in accordance with CCITT V.24 is also described in the American norm RS232/E and in the German DIN 66 020. The interface signals are given on the following page. In practical applications only some of the many signals are analysed. Here is an example for two devices connected with V.24 interfaces:

Without handshake		CTS handshake		DTR handshake		Full handshake	
DTE	DCE	DTE	DCE	DTE	DCE	DTE	DCE
2	2	2	2	2	2	2	2
3 →	3	3 →	3	3 →	3	3 →	3
4 ←	4	4 ←	4	4 ←	4	4 ←	4
5	5	5 →	5	5	5	5 →	5
6	6	6 ←	6	6	6	6 ←	6
7	7	7	7	7	7	7	7
8 —	8	8 —	8	8 —	8	8 —	8
15	15	15	15	15	15	15 —	15
17	17	17	17	17	17	17	17
20	20	20	20	20 —	20	20 →	20
24	24	24	24	24	24	24 →	24

CTS and DTR handshakes may be combined with each other.

Nullmodem		CTS handshake					
DTE	DTE	DTE	DTE	DTE	DTE	DTE	DTE
2	2	2	2	2	2	2	2
3 ⨯	3	3 ⨯	3	3 ⨯	3	3 ⨯	3
4 ⨯	4	4 ⨯	4	4 ⨯	4	4 ⨯	4
5	5	5 ⨯	5	5 ⨯	5	5 ⨯	5
6	6	6 ⨯	6	6 ⨯	6	6 ⨯	6
7	7	7	7	7	7	7	7
8 —	8	8	8	8	8	8	8
15	15	15	15	15	15	15	15
17	17	17	17	17	17	17	17
20	20	20	20	20	20	20	20
24	24	24	24	24	24	24	24

Levels: Mark (1) $-15\,\text{V} < V < -3\,\text{V}$
Space (0) $+15\,\text{V} > V > +3\,\text{V}$

Protocols: The data flow control is carried by the RTS/CTS or DTR signals, or through an exchange of the XON/XOFF (DC1/DC3) or ETX/ACK signals.

The arrows in the following table show the signal direction between

	Computer	Modem
DTE:	*data terminal equipment* DCE:	*data communications equipment*

Interface signals					
Notation			Pin		Description	
CCITT	EIA	DIN		RS232C		[1]
101	AA	E1	1	–	Protective ground	⟷
102	AB	E2	7	–	Signal ground	⟷
103	BA	D1	2	TD	Transmitted data	⟵
104	BB	D2	3	RD	Received data	⟶
105	CA	S2	4	RTS	Request to send	⟵
106	CB	M2	5	CTS	Clear to send	⟶
107	CC	M1	6	DSR	Data set ready	⟶
108.1		S1.1	20	–	Connect data set to line	⟵
108.2	CD	S1.2	20	DTR	Data terminal ready	⟵
125	CE	M3	22	RI	Ring indicator	⟶
109	CF	M5	8	DCD	Data carrier detect	⟶
110	CG	M6	21	SQ	Signal quality detect	⟶
111	CH	S4	23	–	Data signal rate selector (DTE)	⟵
112	CI	M4	23	–	Data signal rate selector (DCE)	⟶
126	CK	S5	11	–	Select transmit frequency	⟵
113	DA	T1	24	–	Transmitter signal element timing	⟵
114	DB	T2	15	–	Transmitter signal element timing	⟶
115	DD	T4	17	RC	Receiver clock	⟶
					secondary channel	
118	SBA	HD1	14	–	transmitted data	⟵
119	SBB	HD2	16	–	received data	⟶
120	SCA	HS2	19	–	request to send	⟵
121	SCB	HM2	13	–	clear to send	⟶
122	SCF	HM5	12	–	carrier detect	⟶

[1] From DTE to DCE.

B.17 Dual-Tone Multi-Frequency

Two sinusoidal waveforms of different frequencies are sent by the telephone when a button is pressed. The frequencies and their order are internationally standardised.

697 Hz	1	2	3	A
770 Hz	4	5	6	B
852 Hz	7	8	9	C
941 Hz	*	0	#	D
	1209 Hz	1336 Hz	1477 Hz	1633 Hz

The keys shown in the last column are only availabe on some telephones.

B.18 ASCII Coding

hex	0	1	2	3	4	5	6	7
0	NUL 0	DLE 16	 32	0 48	@ 64	P 80	` 96	p 112
1	SOH 1	DC1 17	! 33	1 49	A 65	Q 81	a 97	q 113
2	STX 2	DC2 18	" 34	2 50	B 66	R 82	b 98	r 114
3	ETX 3	DC3 19	# 35	3 51	C 67	S 83	c 99	s 115
4	EOT 4	DC4 20	$ 36	4 52	D 68	T 84	d 100	t 116
5	ENQ 5	NAK 21	% 37	5 53	E 69	U 85	e 101	u 117
6	ACK 6	SYN 22	& 38	6 54	F 70	V 86	f 102	v 118
7	BEL 7	ETB 23	' 39	7 55	G 71	W 87	g 103	w 119
8	BS 8	CAN 24	(40	8 56	H 72	X 88	h 104	x 120
9	HT 9	EM 25	) 41	9 57	I 73	Y 89	i 105	y 121
A	LF 10	SUB 26	* 42	: 58	J 74	Z 90	j 106	z 122
B	VT 11	ESC 27	+ 43	; 59	K 75	[/Ä 91	k 107	{/ä 123
C	FF 12	FS 28	, 44	< 60	L 76	\/Ö 92	l 108	—/ö 124
D	CR 13	GS 29	- 45	= 61	M 77	]/Ü 93	m 109	}/ü 125
E	SO 14	RS 30	. 46	> 62	N 78	∧ 94	n 110	∼/ß 126
F	SI 15	US 31	/ 47	? 63	O 79	_ 95	o 111	DEL 127

The American Standard Code for Information Interchange (ASCII) is standardised worldwide as ISO/IEC 646. It allows national special characters in 12 places. In this table the German extensions according to DIN 66 003 are also shown. ASCII is a 7-bit code, but there exist many extensions to 256 characters, which are not necessarily compatible to each other.

The two- and three-letter symbols are acronyms for control codes for data transmission according to ISO. The symbol DC1 is also known as XON, and DC3 as XOFF.

B.19 Resolution and Coding for Analogue-to-Digital Converters

Resolution			
Bits n	Number of steps 2^n	Resolution at 10 V	Dynamic range (dB)
1	2	5 V	6.02
2	4	2.5 V	12.04
4	16	625 mV	24.08
6	64	156 mV	36.12
8	256	39 mV	48.16
10	1 024	9.77 mV	60.21
12	4 096	2.441 mV	72.25
14	16 384	610.352 µV	84.29
16	65 536	152.588 µV	96.33
18	262 144	38.1470 µV	108.37
20	1 048 576	9.53674 µV	120.41

The resolution is given for an input signal range of 10 V. The dynamic range denotes the logarithmic ratio between the largest and the smallest representable signal.

Coding				
Value	Offset binary	Two's complement	One's complement	Sign magnitude
+FS−1 LSB	1111 ... 1111	0111 ... 1111	0111 ... 1111	1111 ... 1111
+1/2 FS	1100 ... 0000	0100 ... 0000	0100 ... 0000	1100 ... 0000
+0	1000 ... 0000	0000 ... 0000	0000 ... 0000	1000 ... 0000
−0	...	...	1111 ... 1111	0000 ... 0000
−1/2 FS	0100 ... 0000	1100 ... 0000	1011 ... 1111	0100 ... 0000
−FS+1 LSB	0000 ... 0001	1000 ... 0001	1000 ... 0000	0111 ... 1111
−FS	0000 ... 0000	1000 ... 0000	−	−

LSB: *least significant bit*

FS: *full scale*, maximum range allowed for the converter. The maximum output value is given by the input voltage $(2^n - 1) \cdot$ FS.

One's complement is derived by inverting each of the bits of the value. The representation as *sign magnitude* uses the highest value bit to show the positive sign. Both representations have two different representations for zero.

B.20 Chemical Elements

Z	Symbol	Element name	Atomic weight	Remark
1	H	Hydrogen	1.008	Gas
2	He	Helium	4.003	Inert gas
3	Li	Lithium	6.941	Alkaline metal
4	Be	Beryllium	9.012	Alkaline earth element
5	B	Boron	10.81	–
6	C	Carbon	12.01	–
7	N	Nitrogen	14.01	Gas
8	O	Oxygen	16.00	Gas
9	F	Flourine	19.00	Halogen
10	Ne	Neon	20.18	Inert gas
11	Na	Sodium	22.99	Alkaline metal
12	Mg	Magnesium	24.31	Light metal
13	Al	Aluminium	26.98	Light metal
14	Si	Silicon	28.09	Semiconductor
15	P	Phosphorus	30.97	–
16	S	Sulfur	32.06	–
17	Cl	Chlorine	35.45	Halogen
18	Ar	Argon	39.95	Inert gas
19	K	Potassium	39.10	Alkaline metal
20	Ca	Calcium	40.08	Alkaline earth element
21	Sc	Scandium	44.96	Metal
22	Ti	Titanium	47.88	Light metal
23	V	Vanadium	50.94	Heavy metal
24	Cr	Chromium	52.00	Heavy metal
25	Mn	Manganese	54.94	Heavy metal
26	Fe	Iron	55.85	Heavy metal
27	Co	Cobalt	58.93	Heavy metal
28	Ni	Nickel	58.69	Heavy metal
29	Cu	Copper	63.55	Heavy metal
30	Zn	Zinc	65.39	Metal
31	Ga	Gallium	69.72	Semiconductor
32	Ge	Germanium	72.59	Semiconductor
33	As	Arsenic	74.92	–
34	Se	Selenium	78.96	Semiconductor
35	Br	Bromine	79.90	Halogen
36	Kr	Krypton	83.80	Inert gas
37	Rb	Rubidium	85.47	Alkaline metal
38	Sr	Strontium	87.62	Alkaline earth element
39	Y	Yttrium	88.91	Metal
40	Zr	Zirconium	91.22	Metal
41	Nb	Niobium	92.91	Metal
42	Mo	Molybdenum	95.94	Metal
43	Tc	Technetium	(98)	Artificial metal
44	Ru	Ruthenium	101.1	Transition metal

Z: atomic number. The atomic weight is given in g/mol.

Z	Symbol	Element name	Atomic weight	Remark
45	Rh	Rhodium	102.9	Precious metal
46	Pd	Palladium	106.4	Precious metal
47	Ag	Silver	107.9	Precious metal
48	Cd	Cadmium	112.4	Metal
49	In	Indium	114.8	Metal
50	Sn	Tin	118.7	Heavy metal
51	Sb	Antimony	121.8	Heavy metal
52	Te	Tellurium	127.6	Semiconductor
53	I	Iodine	126.9	Halogen
54	Xe	Xenon	131.3	Inert gas
55	Cs	Cesium	132.9	Alkaline metal
56	Ba	Barium	137.3	Alkaline earth element
57	La	Lanthanum	138.9	Rare earth element
58	Ce	Cerium	140.1	Rare earth element
59	Pr	Praseodymium	140.9	Rare earth element
60	Nd	Neodymium	144.2	Rare earth element
61	Pm	Promethium	145.0	Rare earth element
62	Sm	Samarium	150.4	Rare earth element
63	Eu	Europium	152.0	Rare earth element
64	Gd	Gadolinium	157.3	Rare earth element
65	Tb	Terbium	158.9	Rare earth element
66	Dy	Dysprosium	162.5	Rare earth element
67	Ho	Holmium	164.9	Rare earth element
68	Er	Erbium	167.3	Rare earth element
69	Tm	Thulium	168.9	Rare earth element
70	Yb	Ytterbium	173.0	Rare earth element
71	Lu	Lutetium	175.0	Rare earth element
72	Hf	Hafnium	178.5	–
73	Ta	Tantalum	180.9	–
74	W	Tungsten	183.9	–
75	Re	Rhenium	186.2	Transition metal
76	Os	Osmium	190.2	Heavy metal
77	Ir	Iridium	192.2	Precious metal
78	Pt	Platinum	195.1	Precious metal
79	Au	Gold	197.0	Precious metal
80	Hg	Mercury	200.6	Liquid metal
81	Tl	Thallium	204.4	–
82	Pb	Lead	207.2	Heavy metal
83	Bi	Bismuth	209.0	Heavy metal
84	Po	Polonium	(209)	–
85	At	Astatine	(210)	–
86	Rn	Radon	(222)	Radioactive inert gas
87	Fr	Francium	223.0	Alkaline metal
88	Ra	Radium	226.0	Radioactive metal

Z: atomic number. The atomic weight is given in g/mol. For unstable elements the atomic mass of the longest-lasting isotope is given in parentheses.

Z	Symbol	Element name	Atomic weight	Remark
89	Ac	Actinium	227.0	Actinoid
90	Th	Thorium	232.0	Actinoid
91	Pa	Protactinium	231.0	Actinoid
92	U	Uranium	238.0	Actinoid
93	Np	Neptunium	237.0	Transuranic ↓
94	Pu	Plutonium	(244)	
95	Am	Americium	(243)	
96	Cm	Curium	247	
97	Bk	Berkelium	(247)	
98	Cf	Californium	(251)	
99	Es	Einsteinium	(252)	
100	Fm	Fermium	257	
101	Md	Mendelevium	(258)	
102	No	Nobelium	(259)	
103	Lr	Lawrencium	(260)	
104	Rf	Rutherfordium	(261)	
105	Db	Dubnium	(262)	
106	Sg	Seaborgium	(263)	also Unh, Unnilhexium
107	Bh	Bohrium	(262)	
108	Hs	Hassium	(265)	
109	Mt	Meitnerium		
110	Uun	Ununnilium	(269)	discovered 1994
111	Uuu	Ununium	(272)	discovered 1994
112	Uub	Ununbium	(277)	discovered 1996
113	Uut			
114	Uuq		(285)	discovered 1999
115	Uup			
116	Uuh		(289)	
117	Uus			
118	Uuo		(293)	discovered 1999, announcement retracted 2001
...	??	as of 2001		

Z: atomic number. The atomic weight is given in g/mol. For unstable elements the atomic mass of the longest-lasting isotope is given in parentheses.

B.21 Materials

Material	Chem. symbol	Density kg/dm^3	Resistivity μΩm*	Temperature coefficient 10^{-3}K^{-1}
Aluminium	Al	2.70	0.027	4.3
Antimony	Sb	6.68	0.42	3.6
Brass	–	8.4	0.05–0.12	1.5
Bronze	–	8.9	0.02–0.14	0.5
Cadmium	Cd	8.64	0.077	3.8–4.2
Chromium	Cr	7.20	0.13	–
Chromium–Nickel	–	8.3	1–1.1	0.14
Cobalt	Co	8.9	0.06–0.09	3–6
Constantan	–	8.8	0.5	-0.04
Copper	Cu	8.92	0.017	4.3
Germanium (pure)	Ge	5.35	$0.46 \cdot 10^6$	–
German silver	–	8.5	0.33	0.07
Glass	–	2.4–2.6	10^{17}–10^{18}	–
Gold	Au	19.3	0.022	3.8
Iridium	Ir	22.42	0.06–0.08	4.1
Iron	Fe	7.86	0.1	6.5
55% Cu, 44%Ni, 1% Mn Lead	Pb	11.2	0.21	3.9
Magnesium	Mg	1.74	0.045	3.8–5.0
Manganin	–	8.4	0.43	±0.01
86% Cu, 2% Ni, 12% Mn Mercury	Hg	13.55	0.97	0.8
Mica	–	2.6–3.2	10^{19}–10^{21}	–
Molybdenum	Mo	10.2	0.055	3.3
Nickel	Ni	8.9	0.08	6.0
Palladium	Pd	11.97	0.11	3.3
Platinum	Pt	21.45	0.098	3.5
Rhodium	Rh	12.4	0.045	4.4
Selenium	Se	4.8	10^{11}	–
Silver	Ag	10.5	0.016	3.6
Silicon	Si	2.4	0.59	–
Tantalum	Ta	16.6	0.15	3.1–3.5
Tin	Sn	7.23	0.12	4.3
Titanium	Ti	4.43	0.048	–
Tungsten	W	19.3	0.055	4.5–5.7
Water (distilled)	H$_2$O	1.00	$4 \cdot 10^4$	–
Wood's metal	–	9.7	0.53	2.0
Zinc	Zn	7.14	0.061	3.7

*Conversion: $1\ \mu\Omega\,\text{m} = 1\ \Omega\,\text{mm}^2/\text{m} = 10^{-6}\ \Omega\,\text{m}$. The resistivity (specific resistance) is valid in the range from 0 to 100°C. The density is given for 20°C. The temperature coefficient is valid for 0°C; some ranges are given for temperatures from 0 to 500°C.

Resistivity of isolators (Ωm)			
Amber	$> 10^{16}$	Polyethylene	10^{16}
Epoxy resin	10^{13}–10^{15}	Polystyrene	10^{16}
Glass	10^{11}–10^{12}	Porcelain	$< 5 \cdot 10^{12}$
Hard rubber	10^{16}	PVC, hard	10^{15}
Mica	10^{13}–10^{15}	soft	10^{13}
Micanite	10^{15}	Quartz	10^{13}–10^{16}
Paper	10^{15}–10^{16}	Transformer oil	10^{10}–10^{13}
Plexiglas	10^{15}	Wood (dry)	10^{9}–10^{13}

Permittivity Values (Dielectric Constants)			
Air 1at, 0°C, dry	1.000 594	Methyl alcohol	33.5
Amber	2.2–2.9	Mica	4–9
Acetone	21.4	Micanite	4.0–6.0
Argon	1.000 504	Nitrobenzene	35.5
Barium titanate	1000–9000	Nitrogen	1.000 528
Benzene	2.3	Oxygen	1.000 486
Cable joint resin	2.5	Paraffin oil	2.2
Cable paper, impregnated	4–4.3	Pertinax	3.5–5.5
Cable oil	2.25	Phenoplaste	5–7
Carbon dioxide	1.000 985	Plexiglas	3–4
Cellulose	3–7	Polyethylene	2.2–2.7
Ceramics	up to 4000	Polystyrene	1.1–1.4
Condensa	40–80	Porcelain	4.5–6.5
Diethyl ether	4.3	PVC	3.1–3.5
Epoxy resin	3.7	Quartz glass	3.2–4.2
Ethyl alcohol	25.1	Shellac	2.7–4
Germanium	≈ 16	Silicon oil	2.2–2.8
Glass	2–16	Silicon	≈ 12
Glycerine	41.1	Styroflex	2.5
Hard rubber	2.5–5	Teflon	2.1
Helium	1.000 066	Transformer oil	2.2–2.5
Hydrogen	1.000 252	Vacuum	1.000 000
Kerosene	2.2	Water, distilled	81
Marble	8.4–14	Wood	2.5–6.8

The values for materials are for orientation. For a specific application it is recommended to get more information in reference books and tables. In particular, the conditions of the measurement have to be considered carefully.

C Acronyms

Acronym	Stands for
A	
AC	alternating current
ACD	automatic call distribution
ACIA	asynchronous communication-interface adapter
ACK	acknowledge
ACL	access control list
ACTE	Approval Committee for Telecommunications Equipment
ACW	architecture control word (GAL)
A/D	analogue to digital
AD	administrative domain
ADC	analogue-to-digital converter
ADM	add-drop multiplexer
ADPCM	adaptive differential-pulse code modulation
ADSL	asymmetrical digital subscriber line
ADSR	attack–decay–sustain–release (sound generator)
AEA	American Electronics Association
AF	audio frequency
AFC	automatic frequency control
AFT	automatic fine tuning
AGA	alterable gate array
AGC	automatic gain control
AHDL	analogue hardware-description language
ALC	automatic level control
ALERT	advice and problem location for European road traffic (RDS decoder)
ALGOL	algorithmic language
ALS	advanced low-power Schottky
ALU	arithmetic logical unit
AM	amplitude modulation
AMI	alternate mark inversion
AMPS	advanced mobile phone system
AMVSB	amplitude-modulation vestigeous sideband
ANL	automatic noise limiter
ANSI	American National Standards Institute
ANSI	ANSI code
APD	avalanche photodiode

Acronym	Stands for
API	application programming interface
APL	a programming language
AQL	acceptable quality level
ARP	address resolution protocol
ARRL	American Radio Relay League
AS	advanced Schottky
ASA	American Standards Association
ASCII	American Standard Code for Information Interchange
ASIC	application-specific integrated circuit
ASIS	application-specific instruction set
ASK	amplitude shift keying
ASM	algorithmic state machine
ASRA	application-specific resistor array
AT	control language for dial-up modems
ATAPI	AT-attachment packet interface
ATE	automatic test equipment
ATF	automatic track finding
ATM	Adobe Type Manager
ATM	asynchronous transfer mode
AUI	attachment unit interface (Ethernet)
AVC	automatic volume control
avg	average
AVI	audio–video interlace
AWG	American wire gauge
AWGN	additive white Gaussian noise

B

BALUN	balanced/unbalanced
BASIC	beginner's all-purpose instruction code
bbl	barrel
BBS	bulletin board system
BCC	block check character
BCD	binary coded decimal
BCH	Bose–Chaudhuri–Hocquenghem (code)
bd	French: baud
BEAB	British Electrotechnical Approvals Board
BEL	bell
BER	bit error rate
BFO	beat frequency oscillator

Acronym	Stands for
BG	borrow generate
BGA	ball/column grid array
BI	burn-in
BITBLT	bit block transfer
BIBO	bounded input–bounded output
BIFET	bipolar field-effect transistor
BIOS	basic input/output system
B-ISDN	broadband ISDN
BIST	built-in self test
BISYNC	binary synchronous communication
bit	binary digit
BK	black
BLOB	binary large object
bn	billion
BN	brown
BNC	bayonet nut connector, baby n-connector
BO	borrow-out output, ripple borrow output
BOC	Bell Operating Company
BOM	begin of message
BORSCHT	battery, overvoltage protection, ringing, signalling, coding, hybrid and testing
bp	boiling point
BP	borrow propagate
BPL	biphase level (code)
bpp	bits per pixel
BPSK	biphase shift keying
BRA	basic rate access (ISDN)
BS	base station
BS	backspace
BSC	binary synchronous communication
BSI	British Standards Institution
BTLZ	British Telecom Lempel–Ziv algorithm (data compression standard V.42bis)
BU	blue
BW	bandwidth
BWG	Birmingham wire gauge

C

C	ceramic
CA	collision avoidance
CAD	computer-aided design

Acronym	Stands for
CAE	computer-aided engineering
CAI	computer-assisted instruction
CAM	common access method (SCSI)
CAM	content-addressable memory
CAN	cancel
CAN	controller area network
CAN	customer access network
CAPI	common ISDN application programming interface
CAS	column address strobe
CASE	computer-aided software engineering
CAT	computer-aided telephony
CATV	community area television
CAV	constant angular velocity
CAZ	commutating auto zero (amplifier)
CB	citizen band
CB	common base (circuit)
CBDS	connectionless broadband data service
CBMS	computer-based message system
CCC	ceramic chip carrier
CCD	charge-coupled device
CCFL	cold-cathode flourescent light
CCIR	French: Comité Consultatif International de Radiodiffusion
CCITT	French: Comité Consultatif International de Téléphonique et de Télégraphique
CCN	cordless communication network
CCO	current-controlled oscillator
CCS7	common channel signalling system no. 7
CCTV	closed circuit television
ccw	counterclockwise
CD	call deflection (ISDN)
CD	collision detection (Ethernet)
CD	conditioned diphase (pulse frequency shift keying)
CDDI	copper distributed data interface
CD-I	CD interactive
CDIP	ceramic dual in-line package
CDLC	cellular data link control
CDMA	code-division multiple access
CDN	count down
CDRAM	cached dynamic RAM
CD-ROM	compact disk ROM

Acronym	Stands for
CDV	compressed digital video
CE	chip enable
CE	common emitter (circuit)
CE	concurrent engineering
CECC	French: Comité des Composants Electroniques du CENELEC
CELP	code-excited linear predictive coding
CEN	French: Comité Européen de Normalisation
CENELEC	French: Comité Européen de Normalisation Electrotechniques
CEPT	French: Conférence Européenne des Administrations des Postes et des Télécommunications
CERDIP	ceramic dual in-line package
CF	call forwarding
CF	center frequency
CG	carry generate
CGA	color graphics adaptor
CI	carry-in input
CID	charge injection device
CIE	French: Commission International de l'Éclairage
CIM	computer-integrated manufacturing
CIR	committed information rate
CISC	complex-instruction set computer
CIT	computer-integrated telephony
Ck	clock
CLCC	ceramic leaded chip carrier
CLI	command language interpreter
CLIP	calling line identification presentation (ISDN)
CLIR	calling line identification restriction
CLP	configurable logic block
Clr	clear
CLUT	colour look-up table
CLV	constant linear velocity
CM	circular mil
CMI	coded mark inversion
CMIP	common management information protocol
CML	current mode logic
CMOL	CMIP over LLC (logical link control)
CMOP	CMIP over TCP/IP
CMR	common-mode rejection
CMRR	common-mode rejection ratio

Acronym	Stands for
CMV	common-mode voltage
C/N	carrier-to-noise ratio
CNC	computer numerical control
CNR	carrier-to-noise ratio
CO	carry-out output, ripple carry output
COFDM	coded orthogonal frequency-division multiplex
COHO	coherent oscillator
COLP	connected line identification presentation
COMAL	common algorithmic language
COMEL	French: Comité de Coordination des Constructeurs des Machines Tournantes Electriques du Marché Commun
CompuSec	computer security
ComSec	communications security
CONP	connection-oriented protocol
CP	carry propagate (output)
CP/M	control program for microcomputers
CPE	customer premises equipment
CPFSK	continuous phase frequency shift keying
CPGA	ceramic pin grid array
CPM	continuous phase modulation
CPN	customer premises network
cps	characters per second
cps	cycles per second
CPU	central processing unit
CQFP	ceramic quad flat package
CR	carriage return
CRC	cyclic redundancy check
CRO	cathode ray oscilloscope
CRT	cathode ray tube
c/s	client server (application)
CS	chip select
CSA	Canadian Standard Association
CSMA	carrier sense multiple access
CSMA/CA	carrier sense multiple access/collision avoidance
CSMA/CD	carrier sense multiple access/collision detection
CSTA	computer-supported telephony applications
CT	cordless telephone
CTC	counter/timer circuit
CTI	computer telephone integration

Acronym	Stands for
CTR	counter
CTS	clear to send
CUP	count up
CVD	chemical vapour deposition
CW	call wait
cw	clockwise
CW	continuous wave
CWL	continuous wave laser
Cy	carry

D

D	data
D2B	domestic digital bus
D^2MAC	duobinary coded multiplexed analogue components
D/A	digital to analogue
DAB	digital audio broadcast
DAC	digital-to-analogue converter
DASP	digital audio signal processor
DATEC	data telecommunications
DAU	data acquisition unit
DBS	direct broadcast satellite
DC	direct current
dc	don't care
DCE	data circuit-terminating equipment
DCS	digital cellular system
DCT	discrete cosine transform
DCTL	direct-coupled transistor logic
DD	double density
DDC	direct digital control
DECT	digital enhanced cordless telephone
DEMKO	Danish national quality assurance symbol
DES	data encryption standard
DFB	distributed feedback (laser)
DFT	discrete Fourier transform
DIAC	diode alternating current switch
DIL	dual in-line
DIMM	dual in-line memory module
DIP	dual in-line package
DLC	data link control

Acronym	Stands for
DMA	direct memory access
DMM	digital multimeter
DNS	domain name system
DOV	data over voice
dpb	defects per billion
DPDT	double-pole double-throw
dpi	dots per inch
DPLL	digital phase-locked loop
DPM	digital panel meter
DPSK	differential phase shift keying
DPST	double-pole single-throw
DQDB	distributed queued dual bus
DQPSK	differential quadrature phase shift keying
DRAM	dynamic random-access memory
DRO	digital recording oscilloscope
DS	double sided
DSB	double sideband
DSBS	direct sound broadcasting by satellite
DSO	digital storage oscilloscope
DSO	dual in-line package small outline
DSP	digital signal processing/processor
DSR	data set ready
DSR	digital satellite radio
DSS-1	digital subscriber signalling system no. 1
DSSS	direct sequencing spread spectrum
DSU/CSU	digital service unit/channel service unit
DTE	data terminal equipment
DTL	diode transistor logic
DTMF	dial tone multiple frequency
DTR	data terminal ready
DUT	device under test
DVD	digital versatile disk
DVSO	dual in-line package very small outline
dx	duplex
DX	distant (reception)

E

E	extension input
E^2PROM	electrically erasable EPROM

Acronym	Stands for
EAPROM	electrically alterable PROM
EAROM	electrically alterable ROM
EAV	end of active video
EBCDIC	extended binary-coded-decimal interchange code
EBU	European Broadcasting Union
ECC	error checking and correction
ECC	error correcting code
ECCT	enhanced computer-controlled teletext
ECL	emitter-coupled logic
ECM	error-correcting mode
ECMA	European Computer Manufacturers Association
ECMA-6	extended ASCII code
ECQAC	Electronic Components Quality Assurance Committee
ED	extreme density
EDA	electronic design automation
EDC	error-detecting code
EDC	error-detection and correction
EDFA	Erbium-doped fibre amplifier
EDI	electronic data interchange
EDIF	electronic data interchange format
EDO	extended data out
EDP	electronic data processing
EDRAM	enhanced dynamic RAM
EDTV	enhanced definition TV
EE	electrical engineering
EEPLD	electrically erasable PLD
EEPROM	electrically erasable PROM
EFM	eight-to-fourteen (modulation)
EGA	enhanced graphics adapter
EHF	extremely high frequency (30–300 GHz)
EIA	Electronic Industries Association
EIB	European installation bus
E-IDE	enhanced IDE
EIRP	effective isotropic radiated power
EISA	extended industry standard architecture
ELCB	earth leakage circuit breaker
ELD	electroluminescent display
ELF	extremely low frequency (30–300 Hz)
EMC	electromagnetic compatibility

Acronym	Stands for
emf	electromotive force
EMI	electromagnetic interference
EMR	electromagnetic radiation
EMS	expanded memory specification
EN	enable
EN	European norm
ENQ	enquiry
e/o	electro-optical
EOF	end of file
EOR	exclusive Or
EOT	end of tape
EOT	end of transmission
EPAC	electrically programmable analogue circuit
EPLD	electrically programmable logic device
EPO	European Patent Office
EPROM	erasable ROM
EPS	encapsulated PostScript
erf	error function
ERMES	European radio message system
ERP	effective radiated power
ES	European standard
ESC	escape
ESD	electrostatic discharge
ESDI	enhanced small device interface
ESDS	electrostatic discharge sensitive (device)
ESPRIT	European Strategic Programme for Research and Development on Information Technology
ETS	European Telecommunication Standard
ETSI	European Telecommunication Standards Institute
EUT	equipment under test
EXOR	exclusive Or
E1	transmission rate in European multiplex hierachy 2.048 Mbit/s

F

FACT	Fairchild Advanced CMOS Technology
FAMOS	floating gate avalanche-injection MOS
FAQ	frequently asked questions
FAST	Fairchild advanced Schottky TTL
FAT	file allocation table

Acronym	Stands for
FCC	Federal Communications Commission
FCS	frame check sequence
FDC	floppy disk controller
FDD	floppy disk drive
FDDI	fibre-distributed data interface
FDDI-II	FDDI enhancement
FDM	frequency-division multiplexing
FDMA	frequency-division multiple access
FDX	full duplex
FEC	forward error correction
FET	field-effect transistor
FF	form feed
FFT	fast Fourier transform
FH-CDMA	frequency-hopping CDMA
FHSS	frequency-hopping spread spectrum
FIFO	first in-first out
FILO	first in-last out
FIPS	Federal information-processing standard
FIR	finite impulse response
FIT	failures in time
FLOTOX	floating gate tunnel oxide
FM	frequency modulation
FoD	fax on demand
FOR	fax over radio
FORTRAN	formula translator
FOX	fibre-optic transceiver
fp	freezing point
FPA	floating point accelerator
FPDT	four-pole double-throw
FPGA	field-programmable gate array
FPLD	field-programmable logic device
FPLS	field-programmable logic sequencer
fps	frames per second
FPST	four-pole single-throw
FPU	floating point unit
FR	frame relay
FRD	fast recovery diode
FROM	flash ROM
FSD	full scale deflection

Acronym	Stands for
FSK	frequency shift keying
FSM	finite state machine
FSR	force sensitive resistor
FTAM	file transfer, access and management
FTP	file transfer protocol
FTTC	fibre to the curb
FTTD	fibre to the desk
FTTH	fibre to the home

G

Acronym	Stands for
GA	gate array
GaAs	gallium arsenide
GAFET	gallium arsenide FET
GAL	generic array logic
GB	gigabytes
GCD	greatest common divisor
GCR	group-coded recording
GCT	gamma correction table
GD	gold
GDI	graphics device interface
GFLOPS	giga (10^9) floating point operations per second
GIGO	garbage in, garbage out
GMSK	Gaussian minimum shift keying
GN	green
GND	ground
GNYE	green–yellow
GOLD	GSM one-chip logic device
GOPS	giga (10^9) operations per second
GP	general purpose
GPIA	general-purpose interface adapter
GPIB	general-purpose interface bus
GPS	global positioning system
GSM	global system for mobile communications
GTO	gate turn-off (thyristor)
GUI	graphical user interface
GY	grey, gray

Acronym	Stands for
H	
HAL	hardware array logic
HBT	heterojunction bipolar transistor
HC	high-speed CMOS
HCF	highest common factor
HCMOS	high-density complementary metal oxide on silicon
HCT	high-speed CMOS with TTL thresholds
HD	hard disk
HD	high density
HDB3	high-density binary code with 3 zeros substitution
HDCD	high-density compact disk
HDD	hard disk drive
HDL	hardware-description language
HDLC	high-level data-link control
HDMAC	high-definition multiplexed analogue components
HDTV	high-definition TV
HDVS	high-definition video system
HDX	half duplex
HEMT	high electron mobility transistor
HFO	high-frequency oscillator
HHF	hyperhigh frequency (300–3000 GHz)
HiFi	high fidelity
HIP	hex in-line package
HIPO	hierarchy of input–process–output
HLF	hyperlow frequency (below 3 kHz)
HLL	high-level logic
HLL	high-level language
HLLCMOS	high-speed low-voltage low-power CMOS
HMA	high memory area
HNIL	high noise immunity logic
HPIB	general-purpose interface bus
HSB	hue, saturation, brightness
HSI	hue, saturation, intensity
HSV	hue, saturation, value
HTL	high threshold logic
HTL	high-voltage transistor logic (26–33 V)
HTML	hypertext markup language
HTP	high trigger point
HTS	high-temperature superconductor

Acronym	Stands for
http	hypertext transfer protocol
h/w	hardware

I

Acronym	Stands for
IAE	ISDN attachment unit
IARU	International Amateur Radio Union
IC	integrated circuit
ICAP	Interactive Circuit Analysis Program
ICCS	integrated communications cabling system
ICE	in-circuit emulation
ICIS	current-controlled current source
ICT	in-circuit test
ICVS	current-controlled voltage source
IDE	intelligent drive electronics
IDFT	inverse discrete Fourier transform
IDN	integrated digital network
IDTV	improved definition TV
IEC	International Electrotechnical Commission
IECC	International Electronic Components Committee
IECEE	IEC System for Conformity Testing to Standards for Safety of Electrical Equipment
IEEE	Institute of Electrical and Electronics Engineers
IEV	International Electrotechnical Vocabulary
IF	image frequency
IF	intermediate frequency
IFL	integrated fuse logic
IGBT	insulated gate bipolar transistor
IGES	initial graphics exchange specification
IGFET	insulated gate FET
IIL	integrated injection logic
IIR	infinite impulse response
ILF	infralow frequency (0.3–3 kHz)
IM	intermodulation
IMD	intermodulation distortion
IMPATT	impact avalanche transit time
IMQ	Italian national quality sign
INIC	current inverting negative impedance converter
INT	interrupt
I/O	input–output
IP	Internet Protocol

Acronym	Stands for
IP	international protection
IP3	intercept point of third order
IPC	Institute for Interconnecting and Packaging of Electronic Circuits
IPIP	input intercept point
ips	inches per second
IPX	Internetwork Packet Exchange
IR	infrared
IrDA	Infrared Data Association
IRE	Institute for Radio Engineers
IRE	IRE units
IRED	infrared emitting diode
IRQ	interrupt request
ISA	industry standard architecture
ISDN	integrated services digital network
ISI	intersymbol interference
ISO	International Standards Organisation
ISP	Internet service provider
IT	information technology
ITSEC	information technology security evaluation criteria
ITU	International Telecommunications Union
ITU-R	International Telecommunications Union – Radio Communication Sector
ITU-T	see ITU-TSS
ITU-TSS	International Telecommunications Union – Telecom Standardisation Sector
IVR	interactive voice response
IWG	Imperial wire gauge

J

JAN	Joint Army–Navy
JEDEC	Joint Electron Device Engineering Committee
JFET	junction FET
JIT	just in time
JPEG	Joint Photographic Expert Group (picture compression)

K

kbps	kilobits per second
KCL	Kirchhoff's current law
kc/s	kilocycles per second
KIS	keep it simple

Appendix C Acronyms

Acronym	Stands for
KLT	Karhunen–Loéve transform
kMc	kilo megacycles (GHz)
KOPS	kilo-operations per second
ksps	kilosamples per second
KVL	Kirchhoff's voltage law

L

Acronym	Stands for
L	live
LAN	local area network
LAP-M	link access procedure for modems
laser	light emission by stimulated emission of radiation
LCA	logic cell array
LCC	leadless chip carrier
LCD	liquid crystal display
LCM	least common multiple
LCR	least cost routing
LD	laser diode
LDC	long distance carrier
LDR	light-dependent resistor
LDTV	low-definition TV
LE	local exchange
LED	light-emitting diode
LEMP	lightning electromagnetic pulse
LF	line feed
LF	low frequency (30 Hz–300 kHz)
LFO	low-frequency oscillator
LIFO	last in-first out
LISP	list processing (programming language)
LL	leased line
LLC	logical link control
LLLTV	low-level light TV
LMS	least mean square
LNA	low-noise amplifier
LNB	low-noise block converter
LNC	low-noise converter
LO	local oscillator
LOCMOS	local oxide CMOS
LORAN	long-range navigation
LP	low pass (filter)

Acronym	Stands for
LPC	linear predictive coding
LR	loudness rating
LRC	longitudinal redundancy check
LRU	last recently used (memory)
LS	least square
LSB	least significant bit
LSD	least significant digit
LSI	large-scale integration (1000-5000 gates)
LSTTL	low-power Schottky TTL
LTP	lower trigger point
LUT	look-up table
LZW	Lempel–Ziv–Welch (data compression)

M

MAC	media access control
MAC	multiplexed analogue components
MAD	mean absolute difference
MAN	metropolitan area network
MAP	manufacturing automation protocol
MASK	multiple-amplitude shift keying
MAU	medium attachment unit
MAU	multistation access unit
MB	megabytes
Mbps	megabits per second
MBS	mutual broadcasting system
µC	micro-controller
MCA	microchannel architecture
MCM	1000 circular mils
MCT	MOS-controlled thyristor
MDAC	multiplying analogue-to-digital converter
MDR	magnetic field dependent resistor
MDT	mean down time
MECL	Motorola emitter-coupled logic
MET	multiemitter transistor
MF	medium frequency (300 kHz–3 MHz)
MF	microfarad, µF
MFAQ	most frequently asked questions
MFD	microfarad, µF
MFLOPS	mega floating-point operations per second

Acronym	Stands for
MFM	modified frequency modulation
MFSK	multiple-frequency shift keying
MHC	modified Huffmann code
MHS	message-handling system
MIB	management information base
MIDI	musical instrument digital interface
mil	1/1000 inch
MIL	qualified for military use
MIMD	multiple instruction/multiple data (stream)
MIME	multipurpose Internet mail extension
MIPS	million instructions per second
MLE	maximum likelihood estimation/estimator
MLSE	minimum least-square error
MM-CD	mixed-mode compact disk
MMS43	multimode system 4B3T
MMU	memory management unit
MNP	Microcom network protocol
MO	magneto-optical
MOD	magneto-optical drive
modem	modulator/demodulator
MOS	metal-oxide semiconductor
μP	microprocessor
mp	melting point
MPEG	motion picture expert group
MPLD	mask-programmed logic device
MPP	massively parallel processor
MPP	maximum power point
MPPP	multiwatt power plastic package
MPRII	Swedish norm concerning the maximum values for electric fields from PC monitors
MPSK	multiple-phase shift keying
MPU	microprocessing unit
MR	master reset
MSB	most significant bit
MSD	most significant digit
MS-DOS	Microsoft disk operating system
MSE	mean square error
MSI	medium scale integration (10–1000 gates)
MSK	minimum shift keying
MSPS	mega samples per second

Acronym	Stands for
MTBF	mean time between failures
MTF	modulation transfer function
MTTF	mean time to failure
MTTFF	mean time to first failure
MTTR	mean time to repair
MUSE	multiple subsampling encoding
MUSICAM	masking pattern universal sub-band integrated coding and multiplexing
MUT	mean up time
MUX	multiplexer
MW	hectometric waves
MX	multiplex

N

N	neutral
NAK	negative acknowledge
NB	narrowband
NBFM	narrow-band frequency modulation
NBS	National Bureau of Standards (USA)
nc	normally closed
nc	not connected
NCCF	normalized cross-correlation function
NCO	numerically controlled oscillator
NDI	nondestructive inspection
NDT	nondestructive testing
NE	network element
NEC	National Electric Code (USA)
NEMA	National Electrical Manufacturers Association
NEMKO	Norwegian national quality assurance symbol
NEP	noise equivalent power
NF	noise figure
NFB	negative feedback
NFS	network file system
NI	network interface
NIC	negative impedance converter
NIC	network interface card
NIH	not invented here (syndrome)
NIM	nuclear instrumentation module
NIST	US National Institute of Standards and Technology
NLQ	near letter quality

Acronym	Stands for
NMI	nonmaskable interrupt
NN	neural network
no	normally open
NOT	number of turns
Np	neper (log unit = 8.69 dB)
NPV	net present value
NRZ	nonreturn-to-zero
NRZI	nonreturn-to-zero inverted
NT	network terminator (ISDN)
NTC	negative temperature coefficient
NTFS	new technology file system
NTP	normal temperature and pressure
NTSC	National Television System Committee
nv	nonvolatile
NVM	nonvolatile memory

O

Acronym	Stands for
OA	office automation
OA	operational amplifier
OC	open collector
OC	output control
OCCAM	programming language for transputer
OCP	overcurrent protection
OCR	optical character recognition/reader
ODA	open document architecture
ODIF	open document interchange format
ODL	optical data link
o/e	optoelectronic
OEIC	optoelectronic integrated circuit
OEM	original equipment manufacturer
OFA	optical fibre amplifier
OFDM	optical frequency-division multiplex
OFDM	orthogonal frequency-division multiplex
OG	orange
OGM	outgoing message
OHP	overheat protection
OLMC	output logic macro cell
ONT	optical network termination
OOK	on–off keying

Acronym	Stands for
OOP	object-oriented programming
OOS	out of service
OPIP	output intercept point
OSD	on-screen display
OSF	Open Systems Foundation
OSI	open systems interconnection
OTA	operational transconductance amplifier
OTDM	optical time-division multiplex
OTDR	optical time-domain reflectometer
OTP	one-time programmable
ÖVE	Austrian national quality assurance initials
OVP	overvoltage protection

P

Acronym	Stands for
P	plastic
P	proportional (control)
PA	power amplifier
PA	polyamide
PA	public address
PABX	private automatic branch exchange
PAD	packet assembler/disassembler
PAL	phase alternation line
PAL	programmable array logic
PAM	pulse amplitude modulation
PAP	plug and play
PASC	precision adaptive sub-band coding
PASTA	Poisson arrivals see time averages
PBN	private branch network
PBX	private branch exchange
PC	personal computer
PCB	printed circuit board
PCC	plastic chip carrier
PCD	photo compact disk
PCI	peripheral component interconnect
PCL	printer command language
PCM	pulse code modulation
PCMCIA	PC Memory Card International Association
PCN	personal communication network
pcs	pieces

Acronym	Stands for
PCS	plastic cladded silica
PCSF	plastic cladding silica fibre
PCTA	personal computer terminal adapter
PD	proportional differential (control)
PD	public domain
PDA	personal digital assistant
PDCA	plan, do, check, assess
PDH	plesiosynchronous digital hierarchy
PDM	polarization-division multiplex
PDM	pulse-duration modulation
PE	parallel enable
PE	phase encoding
PE	polyethylene
PE	protective earth
PEARL	process and experiment automation real-time language
PECL	pseudo-ECL
PEEL	programmable electrically erasable logic
PEN	protective earth neutral
PERL	practical extraction and report language
PF	power factor $\cos \varphi$
PFC	power factor correction
PFET	power field-effect transistor
PFM	pulse frequency modulation
PGA	pin grid array
PGA	programmable gain amplifier
PHIGS	programmer's hierarchical interactive graphics system
PHL	physical layer
PI	proportional integral (control)
PIA	peripheral interface adapter
PID	process indentifer
PID	proportional integral differential (control)
PIN	personal identification number
PIN	positive–intrinsic–negative
PIO	parallel input/output
PIP	picture in picture
PIPO	parallel in, parallel out
PIR	passive infrared (detector)
PISO	parallel in serial out
PK	pink

Acronym	Stands for
PKC	public key cryptography system
PKI	public key cryptography infrastructure
PLA	programmable logic array
PLC	programmable logic controller
PLCC	plastic leaded chip carrier
PLD	programmable logic device
PLL	phase-locked loop
PLM	pulse length modulation
PL/1	programming language no. 1
PM	phase modulation
PM	polarization maintaining (fibre)
PMF	power MOSFET
POF	polymer optical fibre
POH	power-on hours
POLSK	polarisation shift keying
PON	passive optical network
PON	power on
POS	product of sums
POST	power-on self-test
POTS	plain old telephone service
pp	peak to peak
PP	polypropylene
PPA	push–pull amplifier
ppb	parts per billion
ppm	parts per million
PPP	point-to-point protocol
PQFP	plastic quad flat pack
PRA	primary rate access (ISDN)
PRBS	pseudo-random binary sequence
PRF	pulse repetition frequency
PRN	pseudo-random noise
PROM	programmable read-only memory
PRR	pulse repetition rate
PS	polystyrol
PSDN	packet-switched data network
PSK	phase shift keying
PSRR	power supply rejection ratio
PSSO	plastic shrink small outline (SMD)
PSTN	public switched telephone network

Acronym	Stands for
PSW	program status word
PTC	positive temperature coefficient
PTFE	polytetraflourineethylene (Teflon)
PTO	public telephone operator
PTT	post, telephone and telegraph company
PU	polyurethane
PVC	polyvinyl chloride
PWD	pulse-width distortion
PWM	pulse-width modulation
PWR	power
PWR DWN	power down
PXO	programmable oscillator

Q

QAM	quadrature amplitude modulation
QASK	quadrature amplitude shift keying
QBE	query by example
QCIF	quarter common intermediate format
QDPSK	quadrature differential-phase shift keying
QFP	quad flat package
QFPP	quad flat plastic package
QIC	quarter-inch cartridge
QIP	quad in parallel
QIP	quad-in-line package
QMS	quality management system
QoS	quality of service
QPP	quiescent push–pull amplifier
QPSK	quadrature phase shift keying
QSPI	queued serial peripheral interface
QTY	quantity

R

RAC	rectified alternating current
RACE	Research on Advanced Communications for Europe
RADAR	radio detection and ranging
RAH	row address hold
RAID	redundant array of inexpensive disks
RAIT	redundant array of inexpensive tapes

Acronym	Stands for
RAM	random-access memory
RAMDAC	digital-to-analogue converter with RAM
RAS	row address strobe
RBER	residual bit error rate
RBOC	regional Bell operating company
RCO	ripple counter output
RCT	reduced contact test
RCTL	resistor-coupled transistor logic
RCV	receive
R&D	research and development
RD	receive data
RD	red
RDBMS	relational database management system
RDS	running digital sum
RDY	ready
RF	radio frequency (3–30 MHz)
RF	reactive factor $\sin\varphi$
RFA	radio-frequency amplifier
RFI	radio-frequency interference
RGB	red, green, blue
RIP	remote image processing
RISC	reduced instruction set computer
RJ45	8-pin connector for network/telecommunications applications
RLC	resistor, inductance, capacitor (filter)
RLE	run-length encoding
RLLE	run-length-limited encoding
RMS	root mean square
RNIS	French: Réseau Numérique Intégration de Services (ISDN)
ROC	region of convergence
ROD	rewritable optical disk
ROM	read-only memory
RPM	revolutions per minute
RPN	reverse polish notation
RPS	revolutions per second
RS	Reed–Solomon (code)
RS232	American interface standard similar to V.24
RSA	Rivest–Shamir–Adleman (code)
RSC	Reed–Solomon code
RS-PG	Reed–Solomon product code

Acronym	Stands for
RTC	real-time convolver
RTD	resistive temperature device (thermistor)
RTF	rich text format
RTL	resistor transistor logic
RTS	request to send
RTTY	radio teletype
R/W	read/write
RX	receiver
RZ	return to zero

S

S_0	ISDN subscriber interface
SAA	standard application architecture
SAH	stuck at high
SAL	stuck at low
SAM	sequential access memory
SAV	start of active video
SAW	surface acoustic wave (filter)
SAWR	surface acoustic wave resonator
SBC	single-board computer
SC	switched capacitor (filter)
SCAM	suppressed carrier amplitude modulation
SCM	subcarrier modulation
SCP	serial communication port
SCR	silicon-controlled rectifier
SCS	silicon-controlled switch
SCSI	small computer systems interconnect (pronounce: scuzzy)
$\Sigma-\Delta$	sigma delta converter
SD	single density
SDH	synchronous digital hierarchy
SDIP	shrink dual in-line package
SDLC	synchronous data-link control
SDRAM	synchronous dynamic RAM
SDTV	standard-definition TV
SECAM	French: Séquential á mémoire (TV)
SEM	scanning electron microscope
SEMKO	Swedish national quality assurance initials
SET	single-electron transistor
SETI	Finnish national quality assurance initials

Acronym	Stands for
SEV	Swiss national quality assurance initials
SFN	single-frequency network
SG	signal ground
SGML	standardized generalized markup language
S/H	sample and hold
SHA	sample-and-hold amplifier
SHF	superhigh frequency (3–30 GHz)
S/I	signal to interference (ratio)
SIA	Semiconductor Industry Association
SIL	single in line
SIMD	single-instruction multiple data (stream)
SIMM	single in-line memory modules
SIO	serial input/output
SIP	single in-line package
SIPO	serial in, parallel out
SISD	single instruction single data
SISO	serial in, serial out
SLALOM	semiconductor laser amplifier in a loop mirror
SLF	superlow frequency ($<$ 3 kHz)
SLIC	subscriber line interface circuit
SMAC	state machine atomic cell
SMC	surface-mounted component
SMD	surface-mounted device
SMDS	switched multimegabit data service
SMPS	switched-mode power supply
SMPTE	Society of Motion Picture and Television Engineers
SMT	surface-mount technology
SMTP	simple mail transfer protocol
S/N	signal to noise (ratio)
SNA	systems network architecture
SNMP	simple network management protocol
SNR	signal-to-noise ratio
SO	serial output
SO	small outline
SOA	safe operating area
SOG	small-outline gull-wing
SOH	start of heading
SOHO	small office, home office
SOIC	small-outline IC

Acronym	Stands for
SOJ	small-outline J (IC housing)
SOP	state of polarisation
SOP	sum of products
SOP	small-outline package
SOS	silicon on sapphire
SOT	small-outline transistor
SP	signal processor
SP	stack pointer
SP	surge protector
SPARC	scalable processor architecture
SPC	stored program control
SPDT	single-pole double-throw
SPE	subscriber premises equipment
sp gr	specific gravity
SPICE	simulation program with IC emphasis
SPN	subscriber premises network
SPST	single-pole single-throw
SPX	sequenced packet exchange
SQFT	shrink quad flat package
SQL	structured query language
SR	shift register
SR	silver
SRAM	static RAM
SRD	step recovery diode
SSB	single sideband (modulation)
SSBSC	single sideband suppressed carrier (modulation)
SSD	solid-state disk
SS/DD	single side double density
SSI	small-scale integration
SSMA	spread spectrum multiple access
SSN7	signalling system no. 7
SSOP	shrink small-outline package
SSPA	solid-state power amplifier
SSR	solid-state relay
SSSC	single sideband suppressed carrier
SSTV	slow-scan television
SS#7	signalling system no. 7
SS7	signalling system no. 7
STDM	synchronous time-division multiplex

Acronym	Stands for
STEP	standard for the exchange of product and model data
STM	synchronous transfer mode
STM-1	synchronous transport module (ISS Mbps)
STP	shielded twisted pair
STP	standard temperature and pressure
STX	start of text
s-VHS	super-video home system
SVP	surge voltage protector
s/w	software
SW	short wave
SWG	Imperial standard wire gauge
SWR	standing-wave ratio
SYN	synchronous idle

T

TAM	telephone answering machine
TAP	terminal access point
TASI	time-assignment speech interpolation
TAT	transatlantic tube
TAZ	transient absorption zener (diode)
TB	terminal block
TC	temperature coefficient
TC	terminal count
TC	two's complement
TCM	Trellis coded modulation
TCO92	Swedish standard for electric fields from PC monitors
TCO95	Swedish standard for power consumption of PC monitors
TCP/IP	transmission control protocol/internet protocol
TD	transmit data
TDD	time-division duplex
TDM	time-division multiplexing
TDMA	time-division multiple access
TDR	time-domain reflectometer
TE	transversal electrical (wave)
TE	terminal equipment
TEM	transversal electromagnetic (wave)
TEMPEST	test for electromagnetic propagation emission and secure transmission
TETRA	trans-European trunked radio
TFP	thin flat package

Acronym	Stands for
TFT	thin-film transistor (LCDs)
T/H	terminal host (application)
T/H	track and hold
THD	total harmonic distortion
THZ	terahertz (10^{15} Hz)
TIFF	tagged image file format
TIM	transient intermodulation
TLA	three-letter acronym
TM	transversal magnetic (wave)
TN-C	terra, neutral, common (protective conductor also serves as neutral)
TN-C-S	terra, neutral, common, separated (contains both combined and separate neutral and protective conductors)
TOC	table of contents
TOR	telex over radio
TPDDI	twisted-pair distributed data interface
TPE	twisted pair Ethernet
tpi	tracks per inch
TQ	turquoise
TQFP	thin quad flat package
TQM	total quality management
TRIAC	triode alternating current switch
Triple nickel	lab-slang for the timer-IC 555
TSR	terminate and stay resident
TSSOP	thin shrink small-outline package
TT	true type
TTL	transistor–transistor logic
TTY	teletypewriter
TÜV	German national quality assurance initals
TV	television
TWAIN	technology without an interesting name
TWT	travelling wave tube
TX	transmitter
T1	transmission rate in US multiplex hierarchy, 1.5 Mbit/s
T3	transmission rate in US multiplex hierarchy, 45 Mbit/s

U

UART	universal asynchronous receiver/transmitter
UDP	user datagram protocol
UDTV	ultradefinition TV

Acronym	Stands for
UEP	unequal error protection
uF	microfarad, μF
uH	microhenry, μH
UHF	ultrahigh frequency (300 MHz–3 GHz)
UI	unit interval
UIT	French: Union International des Télécommunications (ITU)
UJT	unijunction transistor
UKW	German abbreviation for VHF, metric waves
UL	American national quality assurance initials
ULA	uncommitted logic arrays (ASICs)
ULF	ultralow frequency (300 Hz–3 kHz)
ULSI	ultralarge scale integration
UMA	upper memory area
UMB	upper memory blocks
UNIC	voltage-inverting negative impedance converter
UNIX	widely used multiuser operating system for powerful workstations
UPC	universal product code
UPS	uninterruptible power supply
US	unavailable seconds
USART	universal synchronous/asynchronous receiver/transmitter
USAT	ultrasmall-aperture terminal
USD	US dollars
UTC	universal time coordinated
UTP	unshielded twisted pair (cable)
UV	ultraviolet ($\lambda < 400$ nm)

V

VAC	volts alternating current
VANS	value-added network service
VAR	value-added reseller
VAS	value-added service
VC	virtual channel
VCA	voltage-controlled amplifier
V_{cc}	supply voltage
VCD	variable-capacitance diode
VCF	voltage-controlled filter
VCI	virtual channel identifier
VCIS	voltage-controlled current source
VCO	voltage-controlled oscillator

Acronym	Stands for
VCR	videocassette recorder
VCVS	voltage-controlled voltage source
VCXO	voltage-controlled crystal oscillator
VDC	volts direct current
VDE	German national quality assurance initals
VDR	voltage-dependent resistor (varistor)
VDT	video display terminal
VDU	video display unit
VESA	Video Electronics Standards Association
VF	voice frequency (16 Hz–20 kHz)
VFC	voltage-to-frequency converter
VFO	variable-frequency oscillator
VHDL	VHSIC hardware description language
VHF	very high frequency (30–300 MHz)
VHSIC	very high speed integrated circuit
VIA	versatile interface adapter
V_{IH}	high-level input voltage
V_{IL}	low-level input voltage
VIL	vertical in-line
VLB	VESA local bus
VLF	very low frequency (3–30 kHz)
VLSI	very large scale integration (more than 5000 gates)
VLT	video look-up table
VME	Versa Module Eurocard (bus system for microcomputers and workstations)
VMOS	V-groove MOS
VMS	voice mail system
VoD	video on demand
V_{OH}	high-level output voltage
V_{OL}	low-level output voltage
VOM	volt–ohm–milliammeter (multimeter)
VOX	voice-operated transmission
VPI	virtual path identifier
VPN	virtual private network
VR	voltage regulator
VRAM	video random-access memory
VRC	vertical redundancy check
VSAT	very small aperture terminal
VSO	very small outline
V_{ss}	ground

Acronym	Stands for
VSW	very short waves (10–1 m, UKW)
VSWR	voltage standing-wave ratio
VT	vertical tabulator
VT	violet
VTF	voltage-tunable filter
VTR	videotape recorder
VTVM	vacuum voltmeter
VXI	VME bus extension for instrumentation
VXO	variable-frequency crystal oscillator
V.24	interface standard of the CCITT

W

W^3	World Wide Web
WAN	wide area network
WARC	World Administrative Radio Conference
WB	wide-band
WBFM	wide-band frequency modulation
WCS	writable control store
WDM	wavelength-division multiplexing
WDT	watchdog timer
WE	write enable
WH	white
WISCA	why isn't Sam coding anything?
WLAN	wireless LAN
WORM	write once read multiple
wpc	watts per candle (light power)
wrt	with respect to
WSI	wafer-scale integration
wt	weight
WWW	World Wide Web
WYSIWYG	what you see is what you get

X

XBIOS	extended basic input/output system
XMS	extended memory specification
XMT	transmit
XOR	exclusive Or
XTAL	crystal

Acronym	Stands for
Y	
yd	yard (0.9 m)
YE	yellow
YIG	yttrium–iron–garnet
YIQ	luminance, in-phase, quadrature
YUV	colour coordinates of the European PAL system (Y = luminance, UV = chrominance)
Z	
Z	zero bit
Z80	highly popular 8-bit microprocessor
ZCS	zero code suppression
ZCS	zero current switching
ZD	zero defects
ZIF	zero insertion force (ICs)
ZIP	zigzag in-line package
ZM	dual sideband modulation
ZTAT	zero turnaround time
ZVS	zero voltage switching
4PDT	four-pole double-throw
4PST	four-pole single-throw
555	triple nickel (nickel: 5-cent coin)

D Circuit Symbols (Selected)

Resistors			
—▭—	Resistor, general	⌀ -Θ	Thermal resistor, NTC
⌀	Adjustable resistor	⌀ Θ	Thermal resistor, PTC
⌐▭	Potentiometer with movable contact	⌀ B	Magnetic-dependant resistor
⌀ V	Voltage-dependent resistor, varistor	⌀ X	Magnetoresistor, linear
—▭—	Resistor with fixed taps	/	Adjustability, general
—▭—	Shunt	/	Adjustability, nonlinear
—▥—	Heating element	/	Variability, inherent, general
		/	Variability, inherent, nonlinear

Capacitors			
⊣⊢	Capacitor, general	⌿	Capacitor, adjustable
⊣⊢	Polarised capacitor	⊤	Lead-through capacitor feed-through capacitor

Inductors

⌒⌒⌒	Inductor, Coil general	⌒⌒⌒	Continuously variable inductor
⌒⌒⌒	Inductor with magnetic core	⌒⌒⌒	Inductor with fixed taps
⌒⌒⌒	Inductor with gap in magnetic core	—⎕—	Ferrit bead, shown on conductor

Transformers

⌒⌒⌒	Transformer with 2 windings	⌒⌒⌒	Transformer with 3 windings
⌒⌒⌒	Transformer with 2 windings and identical voltage polarity	⌒⌒⌒	Autotransformer
⌒⌒⌒	Transformer with 2 windings and opposing voltage polarity	⌒⌒⌒	Pulse transformer

Voltage Sources, Current Sources

⌽	Ideal voltage source	⊖	Ideal current source
⊙	AC voltage source, technical frequency	⊛	AC voltage source, high frequency
⊛	AC voltage source, audio frequency	⏚	Protective earth, Protective ground
⏊	Earth, general Ground, general	⊥ ⏚	Ground, Chassis
—⎍—	Fuse		

Old Symbols (no longer used!)

—⋀⋀—	Resistor, general	⊣⊢	Electrolytic capacitor
⊣⊢	Capacitor, general	—▬—	Inductor
⊣⊢	Polarized capacitor	▬▬	Transformer

Semiconductor Diodes

⩟	Semiconductor diode, general	⩟	Variable capacitance diode Varactor
⩟ ⋆	Light emitting diode, LED general	⩟	Breakdown diode, unidirectional Zener diode
⩟Θ	Temperature sensing diode	⩟	Breakdown diode, bidirectional
⩟	Photodiode	–	–

Thyristors

⩟	Bidirectional diode thyristor DIAC	⩟	Bidirectional triode thyristor TRIAC
⩟	Thyristor	⩟	Turn-off triode thyristor GTO

Transistors			
⊣<	npn-transistor	⊥⊨	Insulated gate field-effect transistor IGFET, enhancement type, p-type channel
⊣<	pnp-transistor	⊥⊨	IGFET, enhancement type, n-type channel
⊣⊘	npn-transistor, collector connected to housing	⊥⊨	IGFET, enhancement type, n-type channel
⊥⊨	Junction field-effect transistor JFET, n-type channel	⊥⊨	IGFET, depletion type, n-type channel
⊥⊨	JFET, p-type channel	⊥⊨	IGFET, depletion type, p-type channel
⊄	Insulated gate bipolar transisor IGBT, enhancement type, n-type channel	⋎	Phototransistor, pnp-type

Measurement Instruments			
○	Indicating instrument	□	Recording instrument
▯	Integrating instrument	○---	Counter
Ⓥ	Voltmeter	W	Recording wattmeter
(cos φ)	Power-factor meter	Wh	Watt-hour meter
Ⓦ	Wattmeter	h	Hour meter Hour counter
Θ	Thermometer	▭○	Pulse meter
-U+	Thermocouple, with polarity	⊥	Thermoelement with noninsulated heating element
Movement symbols, see Sect. 4.1.6			

Switches, Relays			
╲	Switch, make contact	╤	Operating device Relay coil, general
╱	Switch, break contact	■╤	Relay coil of a polarised relay
╱Θ	Temperature-sensitive switch, break contact	╱╤	Relay coil of a remanent relay
⌐╱	Self-operating thermal switch, break contact	╤┐	Operating device of a thermal relay
⊢-╲	Manually operated switch, general	╒◁	Operating device of an electronic relay

Connections, Connectors

Symbol	Description	Symbol	Description
——	Connection, general	•	Junction, connection point
—//—/3	3 connections	⊤⊤	T-connection
—⌒—	Screened conductor	+⊥⊤	Double junction of conductors
=⧸=	Twisted connection, 2 connections	+	Not connected
⊖⊖	Coaxial pair	—(■—	Plug and socket
○	Terminal	—(	Female contact, socket
⊏⊐	Connecting link, closed	■—	Male contact, socket

Sensors

Symbol	Description	Symbol	Description
S N	Permanent magnet	—⌀—	Hall generator
↘◻	Light-dependent resistor	↘⧸	Photovoltaic cell
⌐⌐	Piezoelectric crystal, quartz	–	–

Integrated Circuits

Symbol	Description	Symbol	Description
⚡⇒◁	Optocoupler	▷∞	Operational amplifier
⚡⇒◫	Optical coupling device with slot for light barrier	⚡⇒⚡	Opto-TRIAC

Index

8421-code 452

absolute value 102–104
absolute value squared 102
absolute values of sums 105
absolute voltage level 549
AC bridges *see* bridge circuits
AC current gain 274
AC equivalent circuit 262
AC measurement
   current 177
   voltage 177
AC power
   overview 156
AC voltage amplifier 343
access time 440
acronyms 561
active 265
active filter *see* filter
actual value 99
addition of vectors 105
addition theorems
   of hyperbolic functions 518
   of trigonometric functions 514
address 439
address access time 440
address inputs 424
addressable memory 439
adjacent terms
   logic algebra 407
admittance 119
admittance parameters 266
admittance plane
   complex 120
advanced low-power Schottky TTL series 416
advanced Schottky TTL series 416
AGA 446
air gap 500
   value of A_L 73
all-pass filter 197, 237, 348
alternating quantity 112
alternative phase-shifting circuits 148
A_L-value 500
American units 535
American wire gauge 545
amount of feedback 323
ampere 1
   definition 64

ampere turns 71
Ampere's law 71, 95
amplitude 99, 113
amplitude spectrum 212
amplitude–phase form 211
analogue circuit design 261 ff
   methods of analysis 261
analogue signals 261
analogue stabilisation with transistor 474
analogue-to-digital converter
   resolution and coding 555
AND 444
AND function 392
AND gate 394
angular frequency 99, 106, 113
antisymmetric function 209
apparent power 155, 156
application of the Fourier series 217
approval sign 469
arbiter 442
arc-functions 517
   principal value 517
Argand diagram 103
argument 103, 104
arithmetic mean 114
Aron-circuit 186
Arrhenius-law 372
artificial mains network 508
artificial zero-point 185
ASCII coding 554
ASCII table 554
associative law 397
asymmetric radio-frequency interference voltage 508
asymmetrical 265
asynchronous counters 447, 450
attenuation constant 237
attenuation distortion 237
attenuation factor 194
atto 531
average 114
average power 7, 152
AWG 545
axial symmetry 209

baby cell 547
balanced 158
balanced systems 165

Index

balancing
  of AC bridges 149
balancing condition
  of AC bridges 149
bandpass filter 196, 204, 361
  bandwidth 361
  centre frequency 361
  circuit 362
  frequency response 361
  higher order 362
  ideal 240
bandwidth 26, 196, 240, 361
  definition 240
Barkhausen criterion 364
barkhausencriterion 328
basics of differential calculus 518
basics of integral calculus 519
batteries 547
BCD 452
BCD counter 447
BCD-decimal decoder 426
Bessel-filter 351
bias voltage production 385
bimetallic instrument 171
binary coded decimal 452
binary counter 447, 450
Biot–Savart 68
Biot–Savart's law 68
bipolar transistor 271 ff
  AC equivalent circuit 277
  basic circuits 280
  characteristics 272
  common-emitter circuit 280
  critical frequency 276
  current gain 274
  equivalent circuit, AC 277
  equivalent circuit, static 276
  Giacoletto equivalent circuit 278
  input resistance 275, 279
  output resistance 275
  overview: basic circuits 296
  reverse voltage transfer ratio 276
  static equivalent circuit 276
  thermal voltage drift 275
  unity gain frequency 276
  voltages and currents 272
bipolar transistor current sources 296
bipolar transistor differential amplifier 298, 300
  differential mode gain 300
black-box 192, 220
block diagram 267

Bode plot 269
Boltzmann-constant 533
boost converter 479, 494
bootstrap 384
branch 6
bridge circuits 149
  balancing condition 149
bridge rectification 472
bridge rectifier 472
bridges *see* bridge circuits
buck converter 477, 494
buck-boost converter 481, 494
Butterworth-filter 351

calculation methods for linear circuits 29 ff
capacitance 4, 46
capacitive divider 19
capacitive reactance 123
capacitor 5, 122
CAS 442
cascade circuit 202
cascading counters 456
causal signals 208
causal systems 221
CE sign 469
centre frequency 196, 361
chain rule 518
characterisation of nonlinear systems 253
characteristic equation
  of nonlinear systems 253
characteristic expression 435
characteristics of nonsinusoidal waveforms 115
charge 1
  electric 1
Chebyshev-filter 351
chemical elements 556
choke 4
  current compensated 509
chokes 91
circuit duality 139
circuit symbols 595–600
  capacitors 595
  connections 600
  connectors 600
  current sources 596
  diodes 597
  inductors 596
  integrated circuits 600
  measurement instruments 598
  old symbols 597

relays 598
resistors 595
sensors 600
switches 598
thysistors 597
transformers 596
transistors 598
voltage sources 596
circuits
  equivalent 135
circuits with operational amplifiers 335, 336
  impedance converter 336
  noninverting amplifier 336
class A operation 379
class AB operation 383
  biasing 384
class B operation 382
class C operation 382
classes of precision 188
closed-loop gain 323
closed-loop system 323
CMOS 417
CMOS counters 459
CMOS devices
  technical data 417
CMRR 301, 332
code
  8421-code 452
coercivity 76
coherent units 530
coil 4
colour code
  resistor 542
Colpitts oscillator 368
combinational circuit 408, 423
combinational logic 423
common mode 299
common-base circuit 294, 297
  AC equivalent circuit 295
  AC voltage gain 296
  high frequencies 296
  input impedance 295
  output impedance 295
common-collector circuit 291, 297
  AC current gain 294
  AC equivalent circuit 292
  high frequencies 294
  input impedance 293
  output impedance 293
  voltage gain 291
common-drain circuit 316, 318

common-emitter circuit 280 ff, 297
  AC equivalent circuit 282
  AC voltage gain 285
  at high frequencies 291
  input impedance 283
  load line 290
  operating point 286
  output impedance 284
  thermal voltage drift gain 289
  two-port network equations 281
  two-port network parameters 281
common-gate circuit 317, 318
common-mode gain 299, 301, 320, 331
common-mode input resistance 302
common-mode input swing 331
common-mode radio-frequency
  interference voltage 508
common-mode rejection ratio 301, 321, 332
common-source circuit 310–313, 318
  AC equivalent circuit 312
  feedback capacitance 314
  gain 314
  input impedance 313
  operating point 314
  output impedance 313
  two-port parameters 311
  y-parameter 311
commutative law 396
comparators 335
compass needle 66
compensated voltage divider 141
compensation
  of reactive current 156
complementary emitter follower 379
  biasing 384
  bootstrap 384
  class AB operation 383
  class B operation 379, 382
  class C operation 382
  current-limiting 386
  Darlington pair 385
  efficiency 381
  feedback 387
  input and output impedance 380
  input signal injection 386
  operation classes 383
  oscillation 387
  output power 380
  output voltage limit 380
  power dissipation 381
  pseudo-Darlington circuit 385

zero stability 387
complex admittance plane 120
complex amplitude 110
complex calculus 105
complex conjugate 101, 103, 104
complex exponential function 104
complex Fourier coefficients 212
complex frequency 267
complex function of time 109
complex impedance 116
complex impedances
    overview 121
complex normal form 212
complex number arithmetic
    overview 107
complex numbers 101 ff
    addition 102
    Cartesian form 103
    division 102
    Euler formula 109
    exponential form 104
    multiplication 102, 106
    notation convention 101
    polar form 103
    representations 103, 105
    subtraction 102
    trigonometric form 103
complex plane 103
complex power 155
complex RMS value 110
complex spectrum 212
composite signal spectrum 219
compression point 257
conductance 3, 57, 119, 122
    complex 119
    magnetic 500
conducted-mode interference 508
conductivity 56
confidence intervals 528
constant quantity 112
consumer pointer system 65
continuous mode 478
continuous-mode operation 477
control
    current-mode 496
    of SMPS 496
    PI controller 498
    power factor pre-regulator 506
    voltage-mode 496
controllable resistor 321
converter
    inverting 481, 494

convolution 228
    rules 228–230
convolution integral 228
convolution product 228
core cross-section
    magnetic 499
core length
    magnetic 499
corkscrew rule 64, 66
corner frequency 147, 198, 202
correct current measurement 181
correct voltage measurement 181
cosine function 99, 109
    basic terms 99
    graph 513
    with complex argument 109
cosinusoidal waveforms
    sum of 100
cotan function
    graph 513
coulomb 1
Coulomb integral 44
Coulomb's law 39
counter 447–460
    asynchronous 450
    BCD 450
    binary 450
    cascading 456
    CMOS 458, 459
    decimal 450, 453
    down 453
    overview 458, 459
    partially synchronous 456
    programmable 454
    ripple-through 450
    semisynchronous 456
    TTL 458, 459
    up/down 454
coupling coefficient 88
crest factor 115
critical frequency 147, 348
critical frequency of transconductance 310
critically damped case 26, 27
cross-coil instrument 172
crossover distortion 380, 384
crystal oscillator 368
current 1, 54
    definition 64
    electric 1
current amplifier 324
current compensated double choke 509

current density 55
current direction
　positive 1
current divider 18
　capacitive 19
　complex 140
　inductive 19
current division 140
current error
　with instrument transformers 180
current error circuit 181
current flow
　selecting track dimensions 544
current gain 274
　AC 274
　differential 274
　forward 274
　static 274
current limiting 386
current measurement
　AC 177
　DC 174
current mirror 304
current path 182
current sink 417
current source 5, 13, 417
　conversion into voltage source 13
　ideal 5
　real 12
　voltage-controlled 341
current transformer 179
current-compensated choke 509
current-compensated inductor 510
current-divider rule 18
cutoff frequency 195, 196, 198, 202

D flip-flop 430, 434, 436
D-input 433
D-latch 430
damped oscillation 108
damping ratio 26, 204
　critically damped case 26
　overdamped case 26
　underdamped case 26
Darlington pair 278, 385
data selectors 427
data sheets
　digital technology 412
DC 1
DC measurement
　current 174
　voltage 174
DC part 211

DC systems 1
decibel 193
decimal counter 450, 453
decimal prefixes 531
decoder 426
definition
　linear systems 220
　stable systems 222
　time-invariant systems 221
delay distortion 237
delay time 236
delayed output 432, 433
delta circuit
　transformation into star circuit 17
　transformation to a star circuit 137
　transformation to a wye circuit 137
delta function 224
　spectrum 245
delta–configuration 17
delta–star transformation 137
delta–wye transformation 137
delta-connected generator 161
demagnetisation 77
DeMorgan's rules 398
demultiplexer 427
dependency notation 423–426
depletion 305
derivative
　of step function 225
derivatives 108
　of elementary functions 519
design of the PI controller 498
diamagnetism 69, 74
dielectric constant 44
dielectrics 44
difference amplifier 339
differential amplifier 298
differential amplifier with bipolar
transistors
　common-mode gain 301
　common-mode rejection ratio 301
　examples 303
　input impedance 302
　input offset voltage drift 302
　offset current 302
　offset voltage 302
　output impedance 302
　overview 304
differential amplifier with field-effect
transistors 319, 320
　common-mode gain 320
　common-mode rejection ratio 321

differential mode gain 320
  input impedance 321
  output impedance 321
  overview 321
differential calculus
  chain rule 518
  division rule 518
  product rule 518
differential current gain 274
differential equation, linear 1st-order 19
differential input resistance 275
differential mode 299
differential output resistance 275
differential resistance 262
differential-mode gain 300
differential-mode input impedance 302
differential-mode radio-frequency
  interference voltage 508
differentiator 342
digital circuits
  CMOS family 417
  integration 412
  loading of 410
  noise margin 410
  open collector 420
  power loss 412
  propagation delay 411
  rise time 411
  slew rate 412
  TTL family 414
  voltage levels 409
digital electronics 392 ff
diode 269
  dynamic resistance 271
  parallel combination 271
Dirac impulse 224
direct current 1
discontinuous mode 478
discontinuous-mode 478
display error 187
  classes of precision 188
distortion-free systems 236
distortions
  linear 237
  nonlinear 253–257
distributive law 397
division rule 518
don't care state 405
down counter 449, 453
drain 305
DRAM 441
driver transistors 385

DTL 419
dual 139
dual-tone multi-frequency 553
duality constants 139
duality of circuits 139
duty cycle 477
dynamic component 412
dynamic input 433
dynamic load line 291
dynamic noise margin 411
dynamic RAM 441
dynamic resistance 262
dynamometer 170

EAROM 444
earth leakage current 509
ECC memory 443
ECL 418
edge-triggered flip-flops 434
  synthesis of 436
EEPROM 444
efficiency 9
electric charge 1
electric current 1
electric displacement 43
electric field 39
electric field strength 40, 54
electric flux lines 40
electric induction 42
electric potential 2
electric resistance 2
electric voltage 2
electrical energy 7
electricity meter 172
electrodynamic instrument 170
electrodynamic ratio meter 172
electron
  charge 533
electronic realisation of logic circuits
  409
electrostatic field 39–53
  at a boundary 47
electrostatic induction 42
electrostatic instrument 171
electrostatic movement 171
electrostatic shielding 42
elementary charge 1
elementary signals 222 ff
  delta function 224
  Dirac impulse 224
  Gaussian pulse 223
  impulse function 224
  rectangular pulse 222

step function 222
triangular pulse 223
EMI suppression chokes 510
emitter follower 291, 297, 376
   as power amplifier 376
   complementary 379
   efficiency 379
   input and output impedance 377
   operating limits 377
   output power 377
   power dissipation 378
enable 457
energy 7 ff
   in a capacitor 9
   in an inductor 8
energy in a magnetic field 90
energy in an electrostatic field 49
energy in static steady-state current flow 62
energy signal 208
energy, normalised 208
enhancement 305
EPLD 445
EPROM 443, 445
equipotential surfaces 41, 54
equivalent circuit diagram 6
equivalent circuits 135
equivalent parallel circuit 136
equivalent series circuit 136
error limits 188
EU sign of conformity 469
Euler formula 104, 109
European norm
   EN55022 507
   EN61000 507
   EN61000-3-2 504
even function 209
exa 531
exclusive OR 396
exponential form of complex numbers 104
exponential function
   complex 107–109
   derivations and integrals 108
   with complex exponents 108
   with imaginary exponent 108

failure in time 371
failure rate 371
fall time 411
fan 371
fan-in 410
fan-out 410
farad 4, 46
Faraday cage 42
Faraday's law 95
Faraday's law of induction 84
FAST series 416
feedback 322
feedback capacitance 312
feedback factor 323
feedback, types of 324
femto 531
ferrite 499
ferromagnetics 75
ferromagnetism 75
FET *see* field-effect transistor
FET current sources 319
field constant
   electric 533
   magnetic 533
field lines 66
field-effect transistor 305 ff
   active range 307
   as controllable resistor 321
   basic circuits 310, 311
   critical frequency 310
   depletion type 305
   enhancement type 305
   high frequencies 310
   IGFET 306
   input impedance 309
   insulated gate 305
   JFET 305
   junction 305
   MOSFET 306
   n-channel 305
   output characteristic 307
   output characteristics 307
   output resistance 309
   overview: basic circuits 318
   p-channel 305
   pinch-off voltage 307
   resistive range 307
   symbols 306
   threshold voltage 307
   transconductance 308
   transfer characteristic 307
   voltages and currents 305
FIFO 442
filter 194 ff, 470
   active 348
   all-pass 197, 348
   band-stop 348
   bandpass 196, 361

Bessel 351, 353
Butterworth 351, 352
Chebyshev 351
Chebyshev 0.5 dB 354
Chebeyshev 3 dB 355
coefficients 350
EMI 508
high-pass 195, 359
low-pass 195, 349
normalisation of the transfer function 349
order of a 350
overview 194
passive 349
poles 350
rise time 198
stop-band 197
universal 363
with critical damping 351
filter capacitor 471
filter order 202, 348
filter realisation 206
filters 141
finite state machines 464
first-order systems
impulse response 232
step response 232
fit *see* failure in time
fixed memory *see* ROM
flash 444
flip-flop
circuit symbols 433
overview 434
flip-flop applications 428
flip-flop transition table 435
flip-flops 428
circuit symbols 433
circuits, **overview** 438
clocked SR 430
D 430
edge-triggered 434
JK 432
master–slave 431
overview 434
overview, edge-triggered 434
RS 429
SR 429
syntheses 436
triggering 432
flux density 66
flux linkage 70
flyback converter 477, 482, 494

follower 336
force
at the boundaries 93
at the boundary 50
in a magnetic field 92
in an electrostatic field 50
on a charge 50
on a current-carrying conductor 67, 92
on a moving charge 65
form factor 115
forward converter 477
forward current 270
forward current gain with shorted output 266, 274
forward transconductance 267
forward transconductance with shorted output 267
forward voltage 270
Fourier coefficients 210
complex 212
Fourier series 210
amplitude–phase form 211
application of 217
complex normal form 212
exponential form 212
frequently used 215
overview 213
trigonometric form 210
Fourier transform 242
definition 242
properties, **overview** 244
symmetry 244
Fourier transforms 241
of elementary signals 245–249
FPGA 446
FPLA 446
FPLS 446
free-wheeling diode 24
frequency 99, 113
complex 267
frequency compensation 334
frequency divider 434
frequency domain 242
frequency normalisation 199, 231
frequency response 193, 269
frequently used Fourier series 215
full-bridge push–pull converter 495
full-wave rectification 472
full-wave rectification with dual supply 472
full-wave symmetry 209

GaAs 419
gain criterion 364
gain frequency characteristic 193
gain margin 335
gain response 193
gain-bandwidth product 333
GAL 446
gallium-arsenide 419
galvanometer 169
gate 305
gates 393
Gaussian distribution 527
Gaussian pulse 223
    spectrum 248
Gauss's law 96
Gauss's law of electrostatics 45
general alternating quantity 112
generator 29
generator star point 162
Germanium diode 269
Giacoletto 278
Giacoletto equivalent circuit 278
Gibb's phenomenon 239
Greek symbols 533
group delay 237
guaranteed error limits 188

HAL 445
half-bridge push–pull converter 490, 495
half-wave rectification 472
half-wave symmetry 210
hard iron 77
harmonic 112
harmonic function 99, 108
harmonics 210, 253
Hartley oscillator 367
heating of components 370
heatsink 371
    calculation of 370
henry 4, 73, 74
high frequency transformer 499
    coupling 503
    hysteresis losses 502
    minimum number of primary windings 502
    windings 502
    wire diameter 502
high-frequency transformer 500
high-pass filter 195, 200, 359
    circuits 359
    transfer function 359
high-speed CMOS series 416

higher-order filters 202
hold time 441
homogeneous field 55
hot-wire measuring system 171
h-parameters 265
hybrid-parameters 265
hyperbolic functions 518
    addition theorems 518
hysteresis 422
hysteresis loop 76
hysteresis loss 76
hysteresis-circuit 344

ideal bandpass filter 240
ideal low-pass filter 238
    step response 238
ideal systems 236
IFL 446
IGFET 305
illegal states 467
imaginary numbers 101
imaginary part 101
imaginary unit 101
    powers of 101
impedance 116, 117
    complex 116
impedance converter 292, 336
impedance matching 10, 143
impedance normalisation 231
impedance plane
    complex 117
Imperial units 535
implicant 406
impulse function 224
impulse response 226
    first-order systems 232
    second-order systems 234
impulse response calculation 231
in-phase current 153
in-phase voltage 153
increasing oscillation 108
induced voltage 84
inductance 4, 74
induction 83–90
induction in a moving conductor 83
induction instrument 172
inductive divider 19
inductive reactance 122
inductor 4, 122, 499
    airgap 500
    core 500
    current-compensated 510
    wire diameter 500

input admittance with shorted output 267
input bias current 332
input characteristic 273
input impedance 142, 263, 321
input impedance of the differential amplifier 302
input offset voltage 302
input resistance with shorted output 266
input spectrum 231
input vector 465
instantaneous power 7, 151
   in a three-phase system 165
instantaneous value 99, 113
instrument symbols 173
instrumentation amplifier 340
insulated gate field-effect transistors 305
integrals 108
   basics 519
   definite 524
   involving cosine 523
   involving exponential functions 523
   involving inverse trigonometric functions 524
   involving trigonometric functions 521
   of elementary functions 520
integrator 341
intercept point 257
interference
   conducted-mode 508
intermodulation distortion 255
intermodulation margin 257
internal resistance
   in a voltmeter 176
   voltage-related 176
inverse trigonometric functions 517
inversion 392
inversion laws 398
inverter 394
   controlled 396
inverting amplifier 337
inverting converter 494
inverting Schmitt trigger 344
iron loss 76

JFET 305
JK flip-flop 432, 435
Johnson counter 449
joule 7
junction field-effect transistors 305

Karnaugh map 402
KCL 6, 58
Kirchhoff's laws 6, 58
   current law 6, 58
   first law 6, 58
   mesh law 59
   second law 6, 59
   voltage law 6, 59
Kronecker symbol 215
KVL 6, 59

lagging 113
lagging power factor 154
Laplace frequency domain 267
latch 431
latches 428
LCA 446
leading 113
leading power factor 154
least significant bit 462
Lenz's law 84
levels
   absolute 549
   relative, table 551
lifetime 371
line conductor 159
line current 159
line regulation 496
line spectrum 217
line-to-line voltage 159
linear systems 192, 220
linearisation 261
   operating point 261
load 29
   for instrument transformers 180
   of digital circuits 460
load line 290
load regulation 496
load variation 473
loading
   of digital circuits 410
logic algebra 392 ff
logic circuits
   CMOS family 417
   electronic realisation 409
   integration 412
   loading of 410
   noise margin 410
   open collector 420
   power loss 412
   propagation delay 411
   rise time 411
   slew rate 412

TTL family 414
TTL/CMOS comparison 418
voltage levels 409
logic families 412
logic functions 392
logic high voltage level range 409
logic low voltage level range 409
logic transformations 396
   associative law 397
   commutative law 396
   DeMorgan's rules 398
   distributive law 397
   inversion laws 398
   overview 398
logic variable 392
loop 6
loop analysis 30, 32
loop gain 323
Lorentz force 66, 67
loss factor 26
low-pass filter 195, 197, 349
   calculation 356
   circuits 357
   ideal 238
low-pass filter, 2nd-order 27
   step response 29
low-power Schottky TTL series 416
low-power TTL series 415
lower cutoff frequency 196
LRC low-pass filter 27
LSB 462
LSL 419
LTI systems 222

magnetic circuit 78
magnetic circuit with a permanent magnet 80
magnetic conductance 73, 500
magnetic core length 499
magnetic coupling 88
magnetic coupling coefficient 88, 89
magnetic dipole 66
magnetic field
   direction-pointing convention 64
   force at the boundaries 93
magnetic field strength 69
magnetic fields 64–95
magnetic fields at boundaries 77
magnetic flux 70
magnetic flux density 66
magnetic hysteresis 76
magnetic induction 83
magnetic resistance 73

magnetic saturation 76
magnetic voltage 71
magnetomotive force 71
magnitude 102–104
magnitude frequency characteristic 193
magnitude response 193, 197, 201
mains ripple 298
mask programmable 443
master–slave flip-flop 431
mathematical basics 513
maximum power transfer 10
maxterm 406
Maxwell's equations 95–96
   1st equation 95
   2nd equation 95
   3rd equation 96
   4th equation 96
Maxwell's parallel plates 42
mean time between failures 371
measurement
   AC current 177
   AC power 182
   AC voltage 177
   DC current 174
   DC power 181
   DC voltage 174
   multiphase power 185
   power factor 184
   reactive power 183, 184
   reactive power, multiphase 186
   RMS 180
measurement error 187 ff
   classes of precision 188
   current measurement 176
   random error 187
   systematic error 187
   voltage measurement 177
measurement instruments 169–173
   bi-metallic 171
   cross-coil 172
   electrodynamic 170
   electrodynamic ratio 172
   electrostatic 171
   hot-wire 171
   induction 172
   overview 173
   ratiometer moving-coil- 169
   reed frequency meter 172
   rotary magnet 171
   thermal 171
   vibration 172
measurement methods

overview 190
measurement range extension
　current measurement 174
　voltage measurement 175
　with an instrument transformer 179
Meissner oscillator 367
memory 439
memory access 440
memory cell 439
memory construction 439
mesh 6
mesh analysis 30, 32
mho 57
　unit 534
micro cell 547
microdyn cell 547
mignon cell 547
Miller-capacitance 280, 291
minimum number of primary turns 501
minimum overlap of logic terms 407
minterm 406
mixed quantity 112
MMF 71
mode
　continuous 478
　discontinuous 478
modulo-$(m+1)$ counter 455
mono cell 547
most significant bit 463
moving-coil instrument 169, 174
moving-coil meter 116
moving-iron instrument 171
moving-iron meter 116
MSB 463
MTBF 371
multiplexer 425, 427
multiplexor *see* multiplexer
multiport RAM 442
multivibrator 346, 370
mutual inductance 88, 89
mutual induction 88

n-channel FET 305
n-phase system 158
NAND gate 395
national approval signs 469
natural frequency 26
naturally occurring constants 533
negative feedback 322–329
　closed loop gain 323
　critical frequency 328
　frequency response 327
　gain 328

　input and output impedance 326
　stability 328
negative frequency spectrum 212
negative logic 409
negative-feedback resistor 286
neper 549
network transformations 135–140
networks at variable frequency 192
neutral conductor 159
node 6
node analysis 30, 33
noise margin 410
nominal load 180
noninverting amplifier 336
noninverting Schmitt trigger 345
nonlinear systems 192, 226, 253–257
　characterisation 253
　characteristic equation 253
　definition 253
　THD 254
　total harmonic distortion 254
nonperiodic signals 208
nonreactive 265
NOR gate 395
normalisation of circuits 231
normalised frequency 199
Norton's theorem 33
notch filter 197

odd function 209
offset current 302
offset voltage 302, 330
offset voltage drift 302, 331
ohm 3, 56, 57
Ohm's law 2
one way rectification 472
op-amp 329
open circuit 12
open collector 420
open core 510
open-circuit 11
open-loop gain 323
operating point 261
　linearisation at 261
operating point biasing 286
operating point stabilisation 288
　nonlinear 290
operational amplifier 329 ff
　characteristics 330
　CMRR 332
　common-mode gain 331
　common-mode input swing 331
　common-mode rejection ratio 332

compensation 334
critical frequency 333
equivalent circuit 333
frequency compensation 334
gain margin 335
gain-bandwidth product 333
input bias current 332
input impedance 332
instrumentation amplifier 340
offset voltage 330
output impedance 332
output voltage swing 330
phase margin 335
power supply rejection ratio 332
PSRR 332
rail-to-rail 330
single supply 330
slew rate 333
transit frequency 333
operational amplifier circuits
   AC voltage amplifier 343
   bandpass filter circuit 362
   compensation of the input bias current 338
   current source 341
   difference amplifier 339
   differentiator 342
   follower 336
   high-pass filter circuits 359
   integrator 341
   inverting amplifier 337
   low-pass filter circuits 357
   multivibrator 346
   sawtooth generator 346
   Schmitt-trigger 344
   summing amplifier 338
   triangle- and square-wave generator 345
   voltage setting 343
OR 444
order of a filter 350
OR function 393
OR gate 395
orthogonal 214
oscillation
   amplitude criterion 328
   barkhausencriterion 328
   phase criterion 328
oscillation criterion 328
oscillator 364–370
   Barkhausen criterion 364
   Colpitts 368

crystal 368
feedback loop gain criterion with FET 365
gain criterion 364
Hartley 367
LC 367
Meissner 367
phase criterion 364
phase shifter- 365
Pierce 369
quartz 368
RC oscillators 365
Wien bridge 366
out-of-phase current 154
out-of-phase voltage 154
output admittance with open input 266
output admittance with shorted input 267
output characteristic 273, 307
output impedance 142, 321
output impedance, equivalent source resistance 263
output logic 465
output ROM 467
output spectrum 231
output vector 465
output voltage swing 330
overcompensation 157
overdamped case 26, 27
overload protection
   for moving-coil instruments 176
overshoot 239
overview
   AC power 156
   basic circuits using field-effect transistor 318
   bipolar transistor-basic circuits 296
   capacitances of different geometric configurations 48
   characteristics of a magnetic field 94
   characteristics of a static steady-state current flow 63
   characteristics of an electrostatic field 52
   complex impedances 121
   complex number arithmetic 107
   counter circuits, CMOS 459
   counter circuits, TTL 459
   counters 458
   dependency notation 425
   differential amplifier with bipolar transistors 304

differential amplifier with field-effect transistors 321
filters 194
flip-flop circuits 438
flip-flops 434
flip-flops, edge-triggered 434
Fourier series 213
inductances of different geometric configurations 82
instruments 173
logic transformations 398
measurement methods 190
notation in data sheets for digital circuits 412
properties of the Fourier transform 244
resistances of geometric configurations 61
series and parallel circuits 134
switched-mode power supplies 494
symbols on measurement instruments 188
three-phase system 164

p-channel FET 305
PAL 444, 445
  output circuits 446
PAL assembler 463
parallel circuits
  transformation to series circuits 135
parallel combination 13–16, 130
  of R and C 22, 130
  of R and L 129
  of R, C and L 132
  of capacitances 16, 19
  of conductances 14
  of inductances 15, 19
  of resistors 13
parallel in serial out 447
parallel-equivalent circuit
  of a passive component 157
parallel-resonant circuit 132
paramagnetism 69, 74
partially synchronous 457
pascal
  unit 534
pass-band 194–196
passive 265
passive components 123
  parallel combinations 128 ff
  series combination of 123–128
passive elements
  dual 139

peak magnitude 99
peak value 113, 115
period 99, 112, 113, 208
periodic
  definition 112
periodic quantity 112
periodic signals 208
permanent magnet 80
  designing a 81
permeability 69
  relative 69
permeability of free space 69
permittivity 44
  table 560
peta 531
phase 99, 104
phase constant 237
phase criterion 364
phase current 159
phase delay 237
phase distortion 237
phase error
  for instrument transformers 180
phase factor 194
phase frequency characteristic 193
phase margin 335
phase position 113
phase response 193, 197, 201, 269
phase shift 99, 113
  lagging 113
  leading 113
phase shifter 146
phase shifter oscillator 365
phase shifting
  circuits for 146–149
phase spectrum 212
phase voltage 159
phasor
  rotating 110
  rotation of 106
phasor diagram 110
phasors 110
  multiplication with real number 106
  sum of 111
Π-configuration 145
Pierce oscillator 369
pinch-off voltage 307
PISO 447
PLA 445
PLD 444
PLD types 445
  overview 446

point symmetry 209
poles of a filter function 350
polyphase systems 158
POS 401
positive current direction 1
positive edge-triggered 434
positive feedback 322
positive frequency spectrum 212
positive logic 409
potential 2, 41, 53
powder core 92
powder-core choke 510
power 7
   average 7
   average value 152
   in a reactive element 151
   in a resistor 7
   instantaneous 7
   measurement in a multiphase system 185
power amplifier 376–388
   AC voltage gain 387
   current-limiting 386
   input signal injection 386
   negative feedback 386
power attenuation 549
power factor 153, 156
   measurement 184
power factor control 504
power factor correction 157
power factor preregulator 477
power in a three-phase system 165
power in static steady-state current flow 62
power loss
   in digital logic circuits 412
   transistor 273
power matching 264
power measurement 181
   AC circuit 182
   DC circuit 181
   multiphase circuit 185
power signal 208
power supplies 469–475
power supply rejection ratio 332
power transformer 469
   internal resistance 470
   loss factor 470
   no-load voltage 470
   primary winding 470
   protection 470
   rated power 470
   rated voltage 470
   secondary winding 470
   short circuit protection 470
power transistors 385
power, normalised 208
preparatory inputs 432
presettable counter 454
primary switched SMPS 477, 482
prime implicant 406
principal value 103, 517
principle of superposition 31, 192, 220, 262
probe 141
product of sums 400, 401, 405
product rule 518
program ROM 467
programmable counter 449, 454
programmable logic device 439
programmable logic devices 444–448
PROM 443–445
propagation delay time 411
pseudo-Darlington circuit 385
pseudo-Darlington pair 279
pseudostatic RAM 442
PSRR 332
pulsating quantity 112
pulse width 240
   definition 240
   modulator 346, 388, 496
push–pull converter 489
push–pull converter with common based transistors 496
push-pull amplifier 379
PWM 346, 388, 496

Q-factor 26, 196
quality 26
quality factor 196
quartz oscillator 368
quiescent current 384
Quine–McCluskey minimisation 406
Quine–McCluskey technique 406

radian frequency 113
radio frequency interference radiation 507
radio frequency interference suppression 507
radio frequency interference voltage symmetric 509
radio frequency ranges 548
radio noise field strength 507
radio-frequency interference voltage

differential-mode 508
radio-frequency interference 507
radio-frequency interference filter 508, 510
radio-frequency interference meter 508
radio-frequency interference voltage
    asymmetric 508
    symmetric 508
    unsymmetric 508
RAM 439
    arbiter 442
    ECC memory 443
    multiport RAM 442
    pseudostatic RAM 442
    ring memory 442
    variations 442
RAM controller 442
RAS 442
ratio instrument 169
    electrodynamic 172
ratiometer moving-coil instrument 169
RC combinations 19
RC phase shifter 146
RCL combinations 25
reactance 117, 122, 123
    capacitive 123
    inductive 122
reactive component 152
reactive impedance 117
reactive power 154, 156
    from 3-voltmeter measurement 184
    measurement in a multiphase system 186
    measurement in an AC circuit 183
reactive voltage 154
read-only memory 443
read-write memory *see* RAM
real current 153
real part 101
real power 153, 156
    measurement in a multiphase system 185
real voltage 153
reciprocal, reversible 265
rectangular pulse 222
    spectrum 247
rectangular signal
    spectrum 217
rectifier 470
reduced products of sums 404
reduced sum of products 402
reduction of logic functions 402

Karnaugh map 402
Quine–McCluskey technique 406
reed frequency meter 172
register 447
relative bandwidth 196
relative permeability 69
relative permittivity 44
relative quantities 532
reliability 371
remanent flux density 76
reset 434
resistance 3, 57, 117
    temperature dependency of 3
resistive component 152
resistivity 56, 559
resistor standard series 541
resonance 128
resonant circuit 26
resonant converter 477, 491
resonant frequency 26, 128, 133
reverse current 270
reverse transconductance with shorted input 267
reverse voltage transfer ratio with open input 266
reverse voltage-transfer ratio 276
right-hand rule 66
ring memory 442
ripple voltage 471
ripple-through counter 450
rise time 199, 411
RL combinations 19
RMS 170
RMS measurement 180
RMS value 114, 115
ROM 439, 443
root mean square (RMS) 115
rotary magnet instrument 171
rotating phasor 106
rotation 106, 108
RS flip-flop 429
RS232 552
RTL 419

saturation flux density 76
sawtooth generator 346
sawtooth signal
    spectrum 218
Schmitt trigger 422
schmitttrigger 344
Schottky TTL series 415
second-order systems
    impulse response 234

Index

step response 234
secondary switched SMPS 477
selecting track dimensions for current flow 544
self reciprocal function 249
self-induction 87
semi-synchronous counters 447
semiconductor memory 439
semisynchronous 457
sequential circuit 423
sequential logic 423
   synthesis of 460
serial in parallel out 447
series and parallel circuits
   overview 134
series circuits
   transformation to parallel circuits 135
series combination 13–16
   of R and C 21, 125
   of R and L 124
   of R, C and L 126
   of R, L and C 26
   of capacitances 16, 19
   of conductances 14
   of inductances 15, 19
   of resistors 13
series combination of AC impedances 123
series equivalent circuit
   of a passive component 157
series-resonant circuit 126
set 434
settling processes 19 ff, 25 ff
settling time 239
seven-segment code 458
shape factor 196
shielding 42
   electrostatic 42
   magnetic 75
shift register 447
short circuit 11
short-circuit current 11
shortening minterms 407
shunt resistor 174
SI base units
   definition 530
Si function 239
SI units in electrical engineering 532
siemens 3, 56, 57
signal-to-intermodulation ratio 255
signals and systems 208

signum function 246
   spectrum 246
Silicon diode 269
sinc function
   definition 238
sine function 99, 109
   basic terms 99
   with complex argument 109
single-transistor forward converter 486, 495
sinusoidal quantity 112
sinusoidal waveforms
   sum of 100
SI system 530
skin effect 503
slew rate 333, 412
small signal 261
small-signal amplifier 271, 305
small-signal current gain 274
small-signal equivalent circuit 262
SMPS 476
snubber circuit 484
soft iron 77
soft-iron instrument 171
SOP 400
source 305
source field 40
source follower 316, 318
source pointer system 64
source resistance 11, 12
specific resistance
   table 559
specific thermal capacity 375
spectrum
   composite signal 219
   delta function 245
   Gaussian pulse 248
   of elementary signals 245–249
   of harmonic functions 249
   rectangular pulse 247
   sawtooth signal 218
   signum function 246
   step function 246
   triangular pulse 247
SR flip-flop 429, 434, 436
   clocked 430
SR flip-flop with clock input 430
SRAM 441
standard distribution 527
standard series
   IEC 541
standard TTL series 415

Index

star circuit
   transformation to a delta circuit 137
star–configuration 17
star–connected generator 162
star–delta start 164
star–delta transformation 17, 138
start
   star–delta 165
state memory 465
state vector 465
static component 412
static load line 290
static RAM 441
static steady-state current flow 53–64
   at boundaries 60
step function 222
   spectrum 246
step response 19 ff, 29, 227
   first-order systems 232
   second-order systems 234
step response calculation 231
step-down converter 477
step-up converter 479
stop-band 194–196
stop-band filter 197
sum of products 400
summary of Fourier transforms 250 ff
summation point 267
summing amplifier 338
superposition 31
susceptance 119
switched mode power supplies
   control of 496
   overview 494
   primary switched 477, 482
   radio frequency interference radiation 507
   radio frequency interference suppression 507
   secondary switched 477
switched-capacitor filter 363
switched-mode amplifiers 388
switched-mode power supplies 476
symbols on measurement instruments 188
symmetric function 209
symmetric radio frequency interference voltage 509
symmetric radio-frequency interference voltage 508
symmetrical 265
synchronous counters 447, 455

synchronous sequential logic 464
synthesis of combinational circuits 408
system response 226
   impulse reponse 226
   step response 227
   to arbitrary input signals 228
systems
   linear, definition 220
   stable, definition 222
   time-invariant, definition 221

T flip-flop 434, 437
table
   Fourier expansion into a series 215
   Fourier transforms 250 ff
T-configuration 145
temperature calculation for components 373
temperature coefficient 3
temperature coefficient of the input offset voltage 302
tera 531
tesla 66
   unit 534
THD 254
THD, nth order 255
thermal capacity 374
thermal compound 371
thermal impedance
   transient 375
thermal instruments 171
thermal paste 371
thermal resistance 373
thermal voltage 270
thermal voltage drift 275
thermal voltage drift gain 289
Thévenin's theorem 33
Thévenizing 35
three-ammeter method 183
three-phase supplies 158
three-phase system 159
   overview 164
three-phase systems 159
three-voltmeter method 183
three-wattmeter circuit 187
threshold voltage 270, 307, 409
   temperature dependency 270
time constant 19, 147
time scaling of signals 225
time shift of signals 225
time–bandwidth product 239
time-invariant systems 221
toggle 450

Index

toggle flip-flop 432
total harmonic distortion 116, 254
total harmonic distortion attenuation 255
total harmonic distortion, nth order 255
track dimensions for current flow 544
transconductance 275, 308
transconductance amplifier 324
transfer characteristic 273, 307, 409
transfer function
   192, 197, 200, 230, 231, 267
   definition 230
transformer 90
   rated voltage 471
transient 99
transient thermal impedance 375
transimpedance amplifier 324
transistor characteristics 272
   base 272
   collector 272
   emitter 272
transistor–transistor logic 414
transit frequency 333
transition combinational circuit 465
transparency
   in flip-flops 431
tri-state 422, 424, 447
triangle- and square-wave generator 345
triangular pulse 223
   spectrum 247
triggering of flip-flops 432
trigonometric functions 513
   addition theorems 515
   inverse 517
   products 516
   properties 513
true RMS measurement 180
truth table 392
TTL 414
   basic structure 416
TTL devices
   technical data 415, 416
two-port network 265
   active 265
   asymmetrical 265
   linear 265
   nonreactive 265
   passive 265
   reversible 265
   symmetrical 265
two-port network equations 265
two-port network parameters 265
two-terminal network 264
two-tone signal 255
two-transistor forward converter 488, 495
two-wattmeter method 186

uncertainty principle 240
underdamped case 26, 28
uninterruptable power supplies 469
unit 531
units
   decimal prefixes 531
unity gain frequency 276
universal filter 363
unsymmetric radio-frequency
   interference voltage 508
up counter 449
up/down counter 449, 454
upper cutoff frequency 196
UPS 469
useful Fourier series 215

V.24 552
VAR 154
versor 104
vibration instrument 172
volt–ampere 155
volt–ampere reactive 154
voltage 2, 41, 53
   electric 2
   line 159
   magnetic 71
voltage amplifier 324
voltage attenuation 549
voltage divider 18
   capacitive 19
   compensated 141
   complex 140
   complex loaded 142
   inductive 19
   with def. i/o-resistances 145
voltage division 140
voltage error
   with instrument transformers 180
voltage error circuit 181
voltage gain 291
voltage level 409
voltage measurement
   AC 177
   DC 174
voltage path 182
voltage regulation 475
voltage regulator

for variable output voltage 475
  integrated 475
  pulse width modulated 496
voltage setting with defined slew rate 343
voltage source 5, 13
  conversion into current source 13
  ideal 5
  real 11
voltage stabilisation 473
  analogue 473
  with Zener diode 473
voltage transformer 179
voltage variation 473
voltage-divider rule 18
voltage-related internal resistance 176

walking-ring counter 449
watt 7, 153
weber 70
  unit 534
weighting function 226
Weiss domain 75
Wien bridge 150
Wien bridge oscillator 366
wire diameter 500
wire gauge
  American 545
wired AND 420
wired OR 420
work in static steady-state current flow 62
write cycle time 441
write pulse width 441
wye *see* star
wye circuit
  transformation to a delta circuit 137
wye–delta transformation 17, 137
wye-connected generator 162

X-capacitor 510
XOR 396
XOR gate 396

Y-capacitor 509
y-parameters 266

ZCS push–pull resonant converter 491, 496
ZCS resonant converters 491
Zener diode 473
zero-point
  artificial 185
ZVS resonant converters 491

Learn How to Design Electronic Circuits Fast

U. Tietze, C. Schenk

Electronic Circuits

Handbook of Design and Applications

2nd ed. 2003. Approx. 1700 p. 1800 illus., with CD-ROM.
Hardcover
Approx. € 149.95; approx. sFr 242.50; approx. £ 105
ISBN 3-540-00429-7

Written for the student, practicing engineer and scientist, this reference book covers all important aspects and applications of modern analog and digital circuit design. **Part I** concentrates on basics like analog and digital circuits, on operational amplifiers, combinatorial and sequential logic and memories. **Part II** deals with application. Each chapter offers various solutions to a given problem, thereby enabling the reader to understand ready-made circuits or to proceed quickly from an idea to a working circuit. Usually the design approach is illustrated by an example. Analog applications cover such topics as analog computing circuits. The digital sections deal with AD and DA conversion, digitial computing circuits, microprocessors and digital filters. This edition also contains the basic electronics for mobile communications.

The accompanying CD-ROM contains a PSPICE version that enables the user to simulate analog circuits. It also includes simulation examples and model libraries related to the contents of the book.

Please order through your bookseller

All Euro and GBP prices are net-prices subject to local VAT, e.g. in Germany 7% VAT for books.
Prices and other details are subject to change without notice. d&p · BA 43965/1

Fourth Edition of the
Fundamentals of Mathematics

**I.N. Bronshtein,
K.A. Semendyayev,
G. Musiol, H. Mühlig**

Handbook of Mathematics

4th ed. 2003. Approx. 1000 p.
Hardcover € **59.95**; sFr 99.50; £ 42
ISBN 3-540-43491-7

This guide book to mathematics contains in handbook form the fundamental working knowledge of mathematics which is needed as an everyday guide for working scientists and engineers, as well as for students. Easy to understand, and convenient to use, this guide book gives concisely the information necessary to evaluate most problems which occur in concrete applications. For the 4th edition, the concept of the book has been completely re-arranged. The new emphasis is on those fields of mathematics that became more important for the formulation and modeling of technical and natural processes, namely Numerical Mathematics, Probability Theory and Statistics, as well as Information Processing.

Please order through your bookseller

All Euro and GBP prices are net-prices subject to local VAT, e.g. in Germany 7% VAT for books.
Prices and other details are subject to change without notice. d&p · BA 43965/2

Printing and Binding: Stürtz AG, Würzburg